FRACTURE AND FAILURE OF NATURAL BUILDING STONES

Fracture and Failure of Natural Building Stones

Applications in the Restoration of Ancient Monuments

Edited by

STAVROS K. KOURKOULIS
National Technical University of Athens,
Athens, Greece

Springer

A C.I.P. Catalogue record for this book is available from the Library of Congress.

ISBN-10 1-4020-5076-3 (HB)
ISBN-13 978-1-4020-5076-3 (HB)
ISBN-10 1-4020-5077-1 (e-book)
ISBN-13 978-1-4020-5077-0 (e-book)

Published by Springer,
P.O. Box 17, 3300 AA Dordrecht, The Netherlands.

www.springer.com

Printed on acid-free paper

Contents

Part II: Physico-Chemical and Environmental Aspects

Chapter 5: Weathering

Chapter 6: Freeze-thaw

Chapter 7: Thermal stresses

Contributing Authors

Agioutantis, Z. G., Technical University of Crete, Hania, Hellas.
Aires-Barros, L., Instituto Superior Técnico, Lisboa, Portugal.
Altindag, R., SDU Engineering and Architecture Faculty, Isparta, Turkey.
Álvarez de Buergo, M., Universidad Complutense. 28040 Madrid, Spain.
Antonopoulos, A. K., National Technical University of Athens, Hellas.
Asteris, P. G., National Technical University of Athens, Hellas.
Bakolas, A., National Technical University of Athens, Hellas.
Bellopede, R., Politecnico di Torino, Italy.
Binda, L., Politecnico di Milano, Italy.
Brugnara, M., University of Trento, Italy.
Cardani, G., Politecnico di Milano, Italy.
Chau, K.T., The Hong Kong Polytechnic University, China
Cnudde, V., Ghent University, Belgium.
Della Volpe, C., University of Trento, Italy.
Dierick, M., Ghent University, Belgium.
Dreesen, R., Flemish Institute for Technological Research, Materials Technology, Mol, Belgium.
Eimermacher, R.C., Haifa University, Israel.
Ercoli, L., Università di Palermo, Italy.
Ferrero, A.M., University of Parma, Italy.
Ferretti, A.-S., University of Rome "La Sapienza", Italy.
Figueiredo, C., Instituto Superior Técnico, Lisboa, Portugal.
Fort, R., Universidad Complutense, Madrid, Spain.
Ganniari-Papageorgiou, E., National Technical University of Athens, Hellas.
Giavarini, C., University of Rome "La Sapienza", Rome, Italy.

Gómez-Heras, M., Universidad Complutense, Madrid, Spain.
Güney, A., Mugla University, Turkey.
Hajpál, M., Budapest University of Technology and Economics, Hungary.
Hall, K., University of Pretoria, South Africa. Present address: University of Northern British Columbia, Canada.
Hatzor, Y.H., Ben-Gurion University of the Negev, Beer-Sheva, Israel.
Houck, J., Princeton University, USA.
Jacobs, JS, P., Ghent University, Belgium.
Karagüzel, R., University of Isparta, Turkey.
Konstanti, A., National Technical University of Athens, Hellas.
Kontos, G., Technical University of Crete, Hania, Hellas.
Kourkoulis, S.K., National Technical University of Athens, Hellas.
Kritsotakis, K., Johanes Gutenberg Universitaet, Mainz, Deutschland.
Kwok, K.W., The Hong Kong Polytechnic University, China.
Labropoulos, K., National Technical University of Athens, Hellas.
Lagrou, D., Flemish Institute for Technological Research, Materials Technology, Mol, Belgium.
Larsen, I., SINTEF Petroleum Research, Trondheim, Norway.
Laskaridis, K., IGME - Lithos Laboratory, Peanea, Attiki, Hellas.
Lobovikov-Katz, A., Centre for Conservation and Western Galilee College, Akko; Israel Institute of Technology, Technion City, Haifa, Israel.
Logan, J.M., University of Oregon, USA.
Malaga, K., SP Swedish National Testing and Research Institute, Borås, Sweden.
Manfredotti, L., Politecnico di Torino, Italy.
Maniglio, D., University of Trento, Italy.
Marini, P., Politecnico di Torino, Italy.
Markopoulos, Th., Technical University of Crete, Hania, Hellas.
Masschaele, B., Ghent University, Belgium.
Maurício, A., Instituto Superior Técnico, Lisboa, Portugal.
Mavrouli, O. A., National Technical University of Athens, Hellas.
McKinley, J., Queen's University Belfast, United Kingdom.
Megna, B., Università di Palermo, Italy.
Mentzini, M., Committee for the Conservation of the Acropolis Monuments, Acropolis, Athens, Hellas.
Michailidis, P., National Technical University of Athens, Hellas.
Migliazza, M., University of Parma, Italy.
Modena, C., University of Padova, Italy.
Moggi, G., University of Genoa, Italy.
Moropoulou, A., National Technical University of Athens, Hellas.
Mutlutürk, M., University of Isparta, Turkey.

Myrin, M., Stenkonservatorn Skanska, Department of Environmental Sciences, Göteborg University, Stockholm, Sweden.
Nielsen, P., Flemish Institute for Technological Research, Materials Technology, Mol, Belgium;
Ninis, N.L., Archaeological Museum of Epidauros, Ligourio, Hellas.
Nova, R., Politecnico di Milano, Italy.
Onargan, T., DEU Engineering Faculty, Department of Mining Engineering, Bornova-İzmir, Turkey.
Papamichos, E., Aristotle University of Thessaloniki, Hellas.
Papanicolopulos, S.-A., National Technical University of Athens, Hellas.
Papantonopoulos, C., 10 25th March, 162 33 Vyron, Attiki, Hellas.
Pedemonte, E., University of Genoa, Italy.
Perdikatsis, V., Technical University of Crete, Hania, Hellas.
Poli, T., CNR – ICVBC Sezione di Milano "Gino Bozza", Milano, Italy.
Princi, E., University of Genoa, Italy.
Psycharis, I.N., National Technical University of Athens, Hellas.
Rizzo G., Università di Palermo, Italy.
Roumpopoulos, K., National Technical University of Athens, Hellas.
Saisi, A., Politecnico di Milano, Italy.
Santarelli, M.-L., University of Rome "La Sapienza", Italy.
Scherer, G.W., Princeton University, USA.
Şengün, N., University of Isparta, Turkey.
Siboni, S., University of Trento, Italy
Smith, B.J., Queen's University Belfast, United Kingdom.
Stefanou, I., National Technical University of Athens, Hellas.
Sulem, J., CERMES, Ecole Nationale des Ponts et Chaussées/LCPC, Institut Navier, Paris, France.
Syrmakezis, C. A., National Technical University of Athens, Hellas.
Tedeschi, C., Politecnico of Milan, Italy.
Tiraboschi, C., Politecnico di Milano, Italy.
Toniolo, L., CNR – ICVBC Sezione di Milano "Gino Bozza", Milano, Italy.
Török, Á., Budapest University of Technology and Economics, Hungary.
Tsesarsky, M., The Technion – Israel Institute of Technology, Haifa, Israel.
Valluzzi, M.-R., University of Padova, Italy
Varas José, M., Universidad Complutense, Madrid, Spain.
Vardoulakis, I., National Technical University of Athens, Hellas.
Vicini, S., University of Genoa, Italy.
Warke, P.A., Queen's University Belfast, United Kingdom.
Wong, R.H.C., State University of New York at Stony Brook, USA.
Wong, T.-f., State University of New York at Stony Brook, USA.
Zezza, F., University IUAV of Venice, Italy.

Editor's Preface

The fracture and failure of natural building stones has been for many years the concern of the engineering community and particularly the community of scientists working for the restoration and conservation of stone monuments. The need to protect the authentic stone and the requirement for reversibility of the interventions rendered the in-depth knowledge of the mechanical behaviour of both the authentic material and its substitutes indispensable.

This book contains 36 papers presented at the Symposium on "*Fracture and Failure of Natural Building Stones*" which was organized in the frame of the "*16th European Conference on Fracture (ECF16)*". The Conference took place in Alexandroupolis, Hellas on July 3-7, 2006. To my best knowledge this is the first time that a special Symposium of a European Conference on Fracture is devoted exclusively to the study of the fracture and failure of building stones.

The book consists of invited papers written by leading experts in the field. It contains original contributions concerning the latest developments in the fracture and failure of the natural building stones and their application in the restoration of ancient monuments. It covers a wide range of subjects including purely mechanical aspects, physico-chemical approaches, applications and case studies. The papers are arranged in two parts with a total of nine chapters. Part I is devoted to purely mechanical and structural aspects and applications, while Part II is devoted to the physico-chemical and environmental aspects including thermal effects.

Part I contains four chapters. The first one deals with the behaviour of building stones in the presence of cracks, namely the Fracture Mechanics' point of view. The second chapter is devoted to some special features of the mechanical behaviour and the mechanical properties of building stones. Applications on the behaviour of masonry under static and dynamic loading are included in the third chapter, while the subject of the fourth chapter is the methodological approach to the conservation of stone monuments through particular applications and case studies.

Part II includes five chapters. The first one (fifth chapter of the volume) deals with the problem of weathering, either natural or accelerated, of natural building stones. The freeze-thaw problem, which is a special case of weathering, is the subject of the sixth chapter, while chapter seven deals with the influence of heat and fire on the behaviour of building stones. The petrographical approach and the relation between the fabric and the physico-mechanical properties is the subject of the eighth chapter. Finally, the ninth chapter deals with the protection of natural building stones from weathering with the aid of suitable surface treatment techniques.

I consider it an honour and a privilege that I have had the opportunity to edit this book. I wish to thank sincerely the authors who have contributed to this volume and all those who participated in the Symposium. Also, I would like to thank the reviewers of the papers who assured the scientific quality and originality of the papers of the volume. In addition I would like to thank Mr. Nikolaos Ninis, of the "Team for the Restoration of the Epidauros' Monuments", who spend a lot of time reading the papers as well as for his valuable suggestions concerning the allocation of papers per chapter, the structure of the volume and the sessions of the Symposium.

I would like to thank the PhD student Margarita Satraki for her invaluable help in editing the volume, my PhD students Evangelia Ganniari-Papageorgiou and Panagiotis Chatzistergos and my MSc student Pavlos Tsirigas for proof reading the edited manuscripts. Also, Mr. Iordanis Naziris for the design of the cover page.

Finally, a special word of thanks goes to Mrs. Natalie Jacobs and Mrs. Anneke Pot of Springer for their interest in publishing this work and for their collaboration, support and patience.

February 2006

Athens, Hellas

Stavros K. Kourkoulis
Editor

Foreword

The "*16th European Conference of Fracture* (ECF16)", was held in the beautiful town of Alexandroupolis, Greece, site of the Democritus University of Thrace, July 3-7, 2006. Within the context of ECF16 forty six special symposia and sessions were organized by renowned experts from around the world. The present volume is devoted to the symposium on "*Fracture and Failure of Natural Building Stones - Applications in the Restoration of Ancient Monuments*," organized by Dr. Stavros Kourkoulis of the National Technical University of Athens, Hellas. I am greatly indebted to Stavros who undertook the difficult task to organize this symposium and edit the symposium volume.

Started in 1976, the European Conference of Fracture (ECF) takes place every two years in a European country. Its scope is to promote world-wide cooperation among scientists and engineers concerned with fracture and fatigue of solids. ECF16 was under the auspices of the European Structural Integrity Society (ESIS) and was sponsored by the American Society of Testing and Materials, the British Society for Stain Measurement, the Society of Experimental Mechanics, the Italian Society for Experimental Mechanics and the Japanese Society of Mechanical Engineers. ECF16 focused in all aspects of structural integrity with the objective of improving the safety and performance of engineering structures, components, systems and their associated materials. Emphasis was given to the failure of nanostructured materials and nanostructures and micro- and nanoelectromechanical systems (MEMS and NEMS). The technical program of ECF16 was the product of hard work and dedication of the members of the Scientific Advisory Board, the pillars of ECF16, to whom I am greatly indebted. As chairman of ECF16 I am honored to have them on the Board and work closely with them for the success of ECF16.

ECF16 has been attended by more than nine hundred participants, while more than eight hundred papers have been presented, far more than any other previous ECF over a thirty year period. I am happy and proud to have welcomed in Alexandroupolis well-known experts, colleague, friends, old and new acquaintances who came from around the world to discuss problems related to the analysis and prevention of failurc in structures. The tranquility and peacefulness of the small town of Alexandroupolis provided an ideal environment for a group of scientists and engineers to gather and interact on a personal basis.

I wish to thank very sincerely the editor Dr. S. K. Kourkoulis for the excellent appearance of this volume and the authors for their valuable contributions. Finally, a special word of thanks goes to Mrs. Nathalie Jacobs of Springer who accepted my proposal to publish this special volume and her kind and continuous collaboration and support.

February 2006

Xanthi, Hellas

Emmanuel. E. Gdoutos
ECF16 Chairman

PART I:

Mechanical and Structural Aspects

Chapter 1: Fracture

Chapter 1.1

SUBCRITICAL CRACKING: A CAUSE OF ROCK PANEL FAILURE IN BUILDINGS

K.T. Chau[1], R.H.C. Wong[1], and T.-f. Wong[2]
[1]Department of Civil and Structural Engineering, The Hong Kong Polytechnic University, H.K., China; [2]Departments of Geosciences and Mechanical Engineering, State University of New York at Stony Brook, USA

Abstract: Stone and rock are among the most popular natural construction materials on earth. Dimension stones have been used in many historical buildings and rock panels have been used on the façade in most modern buildings. One of the major problems with stones and rocks is that pre-existing cracks and joints are naturally found in them. These cracks may grow after being exposed to prolonged actions of wind, sunshine and rain. Although rock panels are prone to cracking, fracture analysis has not been considered in their design. In this paper some recent efforts in understanding and modeling the cracking problem of rock panels on exterior cladding walls and façades using fracture mechanics analysis are summarized. This framework includes the use of Fracture Mechanics and sub-critical cracking concept (that is, crack growth is considered as a slow but ongoing process even when the local stress intensity factor at crack tip is less than the so-called fracture toughness). A short review of the problem is first given, the use of the concept of sub-critical cracking and Fracture Mechanics in modeling the cracking of façade rock panels is then discussed, and experimental efforts in testing the cracking process in these rock panels are finally mentioned. It is believed that the analysis of façade rock panels based on Fracture Mechanics will eventually become an indispensable part of rock panel design in near future.

Key words: façade rock panel; subcritical crack growth; fracture mechanics analysis.

1. INTRODUCTION

Dimensional stones (or stone blocks) have been used for buildings and structures since the ancient times. The oldest free-standing stone building is believed the temple of Gozo in Malta, built some 4,500 years ago. Thousand of years ago the Egyptians used blocks of stones in building the Pyramids, which are still considered as one of the wonders of the ancient world; and

S.K. Kourkoulis (ed.), Fracture and Failure of Natural Building Stones, 5–18.

the Chinese used dimensional stones in building the Great Wall of 2,150 miles long, which is arguably the only man-made structure on earth visible from the space. The famous Leaning Tower of Pisa is also built with marble stones. The Acropolis at Athens and the Colosseum of Rome are further known examples. The list continues.

Figure 1. Some famous ancient stone structures: Acropolis, Pyramids, Colosseum and Great Wall.

Although many new building materials have been invented in the last few thousand years, stone and rock still remain one of the most common choices for building, especially for cladding wall design, because of its shinning surface and aesthetic look. All kinds of rocks have been used in building façade, including granite, marble, limestone, sandstone, and quartz, to name a few. Figure 2 shows some modern tall buildings using rock panels on their façades. In modern buildings, rock panels or slabs are considerably thinner than those dimensional rock used in the ancient times. Typically, rock slabs or panels are seldom thicker than 30 mm. Although they are not load-bearing, they have to resist wind pressure on the building surface, they are subjected to daily sunshine, and they are prone to chemical corrosion due to acid rain. It is also because of this, that the façade rock slabs and panels of modern buildings are more conducive to weathering and cracking than the historical ones.

There have been numerous incidences of rock panel failure reported. One of the most famous rock panel failures is the notable example of 344 m tall Amoco Building (now renamed as Aon Center) in Chicago (Figure 3); and this case was well documented and has arouse international awareness on the safety and problem of cladding wall design (Trewhitt and Tuchmann, 1988;

Figure 2. Some famous modern buildings with rock panels on façade, including the Two Prudential Plaza in Chicago, Empire State Building in New York, MLC Center in Sydney, Society for Savings Building in Cleveland.

Anonymous, 1989; Ridout, 1989; Kent, 1990; Logan et al., 1993; Hook, 1994; Rudnicki, 2000). In 1985 bowing and cracking occurred on some 43,000 slabs of Italian Carrara marble of size 1.219 m × 0.9144 m on the external façade of this 80-story building. The marble slabs were completely replaced by granite slabs and the renovation cost in 1991 was US$ 80 millions, which is about half of the total cost of the whole building about 20 years ago. There are numerous examples of problematic façade rock panels, including the cases of Finlandia Hall in Helsinki (repairing cost of 4 millions Euros in 2000) (Royer-Carfagni, 1999a-b), and Lincoln First Tower in Rochester, New York (repairing cost of 20 million US$, in 1988). In Hong Kong, serious spalling and cracking appeared at the granite cladding to the 23-story Bank of East Asia Head-quarter building at Central in 1993, ten years after the building was completed. To prevent the granite slabs from falling off and endangering pedestrians, the Bank replaced the entire cladding at a cost of about HK$ 38 millions. The lawsuit of the Bank against the Architect and Sub-Consultant went all the way to the Court of Final Appeal in Hong Kong. Other examples of cracking problems in cladding can be found in the textbooks by Winkler (1975, 1994). For more information and potential problems on the use of dimensional stone in structures, the readers are referred to Winkle (1975, 1994), Lewis (1995), Franzini (1995), Chew et al. (1998), Smith (1999), Chacon (1999), Gauri and Bandyopadhyay (1999), and Bradley (2001).

The current design codes in the world for façade dimension stone do not include the consideration of potential cracking, which is inevitable in natural

Figure 3. Cracking in the Amoco Building in Chicago and other cracking observed on rock panel cladding.

stones; therefore the cracking problem of rock panels on commercial buildings deserves more detailed investigations. To date, no comprehensive study has appeared. In view of this, Chau and Shao (2006) proposed the use of Fracture Mechanics in modeling the time dependent problem of cracking in rock panels. As discussed in a later section, a simple analytical crack model has been proposed to investigate the crack growth mechanism, by incorporating sub-critical crack growth.

2. ROCK PANELS DESIGN FOR CURTAIN WALLS

Although stone or rock panels have been widely used in cladding all over the world, no international design standard exists (Cohen and Monteiro, 1991; Ruggiero, 1995). Currently, there are various design codes for testing stones for cladding panels, such as ASTM standard in USA, DIN standard in Germany, prEN standard in Italy, EN and WI00246 standards for European countries, BS standards in UK, and CSIRO "BEST" tests in Australia (Quick, 1998). These tests include compressive strength, flexural strength, modulus of elasticity, density, absorption, thermal conductivity, coefficient of thermal expansion, creep deflection, and resistance to chemical agents. Clearly, many of the stone panel failures on façade involve cracking, however none of these codes require the test of fracture properties of rock.

The anchoring system used for holding the stones is also of vital importance to the stability and cracking of rock panels, but no standardized anchoring system design has been adopted internationally. Each supplier has its own design details and concept for the anchoring system. Figure 4 shows some of the potential cracking related to different anchor systems. The main consideration for the design of anchors depends on the ability to resist wind

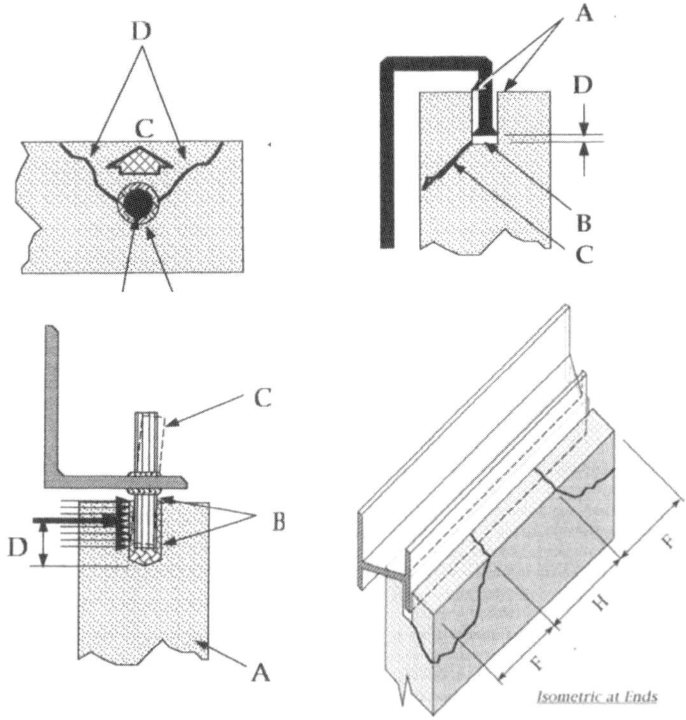

Figure 4. Some cracking patterns emerging from different anchor system of rock panels.

and seismic loads, and avoid bowing problems (Cohen and Monteiro, 1991). Stress concentration between the connection and the rock panels is known to be highly dependent on the details of anchoring system (Ho and Chau, 1997, 1999; Chau and Wei, 2001).

Although cracking is known to appear in cladding and dimensional stones (e.g. Simmons and Richter, 1993; Ayling, 2002), there is no theoretical study analyzing the problem under normal working conditions (i.e. the daily solar heating and wind load). Therefore, Chau and Shao (2006) considered a simple Fracture Mechanics analysis for panels subjected to periodic temperature variations on one surface whereas the other is kept at constant temperature. The end conditions of the panels can be either fixed or simply supported.

3. THEORETICAL MODELING BY CHAU AND SHAO (2006)

3.1 Concept of sub-critical cracking

The theoretical analysis by Chau and Shao (2006) will be briefly summarized in this section. The most likely failure mode of brittle rock panels is tensile cracking. The classical linear elastic fracture mechanics predict that as long as the stress intensity factor (SIF) at a crack tip is less than a critical value called fracture toughness K_{IC}, the crack is stable and no crack propagation will occur. However, failure cases (such as the Amoco case in Chicago and the Bank of East Asia case in Hong Kong) and creeping experiments on rock specimens clearly show that crack propagation did occur even when the sustained SIF is less than K_{IC} as long as a threshold value is exceeded. This phenomenon is known as sub-critical cracking in geophysics and can occur in rocks under high pressure and temperature (Atkinson, 1984; Atkinson and Meredith, 1987). Clearly, this kind of sub-critical crack growth can provide a theoretical basis for long term cracking phenomenon of rock panels on cladding under serviceability condition. For example, cracking of the Bank of East Asia cladding appeared only 10 years after its completion. Therefore, subcritical crack growth in rock panels in many existing and new structures deserves more detailed investigation.

To examine the time-dependent cracking and failure of rock panels on cladding wall, subcritical cracking resulted from periodic solar heating and wind loads was considered by Chau and Shao (2006). They considered the subcritical cracking of either an edge or a center crack in an elastic strip of finite thickness with both free and fully constrained end boundaries subject to periodic temperature variation on one surface (i.e. simulated solar heating on rock panels) while the other is kept at a constant temperature (i.e. simulated constant indoor temperature in the building), as shown in Figure 5. Both of these edge and center cracks are assumed perpendicular to the surface of the elastic strip since this appears to be the most crucial situation. Physically, if a crack (either edge or center) is inclined to the strip surface, the temperature field is being disturbed across the thickness, at least around the crack, such that the temperature field is no longer one-dimensional. It is because a layer of air is expected to be trapped in the crack, which changes the uniformity of the conductivity across the thickness. Thus, the assumption of perpendicular crack reflects the most crucial situation, and at the same time simplifies the problem mathematically.

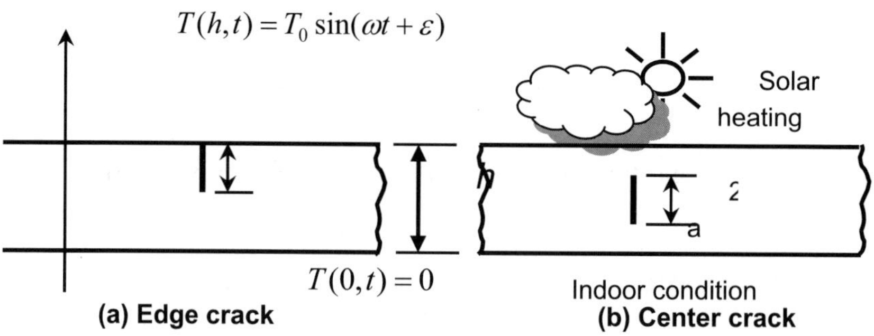

Figure 5. Rock slabs subject to constant indoor temperature and periodic outdoor solar heating: (a) An edge crack of size a; (b) A center crack of size $2a$.

3.2 Temperature and stress fields in the slab subject to periodic heating

Consider a finite slab of thickness h subject to periodic heating on the surface $x=h$ while the temperature is kept at constant on $x = 0$ (see Figure 5). If the coupling between the temperature field and the deformation is negligible, the heat conduction within the slab is governed by the standard diffusion equation and the solution of this problem is given in Section 3.6 of Carslaw and Jaeger (1959) as:

$$T(x,t) = T_0 \psi(x,t) \tag{1}$$

where

$$\psi(x,t) = \Omega(x)\sin[\omega t + \varepsilon + \phi(x)] + \frac{2\pi\kappa}{h^2} \times \sum_{n=1}^{\infty} \frac{n(-1)^n(\alpha_n \sin\varepsilon - \omega\cos\varepsilon)}{\alpha_n^2 + \omega^2} \sin\left(\frac{n\pi x}{h}\right) \exp(-\alpha_n t) \tag{2}$$

where $\alpha_n = \kappa n^2 \pi^2 / h^2$ and the amplitude $\Omega(x)$ and phase $\phi(x)$ of the temperature oscillations at point x are given by:

$$\Omega(x) = \left\{\frac{\cosh(2kx) - \cos(2kx)}{\cosh(2kh) - \cos(skh)}\right\}^{1/2}, \quad \phi = \arg\left\{\frac{\sinh kx(1+i)}{\sinh kh(1+i)}\right\} \tag{3}$$

In addition, the heat wave number k and the imaginary constant are denoted by $[\omega/(2\kappa)]^{1/2}$ and i respectively. The first term on the right hand of Eq.(2) is the periodic steady state solution and the second is the transient term which dies out quickly with the summation index n.

The non zero thermal stress can be shown that is equal to:

$$\sigma_{zz} = C_{11}[A(t)x + B(t)] - \lambda T_0 \psi(x,t), \; \lambda = \frac{(1-\nu)\alpha E}{(1-2\nu)(1+\nu)} \tag{4}$$

where c_{11} is the elastic modulus, E and ν are the Young's modulus and Poisson's ratio, α is the linear coefficient of thermal expansion, and the constants A and B (as a function of time) depends on the end condition of the rock panels. If the slab is constrained, ε_{zz} will be identically zero which in turn leads to $A=B=0$. But, in cladding design, the rock panels are normally separated by sealant, epoxy or cement paste so that a free boundary condition may not be inappropriate. In particular, if the slab is free to expand and is moment free at the end, we have the following expressions for A and B:

$$A(t) = \frac{6T_0\gamma}{h}\left[\frac{2}{h}\int_0^h x\psi(x,t)dx - \int_0^h \psi(x,t)dx\right],$$
$$B(t) = 2T_0\gamma\left[2\int_0^h \psi(x,t)dx - \frac{3}{h}\int_0^h x\psi(x,t)dx\right] \quad (5)$$

where $\gamma=\lambda/(c_{11}h)$. Note that since the temperature field is a function of time, both A and B are functions of time as well. In real situations, the end boundary conditions can be somewhat between free and fixed. Therefore, the end conditions considered here cover the most extreme cases.

Plots for the temperature variation at various normalized depths are shown in Figure 6. It can be seen that the temperature variation is roughly periodic with time. The plots of the corresponding stress were shown in Figure 7.

3.3 Stress intensity factors for a cracked slab subjected to periodic surface heating

The solution for the stress intensity factor can be obtained by using a fundamental solution for the crack problem for both edge crack of size a and for a center crack of size $2a$ (e.g. Rooke and Cartwright, 1976). Details can be found in the paper by Chau and Shao (2006). Applying the principle of superposition, the problem of a crack subjected to a temperature field given in (2) can be decomposed into two associated problems: (I) a non-cracked slab subject to a temperature field given in Eq.(2); and (II) a cracked-slab subjected to an internal stress field which is generated on the position of the crack in Associated Problem I. Since the stress field is not singular anywhere in the slab in Associate Problem I, only the Associated Problem II contributes to the calculation of the stress intensity factor at the crack tip.

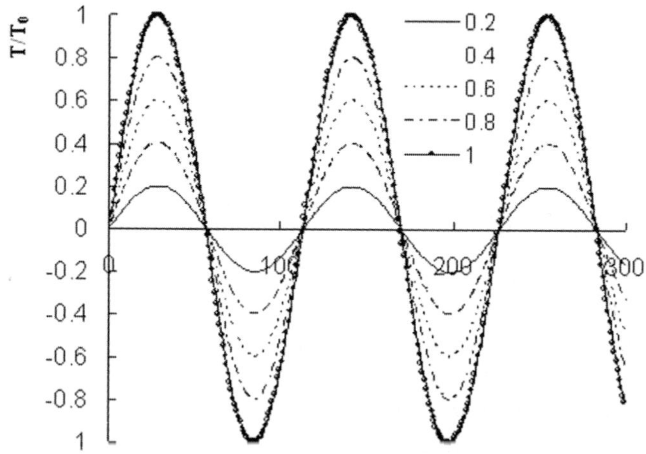

Figure 6. The temperature variation at various depths versus time.

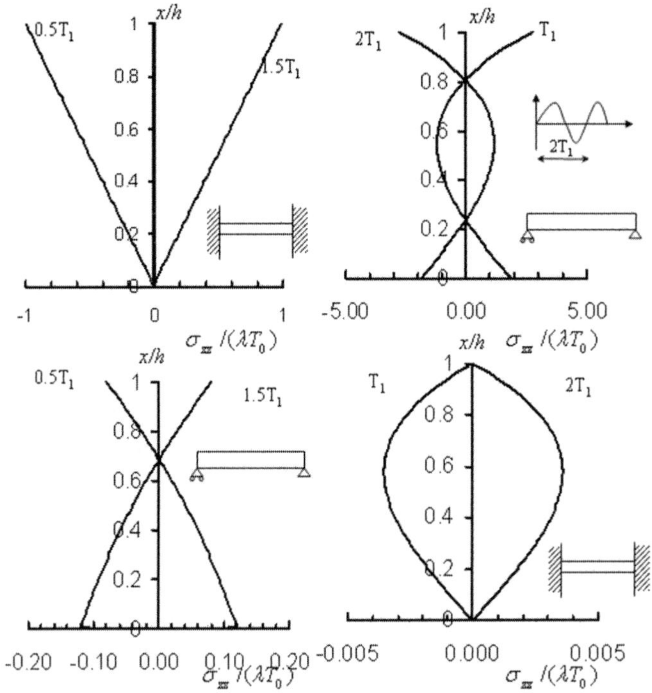

Figure 7. Stress distribution at various time for both fixed and simply support conditions.

For the case of an edge crack, using the fundamental solution due to a pair of point loads *P* applied on the crack faces given by Tada et al. (1985), the mode I stress intensity factor can be obtained as:

$$K_I(\eta) = 2\sqrt{\frac{a}{\pi}} \int_0^1 \sigma_{zz}^*(\xi, t) F_I(\xi, \eta) d\xi \tag{6}$$

where ξ=c/α, η=α/h and

$$\begin{aligned} F_I(\xi,\eta) &= \frac{3.52\xi}{(1-\eta)^{3/2}} - \frac{4.35 - 5.28(1-\xi)}{\sqrt{1-\eta}} \\ &+ \left\{ \frac{1.3 - 0.3(1-\xi)^{3/2}}{\sqrt{\xi(2-\xi)}} + 0.83 - 1.76(1-\xi) \right\}(1-\xi\eta) \end{aligned} \tag{7}$$

$$\begin{aligned} \sigma_{zz}^*(\xi,t) = T_0\varphi(\xi,t) &= \sigma_{zz}(\xi,t) \qquad && \sigma_{zz} > 0 \\ &= 0 && \sigma_{zz} \le 0 \end{aligned} \tag{8}$$

A similar expression can be also obtained for the case of the center crack shown in Figure 2b (details in Chau and Shao, 2006). The integration of Eq.7 can be done by following a standard procedure using Simpson's rule with error control (e.g. Press et al., 1992). A typical plot for the crack length vs. time is given in Figure 8 for both cases of center and edge cracks subjected to both free and fixed end conditions. It can be seen that crack growth is the slowest for the case of a center crack subjected to fixed condition whereas a centered crack subjected to free end conditions is most conducive to crack growth. A detailed parametric study is carried out by Chau and Shao (2006).

4. EXPERIMENTAL STUDIES

Early experiments on the performance of marble slabs can trace back to Rayleigh (1934). In the literature, either fatigue crack test or subcritical crack test has been conducted on rocks, but not both. As discussed by Hertzberg (1996), it is possible to incorporate both fatigue and subcritical crack growths:

$$\left(\frac{da}{dN}\right)_T = \left(\frac{da}{dN}\right)_{Fat} + \int \frac{da}{dt} K(t) dt \tag{9}$$

where N is the number of loading cycles.

4.1 Subcritical crack growth test

The parameters for sub-critical and fatigue crack growth have been determined experimentally in an accompanying paper by Kwok et al (2006). First of all, the usual fracture toughness test will be conducted in order to de-

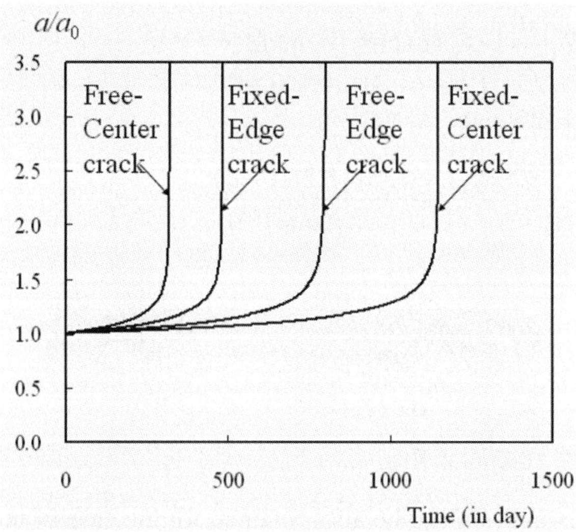

Figure 8. Crack growth versus time for both edge and center subjected to free and fixed boundary conditions.

termine the critical stress intensity factor. For subcritical cracking parameters, the effect of corrosive stress will be simulated by immersing a four-point-loaded beam with a central crack or notch in a solution of diluted sulphuric acid, with the crack growth rate and opening displacement rate being monitored as a function of time. The loading for various levels of stress intensity factor (K) will be applied so that the threshold K value for the onset of subcritical crack growth can be established. A typical set-up is shown in Figure 9.

4.2 Fatigue crack growth test

Fatigue tests on marbles have been done by Royer-Carfagni and Salvatore (2000) and Pino et al. (1999), but no test on fatigue crack growth has been conducted. For fatigue crack growth, the loading on the four-point-bending cracked beam will be added and removed periodically. The crack length will be monitored as a function of time. To separate the effect of subcritical cracking, the experiments will be repeated but with the beam being immersed in either carbonic acid or sulfuric acid. These experiments are still in progress.

4.3 Thermal stress test

The experiments described above are only for fatigue crack propagation under pure mechanical loading, such as those induced by wind loads. In addition, rock panels on cladding walls are also subjected to periodic (or more precisely close to periodic) heating from the sun. Therefore, the fatigue crack experiments will be repeated by combining both the effect of mechanical load and temperature (e.g. Mahmutoglu, 1998). In this test, the mechanical load will be kept constant while periodic heating will be applied to the cracked side of the beam and the other side will be kept at a constant temperature (e.g. using a wet sand bed). The results of these experiments will be reported in our forthcoming publications.

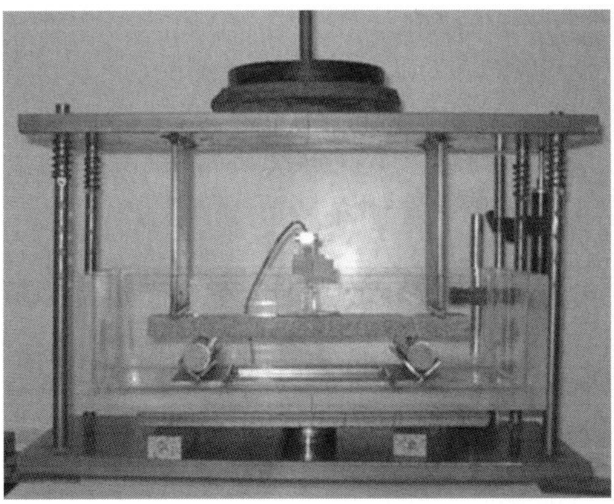

Figure 9. Typical experimental set-up for sub-critical cracking.

5. CONCLUSIONS

In this paper, a systematic Fracture Mechanics framework has been summarized, which can be used in analyzing the fatigue life of rock panels subjected to periodic sunshine, which is prescribed as periodic surface temperature variations. In this analysis, pre-existing micro-cracks are modeled as either edge cracks or center cracks. To cater for the most extreme anchor system in real situations, both fixed and free end conditions have been considered. First, temperature variation in the rock panels is obtained. The corresponding thermal stress is then calculated. Using fundamental solution for the crack problem and the principle of superposition, the stress intensity factor versus time can be estimated. This result can be used to integrate for the sub-critical crack growth. Therefore, understanding of the cracking problem of rock panels on exterior cladding walls and façades as a function of time is achieved by fracture mechanics approach. This frame-work includes the use of Fracture Mechanics and sub-critical cracking concept (that is, crack growth is considered as a slow but ongoing process even when the local stress intensity factor at crack tip is less than the so-called fracture toughness). In addition, series of experimental efforts in testing the parameters for sub-critical cracking have been summarized.

Currently the authors are carrying out experiments on marbles (Wong et al., 1995, 1996) and other rocks, extending the Fracture Mechanics analysis to slabs containing three-dimensional cracks. It is believed that Fracture Mechanics analysis for façade rock panels will eventually become an indispensable part of rock panel design in the future.

ACKNOWLEDGEMENTS

The work described in this paper was fully supported by a grant from the Research Grants Council of the Hong Kong Special Administrative Region, China (Project No. PolyU 5166/04E). The present work was motivated by a review article by Rudnicki (2000) in geomechanics.

REFERENCES

Anonymous, 1989, Granite will replace old marble on Amoco façade, *Civil Engineering*, **59**(6): 19-20.

Atkinson, B.K., 1984, Subcritical crack growth in geological materials, *Journal of Geophysical Research*, **89**(B6): 4077-4114.

Atkinson, B. K., and Meredith, P. G., 1987, The theory of subcritical crack growth with applications to minerals and rocks, In: Atkinson, B. K., ed., *Fracture Mechanics of Rocks*, Academic Press, London, pp. 111-166.

Ayling, L., 2002, A solution for the stone repair of a cracked primary column at the Wellington Arch, Hyde Park Corner, London, in *Stone-Stone Building Materials, Construction and Associated Component Systems: Their Decay and Treatment,* ed. J. Fidler, James & James, London.

Bradley, F., 2001. *Fine Marble in Architecture*, WW Norton, New York.

Carslaw, H.S., and Jaeger, J.C., 1959, *Conduction of Heat in Solids*, second edition, Oxford University Press, Oxford.

Chacon, M.A., 1999, *Architectural Stone: Fabrication, Installation, and Selection*, Wiley, New York.

Chau, K.T., and Wei X.X., 2001, Stress concentration reduction at a reinforced hole loaded by a bonded circular inclusion, *Journal of Applied Mechanics ASME*, **68**(3): 405-411.

Chau, K. T. and Shao J. F. 2006. Subcritical crack growth of edge and center cracks in façade rock panels subject to periodic surface temperature variations, *International Journal of Solids and Structures,* **43**, 807-827 (2006).

Chew, M.Y.L., Wong, C.W., and Kang, L.H. 1998, *Building Façade: A Guide to Common Defects in Tropical Climates,* World Scientific, Singapore.

Cohen, J. M., and Monteiro, P.J.M. 1991. Durability and integrity of marble cladding: A state-of-the-art review, *ASCE, Journal of Performance and Constructed Facilities* **5**(2): 113-124.

Franzini M., 1995. Stones in monuments - natural and anthropogenic deterioration of marble artefacts, *European Journal of Mineral*, **7**(4): 735-743.

Gauri, K.L., and Bandyopadhyay, J.K., 1999. *Carbonate Stone: Chemical Behavior, Durability, and Conservation*, Wiley, New York.

Hertzberg, R.W., 1996, *Deformation and Fracture Mechanics of Engineering Materials*, fourth ed., Wiley, New York.

Ho, K.C., and Chau, K.T., 1997, An infinite plane loaded by a rivet of a different material, *International Journal of Solids and Structures*, **34**(19): 2477-2496.

Ho, K.C., and Chau, K.T., 1999. A finite strip loaded by a bonded-rivet of a different material, Computers and Structures, **70**(2): 203-218.

Hook, G., 1994.,Look out below- the Amoco buildings cladding failure: what went wrong, how it was fixed and the implications for thin-stone veneers, *Progressive Architecture* **75** (2): 58-61.

Jaeger, J.C., and Cook, N.G.W., 1976, *Fundamentals of Rock Mechanics*, second edition, Halsted Press, New York.

Kent, C., 1990, Skin graft for Amoco Building, *Progressive Architecture* **71**(10): 23.

Kwok, K.W., Wong, R.H.C., Chau, K.T. and Wong T.-f., 2006, Studies on sub-critical crack growth in façade rick panel using four-point bending, paper in this volume.

Lewis, M.D., 1995, *Modern Stone Cladding: Design and Installation of Exterior Dimension Stone Systems,* ASTM Manual Series: MNL 21, ASTM Philadelphia.

Logan, J.M., Hastedt, M., Lehnert, D., and Denton, M., 1993. A case study of the properties of marble as building veneer, *Int. J. Rock Mech. Min. Sci. & Geomechanics Abstr.* **30**(7): 1531-1537.

Mahmutoglu, Y., 1998, Mechanical behaviour of cyclically heated fine grained rock, *Rock Mechanics and Rock Engineering*, **31**(3): 169-179.

Press, W.H., Flannery, B.P., Teukolsky, S.A., and Vetterling, W.T., 1992, *Numerical Recipes: The Art of Scientific Computing,* second edition, Cambridge University Press, New York.

Quick, G.W., 1998. CSIRO Methods for Testing Stone Tiles, *CSIRO BCE* Doc. 98/016.

Pino, R.D., Narducci P. and Royer-Carfagni, G.R. 1999, A SEM investigation on fatigue damage of marble. *Journal of Materials Science Letters,* **18**: 1619-1622.

Rayleigh, Lord, 1934, The bending of marble, *Proceedings of the Royal Society of London. Series A*, **144**(852): 266-279.

Ridout, G., 1989, Losing its marble, *Building*, April 28, 1989,**17**: 61.

Rooke, D.P., Cartwright, D.J., 1976. Compendium of Stress Intensity Factors. Her Majesty's Stationery Office, London.

Royer-Carfagni, G., 1999a. Some considerations on the warping of marble facades: the example of Alvar Aalto's Finland Hall in Helsinki, *Construction and Building Materials*, **13**(8): 449-457.

Royer-Carfagni G.F., 1999b. On the thermal degradation of marble, *International Journal of Rock Mechanics and Mining Sciences*, **36**(1): 119-126.

Royer-Carfagni C, and Salvatore W., 2000, The characterization of marble by cyclic compression loading: experimental results, *Mechanics of Cohesive and Frictional Materials*, **5**(7): 535-563.

Rudnicki, J.W. 2000. Geomechanics. *Int. J. Solids and Struct.*, **37**(1-2): 349-358.

Ruggiero, S., 1995. Cladding's ticking time bombs, *Progressive Architecture* **76**(12): 90-94.

Simmons, G., and Richter, D., 1993. Cracks in building stone, *Int. J. Rock Mech. Min. Sci. & Geomechanics Abstr.* **30**(7): 1553-1557.

Smith, M.R., 1999. *Stone: Building Stone, Rock Fill and Armourstone in Construction*, Geological Society Engineering Geology Special Publication No. 16, The Geological Society, London.

Tada, H., Paris, P.C., and Irwin, G.R., 1985, *The Stress Analysis of Crack Handbook,* Del Research Corporation, St. Louis.

Trewhitt, J., Tuchmann, J., 1988. Amoco may replace marble on Chicago headquarters, *Engineering News Record*, March 24, 11-12.

Winkler E.M., 1975, *Stone: Properties, Durability in Man's Environment*, 2nd edition, Springer, Wien.

Winkler E.M., 1994, *Stone in Architecture,* 3rd edition, Springer, Berlin.

Wong, R.H.C., Chau, K.T., and Wang, P., 1996, Microcracking and grain size effect in Yuen Long marbles, *Int. J. Rock Mech. Min. Sci. & Geomechanics Abstr.* **33**(5): 479-485.

Wong, R.H.C., Chau, K.T., and Wang, P., 1995, Microcracking in coarse and fine grain marbles, in: *Rock Mechanics, Proceedings of the 35th U.S. Symposium*, pp. 477-482.

Chapter 1.2

STUDIES ON SUBCRITICAL CRACK GROWTH IN FAÇADE ROCK PANEL USING FOUR-POINT BENDING

K.W. Kwok[1], R.H.C. Wong[1], K.T. Chau[1] and T.-f. Wong[2]
[1]*Dept. of Civil and Structural Engng, The Hong Kong Polytechnic University, H.K., China;*
[2]*Depts. of Geosciences and Mech. Engng, State University of New York at Stony Brook, USA*

Abstract: Thin rock panels are commonly used on exterior cladding walls in high-rise buildings. However, it has been found that these rock panels on exterior cladding may develop long-term time-dependent cracking, due to stress concentrations induced by periodic wind load and sunshine, and stress corrosion due to environmental effects (such as acid rain). This kind of cracking at stress level lower than that required for overcoming fracture toughness can be explained by using the concept of subcritical crack growth (SCG) of the pre-existing microcracks. However, existing design requirements of rock panel do not account for SCG. The standard testing method for SCG is the double torsion test (DTT), which does not simulate the failure condition of rock panel under wind and sunshine. Thus, a new method is introduced to investigate the SCG in rock panel under different environmental conditions: the four-point bending (4PB) test. Rock panel specimens containing a central notch were immersed in water, acid and air buffer during the loading test. For comparison DTTs were also conducted. It was found that the crack growth rate increases drastically if the cracked specimen is moved from a water buffer to a set-up with an acidic buffer. The SCG index n (a larger value indicates a faster crack growth) obtained by 4PB test is found consistently lower than that determined by DTT. This may be due to the different fracture mechanisms activated. One distinct feature of the 4PB test is that the SCG in rock panel is monitored through the crack length measurement whereas no such measurement is possible in DTT.

Key words: façade rock panel; subcritical crack growth; four-point bending test; granite; double torsion test.

1. INTRODUCTION

Traditionally, rock panels have been used for façade cladding for their aesthetic appearance. However, rock panels on exterior cladding walls are subjected to mechanical loads due to wind pressure, thermal loads due to sun-

S.K. Kourkoulis (ed.), Fracture and Failure of Natural Building Stones, 19–34.

Figure 1. Typical cracking on granite cladding panels (Chew et al.[1]).

shine, and chemical corrosions due to polluted moisture air. Due to these loading and corrosion, cracking is not uncommon on rock panels in exterior cladding after several years of exposure to these external effects and loads. This is particular so for the case of thin rock slabs (< 30mm). Figure 1 shows some examples of cracking of rock panels on building façade.

Although rock panels have been widely used on exterior cladding walls and cracking in rock panels has been reported repeatedly worldwide, there is still no internationally accepted design standard for rock panels or slabs[2]. There are, however, various design codes in different countries for testing the mechanical properties of dimensional stones for cladding installation, but none of these codes requires testing the fracture properties of rock. In the conclusion of Bayer and Clift[3], they mentioned that the use of Fracture Mechanics might be important for designing rock cladding wall. Motivated by this application, Chau and Shao[4] developed the first Fracture Mechanics based model in predicting the life of rock panels subjected to periodic sunshine using the concept of subcritical crack growth.

According to the classical theory of linear Fracture Mechanics, a crack is considered to be stable and does not grow when the stress intensity factor K_I at a crack tip is smaller than the critical stress intensity factor K_{IC} or the fracture toughness. However, as summarized by Atkinson[5] a crack can propagate even at a value of K_I smaller than that of the critical value (i.e. fracture toughness, K_{IC}). In geophysics, such crack growth phenomenon is referred as subcritical crack growth (SCG). Crack growth in rock panel appears only years after its completion and crack propagation appears to occur under service loadings. This indicates that crack growth in rock panels induced by periodic thermal loading due to sunshine and corrosion may be considered as a kind of SCG. This is the idea behind the model proposed by Chau and Shao[4]. However, the existing SCG parameters have been obtained mainly under the double torsion test which is quite different from the cracking mode in external cladding. Therefore, it is worthwhile to investigate the fracture behavior of rock panels under bending, which is believed to be the main loading mode on rock panels.

The time for rock panel to crack substantially depends on the rate of SCG, *da/dt* where *da* is the crack growth length and *dt* is the corresponding time interval. The relation between the SCG rate and the K_I can be determined by the power law suggested by Charles[6] and modified by Miura et al.[7] and others:

$$da/dt = v_0 (K_I/K_0)^n \tag{1}$$

where v_0 and n the crack velocity at K_I=K_0 and the SCG index respectively. These two important parameters depend on the type of rock and on environmental effects such as acidity and temperature of the atmosphere. Previous experimental data[8] suggested that there are three regions of SCG for most homogenous materials like ceramics and glass (Figure 2). The main factor affecting the crack growth was explained by the chemical reaction near the crack tip at low K_I (region I), the diffusion of corrosive species to the crack tip at intermediate K_I (region II) and mechanical rupture at large K_I (region III).

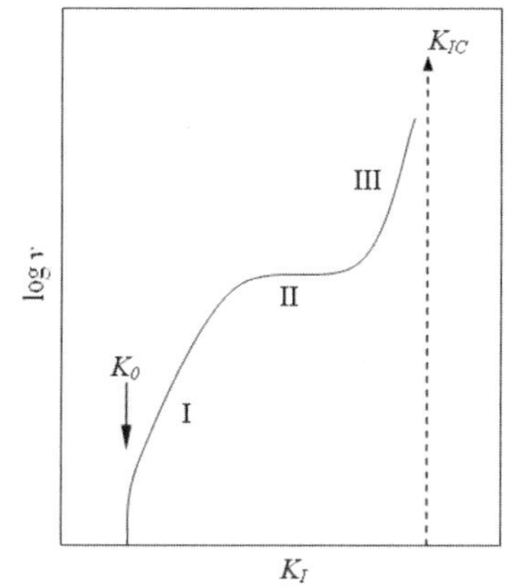

Figure 2. Schematic stress intensity factor/crack velocity diagram (Evans[8]).

Atkinson[5] summarized a wide range of SCG studies on granitic rocks using DTT method and the majority of the results indicated a linear log K_I – log v graph, which is different from that with three regions shown in Figure 2. This crack growth behavior is investigated in the present study.

Although double torsion test has been adopted by many researchers[5, 8-14], for testing the subcritical cracking parameters, as remarked earlier rock panels

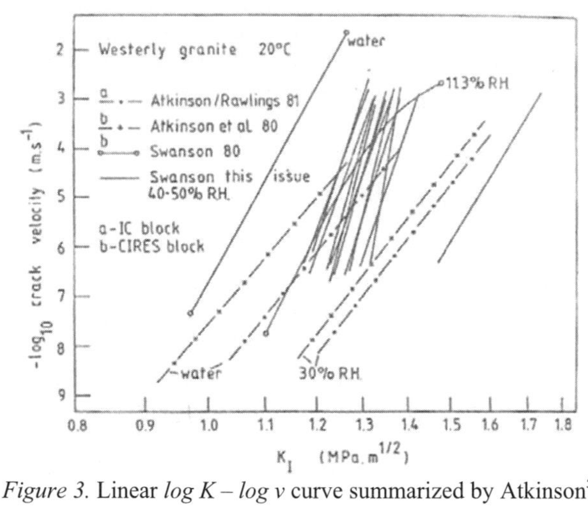

Figure 3. Linear *log K – log v* curve summarized by Atkinson[5].

are likely to bend under both thermal stress induced by sunshine and wind loads. Thus, to simulate a more realistic condition of cracking, 4PB test is proposed for subcritical cracking test in this study. It should be noted that the bending stress is constant in the middle one-third of the rock panel under four-point bending test. For comparison reasons, DTTs were also carried out. Both the $K - v$ curve and the SCG index n are compared.

2. EXPERIMENTAL DETAILS

In this section, the 4PB test method is first presented. Both the theory and the experimental setup are described in details in the following sections.

2.1 Four-point bending test method

2.1.1 Specimen preparation

The rock samples used in this study are fine-grained granite purchased from Fujian, China. The rock contains about 50-55% feldspar, 30-35% quartz, 8-10% mica, and 1-2% opaque minerals. The physical properties of the rock specimen are: density, ρ=2630 kg/m^3; compressive strength, σ_c=197 MPa; indirect tensile strength, σ_t=9.6 MPa; Young's modulus, E=54 GPa; Poisson's ratio, υ=0.387; and porosity=0.9%. The dimensions of the specimens are 300 mm × 30mm × 20mm. Both top and bottom surfaces of all samples are grinded and polished to within a deviation of less than 0.05 mm from the target size using grinding machine. A notch is introduced by a cutting saw. The notch depth (a_0) to thickness (a/W) ratio is 0.5 and the notch width is approximately 1 mm (note that a/W ratio should not be greater than 0.7 according to Murakami[15]). Pre-cracking (a small natural crack is first created at the notch by a small loading) is a general practice used in the double torsion test in studying subcritical crack growth. However, as it was difficult to obtain a stable pre-crack in the 4PB test (it is an inherent unstable test), the specimen is not pre-cracked in this study.

2.1.2 Experimental setup

4PB tests are performed using a four-point bending fixture with inner and outer spans of 200mm and 240mm respectively (Figure 4). Both for the loading points and the supports steel rollers are used. The load P is induced as dead weight and it is recorded by a load cell. A clip gauge is used to record the crack mouth opening displacement. Then the growth of crack length, Δa can be determined by compliance calibration (which will be discussed in §2.1.4). The experimental setup of the 4PB test is shown in Figure 5.

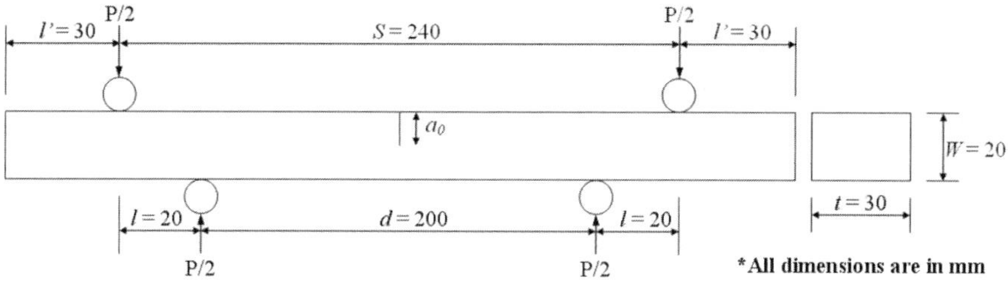

Figure 4. Schematic diagram of rock specimen subjected to 4PB test.

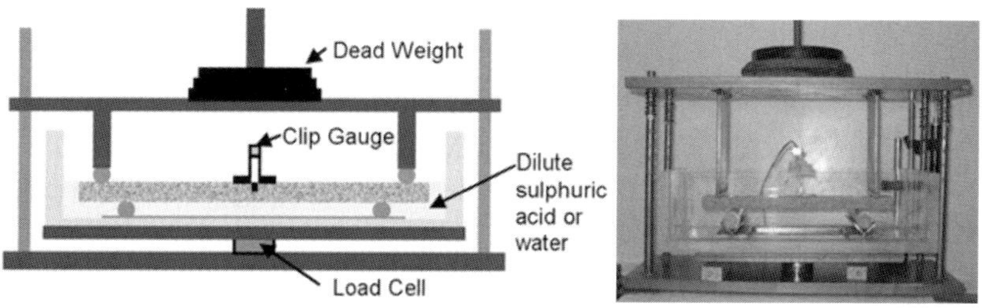

Figure 5. Experimental setup of 4PB test.

2.1.3 Determination of stress intensity factor (K_I)

The Stress Intensity Factor (SIF) K_I at the central notch tip can be calculated in terms of the applied load *P* by the following formula (Murakami[15]):

$$K_I = (3Pl/tW^2) \times (\pi a)^{1/2} \times F_{IM}(\alpha) \tag{2}$$

$$F_{IM}(\alpha) = 1.122 - 1.121\alpha + 3.74\alpha^2 + 3.873\alpha^3 - 19.05\alpha^4 + 22.55\alpha^5 \tag{3}$$

where $\alpha = a/W$ and a is the current crack length defined as the sum of the crack growth (Δa) and the notch length (a_0). The geometric parameters l and t are defined in Figure 4. The fracture toughness can be determined by substituting the critical loading P_c (when crack growth starts) in place of P in Eq.(2) and setting K_I to K_{IC}.

2.1.4 Crack length determination by using compliance calibration

The growth of crack length is monitored by the crack mouth opening displacement (CMOD). The CMOD is measured as a function of time by a clip gauge mounted at the two edges of the notch (Figure 5). The calibration between CMOD and crack length is conducted according to the calibration method proposed by Schmidt[16].

The specimen with a notch length a_1 is loaded with dead weight using the setup mentioned in §2.1.2. The CMOD is measured by a clip gauge and the corresponding load is measured by a load cell. The initial slope of the load vs. CMOD curve, called compliance, is then obtained (see Figure 6 for various *a/W* ratios). The notch is then cut deeper to a value a_2 and the process is repeated. The compliance is then determined for each *a/W* ratio and a calibration curve is plotted (Figure 7). Using this calibration curve, crack growth length under 4PB can be determined by measuring the specimen compliance.

2.1.5 Subcritical crack growth (SCG) test

The fracture toughness of granite obtained by the method described in §2.1.3 is 0.766 MPam$^{1/2}$. This value is used to estimate the pre-determined load for subcritical crack growth test. Dead weight leading to 90% K_{IC} is applied to the specimen. The CMOD is recorded as a function of time throughout the duration of the test. The compliance (mm/kN) can be calculated using the applied load and the corresponding CMOD. Thus, the crack length growth (*a/W*) at the corresponding time can be determined using Figure 7. The crack growth rate *v* at each time interval can be calculated using Eq. (4).

$$v = da/dt \approx [(a_1 - a_2)/(t_1 - t_2)] \quad (4)$$

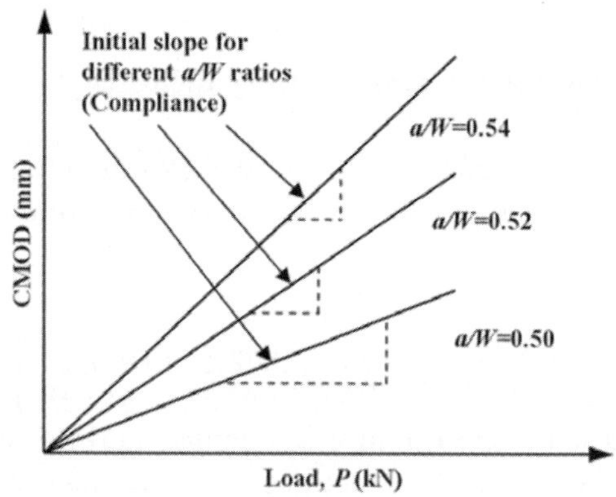

Figure 6. Determination of the initial slope of the load versus CMOD curve.

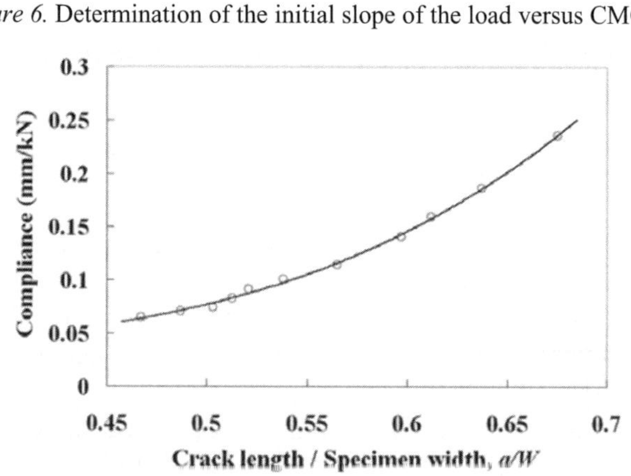

Figure 7. Compliance calibration curve of granite specimen.

where a_1 and a_2 are the crack lengths at times t_1 and t_2 respectively. The SIF, K_I is determined from the pre-determined load using Eqs.(2) and (3). Therefore, K_I/K_{IC} versus v can be plotted. To study the environmental effects, the experiments will be repeated by testing the specimens in water and in dilute sulphuric acid (H_2SO_4) and air, respectively. The experimental results of the environmental effect will be discussed in §3.1.

2.2 Double Torsion Test (DTT) method

2.2.1 Specimen preparation

The dimensions of granite specimens for the DTTs are: length 150 mm, width 60 mm and thickness 4 mm, as shown in Figure 8. Both top and bottom surfaces of the samples are grinded and polished to within a maximum tolerance of 0.05 mm of the specified size using a grinding machine according to the criteria suggested by Atkinson[9]. They were parallel within 0.025 mm.

A central axial 1 mm width groove is cut along the length of each specimen to approximately one-third of its thickness, t, using a table-cutting saw. A notch of about 10 mm long is cut in one end of each specimen along the line of axial groove. An initial crack (or so-called pre-crack) from the notch is first induced by using a displacement rate of 0.05 mm/min through a displacement control loading machine. The onset of cracking is indicated by an instantaneous slope change of the load-time curve.

2.2.2 Subcritical Crack Growth (SCG) test

SCG parameters from DTT are obtained from the load relaxation test. In this test, a pre-determined loading P_i (~90% P_c) is applied at 1.0 mm/min so that no significant crack extension occurs before relaxation. When the load P_i is reached and the crosshead is stopped, the load relaxation is recorded vs. time (Figure 9). A more detailed description of the load relaxation test is given by Henry et al.[10] and Meredith & Atkinson[11]. The SIF K_I at the crack tip is given by Williams and Evans[14] as:

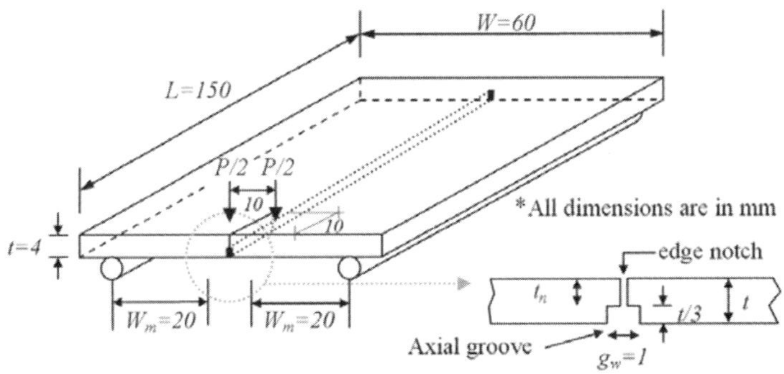

Figure 8. Schematic drawing of the double torsion specimen.

$$K_I = PW_m[3(1 + \upsilon)/Wt^3t_n]^{1/2} \tag{5}$$

where P is the corresponding load, W_m is the semi-moment arm of the specimen, υ is Poisson's ratio, W is the specimen width, t is the specimen thickness, and t_n is the reduced thickness in the plane of the groove (see Figure 8). The crack growth rate, v, is obtained as:

$$v = -(a_0P_0/P^2) \times (dP/dt) \tag{6}$$

where dP/dt is the slope of the relaxation curve at load P. The constants a_0 and P_0 are the initial crack length and the corresponding load. It is noted that the determination of both K_I and v is independent of crack length. This is regarded as one of the advantages of the double torsion test. However, it may also be viewed as a serious limitation as it is clear that stress intensity factor must depend on the crack size, no matter how small is the change.

After the relaxation test, the specimen is fractured at a rapid speed of 10 mm/min. The critical load P_C for fracturing the specimen is used to determine the fracture toughness K_{IC} from Eq.(5). The experimental results of double torsion test will be discussed in Section 3.2.

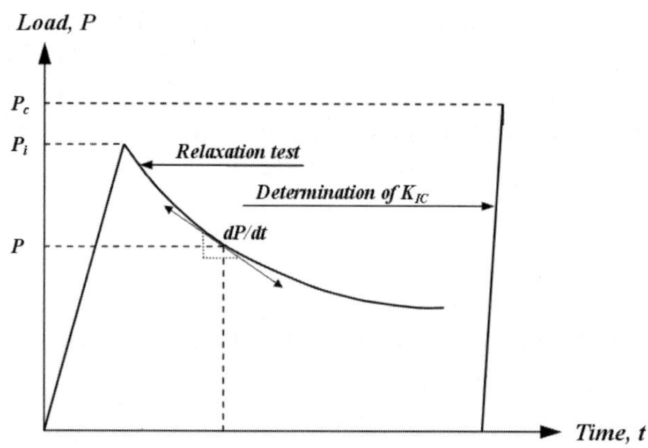

Figure 9. Schematic representation of load relaxation test. (Henry et al.[10]).

3. RESULTS AND DISCUSSION

In this section, the SCG results of both four-point bending and double torsion test are presented and compared.

3.1 Four-point bending test

The log-log plot of the crack growth rate, v, versus normalized SIF, K_I/K_{IC}, of 4PB tests on granite specimens in the environments of air, water and dilute sulphuric acid (pH value of 2) are shown in Figure 10.

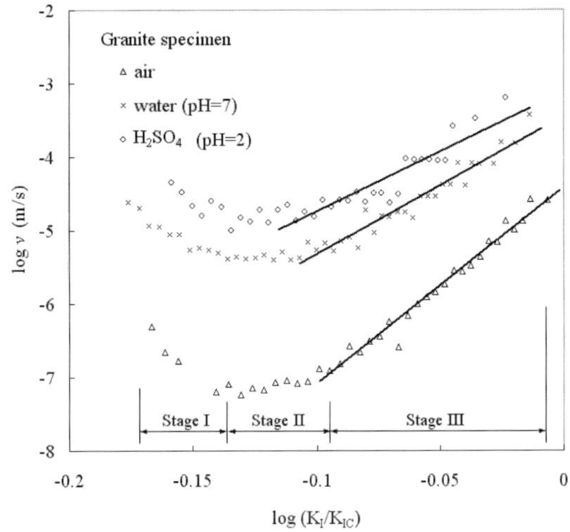

Figure 10. Log (K_I/K_{IC}) – log v curve for granite using four-point bending test.

Three stages are observed in granite specimen under different testing conditions from Figure 10. In Stage I, the crack growth rate v decreases after the application of pre-determined dead weight has been just completed. Then, v attained a minimum value when the normalized SIF (K_I/K_{IC}) reaches about 0.73, that is, log (K_I/K_{IC})= –0.137 in Figure 10. In Stage II, v becomes roughly constant over a certain range of K_I/K_{IC} while the range of K_I/K_{IC} for constant v is shorter in acidic condition (Figure 10). In Stage III, v increases and the crack propagates towards the lower edge, splitting the specimen into two pieces. To the best knowledge of the authors such a crack growth behavior with three different stages has not been reported in the literature on SCG studies using the DTT method (Figure 2). This may be due to different loading processes between 4PB test (constant loading during testing) and DTT (load relaxation during testing).

To further examine the crack growth behavior in 4PB tests, the curves of both CMODs against time are plotted (Figures 11–13), marking the three corresponding stages. It is observed that CMOD increases rapidly when applying dead weight. After weight has been applied, the CMOD increases in a smoother manner than before. The smoother displacement rate of CMOD will cause a decrease rate in crack growth velocity v in Stage I. Moreover, for testing condition in air (Figure 11b) and acid (Figures 12, 13), the difference of CMOD growth rate during and after loading becomes smaller. This is due to crack propagation during applying dead weight in both water and acidic conditions (Figures 12 and 13). Thus, the decrease level of v in Stage I is smaller in acidic condition than that in air condition. In Stage II, the CMOD grows in a nearly constant rate. However, it was found that such time for growth rate of CMOD in acidic condition is very short (Figure 13). When the crack grows to a certain length (a/W=0.56~0.59), the increase of CMOD becomes faster and such increase causes the linear increase of v in Stage III.

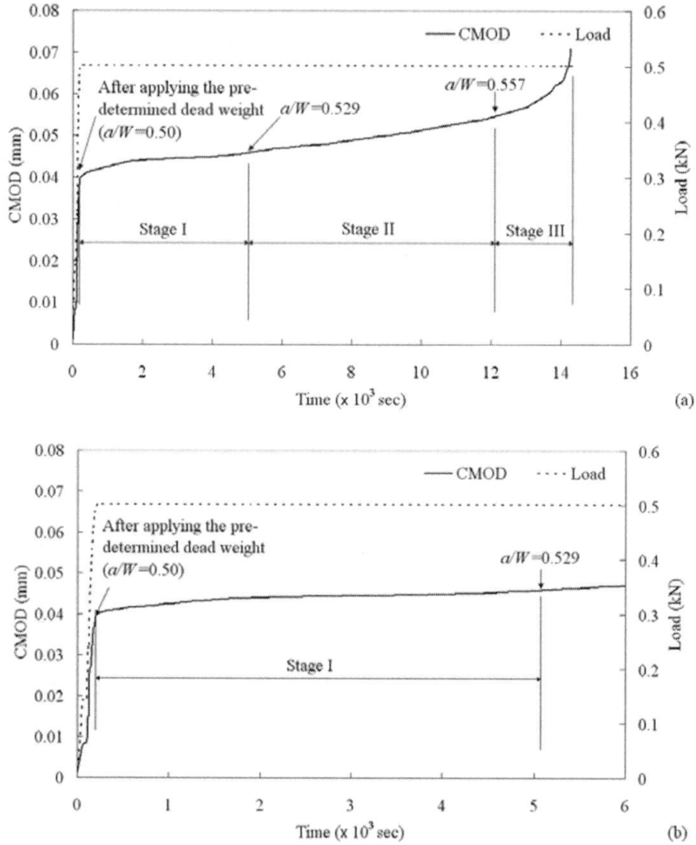

Figure 11. Curve of CMOD against time of granite specimen in air.

From Figures 11-13, it is found that the time to failure of granite specimens becomes shorter when the pH value decreases. According to Plummer et al.[17], the effects of acidic condition may be explained by the chemical weathering of feldspar mineral changing to clay mineral at the crack tip as follows:

$$H_2SO_4 \Leftrightarrow 2H^+ + SO_4^{2-} \quad \text{(Acidic condition)}$$

$$\underbrace{2KAlSi_3O_8}_{\text{Aluminosilicate minerals (feldspar)}} + H_2O + 2H^+ \rightarrow \underbrace{Al_2Si_2O_5(OH)_4}_{\text{Kaolinite (clay)}} + \underbrace{4SiO_2}_{\text{Soluble silica}} + 2K^+$$

Figure 14 shows the crack tip of the specimens after 4PB tests under microscope. By comparing the specimens it can be concluded that in air conditions (Figure14a), the minerals along the tips are weathered in a lower degree compared to those under water conditions (Figure 14b). The white colour along the crack tip is the weathered mineral (Kaolinite). On the contrary under acidic condition (Figure 14c), most of the minerals near the crack tip are weathered and changed in white colour (dotted region in Figure 14c).

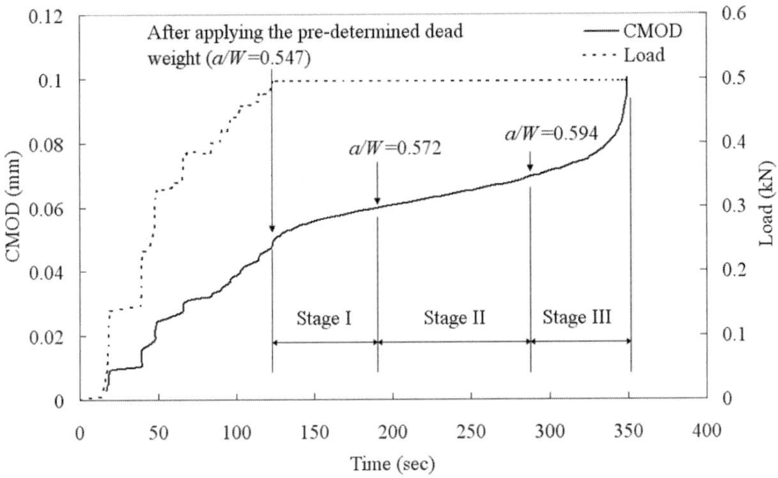

Figure 12. Curve of CMOD against time of granite specimen in water.

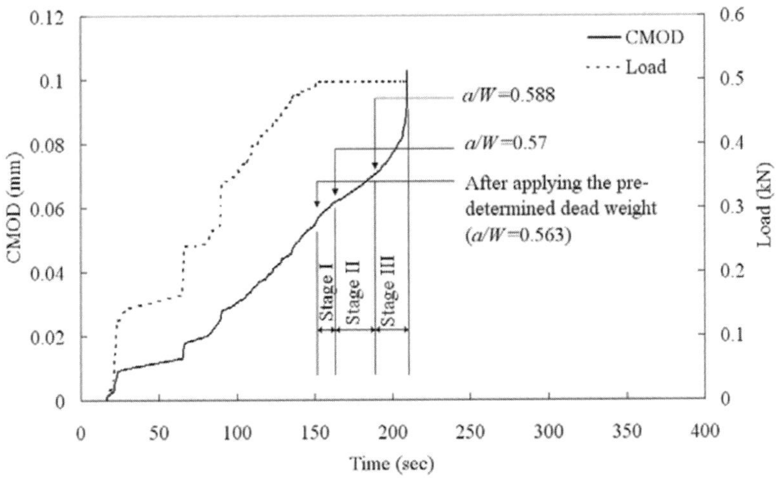

Figure 13. Curve of CMOD against time of granite specimen in H_2SO_4 (pH=2).

To determine the SCG indices *n* of granite specimens, the slope of the linear fitting in Stage III (Figure 10) is used. Table 1 compiles the *n* values of granite specimen under different environmental conditions from the 4PB test.

Table 1. The subcritical crack growth index *n* of granite using four-point bending test

Testing Conditions	Subcritical crack growth index, *n*
Air	27
Water (pH=7.0)	19
H_2SO_4 (pH=2.0)	16

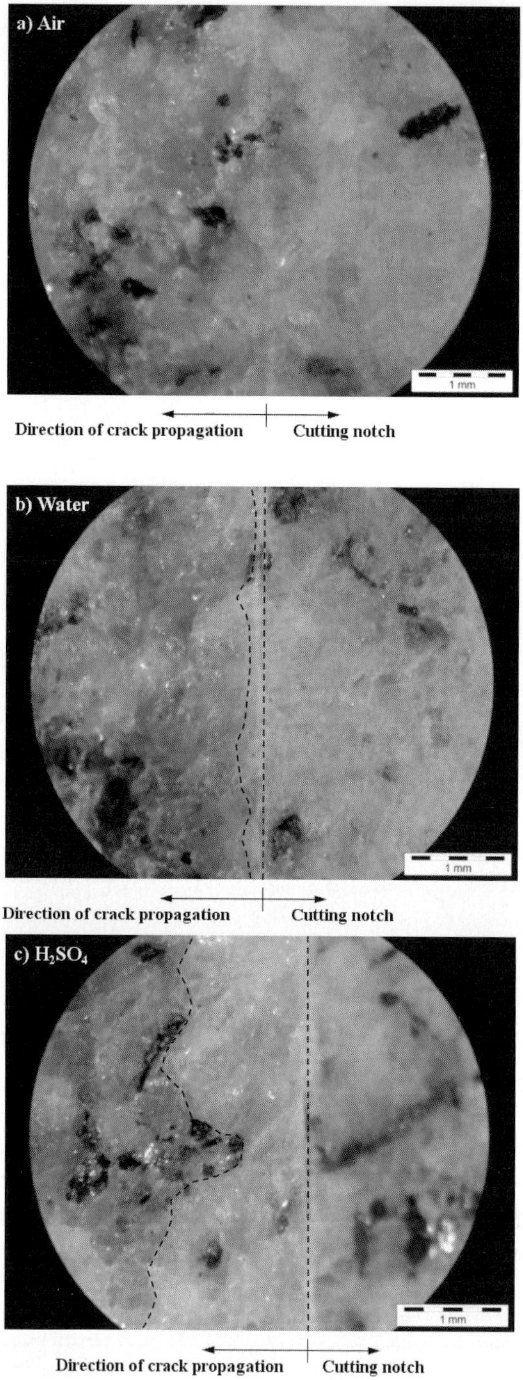

Figure 14. The crack tip of granite specimen after four-point bending test in a) air condition, b) water condition (pH=7), and c) H_2SO_4 (pH=2). Dotted region shows the weathering zone (Kaolinite) of granite specimen.

3.2 Double Torsion Test

The SCG data of load relaxation experiments on granite specimens under air, and dilute sulphuric acid (pH value of 2) is presented by a log-log plot of the crack growth rate v versus normalized K_I/K_{IC} (Figure 15).

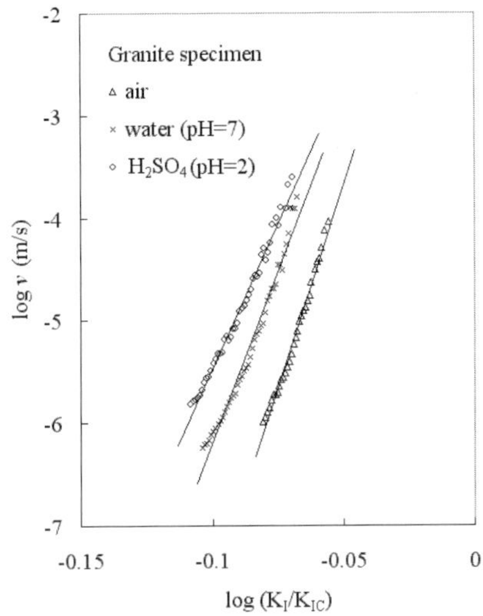

Figure 15. Log K_I – log v curve for Granite using double torsion test.

Based on the log (K_I/K_{IC}) – log v curve, it is concluded the crack growth behavior in granite rock panel (Figure 15) is similar to the behavior of Westerly granite (Figure 3) the v of which decreases during load relaxation. Furthermore, similarly to the result obtained from 4PB tests, it is found that the pH value of the aqueous environment can influence the crack growth rate at the same normalized SIF. To obtain the SCG indices n of granite specimens under different environmental conditions, the slope of the linear part of log-log plot is used (Figure 15). The SCG indices of the specimens are shown in Table 2. From this table it is concluded that the DTT data are in reasonable agreement with the upper bound of data for granitic rocks summarized by Atkinson & Meredith[18].

On the contrary, by comparing the SCG data obtained from 4PB with those from DTT, it is seen that the index n obtained from the latter is higher than that obtained from the 4PB tests by a factor of around 3.5 (Table 3). This may be due to the different fracture mechanisms activated in these test-

Table 2. The subcritical crack growth index n of granite using double torsion test

Testing Conditions	SCG index, n in this study	SCG index, n of granitic rock summarized by Atkinson & Meredith[18]
Air	92	30-90
Water pH=7.0)	69	37-68
H_2SO_4(pH=2.0)	49	Not available

Table 3. Comparison of the SCG index n between four-point bending and double torsion tests

Testing Conditions	SCG index, n		Factor
	4PB test, n_{4PT}	DT test, n_{DT}	(n_{DT} / n_{4PT})
Air	27	92	3.4
Water (pH=7.0)	19	69	3.6
H_2SO_4 (pH=2.0)	16	49	3.1

ing methods. Furthermore, as the determination of the SIF K_I from the DTT is independent of the crack length (as shown in Eq.(5)), K_I may be overestimated or underestimated for different crack lengths[19], since the crack growth length is an important parameter to determine K_I. Thus, it is worth investigating further the influence of the crack length on the results of the DTT.

4. CONCLUSIONS

Both 4PB and DTT were used to investigate subcritical crack growth of granite of rock panel under different environmental conditions. Furthermore, this is the first study to investigate SCG of granite rock panel using four-point bending test. The crack growth behavior and the SCG index n of both testing methods were discussed. Based on the observation and findings of this study, the following conclusions can be drawn:

- 4PB tests can be used for the study of subcritical crack growth in rock panels. The advantage of this method is that the crack growth length can be monitored by clip gauge throughout the whole test and the real situation of rock panel can be simulated.
- Three stages of K–v curve are observed in granite specimens under four-point bending test. In Stage I, the crack growth rate v decreased after the pre-determined dead weight was applied and v attained a minimum value when the normalized stress intensity factor K_I/K_{IC} reached a value of about 0.73. In Stage II, v became roughly constant over a certain range of K_I/K_{IC}. Finally, v increased and the crack propagated towards the lower edge, splitting the specimen into two pieces in Stage III.
- The crack growth behavior of granite specimens as obtained from DTT is similar to that of Westerly granite presented in previous studies.
- The pH value of the aqueous environment can influence the crack growth rate at the same normalized SIF according to the result obtained from both testing methods. Under acidic condition, crack growth rate v becomes faster than under air condition. Moreover, the time to failure of granite four-point bending specimen becomes shorter when the pH value decreases. These effects may be due to the chemical weathering of feldspar mineral changing to clay mineral at the crack tip.
- Although the SCG index n obtained from DTT is in reasonable agreement with the data for granitic rocks summarized in previous studies, this

value is higher than that obtained from 4PB tests, by a factor of about 3 to 4. This may be due to the different fracture mechanism activated in the two methods and the fact that the influence of the crack length on K_I is disregarded in the DTT. Thus, it is worth to investigate further the influence of crack length on the results of the DTTs.

ACKNOWLEDGEMENTS

The work described in this paper was fully supported by a grant from the Research Grants Council of the Hong Kong Special Administrative Region, China (Project No. PolyU 5166/04E). This study is part of the Master study of the first author, K.W. Kwok.

REFERENCES

1. M. Y. L. Chew, C. W. Wong and L. H. Kang, *Building Facades: A guide to common defects in tropical climates*, World Scientific Publishing Co. Pte. Ltd., pp. 12-13 (1998).
2. J. Trewhitt and J. Tuchmann, Amoco may replace marble on Chicago headquarters, *Engineering News Record* **24**, 11-12 (March 1988).
3. J. A. Bayer and C. D. Clift, Use of physical property data and testing in the design of curtain wall, *International Journal of Rock Mechanics and Mining Sciences* **30**(7), 1563-1566 (1993).
4. K. T. Chau and J. F. Shao, Subcritical crack growth of edge and center cracks in façade rock panels subject to periodic surface temperature variations, *International Journal of Solids and Structures,* **43**, 807-827 (2006).
5. B. K. Atkinson, Subcritical crack growth in geological materials, *Journal of Geophysical Research* **89**(B6), 4077-4114 (1984).
6. R. J. Charles, Static fatigue of glass, *Journal of Applied Physics* **29**, 1549-1560 (1958).
7. K. Miura, Y. Okui and H. Horii, Micromechanics-based prediction of creep failure of hard rock for long-term safety of high-level radioactive waste disposal system, *Mechanics of Materials* **35**, 587-601 (2003).
8. A. G. Evans, A simple method for evaluating slow crack growth in brittle materials, *International Journal of Fracture* **9**(3), 267-275 (1973).
9. B.K. Atkinson, Fracture Toughness of Tennessee Sandstone and Carrara Marble using the Double Torsion testing method, *International Journal of Rock Mechanics and Mining Sciences Abstracts* **16**, 49-53 (1979).
10. J. P. Henry, J. Paquet, and J. P. Tancrez, Experimental Study of Crack propagation in Calcite Rocks, *Int. J. Rock Mech. Min. Sci. & Geomech. Abstr.* **14**, 85-91 (1977).
11. P. G. Meredith, B. K. Atkinson, Fracture toughness and subcritical crack growth during high-temperature tensile deformation of Westerly granite and Black gabbro", *Physics of the Earth and Planetary Interiors* **39**, 33-51 (1985).
12. P. L. Swanson, Subcritical Crack Propagation in Westerly Granite: An Investigation into the Double Torsion Method, *Int. J. Rock Mech. Min. Sci. & Geomech. Abstr.* **18,** 445-449 (1981).
13. P. L. Swanson, Subcritical crack growth and other time- and environment-dependent behavior in crustal rocks, *Journal of Geophysical Research* **89**(B6), 4137-4152 (1984).
14. D. P. Williams and A. G. Evans, A simple method for studying slow crack growth, *Journal of Testing and Evaluation* **1**, 264-270 (1973).

15. Y. Murakami, *Stress Intensity Factor Handbook*, Vol. 1, Pergamon Press, pp. 16-17 (1987).
16. R. A. Schmidt, Fracture-toughness testing of limestone, *Experimental Mechanics* **16**(5), 161-167 (1976).
17. C. C. Plummer, D. McGeary and D. H. Carlson, Physical Geology, New York: McGraw-Hill, 9th ed., pp. 111 (2003).
18. B. K. Atkinson and P.G. Meredith, *Fracture Mechanics of Rock*, Academic Press Inc. Ltd. (London), pp. 477-525 (1987).
19. G. G. Trantina, Stress analysis of the double torsion specimen, Journal of the American Ceramic Society **60**, 338-341 (1977).

Chapter 1.3

NOTCHED MARBLE SPECIMENS UNDER DIRECT TENSION: THE INFLUENCE OF THE SHAPE OF THE NOTCH

Zacharias G. Agioutantis[1], Stavros K. Kourkoulis[2] and George Kontos[1]
[1]Department of Mineral Resources Engineering, Technical University of Crete, 73100 Hania, Crete, Hellas, zach@ mred.tuc.gr; [2]School of Applied Sciences, Department of Mechanics, National Technical University of Athens, 5 Heroes of Polytechnion Avenue, 157 73 Zografou, Attiki, Hellas, stakkour@central.ntua.gr

Abstract: In this paper the influence of the shape of the notch on the intensity of the strain field developed in marble plates subjected to direct tension was studied both experimentally and numerically. The specimens were made from Dionysos marble, the material used for the restoration project of the Parthenon Temple of Athens. The experimental results from specimens with different notch shape were compared to each other and, also, with the respective values obtained from a numerical analysis. The material was modeled as linear elastic either isotropic or transversely isotropic. The comparison between experimental and numerical results is satisfactory concerning the specimens with the semi-circular notches. Deviations are observed for the U-shaped notch, which may be attributed to the magnitude of the process zone developed, that renders the elasticity hypothesis, adopted in the numerical analysis, invalid.

Key words: natural building stones; Dionysos marble; uniaxial tension; double edge notched specimens; finite element method.

1. INTRODUCTION

The purpose of the present study is to explore the influence of the shape of the notch on the strain field developed in double edge notched specimens made from marble when subjected to uniaxial tension. The distribution of stresses and strains in the presence of notches of various shapes concerns the engineering community for long since the stress concentration generated in the vicinity of notches is of crucial importance for the safe design of structu-

S.K. Kourkoulis (ed.), Fracture and Failure of Natural Building Stones, 35–52.

res of any kind[1]. Already from the beginning of the previous century the solutions for the stress concentration around circular and elliptic notches within an infinite plate have been introduced by Kirsch[2] and Inglis[3] respectively; however the problem in its general form is by no means closed. Particularly for cases where the assumption of infinite dimensions of the structure is not applicable, i.e. the dimensions of the plate containing the notch are of the same order of magnitude with those of the notch itself, the situation becomes extremely complicated and the analytical solutions available nowadays correspond to a limited number of special configurations. Things become worse in case the notched plate is cut from a non-isotropic material. Then the general analytical solution becomes prohibitively difficult even for the simplest degree of anisotropy, i.e. that of transverse isotropy.

Fortunately the rapid development of numerical techniques during the last two or three decades offer to the design engineers a useful tool for an at least qualitative solution of the problem, independently of the degree of complexity of the geometry analyzed. Nevertheless, as it is pointed out by Filippi and Lazzarin[4], the numerical approximations "*...lead to a typical sparse data output which is much less manageable than analytical results and, moreover, make uneasy to understand the role played by all geometrical parameters involved*". Therefore it looks like the experimental study remains the only reliable tool that can provide data necessary for the calibration and assessment of numerical models as well as for the validation of analytical attempts like the one recently introduced by Kotousov and Wang[5].

In this direction the variation of the strain-field in plates made of Dionysos marble (a typical transversely isotropic material) with either semicircular or U-shaped notches is studied here experimentally in an effort to determine the influence of the shape of the notch on the strain concentration in the vicinity of the notch tip and quantify the strain concentration factor. The results of the experimental program are then used for the calibration and assessment of a numerical model solved with the aid of the Finite Element Method and commercially available software.

2. EXPERIMENTAL PROCEDURE AND RESULTS

2.1 The material

The specimens used for the present study were cut from recently quarried marble blocks from the Dionysos Mountain in Attica, Greece. This material is used for the restoration project of the Parthenon Temple on the Acropolis of Athens because it has properties similar to those of the Pentelik marble, the authentic building material of the monument[6]. The need to study the behaviour of Dionysos marble under tension in presence of notches emanates from the fact that a number of structural elements of the Temple (particularly architraves and capitals) are partly cracked. In addition as it was indicated by Theocaris and Koroneos[7] "*the critical stresses in case of seismic loading... are the tensile ones... at the upper part of the columns...*".

Dionysos marble is composed by 98% of calcite and it contains very small amounts of muscovite, sericite, quartz and chlorite. Its density is 2730 kg/m^3. The porosity is very low (between 0.3% in the virgin state to 0.7% after the action of natural weathering and corrosive agents). The grain size varies around 0.43x10^{-3} m and the crystals are of polygonic shape and of almost uniform size. It is of white colour with a few thin ash-green veins following the schistosity of marble and containing locally silver areas due to the existence of chlorite and muscovite[8].

From the mechanical point of view it is an orthotropic material. Series of direct tension tests[9] with cylindrical specimens cut and loaded along the three anisotropy axes indicated that the mechanical properties along the two of them are very similar to each other. Thus the material is considered as transversely isotropic described with the aid of five constants: the elastic moduli E and E′, in the plane of isotropy and normal to it, respectively, the Poisson's ratios ν and ν′ characterizing the lateral strain response in the plane of isotropy to a tensile stress acting parallel and normal to it, respectively, and G′ the shear modulus in planes normal to the plane of isotropy. The values of the elastic moduli and Poisson's ratios are shown in Table 1, together with the values of the tensile strength, σ_f.

Table 1. Mechanical constants of Dionysos marble

	E [GPa]	ν	σ_f [MPa]
Strong Direction	84.5	0.26	10.8
Intermediate Direction	79.5	0.26	9.5
Weak Direction	50.0	0.11	5.3

The axial stress-axial strain curve of Dionysos marble as obtained from the above experiments is shown in Figure 1. The material appears to be slightly bimodular, i.e. its elastic moduli in tension and compression are not exactly equal. Also, it appears to be slightly non-linear both in the tension and in the compression regime. Especially for the tension regime it was found that the stress-strain curve is accurately described by Gerstner's parabolic law[10]:

$$\sigma=E(\varepsilon-m\varepsilon^2) \qquad (1)$$

where m is a numerically determined constant, equal to about 1880. The modulus of elasticity is a linearly decreasing function of strain of the form:

$$E(\varepsilon)=(1-D)Eo \qquad (2)$$

In Eq.(2) D is a constant and E_o the elastic modulus at the undamaged state.

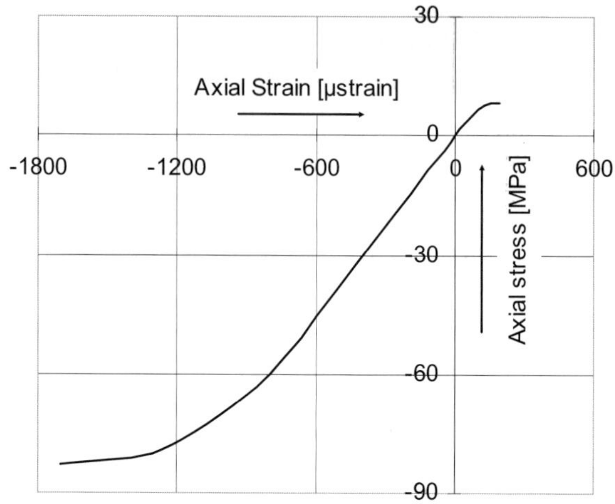

Figure 1. The axial stress - axial strain curve of Dionysos marble[9].

2.2 The specimens and the experimental procedure

Two classes of specimens were prepared. The first class included plates with two symmetric semi-circular notches of radius ρ=10 mm machined at both sides of the specimens. The geometry and dimensions are shown in Figure 2. The second class included specimens with two symmetric U-shaped notches of length a=10 mm at the vertical edges normal to the loading direction. The geometry and dimensions of these specimens are shown in Figure 3a.

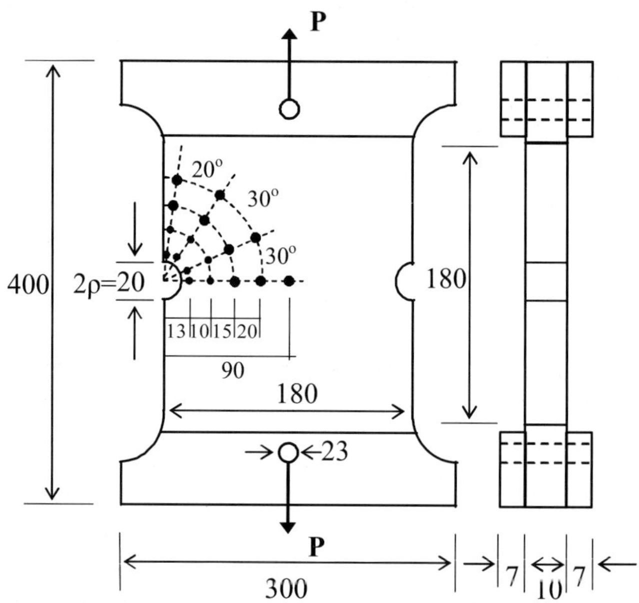

Figure 2. The geometry of the first class of specimens. The dimensions are in mm (not scaled).

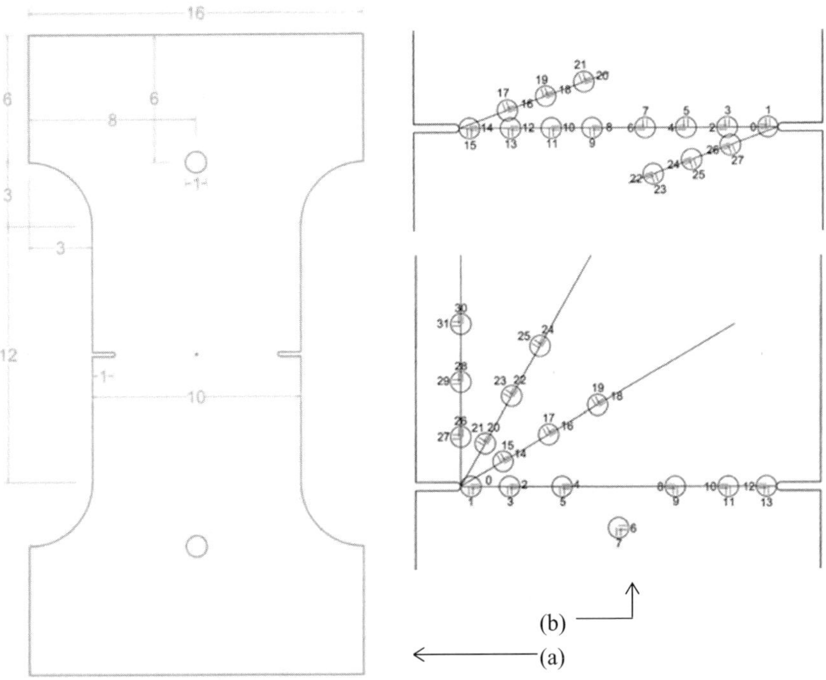

Figure 3. (a) The geometry of the second class of specimens. (b) Detailed view of the two different arrangements of the strain gauges. The dimensions are in cm.

In both cases the load was applied with the aid of two steel pins. For the first class the pins had a diameter equal to 22 mm and were mounted through holes with a diameter of 23 mm while in the second class the respective dimensions were 9 mm and 10 mm. In order to avoid fracture in the vicinity of the load-application region the "ears" of the specimens were reinforced with two pairs of plates made from plexiglas, as it is seen in Figures 2, 4 and 5b. Attention was paid to avoid direct contact of the metallic cylinders with the marble: the load was transferred from the cylinders to the marble through the adhesive material resembling thus a uniformly applied surface shear stress and avoiding stress concentration in the vicinity of the loading pin.

The plates were cut within the plane of isotropy in such a way that the load direction coincided with the strong anisotropy axis. The load was applied statically at a rate equal to 0.02 cm/min with the aid of an electric loading frame with a capacity of 250 kN. The strain components were obtained using a system of 34 strain gauges for the first class of specimens (17 orthogonal rosettes of the Kyowa KFG-X-120-D16 type) and either 28 or 32 strain gauges (14 or 16 orthogonal rosettes) for the second class of specimens. For the rosettes closest to the rim of the semi-circular notch (the ones on the circle with radius 13 mm) and the one on the tip of the U-shaped notch X=1 while for the other ones X=2. A polar system was chosen for the positioning of the strain gauges on the surface of the specimens. The exact coordinates of the application points are shown in Figures 2 and 3b. For the specimens with U-shaped notches two arrangements of the strain gauges were used, as shown in Figure 3b. The angle between the gauge lines was 20° for the first arrangement and 30° for the second. The distance between the gauges was 1 cm.

A typical specimen with semi-circular notches is shown in Figures 4(a,b) just before being tested, while Figure 4c depicts the fracture surfaces of a different specimen. In Figure 5 two typical fractured specimens with U-shaped notches are shown with and without reinforced gripping zones.

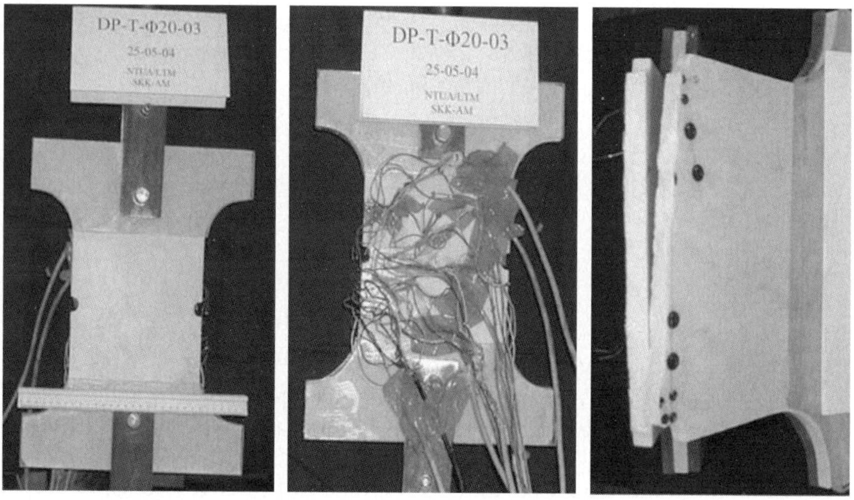

(a) (b) (c)

Figure 4. (a) The front and (b) the rear view of a typical specimen with semi-circular notches just before being tested. (c) The fracture surfaces of a fractured specimen.

(a)

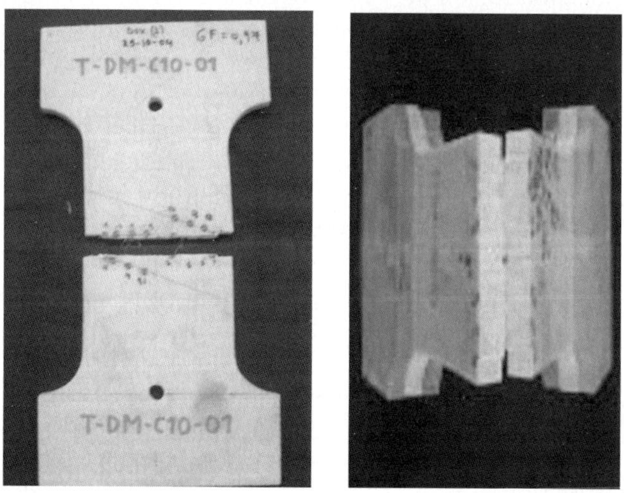

(b)

Figure 5. (a) A typical fractured specimen with U-shaped notch and the first arrangement of strain gauges. (b) Fracture surfaces of a U-notched specimen with reinforced gripping zone.

2.3 Experimental results

2.3.1 Specimens with semi-circular notches

The variation of the tangential (longitudinal for θ=0°) strains developed around the notch versus the nominal stress is plotted in Figure 6a according to the results of the two rosettes positioned at (r/ρ, θ)=(1.3, 0°) and (r/ρ, θ)=(9, 0°), for the ultimate load P≈7 kN. The specific rosettes were selected since they correspond to the point as close as possible to the notch and to the point most distanced from the notch (the center of the specimen). The strains are reduced over the maximum tangential strain developed while the nominal stress is reduced over the tensile fracture stress of the intact marble. It can be concluded from Figure 6a that for the gauge closest to the notch the evolution of the strain field is non-linear from the early loading steps, contrary to what is be expected for a quasi-brittle rock-type material. Although a slight non-linearity is reasonable according to Figure 1, it appears here that the phenomenon is not restricted for loads approaching the fracture load but it rather governs the behaviour of the Dionysos marble under tension in the presence of notches during the whole loading procedure. This non-linearity is attributed to the development of a process zone around the notch crown.

The above process zone has been quantified for Dionysos marble by Kourkoulis et al.[10,11] and its size was found to be of the order of 5-8 mm. Considering the dimensions of the rosettes used it is indeed concluded that

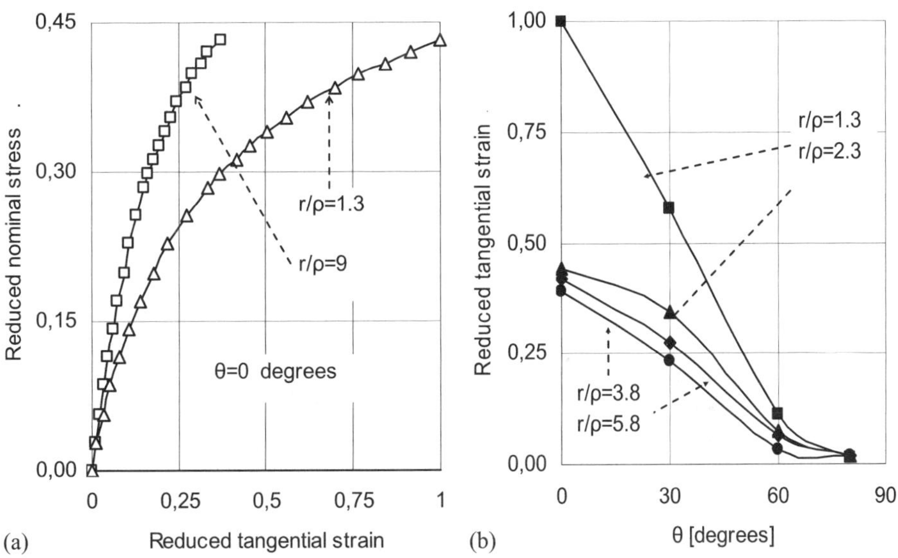

(a) (b)

Figure 6. (a) The variation of the transverse strain vs. the remote nominal stress for the central point of the specimen and the point just ahead the notch tip. (b) The polar distribution of the transverse strain along four semi-cyclic contours with r/ρ=1.3, 2.3, 3.8 and 5.8.

the point under study lies well inside the process zone. On the contrary, at the central point of the specimen the stress-strain relation is linear for the major portion of the loading procedure and the non-linearity observed when approaching the fracture load is attributed to the nature of Dionysos marble (Figure 1). It is also seen from the same figure that the transverse strain developed in the immediate vicinity of the crown of the notch is about 2.7 times the respective strain at the central point of the specimen, providing an indication of the strain field amplification due to the presence of the notch.

In Figure 6b the polar distribution of the tangential strain component, reduced over the respective maximum value, is plotted for all four contours studied, i.e. for values of the r/ρ parameter equal to 1.3, 2.3, 3.8 and 5.8. The graphs correspond again to the ultimate value of the remotely applied load P≈7 kN. It is concluded from this figure that the influence of the notch on the strain field is restricted for r-values smaller than about two times the value of the radius ρ of the notch. In fact the differences of the transverse strains for all three contours with r/ρ=2.3, 3.8 and 5.8 are almost negligible independently from the value of angle θ and only the contour with r/ρ=1.3 diverges.

The dependence of the tangential and radial strains on the distance from the notch tip is plotted in Figures 7(a,b), respectively, where the tangential and radial strains are plotted along the linear contours θ=0°, 30°, 60° and 80°. In both figures the strains have been reduced over the maximum tangential strain developed, while the distance from the origin of the reference system (the center of the notch) is reduced over the notch radius. As it is seen from Figure 7a the transverse strain is strongly affected by the presence of the notch for the lines θ=0° and θ=30°. It is again concluded that the influence of

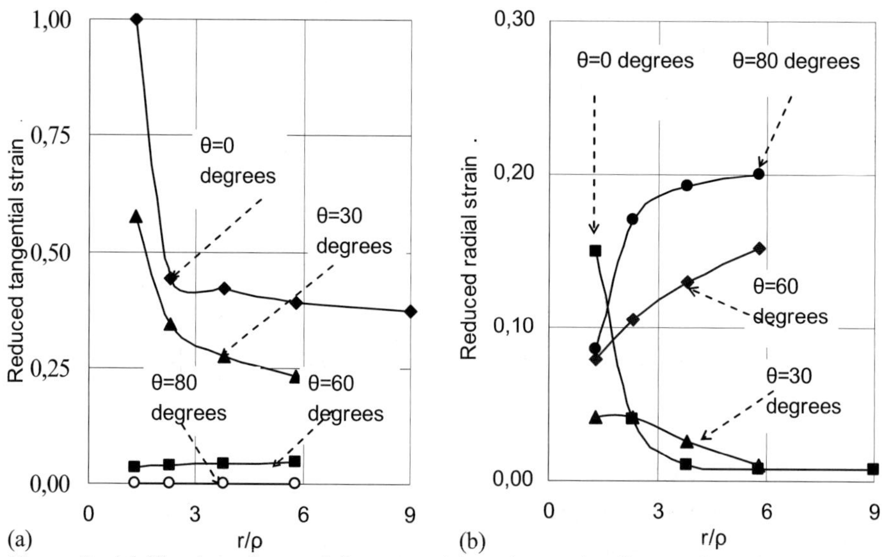

Figure 7 (a) The dependence of the tangential strains on the distance from the tip of the notch (b). The dependence of the radial strains on the distance from the tip of the notch.

the notch on the values of the transverse strain is eliminated very rapidly with increasing distance from the notch tip: indeed, for r/ρ>2.5 the transverse strains remain essentially constant for the whole range of θ. The variation of the radial strains is more complicated; the slope of the curves changes sign depending on the angle θ. For θ=0° the ratio $\varepsilon_{rr}(r/\rho=1.3)/\varepsilon_{rr}(r/\rho=9)$ approaches 20 indicating a tremendous influence of the notch on the radial strain.

2.3.2 Specimens with U-shaped notches

The dependence of the tangential (longitudinal for θ=0°) and radial strains on the axial stress is shown in Figures 8(a,b), according to the data of the rosettes along the line connecting the tips of the two notches. It is again concluded that the results of the gauge closest to the tip of the notch are dramatically influenced by the notch. Indeed for both the transverse and the radial strains all three gauges at distances from the tip equal to 1.5a, 2.5a and 3.5a (a the length of the notch) give almost identical results indicating that at these points the strain field is almost uniform. In addition the stress-strain relation for these gauges is almost linear. On the contrary for the gauge with r=0.5a the stress-strain relation is strongly non-linear for both the tangential and the radial strains. It is concluded that in the case of U-shaped notches the size of the process zone for Dionysos marble varies between 5 mm and 15 mm since for the present series of tests the length of the notch is a=10 mm.

The polar distribution of the tangential and the radial strains around the tip of the notch for three semi-circular paths, with r=a, r=2.5a and r=4a are plotted for a characteristic specimen in Figures 9a and 9b, respectively. It is interesting to observe the negative values attained by the radial strains on the line

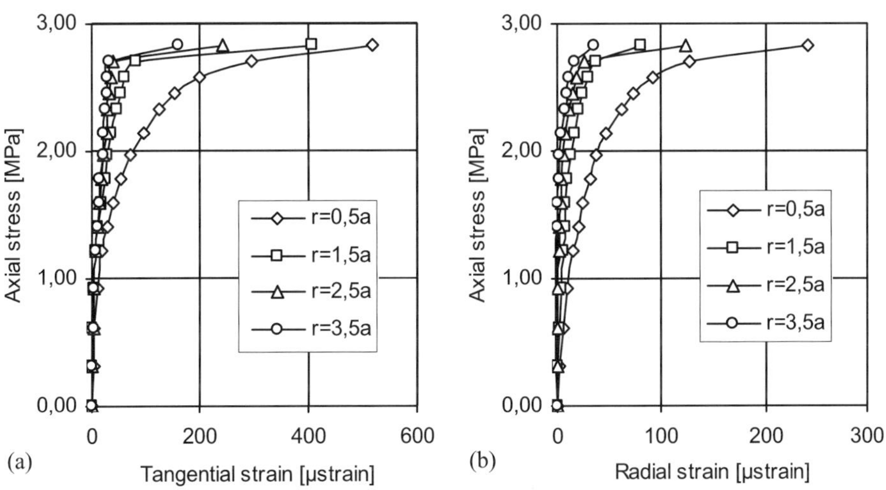

Figure 8. The variation of the transverse (longitudinal) (a) and the radial (b) strains vs. the remote axial stress for specimens with U-shaped notches at the fracture load.

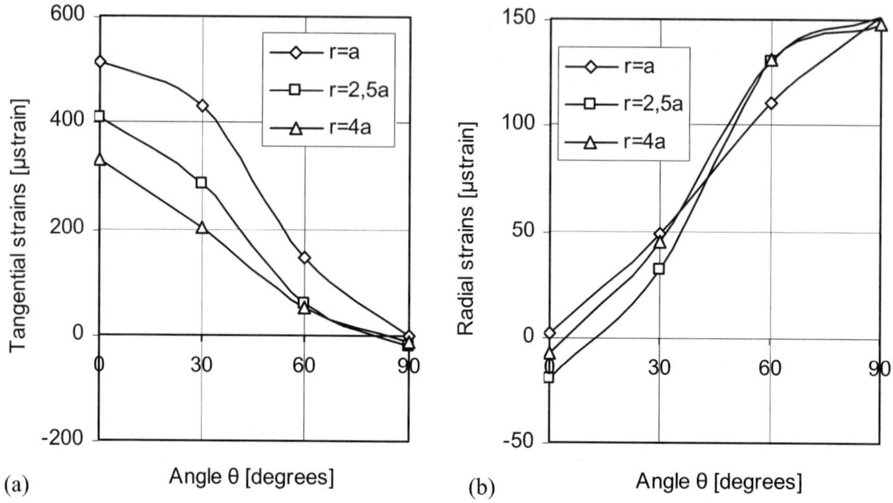

Figure 9. The polar distribution of the tangential (a) and the radial (b) strains for three semi-circular paths around the crown of the notch just before the fracture of the specimen.

with θ=0°. In addition it is stressed out that the values of the radial strains are not significantly affected by the notch, contrary to the tangential ones.

3. NUMERICAL ANALYSIS AND RESULTS

Both type of specimens described in the previous paragraph were modeled numerically in 2-D space in the MSC.Mentat front-end program, and the simulations of the tests were solved by the MSC.Marc Finite Element Analysis program[12,13]. The symmetry of the configurations was not taken into consideration. Therefore the dimensions of the final models matched exactly those of the specimens used for the experimental procedure. The model of the specimen with semi-circular notches consisted of 4502 plane stress elements and 4619 nodes while that of the specimen with U-shaped notches consisted of 7848 triangular and quadrilateral elements and 7883 nodes. In both cases the node positions were defined to coincide with the strain measuring points (the strain gauges) on the specimens in order to avoid interpolation errors.

The boundary conditions imposed include a pin-type restriction of the central node(s) of the specimens as well as distributed load on the edges. The discretization and the boundary conditions are shown in Figures 10(a,b).

The material of the specimens was modeled as linear elastic of either isotropic or orthotropic nature. Its mechanical constants matched those described in section 2.1 for Dionysos marble. The load was applied as edge load on each side of the specimen, and its total value (on both edges) corresponded to a tensile force equal to the average value of the fracture force determined experimentally for each class of specimens.

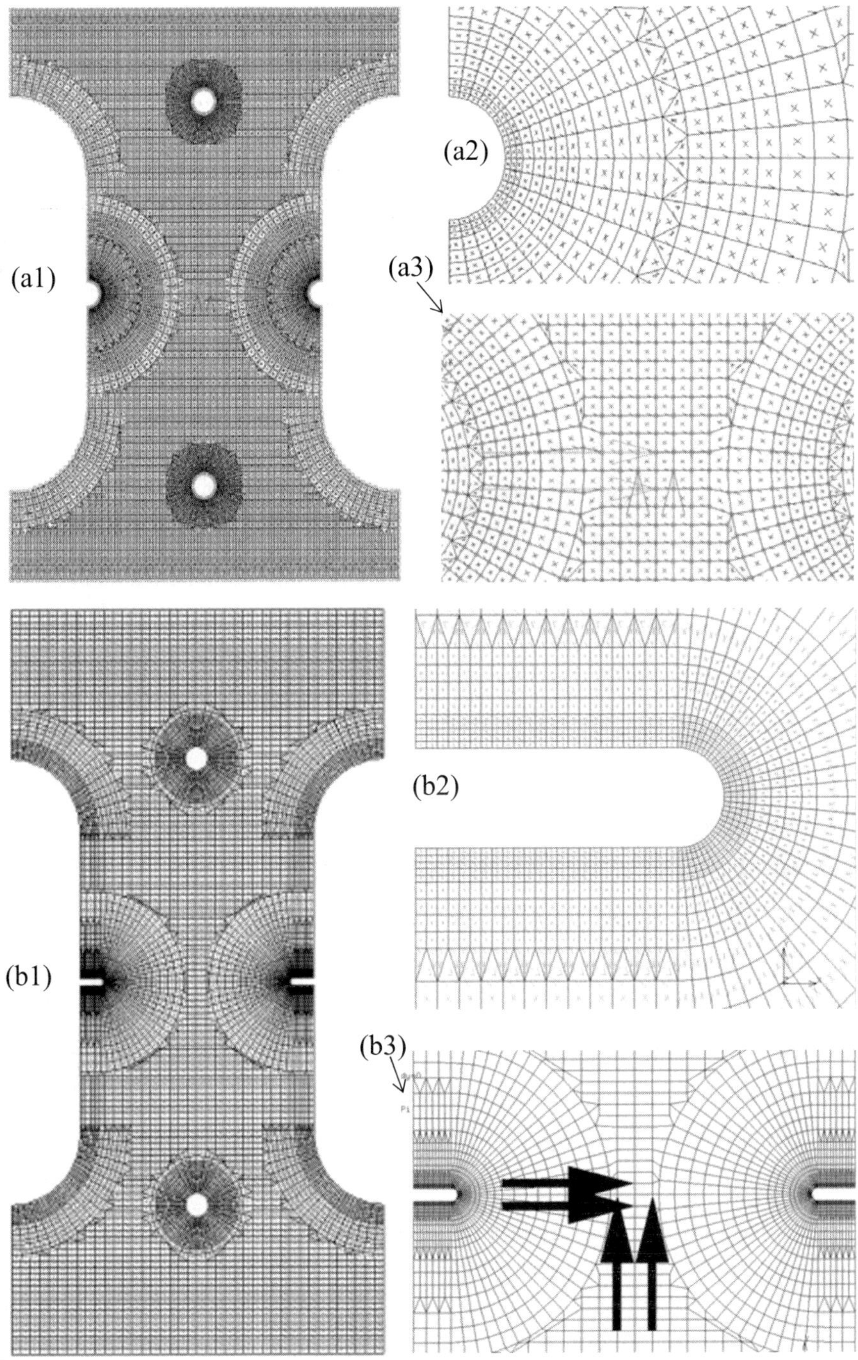

Figure 10. An overview of the numerical models of the specimens with semi-circular (a1) and U-shaped notches (b1). Details of the meshing around the crowns of the notches are also shown (a2, b2) together with the boundary conditions applied (a3, b3).

Characteristic results concerning the variation of the longitudinal (axial) strain, all over the surface of the specimens are presented in Figures 11(a,b).

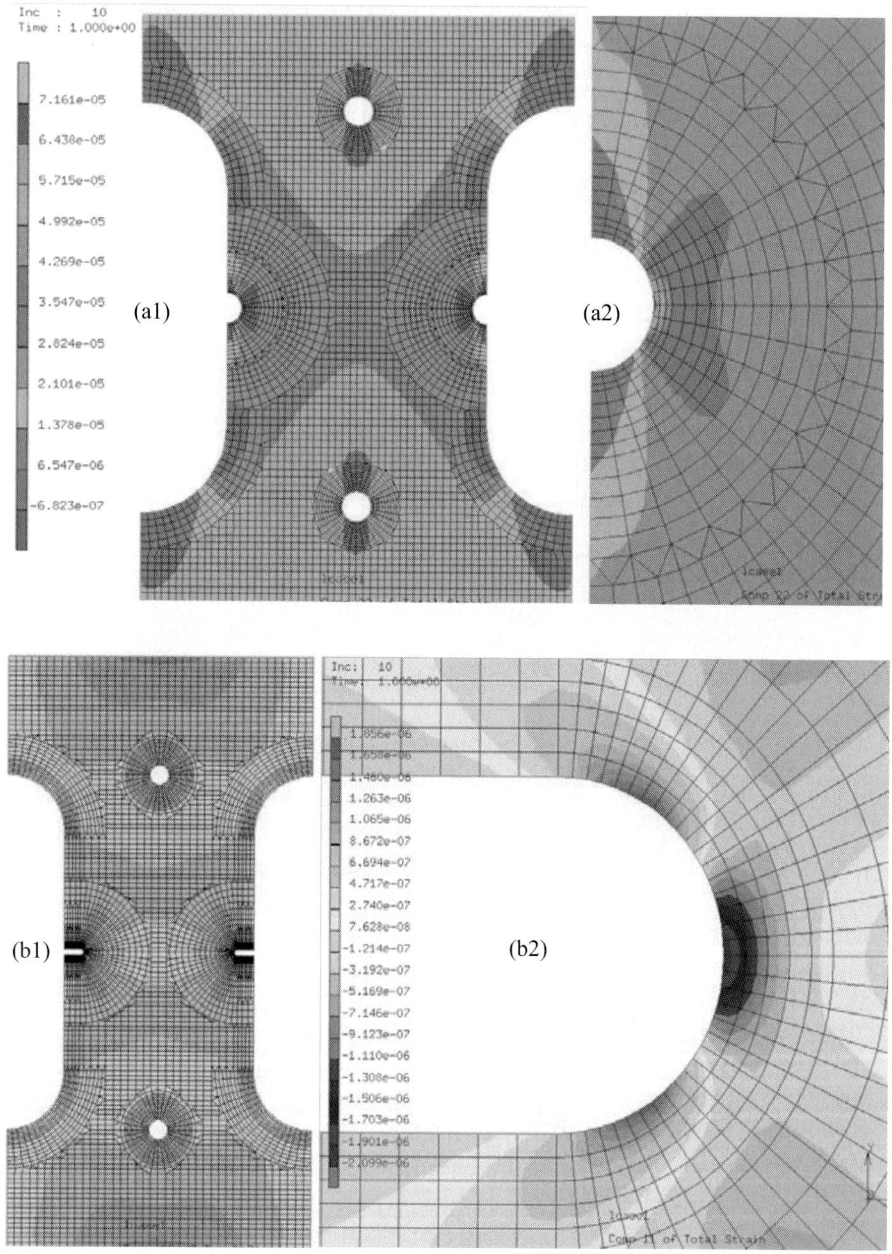

Figure 11. The distribution of the longitudinal (axial) strains ε_{yy} for the specimens with semi-circular (a1) and U-shaped notches (b1) for the isotropic model. Detailed views of the immediate vicinities of the notches are also shown in (b1) and (b2), respectively.

The variation of the normal strain components relatively to Cartesian reference systems with origins at the center of the semi-circular notch and the center of the semi-circular crown of the U-shaped notch is plotted in Figure 12. The axes of the system are parallel to the load and normal to it. For the specimens with semi-circular notches the strains are plotted along a semi-circular contour with r/ρ=1.3 encircling the notch. For the specimens with U-shaped notches the strains are plotted along a path parallel to the border line of the notch at a distance equal to r/a=1.5 from it. Both the axial (square symbols) and the transverse strains (rhombic symbols) are considered. From

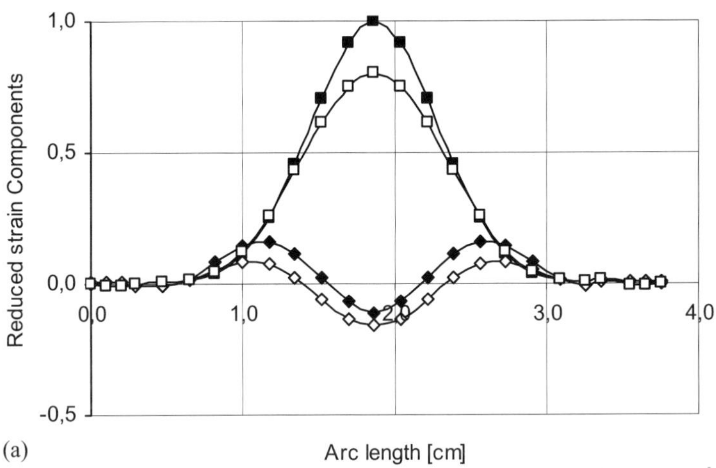

(a)

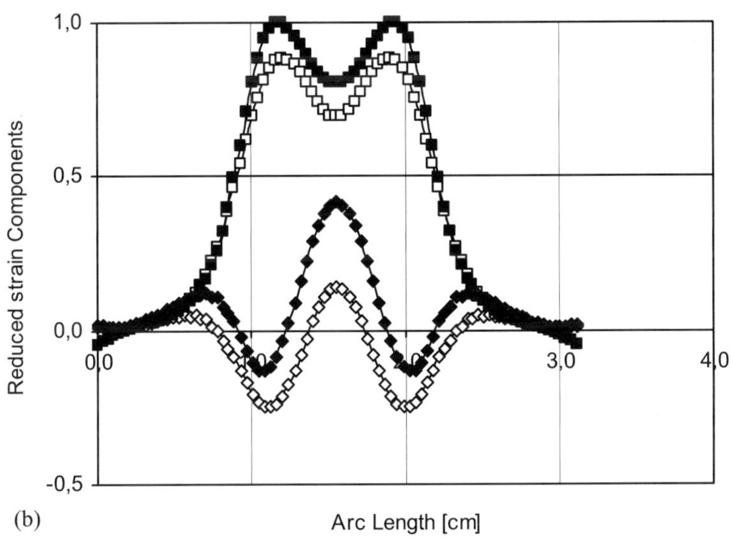

(b)

Figure 12. The variation of the normal strain components for the specimens with semi-circular (a) and U-shaped (b) notches. Filled symbols correspond to the orthotropic model while empty ones to the isotropic model. Square symbols represent the axial strains while rhombic ones represent the transverse strains.

this figure it becomes evident that the strains predicted in case the material is considered as orthotropic (filled symbols) exceed systematically those predicted from the isotropic model (empty symbols). For the specimens with semi-circular notches the difference is more pronounced for the longitudinal strains (about 20%) straight ahead from the notch tip. For the specimens with U-shaped notches the difference is more pronounced for the transverse strains (about 40%), again along the line connecting the tips of the notches.

Comparing Figures 12a and 12b it is observed that the maximum of the axial strains appears along the horizontal axis of symmetry for specimens with semi-circular notches while for the specimens with U-shaped notches two off-axis extrema appear. Although such an observation perhaps contradicts common sense it could be attributed to the finite dimensions of the specimens and the inevitable interaction between the tips of the notches. Further study is required before definite conclusions are drawn and also in order to determine the dimensions of the specimens for which the assumption of not interacting notches is valid.

4. DISCUSSION AND CONCLUSIONS

The radial distribution of the reduced axial strain along the line θ=0° is plotted in Figure 13 as it is obtained both from the experimental measurements (empty symbols and dotted line) as well as from the numerical analysis (filled symbols and continuous lines), for the semi-circular notched specimens. The predictions of the numerical analysis are plotted both for the isotropic and the orthotropic material models (rhombic and triangular symbols,

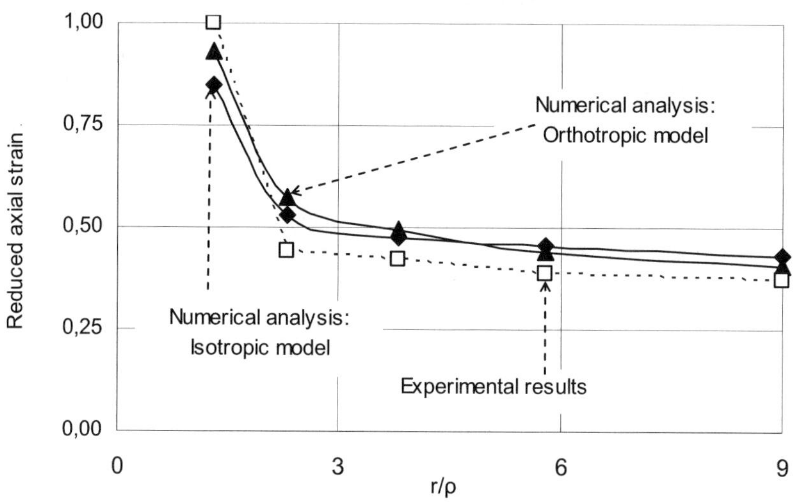

Figure 13. The radial variation of the axial strains along the line with θ=0°, according to the numerical analysis and the experimental results for the semi-circular notched specimens.

respectively). In all three cases the strains have been reduced over the respective maximum value, i.e. the value of the strain rosette at r/ρ=1.3, while the distance from the notch tip is reduced over the radius of the notch. It can be concluded from Figure 11 that the results of the numerical simulations are in very good agreement with those of the experimental study. The numerical results exceed slightly the experimental ones for r/ρ>2. From this point down the experimental results increase much more rapidly. This behaviour can be justified when considering that the linearity assumption adopted by the numerical analysis is not valid for r/ρ<2 since this region lies (partly or as a whole) within the process zone where intense micro-cracking is developed.

The respective results for the axial strain along the line θ=0° for the specimens with U-shaped notches are plotted in Figure 14. Again the strains are reduced over the respective maximum value, i.e. the value of the strain rosette at r/a=0.3. It is seen that the results of the numerical simulations are not in such a good agreement with those of the experimental study, at least compared to the previous case. The experimental results exceed systematically those of the numerical analysis for both models and only at the center of the specimen (r=4a) the experimental strain is the same with that predicted by the isotropic model. These discrepancies could be expected since the strain field is more intense for the case of a U-shaped notch and therefore the non-linearity of the constitutive behaviour of marble influences a wider region around the notch. On the other hand such a result should be studied in conjunction with the conclusions drawn from Figure 8 where it was observed that the influence of the notch is restricted within an area of about 5 to 15 mm (only the results of the strain rosette closest to the tip are non-linear).

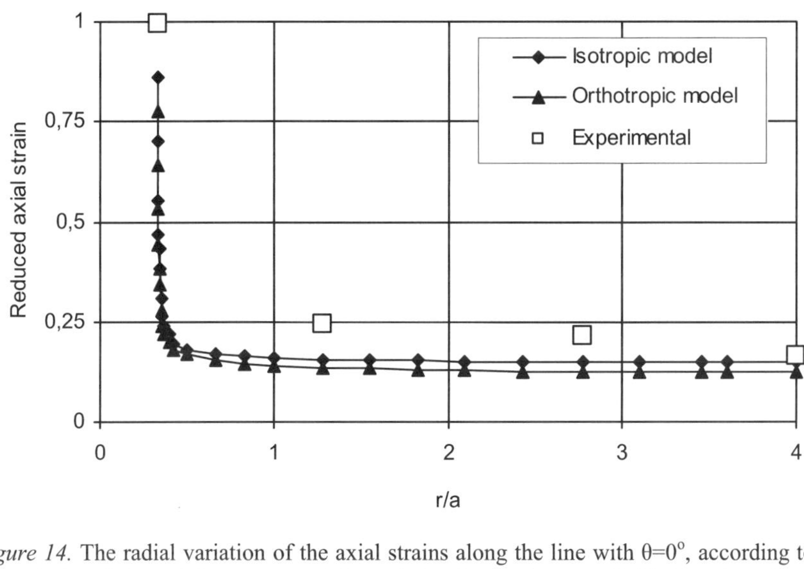

Figure 14. The radial variation of the axial strains along the line with θ=0°, according to the numerical analysis and the experimental results for the specimens with U-shaped notch.

For the specimens with U-shaped notches the distribution of the radial strain along a path parallel to the border-line of the notch at a distance equal to 11.5 mm from it, is plotted in Figure 15, according to both the numerical analysis and the experimental results. It is concluded that the isotropic model predicts strains significantly higher from those of the orthotropic one. The experimental results support the predictions of the latter. Some deviations appear only along the axis of symmetry of the notch.

The comparative study of the amplification of the strain field due to the presence of different types of notches, according to both the isotropic and the orthotropic models can be carried out with the aid of Figures 16(a,b). The comparison is carried for the strain fields developed along the horizontal axis of symmetry of the specimens, i.e. the line connecting the two tips. The origin of the x-axis is the center of the specimens and its end is the tip of the notch. In other words x/r=0 corresponds to the central point of both types of the specimens while x/r=-0.8 and x/r=-0.9 correspond to the tips of the U-shaped and the semi-circular notch, respectively.

The stress concentration factor for the semi-circular notches, defined as the axial strain developed at the center of the specimens over the respective strain at the tip of the notch, is about 2.7, while for the U-shaped notches the respective factor is equal to about 10. The predictions of the orthotropic model are, in general, higher from the respective ones of the isotropic model. The above values should not be compared with the respective ones obtained from analytical solutions since they can not take into consideration neither the fact that the specimens are of finite dimensions permitting interaction of the notches nor the exact shape of the notch.

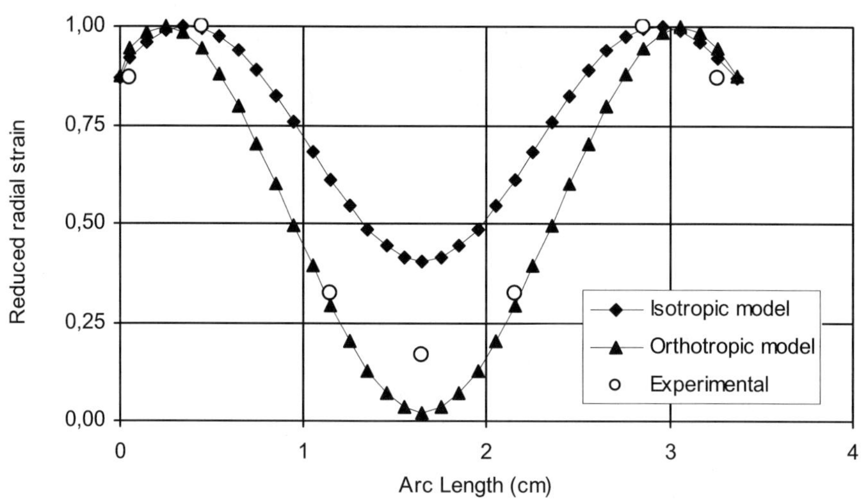

Figure 15. The variation of the radial strains along a path parallel to the border-line of the notch at a distance equal 11.5 mm from it, according to the numerical analysis and the experimental results for the specimens with U-shaped notch.

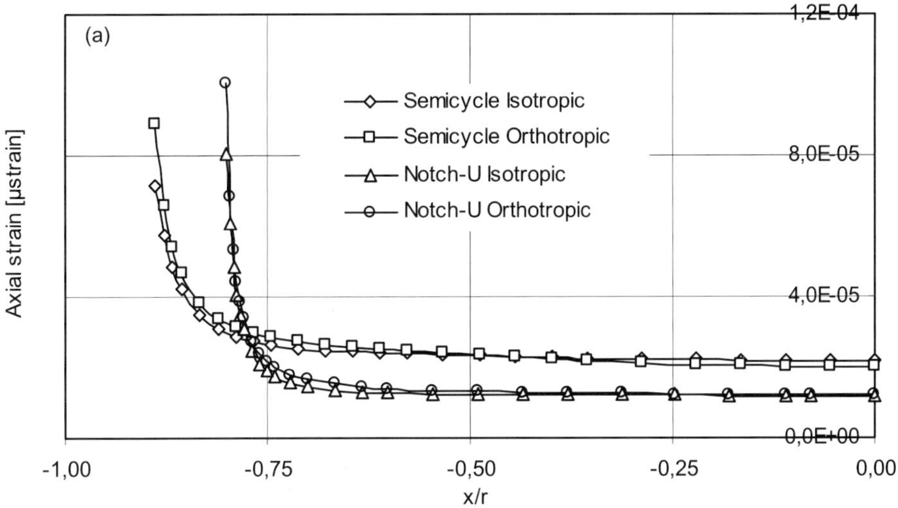

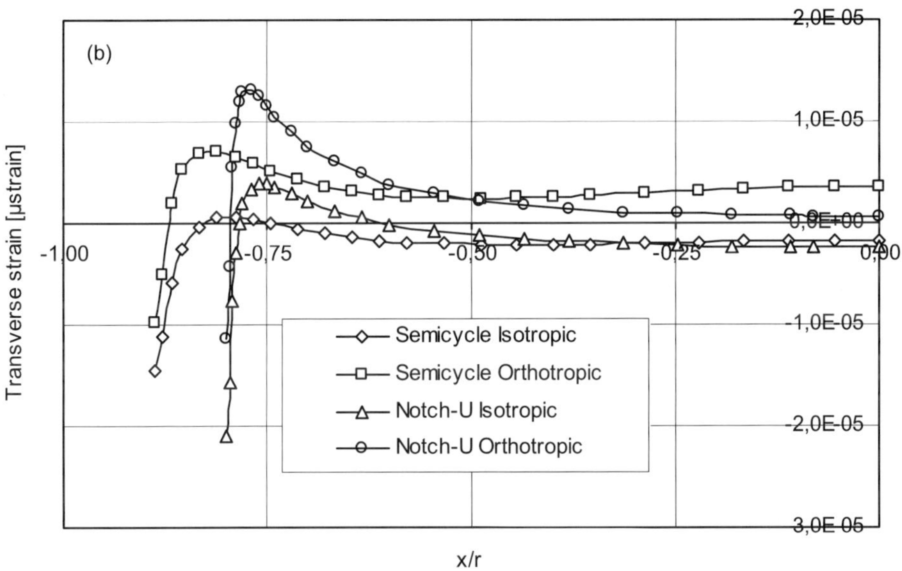

Figure 16. The variation of the axial (a) and the transverse (b) strain components along the line connecting the two notches for the different shape of the notches and for both the isotropic and the orthotropic models according to the results of the numerical analysis.

In summary, the strain field developed in marble plates with semi-circular and U-shaped notches was studied experimentally and numerically. Dogbone specimens with specially designed reinforced gripping zone were subjected to direct tension. The study revealed that:

- A process zone of intense damage (micro-cracking etc.) is developed around the notch in both cases rendering the macroscopic axial stress - axial strain relation non-linear from relatively early loading steps.

- The extent of the process zone varies between 5-8 mm for the specimens with semi-circular notches and 5-15 mm for those with U-shaped notches.
- The influence of the notch is eliminated rapidly: For $r/\rho>2$ (semi-circular notch) and $r/a>1.5$ (for the U-shaped notch) it becomes almost negligible.
- The results of the numerical analysis are in good agreement with the respective experimental ones for the case of specimens with semi-circular notches while for the ones with U-shaped notches the numerical predictions underestimate the experimental reality: Significant discrepancies are observed for points very close or inside the process zone.
- The semi-circular notch amplifies the strain field by almost 2.7 times while the U-shaped one by almost 10 times.

ACKNOWLEDGEMENTS

The authors would like to express their gratitude to Mr N. Moschakis for his help in the preparation of the specimens and the execution of the experiments. The numerical analysis is part of the Master Thesis of the third author.

REFERENCES

1. Z. H. Li and W.L. Guo, Three-dimensional elastic stress fields ahead of blunt V-notches in finite thickness plates, *Int. J. Fracture* **107**, 53-71 (2001).
2. G. Kirsh, Die Theorie der Elasticität und die Bedürfnisse der Festigkeitlehre, *Z. Ver. Dtsch. Ing.* **32**, 797-807 (1898).
3. C. E. Inglis, Stresses in a plate due to the presence of cracks and sharp corners, *Proc. Inst. Nav. Arch.* **55**(1), 219-230 (1913).
4. S. Filippi, P. Lazzarin, Distributions of the elastic principal stress due to notches in finite size plates and rounded bars uniaxially loaded, *Int. J. Fatigue* **26**, 377-391 (2004).
5. Kotousov, A., Wang, C.H., A generalized plane-strain theory for transversally isotropic plates, *Acta Mechanica* **161**, 53-64 (2003).
6. C. Zambas, Structural Repairs to the Monuments of the Acropolis-The Parthenon, *Proc. of the Institution of Civil Engineers - Civil Engineering* **92**, 166-176 (1992) (in Greek).
7. P. S. Theocaris and E. Coroneos, Experimental study of the stability of Parthenon, *Publications of the Academy of Athens* **44**, 1-80 (1979) (in Greek).
8. A. G. Tassogiannopoulos, A Contribution to the Study of the Properties of Structural Natural Stones of Greece, Ph.D. Dissertation (Nat. Techn. Univ. of Athens, 1986) (in Greek).
9. I. Vardoulakis, S. K. Kourkoulis, Mechanical Properties of Dionysos Marble, in: *Final Report of the Environment Project EV5V-CT93-0300* (Nat. Techn. Univ. of Athens, 1997).
10. S. K. Kourkoulis, G. E. Exadaktylos, and I. Vardoulakis, Notched Dionysos-Pentelicon marble in three point bending: The effect of nonlinearity, anisotropy and microstructure, *Int. J. Fracture* **98**, 369-392 (1999).
11. S. K. Kourkoulis, G. E. Exadaktylos and I. Vardoulakis, in: *Proc. 14th European Conf. on Fracture*, edited by A. Neimitz et al. (EMAS Publishing, UK 2004), pp. 243-250.
12. MARC: Analysis Research Corporation, Users' Manuals (1998).
13. Mentat 2: User's guide (1996).

Chapter 2: Mechanical Behaviour and Properties

Chapter 2.1

MODELLING WEATHERING EFFECTS ON THE MECHANICAL BEHAVIOUR OF NATURAL BUILDING STONES

Roberto Nova
Milan University of Technology (Politecnico), P.za L. da Vinci 32, 20133 Milan, Italy

Abstract: The main features of the mechanical behaviour of natural building stones are first recalled. It is shown that a convenient constitutive model can be conceived in the framework of the theory of strain-hardening plasticity, successfully applied to soils and soft rocks. Extension of the theory to chemo-mechanical conditions is presented and the main results of the numerical analyses of two geotechnical problems are shown.

Key words: soft rocks; plasticity; constitutive modelling; weathering; chemo-mechanics.

1. INTRODUCTION

Soft rocks were extensively used in the past as natural building stones. Limestones, tuffs, calcarenites and similar rocks were in fact employed as construction materials for temples, churches and palaces in all times, all over the world. Their mechanical properties are similar to those of concrete, the principal building stone of our age. They are therefore strong enough to bear high loading but at the same time weak enough to be easily quarried. Limestone, for instance, has a tensile strength that is one order of magnitude less than that of granite. They are also widespread so that their cost is limited: an ideal material to work with.

The low tensile strength is due to the weakness of the chemical bonds linking together the mineral grains constituting the rock skeleton. These bonds are attacked by environmental agents such as water, wind, bacteria, so that their strength is progressively reduced with time of exposure to the atmosphere.

The surface degradation of statues and monuments in the polluted air of our metropoleis is an apparent clue of the effect of weathering on natural

S.K. Kourkoulis (ed.), Fracture and Failure of Natural Building Stones, 55–70.

building stones. The decay of the mechanical properties is not only superficial but extends to the core of the stone, so that even its structural efficiency can be jeopardised by the action of atmospheric agents.

It is therefore of some interest also for engineers to know more about the weathering effects on the mechanical behaviour of natural building stones. Furthermore, it is important to clarify on which lines a constitutive model for such materials should be developed in order to cope with the observed phenomena.

In this paper, the fundamental properties of soft rocks are recalled first. The way in which phenomena as yielding and subsequent hardening, brittle ductile transition, effects of chemical attack can be dealt with is sketched next. Comparisons with experimental data in laboratory tests and description of the progressive degradation of structures in engineering problems are then presented. Finally, a simplified model for structural elements is discussed in elementary cases. It is shown how it is possible to calculate the variation of the bearing capacity of a structural element with time.

2. GENERALITIES ABOUT THE MECHANICAL BEHAVIOUR OF SOFT ROCKS

Traditionally, engineers divided natural geomaterials in two classes, depending on the degree of bonding between the mineral grains that constitute the geomaterial skeleton: soils and rocks. The former are characterized by little or no bond strength. Permanent strains take place as soon as loading is applied and failure is controlled mainly by friction. The latter are characterized by large bond strength. Strains are more or less reversible until fractures occur in the rock mass. Even though, for the sake of simplicity, for hard rocks engineers use a plastic failure criterion, characterized by friction and cohesion, it is more the fracture energy that controls the strength and the deformability of such materials.

In the last fifteen years, it was recognized that there exists a wide transitional class of geomaterials that behave in a way intermediate between the former two. At low stress levels their stress strain response is qualitatively similar to that of a hard rock. On the contrary, at high stress levels their behaviour in more similar to that of an unbonded soil. This class of materials was named hard soils-soft rocks. Many natural building stones, especially those characterized by high porosity, belong to it.

Typical examples of behaviour are given by Figures 1 and 2 (after Lagioia and Nova, 1995). Both refer to constant cell pressure triaxial tests on specimens of Gravina calcarenite (Southern Italy). In Figure 1 the behaviour of three specimens under relatively low confining pressures is shown. The initially linear stress - strain behaviour as well as the occurrence of a peak for a certain value of the deviator stress and the following loss of strength and strain uniformity of the specimen can be noticed. The peak is in fact associated to the occurrence of a planar fracture across the specimen. From that point onwards, the specimen deformation is essentially due to the sliding of one rigid block over another and strains, calculated as if they were uniform, are meaningless. In Figure 2 the behaviour of a specimen made of the same material but with a much higher confining pressure is shown. Also in this case the initial stress-strain relationship is linear, characterized by a stiffness that is more or less equal to that of the tests at low confining pressure.

Yielding occurs in a different way, however. After a while in which the displacement controlled specimen deforms at more or less constant load, hardening takes place up to a failure value characterized by large uniform strains. The clearly observed saw-tooth shape of the constant load step is apparently a clear sign that some sort of instability is occurring. Figure 3 shows that the stiffness in unloading-reloading is not affected by the large permanent strains experienced. In unloading-reloading the behaviour can be therefore assumed to be linear elastic.

A convenient conceptual framework to model this behaviour can be constructed, starting from the elasto-plastic strain-hardening model proposed first by the School of Cambridge (Schofield and Wroth, 1968). The fundamental idea on which such models are based is that it is possible to define a domain in the stress space, as shown in Figure 4, in which the behaviour of the rock can be considered linear elastic.

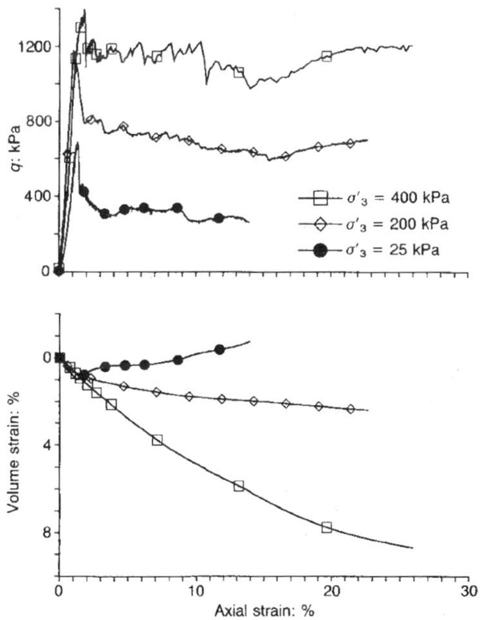

Figure 1. Drained constant cell pressure tests on specimens of Gravina calcarenite consolidated at low values of confining pressure.

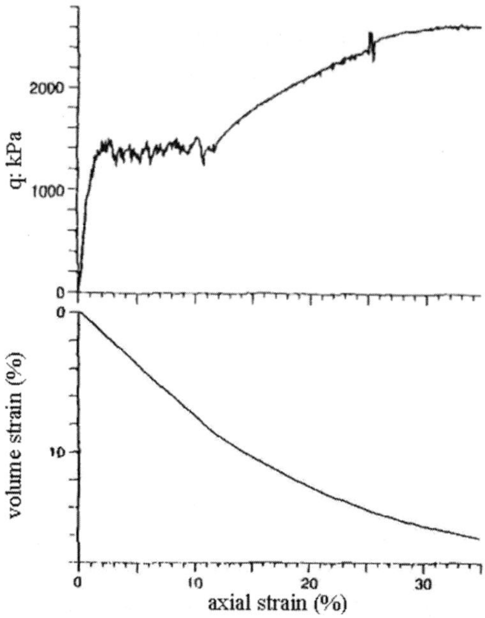

Figure 2. Drained constant cell pressure test ($\sigma'_3 = 900$ kPa) on Gravina calcarenite.

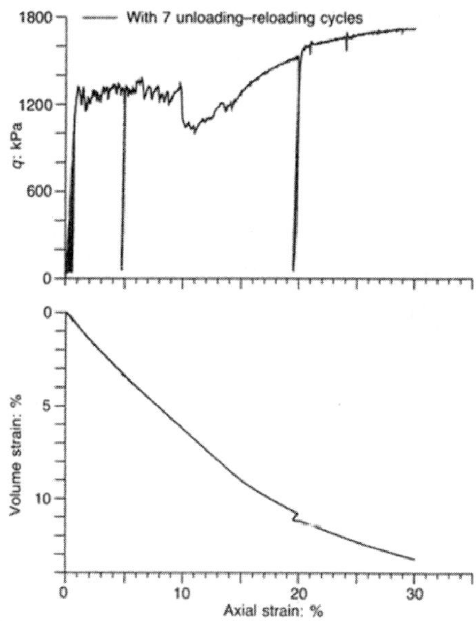

Figure 3. Constant cell pressure compression test ($\sigma'_3 = 600$ kPa): unloading-reloading cycles performed in the elastic phase, the destructuration phase and the hardening phase.

Yielding occurs when the stress point touches the frontier of the domain. The occurrence of plastic strains induces either strain-hardening or strain-softening depending on the sign of plastic volumetric strain rates, which in turn is governed by the so-called normality rule. It is apparent from Figure

4b that such a model can qualitatively describe at least some of the observed features of soft rock behaviour. For instance, it is possible to model the strength increase with the increase of confining pressure, the transition from brittle to ductile behaviour, the occurrence of dilating plastic volumetric strains at small confining pressures, the occurrence of yielding at lower deviator stress levels for higher confining pressures (point H in Figure 4).

To obtain quantitative agreement with experimental tests, many things have to be modified. First of all the shape of the yield locus must be as in Figure 5 (Nova, 1992). The normality rule is no more valid and a plastic potential different from the yield function is introduced. Hardening is controlled by one parameter linked to the grain structure, p_s, as in soil, and two parameters p_m and p_t related to the existence of bonds, (Nova, 1999; Gens and Nova, 1993). Possibly these latter parameters could be considered proportional to each other. In general, hardening depends on both volumetric and deviatoric plastic strains. With these modifications and appropriate choice of constitutive functions, Lagioia and Nova (1995) were able to reproduce the behaviour of Gravina calcarenite with remarkable accuracy in drained tests with low and high confining pressure, in p′ constant and undrained tests, in isotropic and oedometric compression. The behaviour of other materials such as tuff, marl, natural and artificial calcarenite, limestone was also simulated (Nova, 1992; Lagioia and Nova; 1993, Nova and Lagioia, 1996).

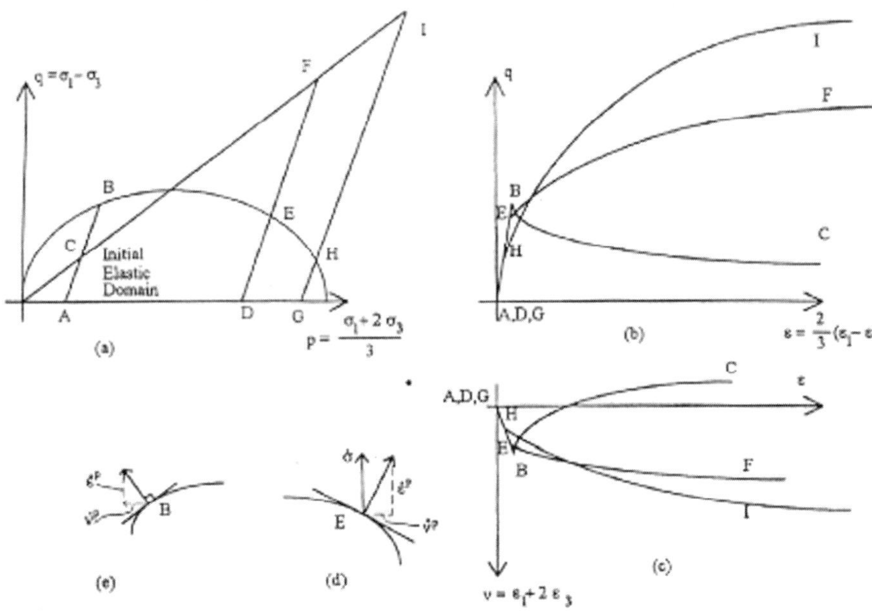

Figure 4. Schematic picture of the Cam Clay model: (a) stress paths in constant cell pressure tests with different overconsolidation ratios; (b) stress-strain behaviour; (c) volumetric strains; (d), (e) normality rule for different stress states.

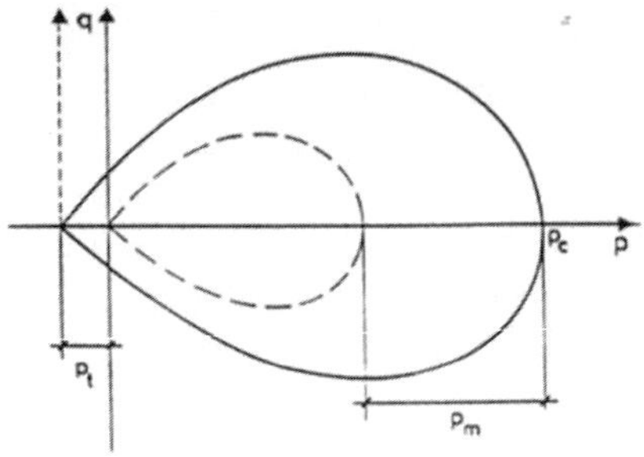

Figure 5. Initial yield surface of cemented (continuous line) and uncemented (dotted line) soil and evolution of the former during destructuration.

3. MODELLING OF WEATHERING EFFECTS

As previously discussed, the action of the atmospheric agents alters the intimate structure of geomaterials and influences their mechanical behaviour. It is mainly the plastic part that is affected, while the elastic properties can be considered constant with time, as a first approximation at least. The results presented in Figure 3 show in fact that the mechanical breaking of the bonds associated to plastic strains does not modify the elastic stiffness. Castellanza and Nova (2004) have shown that the elastic properties of an artificial cemented silica sand are not altered by the chemical aggression of an acid. They showed instead that such a chemical attack transforms a soft rock in an assembly of grains without any bonding.

To model such a behaviour, the following way of reasoning is possible (Nova, 2000): it is assumed that there exists an elastic domain governed by the following expression:

$$f(p, s_{ij}, p_k) \leq 0 \tag{1}$$

where p is the isotropic pressure and s_{ij} is the deviator stress:

$$p \equiv \frac{1}{3}\sigma_{ij}\delta_{ij} \tag{2}$$

$$s_{ij} \equiv \sigma_{ij} - p\delta_{ij} \tag{3}$$

In addition p_k is a set of hardening parameters. To make things simple it is assumed here, making reference to Figure 5, that:

$$p_m = \alpha p_t \tag{4}$$

where α is a constant and

$$p_c^* = p_c + p_s + p_m = p_s + p_t(1+\alpha) \tag{5}$$

The hardening parameter p_s depends on the plastic strains experienced

$$p_s = p_s\left(\varepsilon_{ij}^p\right) \tag{6}$$

while p_t is assumed to depend on a scalar measure of the degradation, x_d, that is linked to the particular process considered. It can be a function of the quartz content in a granite specimen, as in the original definition of Lumb (1962) who studied the weathering of granite. Or it can be a function of the exposure time to an acid attack or to the temperature at which a specimen is tested, and so on. To make things simple it is assumed that:

$$p_t = p_{t0}\left(1-x_d\right)^2 \tag{7}$$

Several experimental data (Park, 1996; Baynes and Dearman, 1978; Castellanza and Nova, 2004) support this hypothesis for different geomaterials.

Clearly, bonds brake also for mechanical reasons, as shown in Figure 2. The value of p_t should also depend on plastic strains (Nova et al., 2003):

$$p_t = p_t\left(\varepsilon_{ij}^p, x_d\right) = p_{t0}P_t\left(\varepsilon_{ij}^p\right)\left(1-x_d\right)^2 \tag{8}$$

For space reasons, it is assumed that P_t is a constant equal to 1. Only the non-mechanical agents are considered to be responsible for the degradation of the mechanical properties. Even with this limitation, it is possible to describe interesting phenomena, as it will be seen in the following.

Since when plastic strains occur f must be equal to zero before and after the "load" increment (an increment in x_d is also considered as a generalized load), the total differential of f must be zero. Therefore

$$df = \frac{\partial f}{\partial p}dp + \frac{\partial f}{\partial s_{ij}}ds_{ij} + \frac{\partial f}{\partial p_s}\frac{\partial p_s}{\partial \varepsilon_{rs}^p}d\varepsilon_{rs}^p + \frac{\partial f}{\partial p_t}\frac{\partial p_t}{\partial x_d}dx_d = 0 \tag{9}$$

Since

$$d\varepsilon_{rs}^p = \Lambda\frac{\partial g}{\partial \sigma_{rs}} \tag{10}$$

where $g(\sigma_{rs})$ is the plastic potential and Λ is a scalar, called plastic multiplier. Taking account of Eqs.(5) and (7) and solving for Λ one gets:

$$\Lambda = -\frac{\dfrac{\partial f}{\partial p}dp + \dfrac{\partial f}{\partial s_{ij}}ds_{ij} - 2\dfrac{\partial f}{\partial p_t}p_{t0}\left(1-x_d\right)dx_d}{\dfrac{\partial f}{\partial p_s}\dfrac{\partial p_s}{\partial \varepsilon_{rs}^p}\dfrac{\partial g}{\partial \sigma_{rs}}} \tag{11}$$

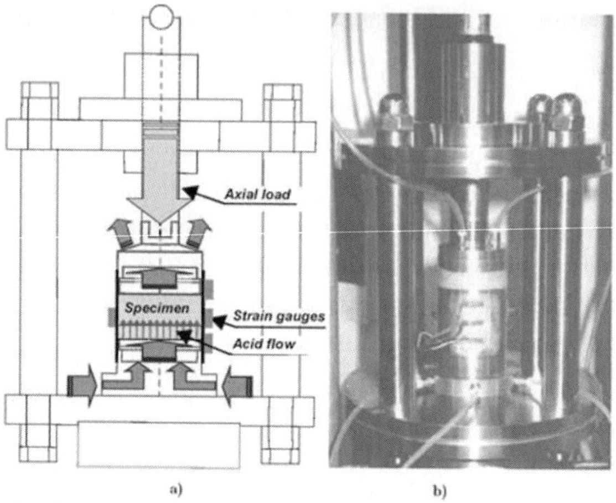

Figure 6. Schematic picture of the weathering test device (a) and photograph of the apparatus (b); after Castellanza (2002).

From Eq.(11) it is apparent that plastic strains are generated both from stress increments and by a variation of x_d. In particular plastic strains can occur even under constant stresses.

Consider as a special case an oedometric test on an artificial specimen of soft rock composed of a mixture of quartz grains and cement constituted by hydraulic lime and water. The test is conducted in the so-called WTD (weathering test device, Figure 6 after Castellanza, 2002). The apparatus consists of a thin metal ring that laterally constrains the geomaterial specimen. Un upper and a lower platen are used to load the specimen in the vertical direction. The two platens are permeable so that a fluid (water or acid) can seep through the specimen. Horizontal strains are almost prevented as in a normal oedometer but not completely, since the ring is slightly deformable. The test is composed of two phases. In the first one the specimen is subject to an increasing vertical loading as in a normal oedometer. In the second one, the load is kept constant to the level achieved in the first phase and an acid is forced to seep into the specimen. With time, the acid dissolves the calcareous bonds so that further vertical strains take place.

A very sensible strain gauge is attached along the circumference of the right section of the ring. This can record the circumferential strains and from those it is possible to determine the radial stresses by means of the Mariotte formula. Therefore the entire stress path can be determined.

Figures 7-10 (after Nova et al., 2003) show the comparisons of the experimental and calculated stress path, radial stresses, vertical settlements and the variation of the calculated internal variables for a specimen of cemented silica sand. The constitutive model used is presented in detail in Nova et al. (2003), where also more complex tests are considered.

It is apparent that the model proposed can describe with sufficient accuracy the effects of weathering on specimens of soft rock.

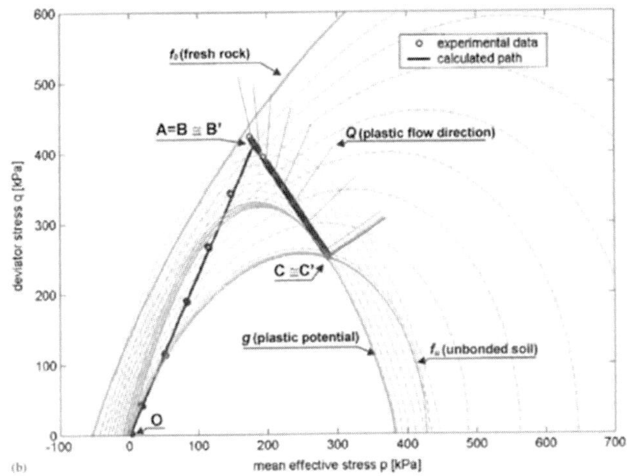

Figure 7. Experimental versus calculated stress paths in a weathering test on cemented silica sand, after Nova et al. (2003).

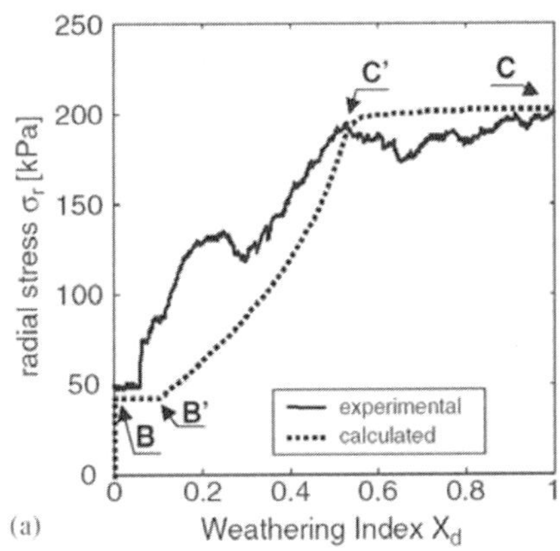

Figure 8. Experimental versus calculated radial stress for the stress path shown in Figure 7, after Nova et al. (2003).

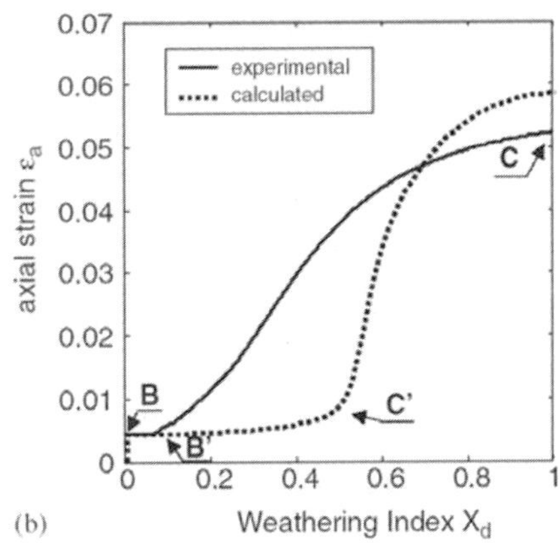

Figure 9. Experimental versus calculated vertical strains for the stress path shown in Figure 7, after Nova et al. (2003).

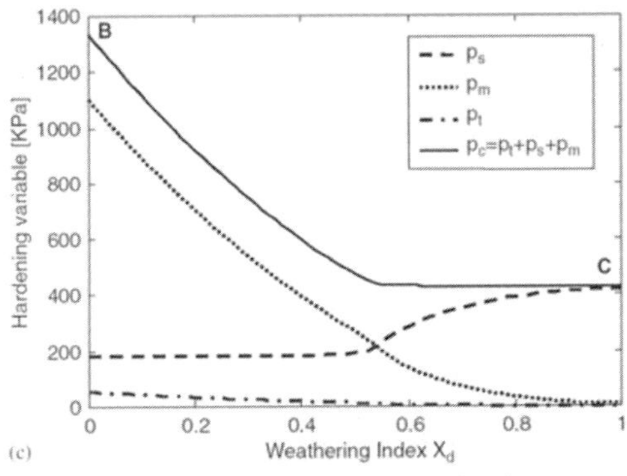

Figure 10. Evolution of the calculated internal variables for the stress path of Figure 7 (after Nova et al. (2003).

4. APPLICATIONS

By employing the model of Nova et al. (2003), Castellanza et al. (2002) showed that it is possible to simulate the progressive failure of geostructures, subjected to chemically induced bond degradation. In Lorraine, for instance, several collapses were caused by the progressive degradation of the pillars of abandoned iron mines, attacked by bacteria, which live in the water that flooded the mine chambers after the end of the mining activities.The damage progression was simulated simply by imposing a degradation law to the pillars. The closer the rock element is to the flooded chamber, the higher is the value of x_d (up to one that corresponds to full degradation). The value of x_d increases with time at a conveniently established rate. Furthermore the zone of the pillar affected by degradation also increases with time at a given rate.

The results obtained are shown in Figure 11. The main effect of the degradation is that of reducing the bearing capacity of the layers that are more exposed to the chemical attack. There is therefore a progressive stress migration towards the centre of the pillar that can eventually collapse under the weight of the overburden that cannot be sustained anymore. Figure 11b shows how relevant can be the weathering induced displacements at the surface with respect to the original ones due solely to mine excavation.

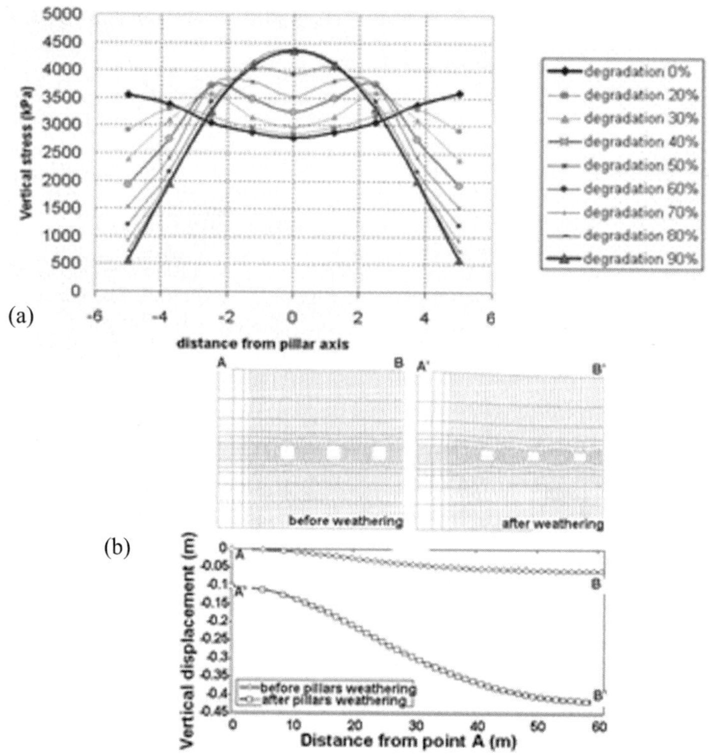

Figure 11. Subsidence induced by the degradation of the pillars of an abandoned iron mine; (a) Stress redistribution within the pillar; (b) Settlements, after Castellanza et al. (2002).

Recently, Parma (2004) considered the effect of a pollutant migrating within a bonded soil mass on the settlement of a shallow foundation. To solve this problem, the contaminant diffusion equations were treated by means of a finite difference computer code combined with GeHoMadrid, the finite element code (Fernandez Merodo et al., 1999) in which the aforementioned Nova et al. model is implemented.

Figure 12 shows the evolution of the contaminant, the associated plastic strains induced by chemical degradation and the progression of the foundation settlements with time. It is apparent that the calculated results look reasonable. A series of experimental tests are presently performed in Milan on a model foundation resting on an artificially cemented soil, to have a benchmark against which checking the validity of the proposed theory.

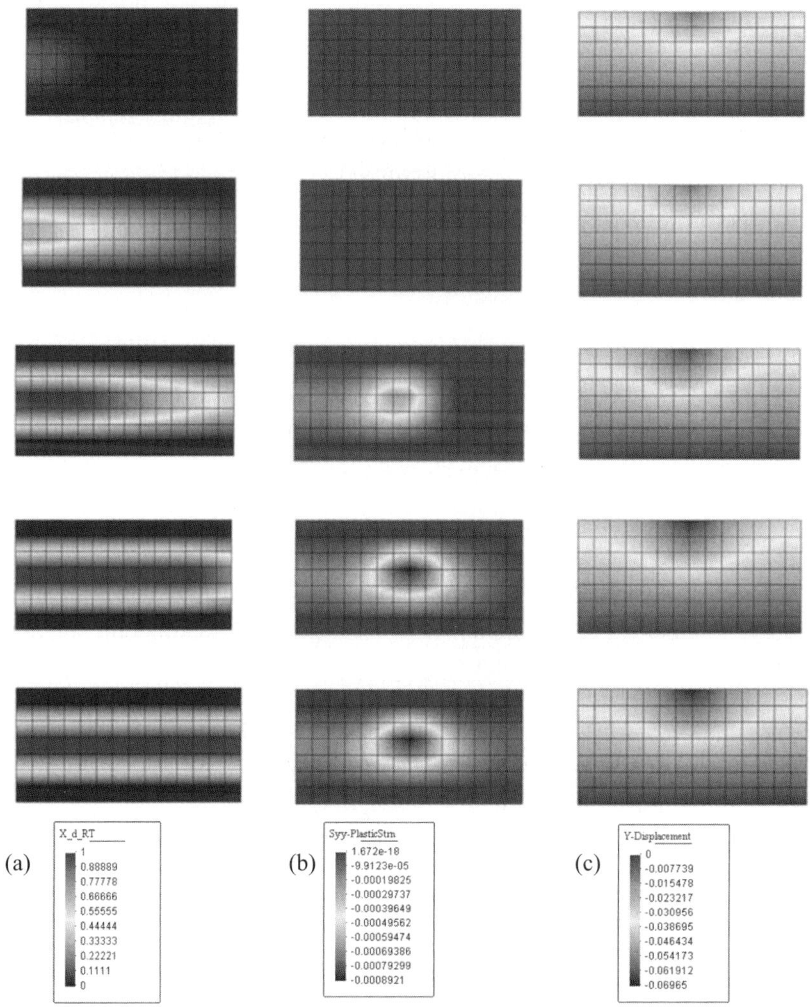

Figure 12. Effects of the diffusion of a contaminant under a shallow foundation; (a) Variation of x_d; (b) Deviatoric plastic strains; (c) Vertical displacements for successive time steps (Parma, 2004).

5. SIMPLIFIED MODEL FOR STRUCTURAL BEHAVIOUR OF NATURAL BUILDING STONES

The conceptual structure of the constitutive model presented so far can be applied to structural problems involving the weathering of natural building stones. To achieve this goal, a simplified model is formulated in this section.

At variance with foundation soils, structural elements are not laterally confined. Therefore the range of interest is that of low confining pressures, where the natural building stones behaviour is qualitatively similar to that of hard rocks. The yield condition is therefore also a failure condition, since yielding is associated to negative hardening values (softening) and a sharp peak occurs in the deviatoric stress-strain law. It is apparent from Figure 13 that such a failure condition can be approximated by a straight line of equation

$$q = mp' + q_m \tag{12}$$

at least for positive p′. It is also shown in Figure 13 that as the degradation proceeds, m remains more or less constant, while q_m decreases more or less in proportion to p_m. One can therefore assume that the unconfined compressive strength of a natural building stone q_C varies with the weathering degree as

$$q_C = q_{C0}\left(1 - x_d\right)^2 \tag{13}$$

or even more simply as

$$q_C = q_{C0}\left(1 - x_d\right) \tag{14}$$

The plastic potential rules the ratio of plastic strains. A constant dilation rate can be assumed again for the sake of simplicity

$$\dot{\varepsilon}_v^p = -\psi \dot{\varepsilon}_d^p \tag{15}$$

where $\dot{\varepsilon}_v^p$ and $\dot{\varepsilon}_d^p$ are the volumetric and deviatoric plastic strain rates, respectively. Furthermore, the value of q_C can be made to depend on plastic strain experienced, for instance as:

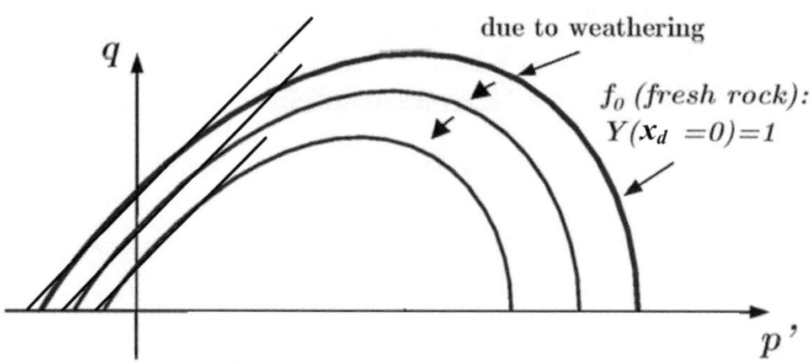

Figure 13. Shrinkage of the yield domain due to weathering.

$$\frac{\partial q_C}{\partial \varepsilon_d^p} = -q_C \rho_C \tag{16}$$

If ρ_c is taken to be zero, a ductile behaviour is assumed after yielding. If ρ_c is taken to be infinite, a perfectly brittle behaviour is considered. This latter possibility is considered here. In practice, this means that when a material element reaches the yielding condition it suddenly looses the ability of bearing stress. The existing stress must be therefore instantaneously transferred to the parts that have not yet yielded to guarantee the fulfilment of equilibrium. Finally, in accordance with many experimental observations, the elastic moduli are considered to be constant, affected neither by weathering nor by the occurrence of plastic strains.

6. AN ANALYSIS OF THE STABILITY OF A CIRCULAR PILLAR SUBJECTED TO WEATHERING

As an application, consider a cylindrical pillar with circular cross section subjected to a constant load Q (consider negligible the pillar own weight). When the pillar is intact at time zero the pillar contraction is given by

$$\delta = \frac{QL}{E\pi R_0^2} \equiv \delta_0 \tag{17}$$

where L is the height of the pillar, R_0 its radius and E the Young's modulus of the material. Assume now that weathering affects the lateral surface of the pillar and progresses towards the centre of the pillar linearly with time. The front of intact rock is therefore at a distance from centre equal to

$$R = R_0 - \alpha t \tag{18}$$

Assuming that the behaviour is perfectly brittle, the rock strength in a point at distance R falls to zero as soon as the front reaches it. The pillar therefore collapses abruptly when:

$$Q = \pi \left(R_0 - \alpha t\right)^2 \sigma_{c0} \tag{19}$$

where σ_{c0} is the compressive strength of the stone. This occurs for a time t_f after weathering begins equal to:

$$t_f = \left\{1 - \sqrt{\frac{Q}{\pi R_0^2 \sigma_{c0}}}\right\} \frac{R_0}{\alpha} \tag{20}$$

The contraction increases with time as:

$$\delta = \frac{QL}{E\pi R_0^2\left(1-\frac{\alpha t}{R_0}\right)^2} = \frac{\delta_0}{\left(1-\frac{\alpha t}{R_0}\right)^2} \tag{21}$$

until the time of failure is achieved and then sudden collapse occurs. The displacement corresponding to t_f according to Eq.(20) is equal to:

$$\delta_f = \frac{Q_0}{Q}\delta_0 \tag{22}$$

where Q_0 is the force the intact pillar can bear. The results are summarized in Figure 14. The values of the parameters of Eqs.(20) and (21) can be determined either by recording the contraction variation with time or by performing appropriate tests in the laboratory. To give an order of magnitude, by assuming that the degradation front progresses at 1 mm/year for a column with a diameter 1 m wide loaded at 0.1 of its bearing capacity in non-weathered conditions, the failure time is 342 years. This time reduces to 146 years if the load is one half of the bearing capacity.

7. CONCLUSIONS

Often old constructions are made of natural materials belonging to the class of soft rocks. Their behaviour is intermediate between that of a hard rock and that of a soil and depends on the confining pressure and the degree

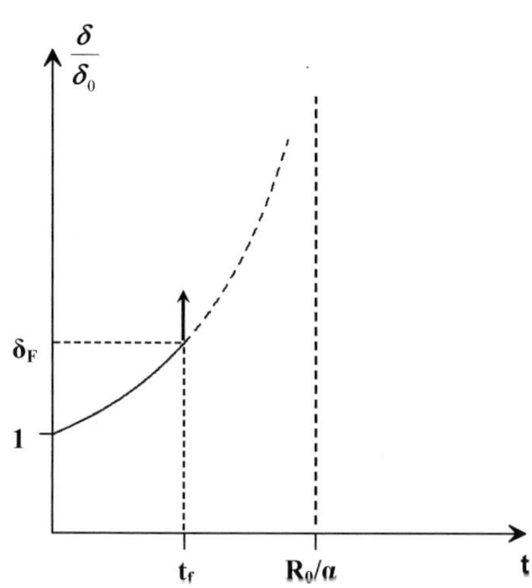

Figure 14. Contraction with time of a natural building stone affected by weathering.

of weathering. At high confining pressures or high weathering degree the rock behaves as a soil, which is as an unbonded material (no cohesion).

By exploiting the conceptual framework of soil models, with appropriate modifications to take the role of bonding into account, it is possible to model the behaviour of rock specimens under mechanical loading and/or chemical attack, with reasonable agreement between experimental data and numerical results. The constitutive model developed can be implemented in a finite element code and, in principle, any type of boundary value problem can be tackled, taking into account the effect of progressive chemical weathering. Two geotechnical problems were discussed.

Also structural problems could be analysed in the same way. However, since in this case the load condition is simple, simplified methods may be used. Such methods can be derived by drawing inspiration from the more realistic 3D model described here. A simple model for evaluating the life-time of a column under centrally loaded subject to environmental degradation was finally proposed.

REFERENCES

Baynes F. J., and Dearman, W. R., 1978, The relationship between the microfabric and the engineering properties of weathered granite, *Bulletin of the International Association of Eng. Geol.* **18**:191-197.

Castellanza R., 2002, Weathering effects on the mechanical behaviour of bonded geomaterials: an experimental, theoretical and numerical study, PhD Thesis, Politecnico di Milano.

Castellanza R., Nova R., 2004, Oedometric Tests on Artificially Weathered Carbonatic Soft Rocks, *J. Geotechnical and Geoenvironmental Engineering* ASCE **130**(7):728-739.

Castellanza R., Nova R., Tamagnini C., 2002, Mechanical effects of chemical degradation of bonded geomaterials in boundary value problems, *Rev. Fran. Génie Civil* **6**(6):1169-1192.

Gens A., Nova R., 1993, Conceptual bases for a constitutive model for bonded soils and weak rocks, in *Proceedings of the International Symposium on Hard soils-soft rocks*, Athens, 485-494.

Lagioia R., Nova R., 1993, A constitutive model for soft rocks, in *Proceedings of the International Symposium on Hard soils-soft rocks,* Athens, pp. 625-632.

Lagioia R, Nova R., 1995, An experimental and theoretical study of the behaviour of a calcarenite in triaxial compression. *Géotechnique* **45**(4):633–648.

Lumb P., 1962, The properties of decomposed granite, *Géotechnique* **12**:226-243.

Nova R., 1992, Mathematical modelling of natural and engineered geomaterials. General lecture 1st E.C.S.M. Munchen, *European J. Mech. A/Solids* **11**(Special issue):135–154.

Nova R., 2000, Modelling weathering effects on the mechanical behaviour of granite, in *Constitutive Modelling of Granular Materials*, D. Kolymbas editor, Springer, Berlin, 397-411.

Nova R., Castellanza R, Tamagnini C., 2003, A constitutive model for bonded geomaterials subject to mechanical and/or chemical degradation. *Int. J. Num. Anal. Meth. Geomech.* **27**(9):705–732.

Nova R., Lagioia R., 1996, Soft Rocks: Behaviour and Modelling" Keynote lecture, Proc. Eurock '96, Turin, Balkema 2000, 3, 1521-1540.

Parma M., 2004, Modellazione dell'accoppiamento chemo-meccanico nei materiali cementati, Tesi di Laurea (in Italian).

Park H. D., 1996, Assessment of the geotechnical properties of the weathered rocks at historical monuments in Korea, in *Proceedings of Eurock '96*, edited by G. Barla, Balkema, Rotterdam, 1413-1416.

Schofield A. N., Wroth C. P., 1968, Critical State Soil Mechanics, McGraw Hill, New York.

Chapter 2.2

MECHANICAL PROPERTIES AND DAMAGE DIAGNOSIS OF NATURAL BUILDING STONES

Euripides Papamichos[1], Stefanos-Aldo Papanicolopulos[2] and Idar Larsen[3]
[1]Aristotle Univ. of Thessaloniki, Dept. of Civil Engineering, GR-54124 Thessaloniki, Greece, epapamic@civil.auth.gr; [2]National Technical University of Athens, Dept. of Applied Mathematical and Physical Sciences, GR-15773 Athens, Greece, stefanos@mechan.ntua.gr; [3]SINTEF Petroleum Research, N-7465 Trondheim, Norway, idar.larsen@iku.sintef.no

Abstract: The indentation technique is tailored here to the needs of a portable tool for in situ diagnosis of mechanical properties and damage of natural building stones. Indentation tests were performed in twelve natural building stones (calcarenites, limestones, sandstones, marbles) and mortars used for restoration. A wide range of mechanical and petrophysical properties with different failure mechanisms in indentation is thus represented. Indicatively, the Unconfined Compressive Strength (*UCS*) ranges between 3.1 MPa and 116 MPa and the tangent Young's modulus E_{50} at 50% of the maximum stress in uniaxial compression ranges between 0.9 GPa and 50 GPa. The tests were performed with three indenter diameters, 1-, 2- and 3-mm, to analyze scale effects in the results, and at five different depths, as the technique will be used not only for surface measurements but also for measurements in the interior, at the bottom of a small drilled hole. Such measurements can provide information on stone damage with depth. The results are used to build correlation functions and databases between indentation parameters and stone stiffness and strength. The technique is applied to two marbles that had been artificially weathered with exposure to moisture and temperature cycles, and to a consolidated mortar, i.e. a mortar treated with a consolidant for improving its weathering characteristics.

Key words: indentation test; modulus of flat indentation; critical transition stress; natural building stones; characterization of mechanical properties; damage diagnosis.

1. INTRODUCTION

Stone monuments represent a significant part of our cultural and historical heritage. However, damage (weathering) of historical buildings, monuments, works of art and other cultural properties due to the aggressive urban environ-

S.K. Kourkoulis (ed.), Fracture and Failure of Natural Building Stones, 71–92.

ment of the last decades and the ambient climatic conditions, is reported from all over the world. This has resulted in a substantial effort from local and governmental authorities and the industry for quick and suitable measures for the preservation of national monuments. Several fundamental questions arise in the decision making of the optimum restoration and preservation strategy of old monumental building stones. Among others, there are questions on stone suitability, effectiveness of consolidation measures, and in situ quantitative measurements of the degree of damage of a part or element of a monument or building. Weathering of natural building stones is also of considerable practical interest to the quarrying and the construction industries.

An effort has therefore been undertaken to develop a portable tool that has been missing from the industry for the in situ, quasi non-destructive, reliable and accurate diagnosis of mechanical properties and damage of natural building stones in structures of cultural heritage, in order to improve preservation strategies.

The work builds on previous work by the authors and others on the determination of mechanical properties and damage through indentation testing. The indentation technique is a common method for measuring hardness in metals, glass and ceramics; also for measuring the elastic parameters of surface coatings[1-3]. In rocks, emphasis has been placed in the search for correlations between indentation measurements and rock parameters such as rock strength and stiffness[4-9]. For our purpose the indentation technique is tailored to natural building stone applications.

2. STONE CHARACTERIZATION TESTS

A mechanical and acoustical characterization of 12 common building stones was performed to obtain their basic mechanical and acoustical properties. These properties are used to obtain relations and correlations with the indirect measurements of mechanical properties from indentation tests. They are obtained from conventional triaxial compression tests with acoustics, i.e. ultrasonic wave transmission measurements, and comprise strength, stiffness and dilation parameters, and P- and S- wave velocities. The bulk density is also calculated from the weight and dimension of the specimens.

The tested stones and their origin are listed in Table 1. They comprise one calcarenite, three limestones, three marbles, two sandstones, one soapstone and two mortars. The mortars Block 1 and Block 2 are prepared as twin blocks and therefore have similar properties. Subsequently Block 2 is treated with a consolidant and characterized again to quantify the effect of the consolidation on the mechanical and acoustical properties. The consolidant was supplied by CNR-ICVBC (Italian Research Council, Institute for the Conservation and Promotion of Cultural Heritage) together with the protocol of treatment.

A total of 13 uniaxial, 27 conventional triaxial compression tests each with two load-unload cycles down to almost the confining stress (to capture the elastic behavior), and 13 uniaxial compaction tests were performed. The stones were divided in two groups, the first group of stones 1 to 6 (see Table 1) comprises the weaker stones with $UCS < 50$ MPa and the second group of stones 7 to 12 the stronger stones with $UCS > 50$ MPa. The stones in the first group were tested in triaxial compression at confining stresses of 4 and 10 MPa, while the stones in the second group were tested at confining stresses 5 and 15 MPa.

The main test results are given in Table 1 which lists the grain size, the porosity ϕ, the bulk density ρ, the *UCS*, the static and dynamic Young's moduli E_{50}, E^d_{50} and Poisson's ratios ν_{50}, ν^d_{50} in unconfined tests at 50% of the peak stress, and the peak friction angle φ and peak cohesion c. The peak axial stress in the uniaxial and triaxial tests is plotted in Figure 1 vs. the confining stress.

The marbles are contractant up to or near peak axial stress and show a rather ductile post-peak behavior even at low or zero confining stress. The sandstones are dilatant after about 70% of the peak axial stress and show a rather brittle post-peak behavior. The calcarenite and Pietra di Lecce limestone exhibit pore collapse due to compaction, the Pietra di Vicenza limestone a brittle to ductile behavior with increasing confining stress and the Portland limestone a brittle behavior. The mortars are ductile. With increasing axial stress, the tangent Young's moduli increase and then decrease as the peak stress is approached. In the post-peak region the modulus becomes negative. The tangent Poisson's ratios increase with increasing stress.

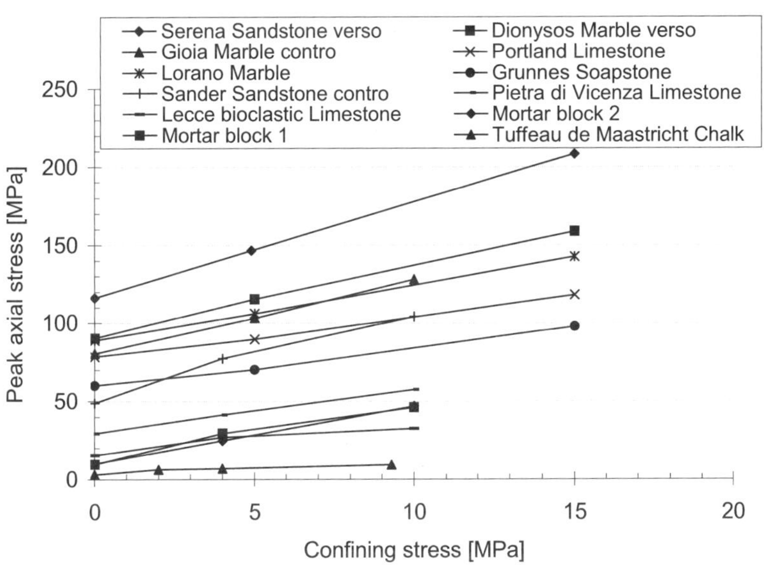

Figure 1. Peak axial stress vs. confining stress in triaxial tests.

Table 1. Stones characterized for their mechanical and acoustical properties

Stone	Origin	Grain size [μm]	ϕ [%]	ρ [g/cm^3]	*UCS* [MPa]	E_{50} [GPa]	ν_{50} [-]	E^d_{50} GPa]	ν^d_{50} [-]	φ [°]	c [MPa]	K_oS *[MPa]*	M_{50} *[GPa]*	K_{o50} [-]
1. Tuffeau de Maastricht Calcarenite	Netherlands	100	50.0	1.304	3.1	0.90	0.09	3.6	0.14	13.3	1.23	8.0	-	-
2a. Mortar Block 1	Italy	1500	33.0	1.735	9.6	3.8	0.27	8.4	0.24	34.2	3.10	-	1.1	0.8
2b. Mortar Block 2	Italy	1500	33.0	1.742	10.0	3.6	0.15	-	-	34.9	2.62	-	-	-
3. Mortar Block 2 consolidated	Italy	1500	-	1.758	6.2	1.7	0.09	-	-	34.7	1.88	-	-	-
4. Pietra di Lecce Limestone	Italy	80-100	47.4	1.470	15.4	4.8	0.23	10.0	0.23	14.2	6.71	29.6	5.2	0.19
5. Pietra di Vicenza Limestone	Italy	400-2000	29.3	1.948	29.5	15.8	0.31	20.2	0.25	28.1	8.93	67.3	15.3	0.15
6. Sander Sandstone	Germany	150-200	19.7	2.195	48.8	8.8	0.36	23.6	0.23	43.6	11.0	-	10.6	0.32
7. Grunnes Soapstone	Norway	800-2000	0.8	2.895	60.1	29.8	0.39	-	-	25.9	18.5	-	44.4	0.27
8. Portland Island Limestone	UK	300	20.1	2.231	78.5	33.1	0.24	41.9	0.27	27.1	23.8	160.3	33.0	0.16
9. Gioia Marble	Carrara, Italy	150-250	1.5	2.708	80.5	42.1	0.42	53.8	0.32	40.7	18.4	-	47.8	0.24
10. Lorano Marble	Carrara, Italy	-	-	2.710	89.0	49.8	0.25	79.9	0.28	34.3	23.4	-	86.0	0.17
11. Dionysos Marble	Athens, Greece	300-400	1.6	2.708	90.3	45.6	0.32	67.0	0.25	39.7	21.5	-	-	
12. Serena Sandstone	Firenzuola, Italy	200-800	6.5	2.588	116.3	22.7	0.30	43.1	0.22	46.0	23.5	-	27.0	0.23

Generally, the P- and S-wave velocities do not change dramatically with increasing axial stress or during the load-unload cycles. A decrease in the velocities was only observed close and after the peak axial stress demonstrating the damage of the stone. Cycles at axial stress close to the peak stress showed that the decrease in velocity and thus the damage is permanent. The dynamic Young's modulus is generally higher than the static but in the more competent stones it coincides well with the static modulus at the start of the unloading cycles. The dynamic and static Poisson's ratios show generally a reasonable agreement.

In addition, Table 1 lists also the main results from the uniaxial compaction tests, i.e. the uniaxial compaction strength K_oS, and the compaction modulus M_{50} and the K_{o50} value at 50% of the peak axial stress in uniaxial compaction. K_o is the ratio of radial to axial stress in the uniaxial compaction test. From the tested stones, compaction failure experienced the Tuffeau de Maastricht calcarenite and the Pietra di Lecce and Pietra di Vicenza limestones. Shear failure experienced the Portland limestone, while the mortar showed a transition to increased compaction. The remaining stones did not show compaction failure in the range of stresses tested.

3. INDENTATION TESTS

3.1 Test description and program

In the indentation test a normal to the specimen force is applied to an indenter, which penetrates into the specimen under constant displacement (or penetration) rate. Figure 2 shows a schematic view and a photograph of the indentation test equipment, which is placed inside a load frame for the application of the load, test control and data acquisition. A load cell and a Linear Variable Differential Transformer (LVDT) measure, respectively, the applied force and the indentation depth. All tests are performed on oven-dried specimens.

Various indenter shapes, such as flat, hemispherical or conical, and various indenter diameters can be used. Measurements with hemispherical or conical indenters are more difficult to interpret because the contact area varies continuously with indentation depth. With respect to indenter diameter, larger indenters provide more reliable results as they are less affected by the stone's grain size and other inhomogeneities in the petrophysical structure of the stones. However, for high indentation strength stones, the diameter of the indenter may be constrained by the load capacity of the equipment (especially for portable equipment), or the yield or buckling load of the indenter, which can lead to indenter failures (Figure 3a). Based on these, flat indenters with diameters 1-, 2- and 3-mm were used. In this way, an optimum indenter diameter can be selected according to the stone and indenter properties. The indenter is constructed from hardened at 60 HRC steel with elastic modulus about 190 GPa and yield strength about 2150 MPa at 20°C. It comprises a 47-mm long and 4-mm diameter stem, a top section for the connection with the load frame holder, and a 2-mm long indentation tip with diameter 1-, 2- or 3-mm (Figure 3b). The stem is necessary in order to perform indentation tests not only at the surface of a stone but also at different depths, i.e. at the bottom of a 5-mm diameter drilled hole. In this way, the mechanical parameters of the stone can be obtained as a function of depth. These parameters in a monumental or other structure may vary due to varying stone degradation.

The indentation tests were performed in stone cubes of 10-cm sides. The vertical orientation of the cubes was selected to coincide with the vertical orientation of the cylindrical specimens tested triaxially for their mechanical

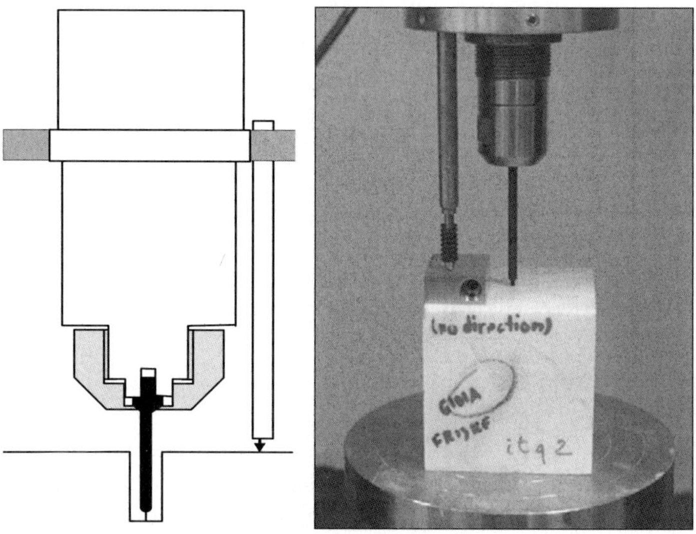

Figure 2. Schematic and photograph of the indentation equipment.

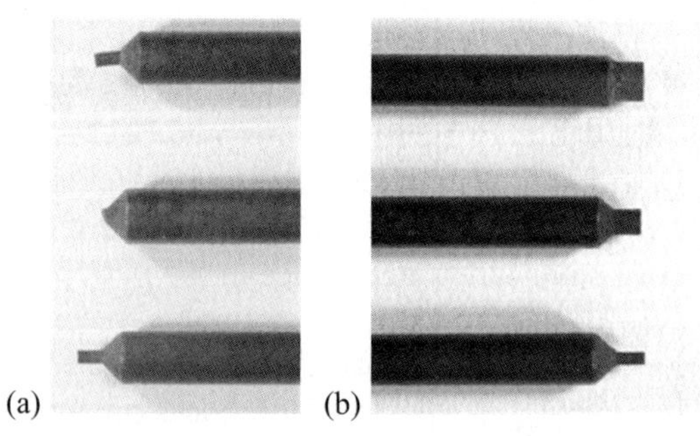

Figure 3. (a) Indenter failures, and (b) indenters with diameter (from top) 3-mm, 2-mm, 1-mm used in testing.

characterization. In each cube, tests were performed with the 1-, 2- and 3-mm indenter in three different locations in a triangular pattern that maximizes the distance between the indenter locations and the corners of the cube to avoid undesirable interaction effects (Figure 4). In each of the three locations, tests were performed at six different depths, i.e. 0-, 5-, 10-, 20-, 30- and 40-mm depths. Finally, the tests were repeated in two more cubes for each stone, cut from the same block. In this way for each of the twelve stones there were three tests for each indenter size and depth, a total of $12 \times 3 \times 3 \times 6 = 648$ tests. Prior to testing and after testing each block, indentation tests were performed in a reference Plexiglas block in order to follow the wear of the indenters in each stone and replace the indenters accordingly. The tests were performed to a maximum indentation depth of 1.5 mm or to a maximum force of 2-, 6- and 9-kN for indenter diameters 1-, 2- and 3-mm, respectively. The indentation rate was about 0.01 mm/s, uncorrected for the indenter deformation.

Figure 5 shows typical test results where the force on the indenter is plotted against the indentation depth for Sander sandstone and Dionysos marble. Generally, the force vs. indentation depth curves show an initial part of increasing slope, a rather linear part and a third part of decreasing slope that may or may not include a peak force. A peak was observed in the marbles, the soapstone and the Serena sandstone, which possibly results from the extensive stone failure around the indenter as shown in Figure 6.

The repeatability of the results is generally satisfactory in the more homogeneous stones but deteriorates as the inhomogeneity increases. Typical example of inhomogeneous material, at least for the scale considered here, is the mortar where different results are obtained when the indentation is performed in a piece of gravel or in the cement (Figure 7).

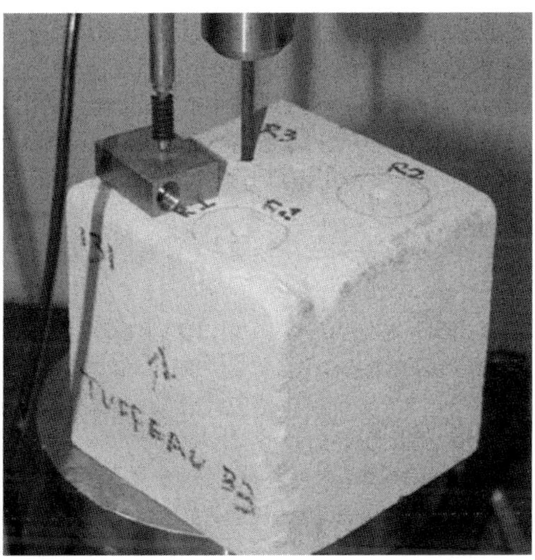

Figure 4. Indentation pattern in a stone block.

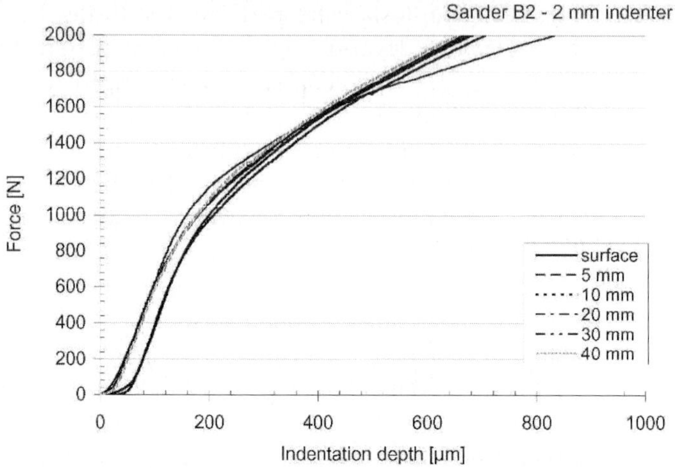

(a)

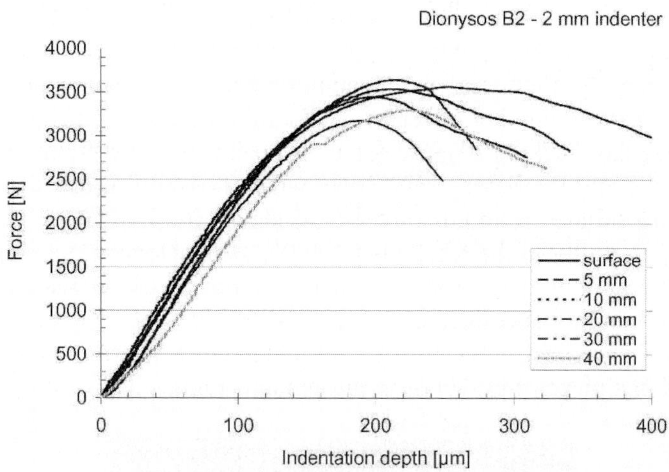

(b)

Figure 5. Typical indentation results. Force vs. indentation depth for (a) Sander sandstone and (b) Dionysos marble with 2-mm diameter indenter. Indentations at surface and at five depths.

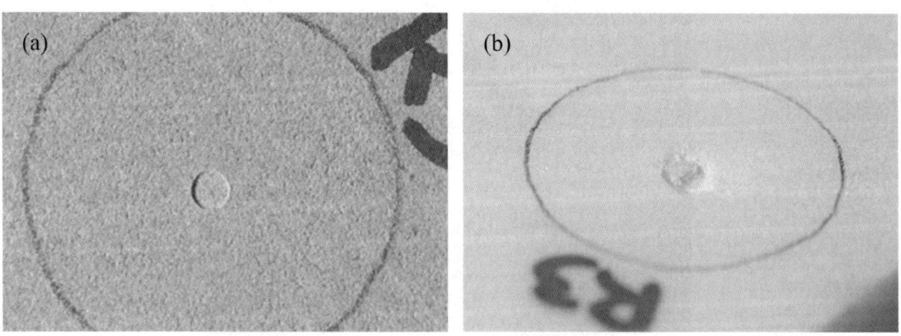

Figure 6. Indentation signature in (a) Sander sandstone and (b) Dionysos marble.

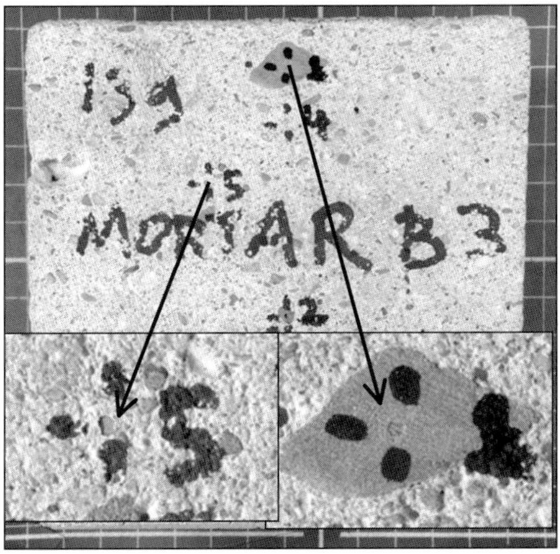

Figure 7. Indentation tests in mortar. Effect of material inhomogeneity on indentation.

3.2 Analysis of indentation results

The results are first corrected for the indenter deformation as a function of force. This is calculated from the indenter's dimensions and elastic modulus. From the corrected results two parameters are calculated, the Scaled Modulus of Flat Indentation and the Critical Transition Stress.

3.2.1 Scaled Modulus of Flat Indentation (*MFIS*)

The Scaled Modulus of Flat Indentation (*MFIS*) with dimensions of stress [Pa], is obtained by scaling the Modulus of Flat Indentation (*MFI*) with the indenter diameter d_i. *MFI* is defined as the slope of the force F vs. indentation depth w curve, at the linear or quasi-linear part of the curve, as shown in Figure 8. The method for calculating *MFI* is to choose first the point in the indentation data where the linear region starts and then finding the linear fit. In cases with particularly noisy data, it is practical to extend the linear region from the best fit until a tolerance in the error function is reached. The slope of this region gives the *MFI*. Alternatively, *MFI* can be calculated from the slope of the region around the inflection point of the curve. This method usually gives a similar *MFI* as the linear region method. However, it was found to be more sensitive to noise in the data or inhomogeneities in the material.

The elastic solution for rigid, flat indentation in a semi-infinite, isotropic, homogeneous material with Young's modulus E and Poisson ratio v is[10]

$$F = \frac{Ed_i}{1 - v^2} w \tag{1}$$

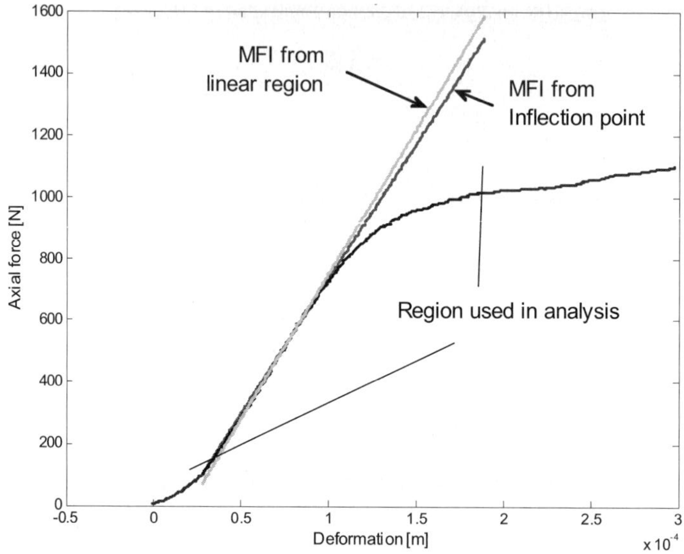

Figure 8. Indentation force vs. indentation depth in a typical indentation test on Dionysos marble. Definition of the region used in the analysis and selection of *MFI* using the linear region method and the inflection point method.

Therefore, according to its definition, the *MFIS* for an isotropic, homogeneous elastic material indented by a rigid, flat indenter is

$$MFIS = \frac{E}{1-\nu^2} \tag{2}$$

and it is independent of the indenter diameter. By extending this expression to inelastic materials, it may be assumed that

$$MFIS \approx \frac{E_{50}}{1-\nu_{50}^2} \tag{3}$$

Therefore, the parameter *MFIS* may be assumed to correlate to E_{50}. The correlations of the calculated values of *MFIS* and the values of E_{50} in uniaxial compression (Table 1) are plotted for each indenter diameter in Figure 9. The results are approximated with power curves of the form

$$E_{50} = b_1 \left(\frac{MFIS}{1\ \text{GPa}} \right)^{b_2} \tag{4}$$

which are also shown in the figure and where the constant b_1 has dimensions of stress [Pa] and the constant b_2 is dimensionless. The values of these constants for the three indenter diameters are given in Table 2 together with the coefficient R^2 of reliability of the approximations. The increase in reliability with indenter diameter shows that small indenters are, as expected,

Table 2. Correlation constants for E_{50} for the 1-, 2- and 3-mm diameter indenters and reliability coefficient R^2

Indenter diameter	Correlation constants b_1 [GPa]	b_2 [-]	Quality of fit R^2 [-]
1-mm	5.0589	0.8014	0.955
2-mm	3.7133	0.7435	0.967
3-mm	3.2639	0.7340	0.988

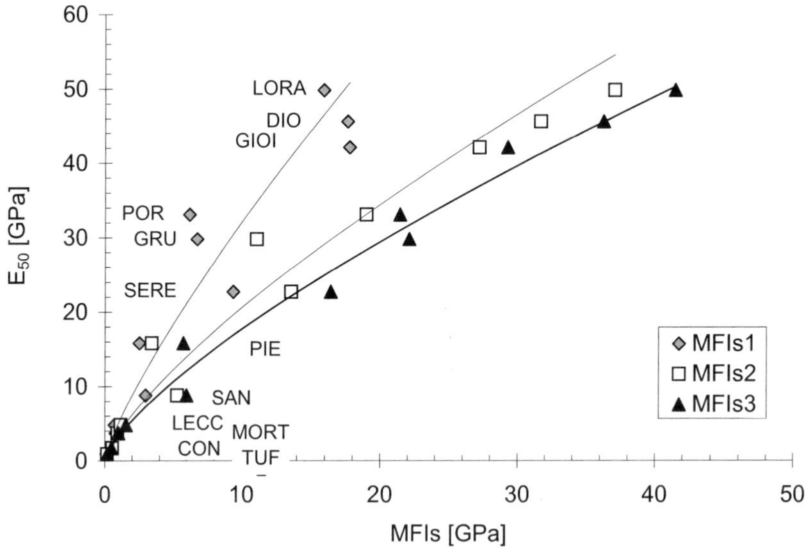

Figure 9. Correlations between uniaxial compression stiffness E_{50} and Scaled Modulus of Flat Indentation *MFIS* for indenter diameter 1-mm (MFIs1), 2-mm (MFIs2) and 3-mm (MFIs3).

more prone to be affected by the internal (petrophysical) structure of the stone. Thus a larger indenter diameter is preferable when the capacity and the specifications of the equipment allow it. The results presented here include data from tests at all depths as the results showed insignificant variations with depth, as it can be seen e.g. in Figure 5.

Reference tests in Plexiglas for indenter diameters 1-, 2- and 3-mm have also shown indenter size-dependency of *MFIS*. For example *MFIS* = 2.79, 3.08 and 3.55 GPa for the 1-mm, 2-mm, and 3-mm indenters, respectively. The indenter size-dependency or scale effect for the tested stones is demonstrated in Figure 10 where the normalized (with the 1-mm indenter diameter values) calibration constants b_{1n} and b_{2n} are plotted vs. the indenter diameter. This size-dependency should be viewed as an average one for all the tested stones whereas individual stones exhibit different size-dependency. A more detail analysis of the size-dependency of *MFIS* for the tested stones is shown in Figure 11, which demonstrates that the *MFIS* is decreasing with decreasing indenter diameter. The size effect varies among the various stones and was measured to be up to approximately 70%.

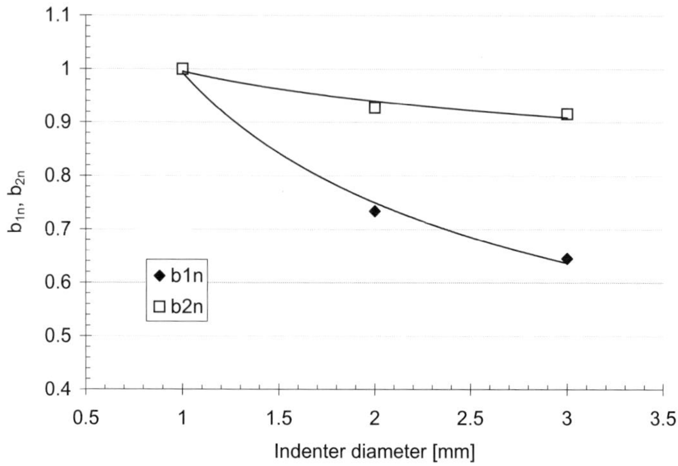

Figure 10. Indenter size-dependency of the calibration constants $b1n$ and $b2n$ for E_{50}.

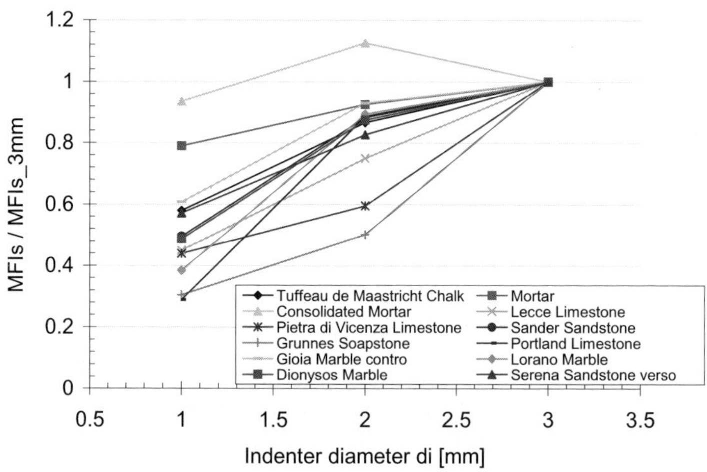

Figure 11. Indenter size-dependency of *MFIS*.

The applicability of Eq.(3) in the indentation of stones may be judged by considering the Elastic Indentation Ratio *EIR* defined as the ratio:

$$EIR = \frac{E_{50} \Big/ \left(1 - \nu_{50}^2\right)}{MFIS} \tag{5}$$

For an elastic material, this ratio is unity, and thus deviations from unity reflect the degree to which the elastic solution deviates from the test results. For the *MFIS* in Figure 9 and the values of E_{50} and ν_{50} in Table 1, the *EIR*s are plotted in Figure 12 vs. E_{50}. It is noted that, in stones, E_{50} and ν_{50} are not necessarily the elastic parameters but include inelastic effects as well. The re-

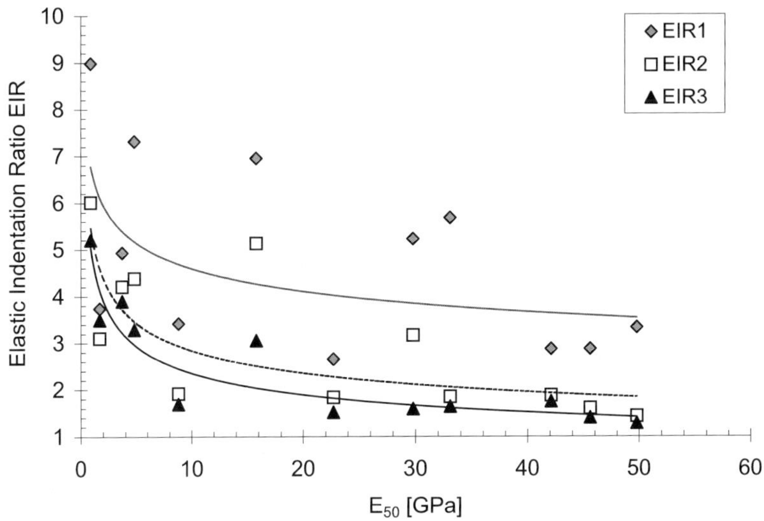

Figure 12. Elastic Indentation Ratio *EIR* vs. E_{50} for indenter diameter 1-mm (EIR1), 2-mm (EIR2) and 3-mm (EIR3).

results show that as the indenter diameter and E_{50} increases, *EIR* approaches unity. This behavior may be interpreted by understanding that with increasing indenter diameter, we move from the softer individual grain and its contacts response to the stiffer response of the stone as a structure. Also the stones with large E_{50} behave more elastically than the low E_{50} stones.

MFIS can also be correlated with the unloading Young's modulus *E*, which can be considered as the elastic Young's modulus of the material since it is calculated from the unloading-reloading cycles of the uniaxial tests. Another correlation can be made with the dynamic Young's moduli E^d_{50} (Table 1), calculated from the P- and S- waves of the uniaxial tests. The correlations for the 3-mm diameter indenter are shown in Figure 13a and are compared with the correlation for the loading Young's modulus E_{50}. All of these correlations show a high fit quality.

Numerical simulations of the flat indentation problem demonstrate that the stress field beneath the indenter is rather one of uniaxial compaction (i.e. zero radial deformation) than one of uniaxial compression (i.e. zero radial stress). Based on this observation it is expected that the indentation stiffness or *MFIS* will correlate well with the uniaxial compaction modulus M_{50} (Table 1), as shown in Figure 13b.

3.2.2 Critical Transition Stress (CTS)

The Critical Transition Stress *CTS*, also with dimensions of stress [Pa], is related to the average stress σ under the indenter, i.e. $\sigma = 4F/\pi d_i^2$, at which the force vs. indentation depth curve enters the region of reduced slope. The material response in this region suggests that some kind of failure takes place. The *CTS* may thus be obtained as the stress at which the slope of the force vs. indenter depth curve differs from the slope that corresponds to *MFIS* with more than a predefined percentage. Alternatively, *CTS* may be calculated from the minimum of the double derivative (or apex) of the force vs. indentation depth in the region of reducing slope (Figure 14).

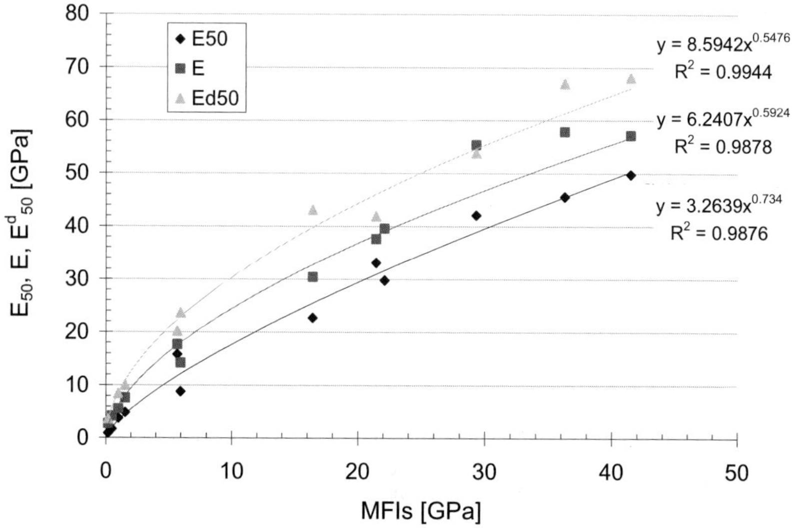

(a)

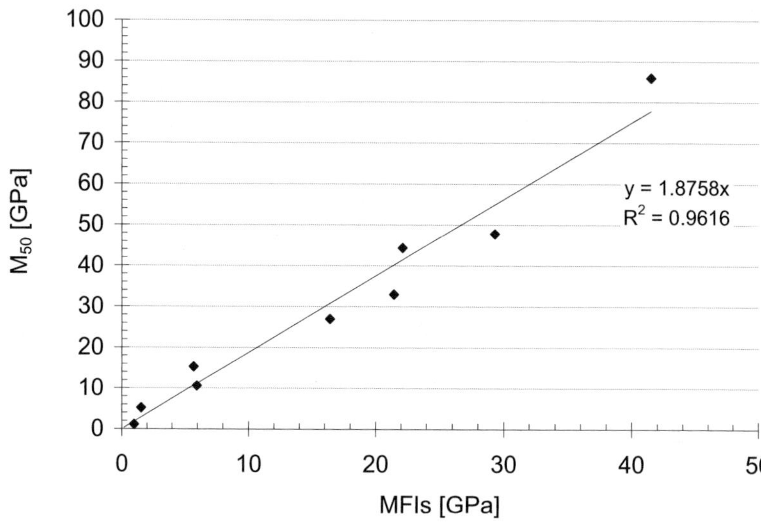

(b)

Figure 13. Correlations between Scaled Modulus of Flat Indentation *MFIS* for 3-mm diameter indenters and (a) uniaxial compression stiffness E_{50}, elastic Young's modulus E, and dynamic Young's modulus E^d_{50}, and (b) uniaxial compaction modulus M_{50}.

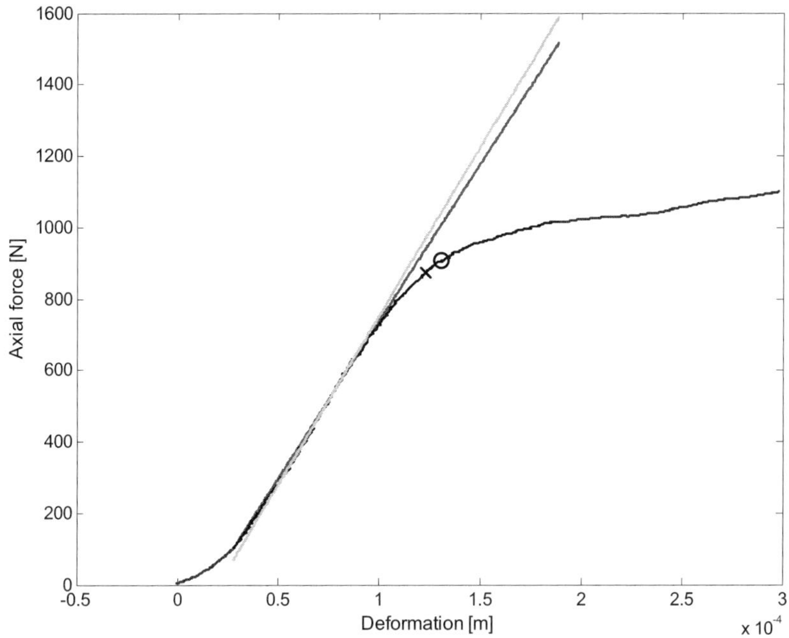

Figure 14. Indentation force versus indentation depth in a typical indentation test on Dionysos marble. The black cross is the *CTS* calculated with a predefined deviation equal to 60% and the black circle is the *CTS* calculated from the curve's minimum of the double derivative.

CTS is correlated with the *UCS* which is a failure strength index in uniaxial compression. This correlation lacks rigorous theoretical background as the stress state in the material under the indenter has a low deviator stress and in reality failure is rather associated with a pressure cap, i.e. grain crushing or pore collapse, than a shear failure surface. This is reflected in the less reliable correlations in Figure 15a between the calculated *CTS* for each indenter diameter and the *UCS* of Table 1. The results are approximated with power curves of the form:

$$UCS = c_1 \left(\frac{CTS}{1\ \text{MPa}} \right)^{c_2} \qquad (6)$$

which are also shown in the figure and where the constant c_1 has dimensions of stress [Pa] and the constant c_2 is dimensionless. The values of these constants for the three indenter diameters are given in Table 3 together with the coefficient R^2 of reliability of the approximation.

A failure associated with the pressure cap is encountered in the uniaxial compaction and thus it is expected that the indentation failure stress or *CTS* will correlate with the uniaxial compaction strength K_oS (Table 1), as shown in Figure 15b.

Table 3. Correlation constants for *UCS* for 1-, 2- and 3-mm diameter indenters and reliability coefficient R^2

Indenter diameter	Correlation constants		Quality of fit
	c_1 [MPa]	c_2 [-]	R^2 [-]
1-mm	0.5375	0.7619	0.811
2-mm	0.5098	0.8196	0.8378
3-mm	0.5776	0.8150	0.8526

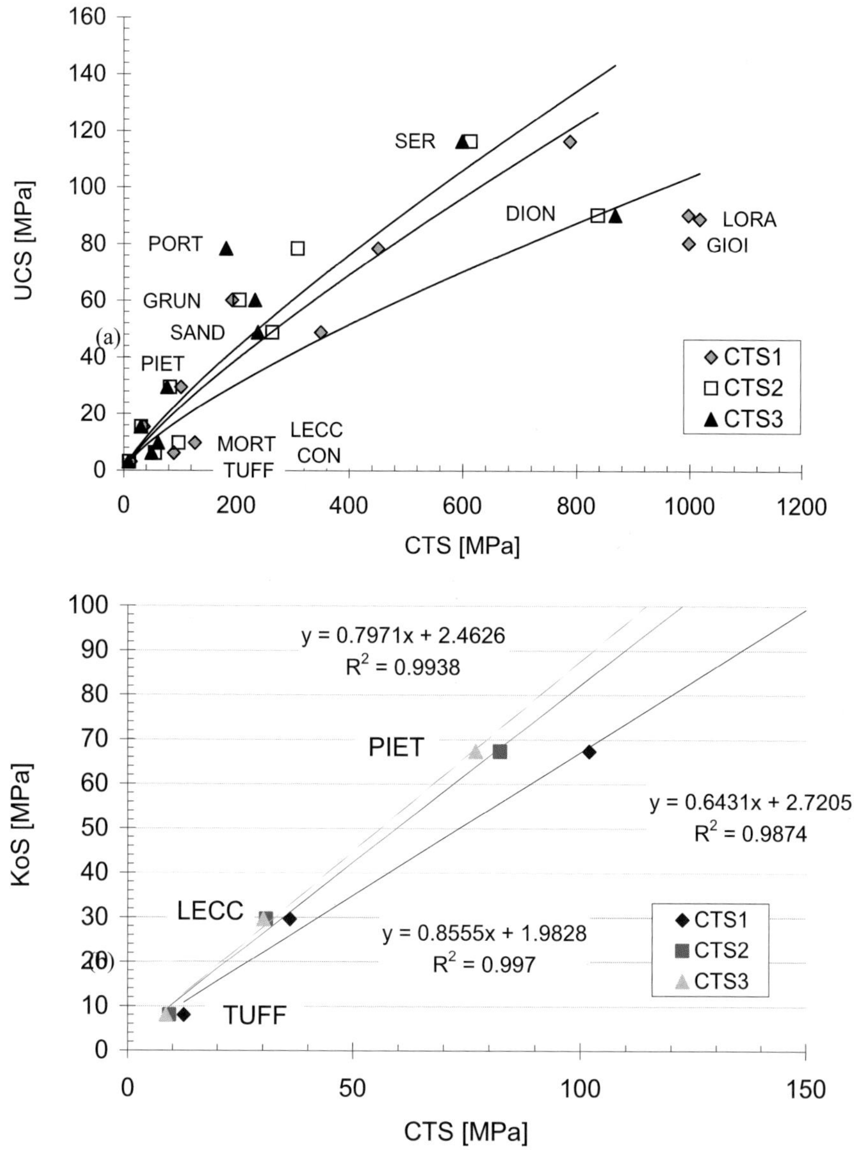

Figure 15. Correlations between (a) *UCS*, and (b) Uniaxial compaction stress K_oS, and Critic al Transition Stress *CTS* for indenter diameter 1-mm (CTS1), 2-mm (CTS2) and 3-mm (CTS3).

The data for *CTS* show also an indenter size-dependency for the tested stones. As it is observed in Figure 16, which shows the size dependency of *CTS* for the various stones, this size-dependency is opposite from the size-dependency observed for *MFIS*, i.e. the *CTS* decreases as the indenter diameter increases. Similarly, size-dependency of *CTS* was also observed in Plexiglas for indenter diameters 1-, 2- and 3-mm. For example *CTS*=335.9 MPa for the 1-mm indenter, 299.8 MPa for the 2-mm indenter and 254.7 MPa for the 3-mm indenter.

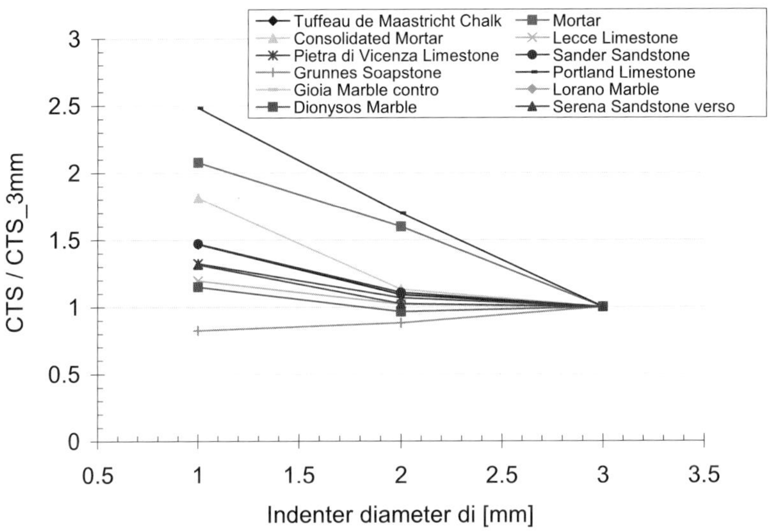

Figure 16. Indenter size-dependency of *CTS*.

4. APPLICATIONS

4.1 Consolidated mortar

The indentation technique was applied for the determination of the effect of a consolidant, as described in Section 2, in the mechanical properties of the mortar. The consolidant was intended to increase the cohesion of monument faces (to a depth of one cm) during restoration works although both triaxial and indentation tests show a deteriorating effect on the mechanical properties and thus it is not appropriate for restoration. From indentation tests with the three indenter diameters and the use of correlations Eqs.(4) and (6), the stiffness E_{50} (in uniaxial compression) and the *UCS* of Mortar and Consolidated Mortar were calculated. These values are compared in Figure 17 with the E_{50} and *UCS* values from uniaxial compression tests (Table 1). The results appear to be satisfactory both quantitatively, i.e. actual values especial-

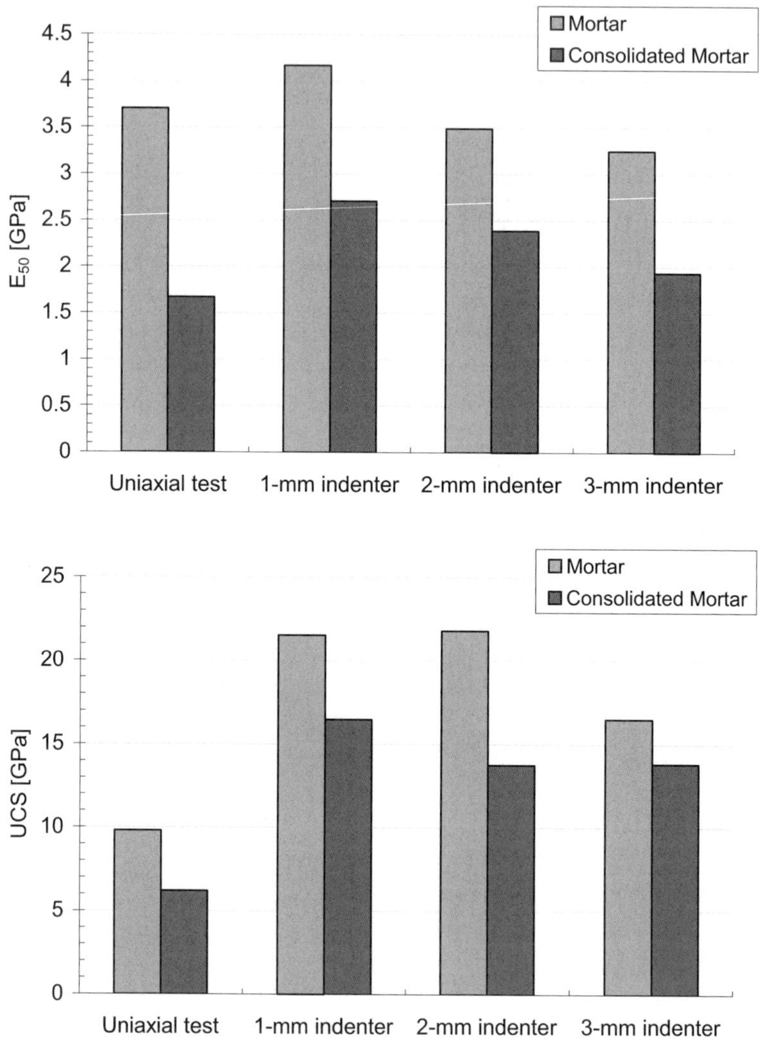

Figure 17. E_{50} (top) and *UCS* (bottom) in Mortar and Consolidated Mortar from uniaxial compression tests and indentation tests with 1-, 2- and 3-mm indenters.

ly for the E_{50}, and qualitatively, i.e. reduction due to consolidant, taken into account the facts that mortar is an inhomogeneous material and in the extreme low of the range of stones tested.

4.2 Artificially-weathered marbles

The indentation technique was also applied for the determination of the effect of artificial weathering of marbles on their mechanical properties. The

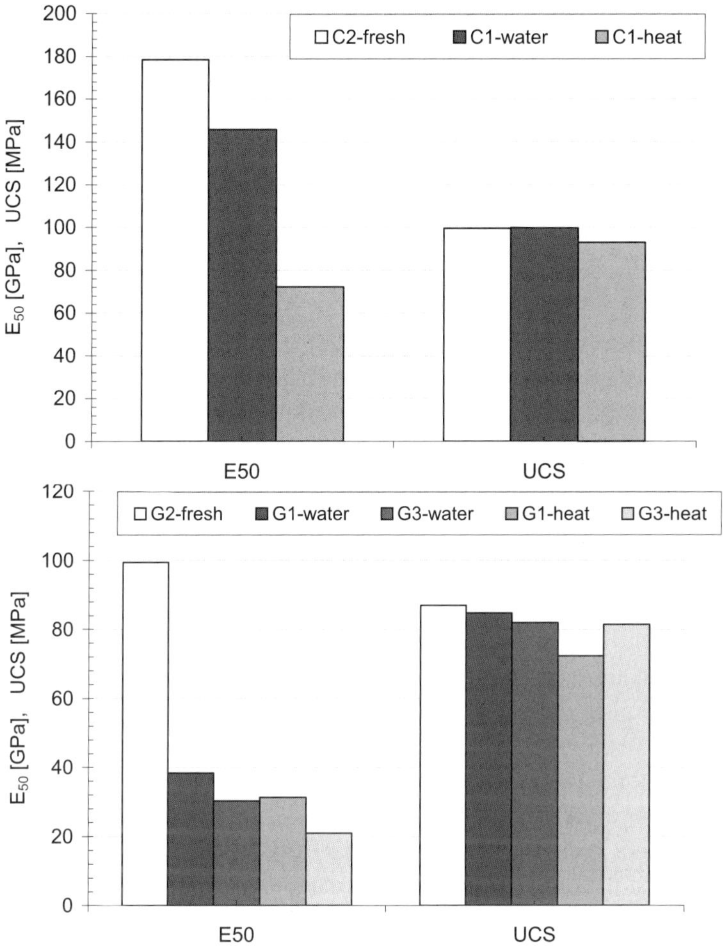

Figure 18. E_{50} and *UCS* in fresh and artificially weathered (heat-treated) Marble 1 (top) and Marble 2 (bottom) from indentation tests.

marbles were weathered using the method NT BUILDT 499 developed to provoke bowing of marbles in an accelerated laboratory environment[11,12]. In this method, a specimen is exposed to water from below and to a number of temperature cycles between 20ºC and 80ºC by heating from above. The specimens are rectangular plates with 30-mm thickness, in accordance with typical panel thickness in claddings. Marble cladding panels are used to cover several modern buildings. At certain buildings the panels start to bow and loose strength after some years. The phenomenon occurs occasionally in crystalline marble exterior claddings[13-14] and is assumed to be controlled by temperature variations and moisture[15].

Twin specimens of two marbles were tested with 1-mm diameter indenters. One of the twin specimens was fresh, i.e. specimens G2 of Marble 1 and C2 of Marble 2, and the others were exposed, i.e. specimens G1 and G3 of Marble 1 and C1 of Marble 2. Of the two marbles, Marble 1 was visibly bowed after exposure while Marble 2 did not show any signs of bowing. On the exposed specimens, both sides, the water- and the heat-side, were indented. From the indentation results and the use of correlations Eqs.(4) and (6), the stiffness E_{50} (in uniaxial compression) and the *UCS* were calculated and plotted in Figure 18. From this figure it is concluded that:

- The degradation effect of moisture and heat cycling is more significant on the stiffness E_{50} than on the *UCS*, and on the heat-side than in the water-side of the specimen.
- Both marbles have suffered degradation of mechanical properties although the degradation appears to be larger in Marble 1, which also was bowed after exposure.
- The E_{50} and *UCS* of fresh Marble 1 (specimen G2) are captured accurately when compared with values in our database.

5. CONCLUSIONS

The indentation technique has been applied for the development of a method for in situ, quasi non-destructive diagnosis of mechanical properties and damage of natural building stones. The method consists of the indentation test specifications, the indentation results analysis, and a correlation database for a wide variety of natural building stones, i.e. calcarenites, limestones, sandstones and marbles, and a mortar used for restoration. From the indentation test data two parameters are calculated, the scaled modulus of flat indentation MFIS and the critical transition stress CTS. From these parameters the following material properties can be derived:

- Tangent stiffness modulus E_{50} in uniaxial compression.
- Elastic Young's modulus E in uniaxial compression.
- Dynamic Young's modulus $E^d{}_{50}$.
- Uniaxial compaction stiffness modulus M_{50}.
- Uniaxial compressive strength *UCS*.
- Uniaxial compaction strength K_oS.

where the subscript 50 refers to the value at 50% of the peak axial stress.

The method has been used in the portable and integrated tool DIAS[16], developed for in situ measurements of mechanical properties of natural building stones, both at the surface and at some depth such that the degradation of stone due to weathering can be quantified. The tool is targeted for the preservation and restoration of monuments, although other applications, such as in the construction industry, can benefit from the technique.

The application of the method to a consolidated mortar and two artificially weathered marbles has shown its potential and limitations. The latter are primarily related to the nature of the indentation test, which being a local test is more reliable in homogeneous stones. As an example, the method is not suited for fracture detection in stones. Future work will extend the method to other stone types and especially granites which have not been tested so far.

ACKNOWLEDGEMENTS

The authors acknowledge the support from the EU projects 'Integrated tool for in situ characterization of effectiveness and durability of conservation techniques in historical structures' (DIAS-EVK4-CT-2002-00080) in the framework of the Development of Innovative Conservation Strategies Program, 'Degradation and Instabilities in Geomaterials with Application to Hazard Mitigation' (DIGA-HPRN-CT-2002-00220) in the framework of the Human Potential Program, Research Training Networks, and the project Pythagoras II (EPEAEK II) co-funded by the European Social Fund (75%) and Greek Resources (25%).

REFERENCES

1. B.R. Lawn and R. Wilshaw, Review on indentation fracture: Principles and applications, *J. Material Sci.* **10**, 1049-1081 (1975).
2. R. Mouginot and D. Maugis, Fracture indentation beneath flat and spherical punches, *J. Material Sci.* **20**, 4354-4376 (1985).
3. H.E. Hintermann, Characterization of surface coatings by scratch adhesion test and by indentation measurements, *Fresenius J. Anal. Chem.* **346**, 45-52 (1993).
4. T.W. Miller and J.B. Cheatham, Analysis of the indentation of a compacting material by a perfectly rough wedge, *Int. J. Rock Mech. Min. Sci.* **9**, 475-492 (1964).
5. A.C. van der Vlis, in: *Proc. 2nd Int. Symp. Rock Mech.* (1970), Paper 3-4.
6. J. Geertsma, Some rock mechanical aspects of oil and gas well completions, *SPE J.*, 848-856 (1985).
7. M. Thiercelin and J. Cook, in: *Proc. 29th U.S. Symp. Rock Mech.*, (Balkema, Rotterdam, 1988), pp. 135-142.
8. R. Suárez-Rivera, Z. Zheng, N.G.W. Cook and G. Cooper, in: *Proc. 31st U.S. Symp. Rock Mech.*, (Balkema, Rotterdam, 1990), pp. 671-678.
9. F.J. Santarelli, J.L. Detienne and J.P. Zundel, in: *Proc. 32nd U.S. Symp. Rock Mech.*, (Balkema, Rotterdam, 1991), pp. 647-655.
10. I.N. Sneddon, The relation between load and penetration in the axisymmetric Boussinesq problem for a punch of arbitrary profile, *Int. J. Engng. Sci.* **3**, 47-57 (1965).
11. B. Schouenborg, B. Grelk, J.-A. Brundin and L. Alnæs, Bow test for facade panels of marble, NORDTEST Project 1443-1499 (2000).
12. B. Grelk, P. Golterman, B. Schouenborg, A. Koch and L. Alnæs, in *Proc. Dimension Stone 2004* (2004).

13. J. Cohen and P.J.M. Monteiro, Durability and integrity of marble cladding: A state of the art review, *ASCE Journal* **5**(2), 113-124 (1991).
14. K Hook, Look out below – The Amoco Building cladding failure, *Progressive Architecture* **75**, 58-62 (1994).
15. T. Yates, J.-A. Brundin, P. Goltermann and B. Grelk, in: *Proc..Dimension Stone 2004* (2004).
16. P. Tiano, G. Exadaktylos, E. Papamichos, E. Valentini, in: *Proc. of the HWC-2006 Heritage Weathering and Conservation Conference* (Balkema, Rotterdam, 2006)

Chapter 2.3

THE MECHANICAL BEHAVIOUR OF COMPOSITE SPECIMENS MADE OF TWO DIFFERENT STONES

Nikolaos L. Ninis[1] and Stavros K. Kourkoulis[2]
[1]Archaeological Museum of Epidauros, 210-52 Ligourio, Hellas; [2]School of Applied Mathematical and Physical Sciences, Department of Mechanics, Laboratory of Testing and Materials, National Technical University of Athens, 5 Heroes of Polytechnion Avenue, 157 73 Zografou, Attiki, Hellas, stakkour@central.ntua.gr

Abstract: The mechanical behaviour of composite specimens made from two mechanically incompatible natural building stones is the subject of the present paper. Attention is focused on a particular case, namely the porous oolitic limestone of Kenchreae used in the erection of the monuments at the Epidaurean Asklepieion in northern Peloponnesus and one of its candidate substitutes, the Alfopetra of Crete. The study is carried out experimentally. It is concluded that the mechanical properties of the composite specimen, considered as a whole, are strongly affected by the inclination of the adhesion plane and the dependence can be described by a non-monotonous relation. In addition strong strain discontinuities are recorded in the vicinity of the adhesion plane, which are responsible for cracking initiation in either of the two materials.

Key words: natural building stones; compatibility; restoration of monuments; uniaxial compression; composite specimens.

1. INTRODUCTION

The compatibility of the original building material with new material inserts used for the completion of damaged structural members is among the most serious aspects of a restoration project in case of stone monuments. Under ideal conditions the two materials should be similar in terms of geological description, mineralogical analysis and characteristics related to colour, texture and workability. They should also have comparable physical properties such as apparent and absolute density, porosity, water absorption,

S.K. Kourkoulis (ed.), Fracture and Failure of Natural Building Stones, 93–106.

permeability and swelling. In addition they should have similar response to mechanical and physicochemical weathering and of course they must be mechanically compatible.

The mechanical compatibility is usually judged on the basis of characteristics obtained from standardized tests. Such quantities are, for example, the compressive and bending strength, the modulus of elasticity, Poisson's ratio, as well as the tensile strength, the characteristics of anisotropy and the behaviour under triaxial testing. However, it is to be emphasized that the direct comparison of the failure stress or the ductility or any other property is not adequate for judging the compatibility and may be misleading. In other words the compatibility of new stone with the original one should not be decided upon the mechanical characteristics of each stone separately. In fact it is the combined behaviour that matters.

It is thus very interesting to know in what way and to what extent the different mechanical properties of the insertion material are going to modify the characteristics of the original stone. In other words the composite behaviour represents the behaviour of a "*new*" non-uniform material. The relative size, the shape and the position of the insert, as well as the bonding strength of the interface, all play an important role in the observed behaviour, and constitute the parameters of the problem. By understanding the function of these parameters one can predict, and hopefully minimize, the effects of introducing a mechanically incompatible stone into an ancient stone structure. Therefore, it becomes obvious that the experimental study of the mechanical behaviour of composite specimens is a unique tool in the hands of the scientists working for a restoration project, which can help in the direction of understanding the effect of this special type of non-uniformity in a systematic and controlled way.

In this direction a series of uniaxial compression tests was carried out in the present study using composite specimens, made from two relatively incompatible stones. The particular aim of the study is to explore the influence of one of the above parameters, i.e. the inclination of the adhesion plane, on the overall mechanical behaviour of the *"new"* material. The original building stone studied here is the porous stone of Kenchreae, which has been used for the construction of various monuments of the Epidaurean Asklepieion in northeastern Peloponnesus, Hellas.

The sanctuary of Asklepios at Epidauros was the most celebrated healing center of the ancient world. Extensive remains of the site have been brought to light in excavations conducted on behalf of the Archaeological Society of Athens in the late nineteenth century. Given its importance in antiquity and the relevance of its healing ideals to the modern world the sanctuary has earned a place in the World Heritage List. Since 1984 the Committee for the Preservation of the Epidaurean Monuments has undertaken a systematic effort for the protection, conservation and restoration of the archaeological site.

2. THE EXPERIMENTAL PROCEDURE

2.1 The materials

The Kenchreae stone is a relatively homogeneous porous oosparite of 'sandstone-like' appearance. Its porosity is relatively high and varies between 35% and 40%. It is characterized by layered structure and complex nets of internal pores and surface vents. The dimensions of the pores and the vents vary between a few millimeters and a few centimeters and sometimes they run through the whole width of the laboratory specimens rendering the scattering of the experimental results unavoidable. Calcite veins or small calcite accumulations appear within it. Its texture ranges from massive to very thin-bedded and the colour varies from whitish-grey to greyish-beige (rarely light yellow).

From the mechanical point of view it is a rather soft and weak material with a ratio of the elasticity modulus, E, over the compressive strength, σ_c, equal to about 400. Typical axial stress - axial strain curves obtained from specimens made of original material (after a special permission of the respective archaeological authorities) and subjected to unconfined compression are shown in Figure 1[1,2]. It is seen from this figure that up to the peak load the constitutive law is almost linearly elastic. Then an abrupt load drop is observed, followed by a load recovery which can even exceed the initial peak stress. From this point on and after a period of strong fluctuations the curves are characterized by a sub-horizontal decline up to the final disintegration of the specimens. The average values of the peak stress, σ_p, the elasticity modulus, E and Poisson's ratio, ν, are summarized in Table 1.

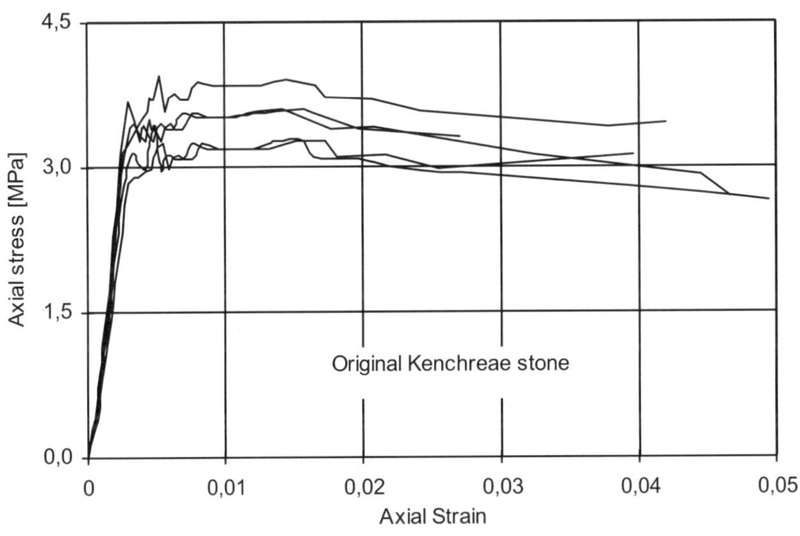

Figure 1. Stress - strain diagrams of the original Kenchreae porous stone.

Table 1. Mechanical properties of the original and the substitute materials

Property→ Material ↓	Peak stress, σ_p [MPa]	Young's modulus, E [GPa]	Poisson's ratio, ν [-]
Ancient (original) Kenhreae stone	3.5	1.5	0.26
Freshly excavated Kenchreae stone	6.2	1.7	0.26
Alfopetra stone of Crete	15.2	2.5	0.27

Unfortunately the access to the ancient quarries of Kenchreae is prohibited and, therefore, for the needs of the restoration program the use of alternatives amongst the commercially available natural building stones became an indispensable demand. The solutions once proposed included, among others, the Alfopetra stone from the Greek island of Crete. Although it was not finally chosen due to its low weathering resistance it has been used in the present research since it represents a material with different mechanical behaviour compared to both the original and the freshly quarried Kenchreae stone. This can be clearly seen from Figure 2, where characteristic axial stress - axial strain curves are plotted for the Alfopetra stone together with the respective curves for freshly quarried Kenchreae stone, as they were obtained from uniaxial compression tests[3]. Comparing, also, Figures 1 and 2, it becomes obvious that both the peak stress and the modulus of elasticity as well as the post peak behaviour of the Alfopetra stone (the slope of which is much more abrupt), are different compared to the respective quantities of either the original or the freshly quarried Kenchreae stone.

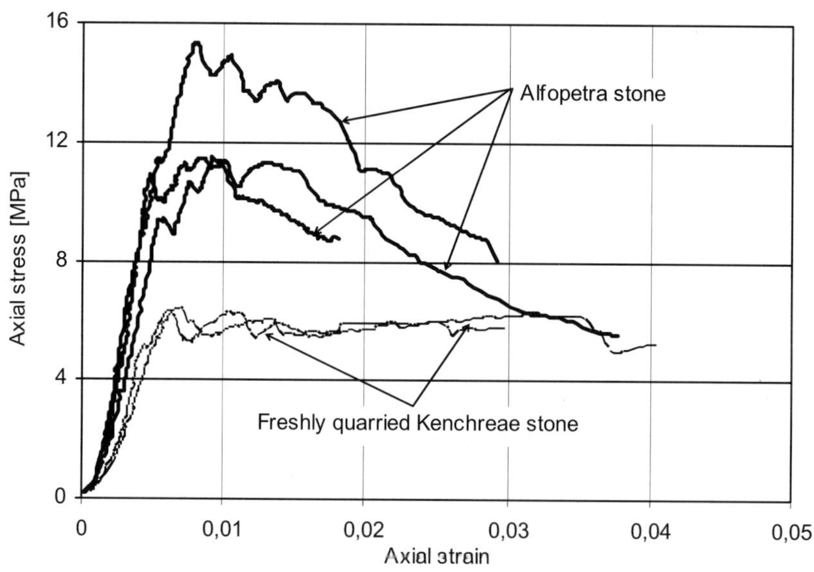

Figure 2. Stress - strain diagrams of the Alfopetra and the freshly quarried Kenchreae stone.

2.2 The composite specimens and the tests

Very carefully prepared composite specimens made from equal parts of freshly quarried Kenchreae stone and Alfopetra stone were used. The specimens were of cubic shape. Their geometry and dimensions are shown in Figure 3. The two parts were bonded together with the aid of a specially prepared epoxy resin. The adhesion plane was inclined with respect to the horizontal by $\theta=0^{o}$ (normal to the load direction), 15°, 30°, 45°, 60°, 75° and 90° (parallel to the load direction). Three specimens were used for each angle θ.

The experiments were carried out without lubricating the bases of the specimens and the loading platens. This is because the conclusions of the present study are to be directly applied by the experts working for the restoration of ancient monuments where the use of lubricants of any type is meaningless because their effectiveness is eliminated with time. Of course it is known that the influence of friction on the results of compression tests is not negligible, especially in the case of geomaterials and natural building stones[4,5]. However, as long as the purpose of the present study is the comparative study of the experimental results obtained from specimens with various inclinations of the bonding plane, it is reasonable to assume that the influence

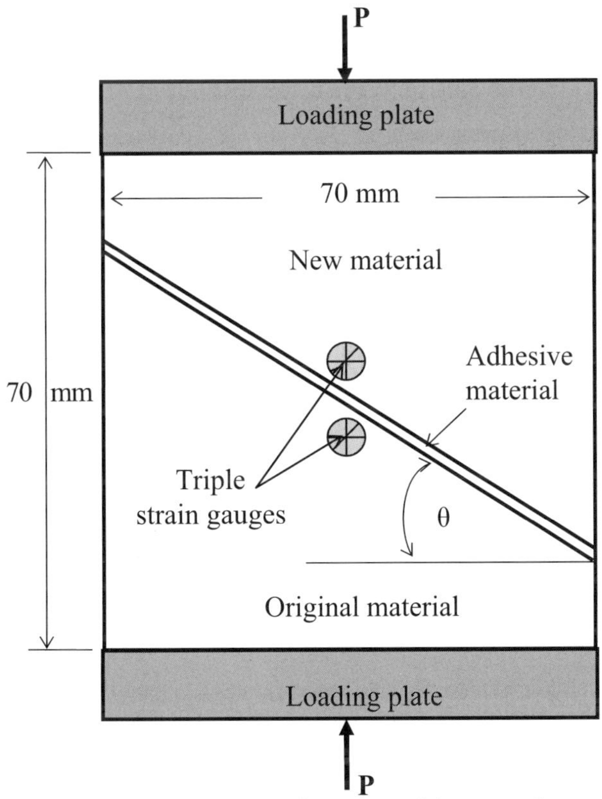

Figure 3. The configuration and the dimensions of the composite specimens.

of friction will not mask the phenomena. The tests were carried out with the aid of a very stiff hydraulic Amsler loading frame of load capacity 1000 kN. Taking into account that the maximum load recorded in the present series of experiments did not exceed in any case 100 kN, it is concluded that the stiffness of the frame can be considered infinite. This is very important in case conclusions are to be drawn concerning the post-peak behaviour of the material. The load was applied statically at a rate of about 0.03 mm/min.

For the measurement of the axial displacements a system of three LVDTs of sensitivity $5x10^{-6}$ m was used. They were placed at 120^0 to each other in order to check the symmetry of the loading. The rotation of the loading plates detected was negligible for the vast majority of specimens. In addition, the components of the strain field developed in the immediate vicinity of the adhesion plane were, also, recorded using a pair of triple strain gauge rosettes suitably positioned, as it is shown in Figure 3.

The quantities measured during the tests were the load, the axial displacements from the LVDTs and the strain components from the strain gauges. Processing properly these data using commercially available software for the elimination of experimental noise and the rejection of statistically unacceptable tests, the overall axial stress - axial strain diagrams were obtained. Successive photographs taken during the experimental process ensured that one could detect the exact moment of appearance of surface cracks, exfoliations or other failure phenomena in order to decide on which of the two materials failed first. Typical composite specimens are shown in Figure 4, after they have been tested. Various failure types can be clearly seen.

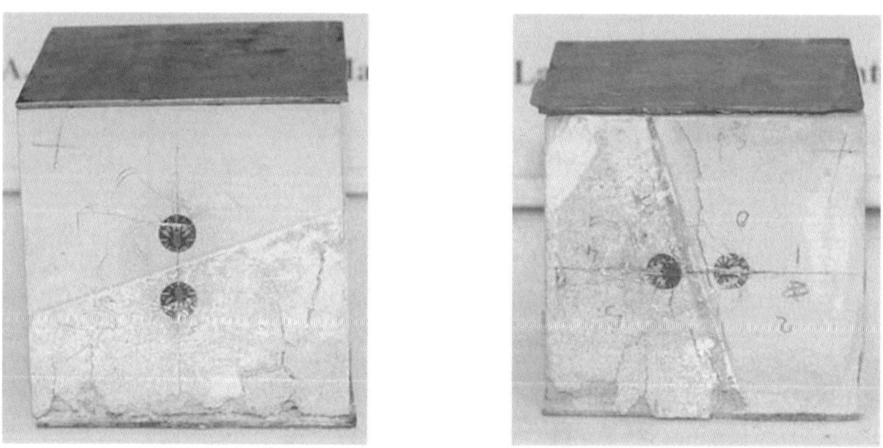

(a) (b)

Figure 4. Typical tested composite specimens. (a) θ=15°. The failure of the original stone (lower part of the specimen) is clearly visible. The substitute material is intact. (b) θ=75°. The failure started from the substitute material (right part of the specimen) very close to the interface. The original material remained almost intact. (Thin lead plates between the specimen and the loading plates ensured optimum contact in both cases).

3. EXPERIMENTAL RESULTS AND DISCUSSION

Characteristic plots concerning the overall macroscopic behaviour of the axial stress - axial strain curves of the composite specimens are shown in Figures 5 and 6 for the whole range of the inclinations of the adhesion plane. It is clearly concluded from these figures that for low values of the inclination of the interface ($\theta<30^{o}$) the behaviour of the composite specimen is dictated by the original material. Indeed, both the elastic regime as well as

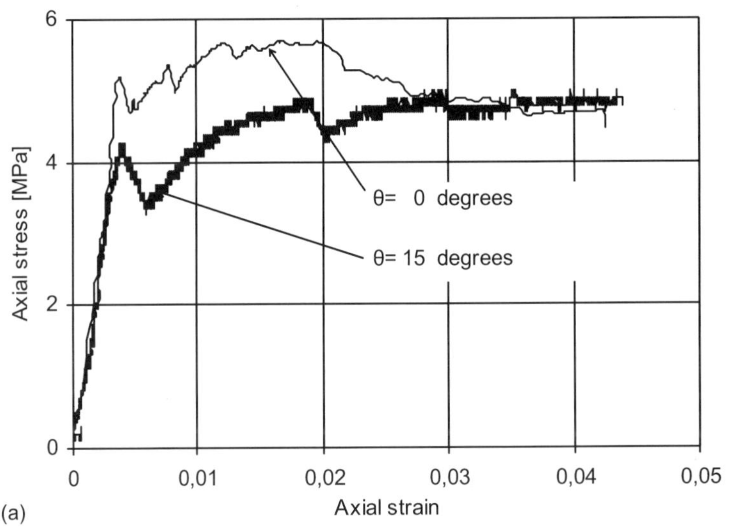

(a)

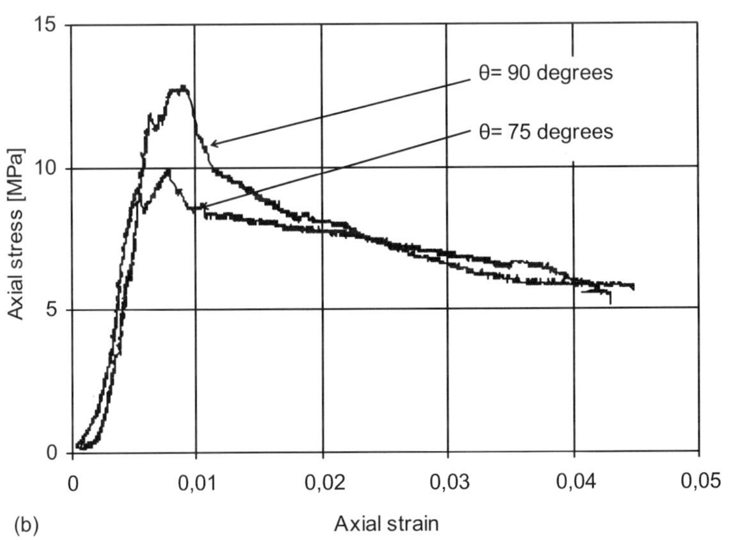

(b)

Figure 5. The behaviour of composite specimens with (a) $\theta=0^{o}$, $\theta=15^{o}$ and (b) $\theta=75^{o}$, $\theta=90^{o}$.

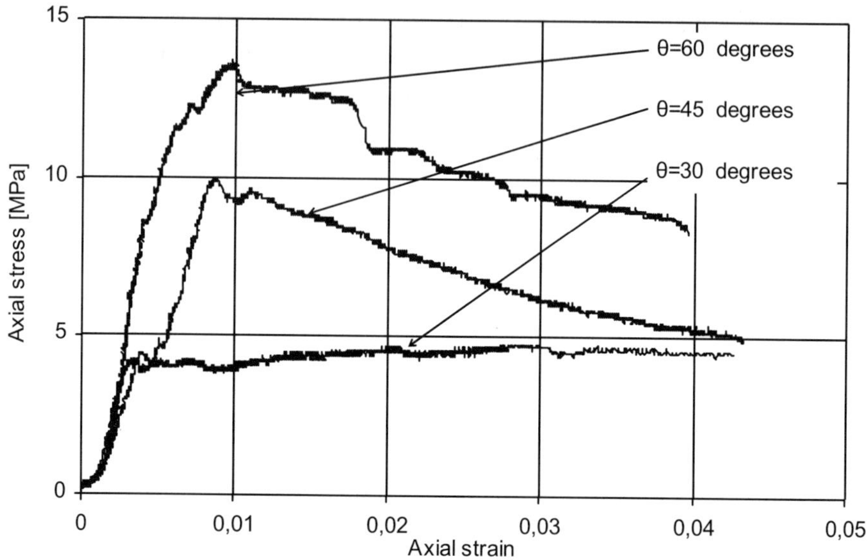

Figure 6. The behaviour of composite specimens with: θ=30°, θ=45° and θ=60°.

the post-peak portion of the graphs are almost identical to the respective ones of the Kenchreae stone. However, it should be emphasized, that besides the striking similarity the maximum stress reached in the case of the composite specimens is systematically lower compared to that of the original Kenchreae stone specimens. On the contrary for θ-values exceeding 75° the form of the graphs is clearly dictated by the new material (Alfopetra stone). Again and besides the similarity of the graphs the maximum stress reached is lower in the case of the composite specimens compared to that of the Alfopetra stone specimens. Finally, for the intermediate interval of θ-values ($30^\circ \leq \theta < 75^\circ$) the graphs gradually depart from that of the Kenchreae stone and approach that of the Alfopetra stone (Figure 6).

Concerning now the question of the mechanical compatibility of the two materials the answer is not unique. It was proved from the present series of tests that the protection of the original material by the new one depends, among others, on the inclination of the adhesion plane, in other words on the geometry of the specimen. For small inclination angles ($\theta < 30^\circ$) the failure stress of the composite specimens was either equal to that of the original stone or in some experiments even lower. For the specific geometries the failure of the composite specimen started systematically from the original material either in the form of axial cracks or in the form of crushing of successive bedding planes. On the other hand for inclination angles between 30° and 90° the failure stress of the composite specimen was systematically higher compared to that of the original material. It is mentioned characteristically that in some cases the failure stress of the composite material even exceeded that of the new material. For these configurations the failure started from the substitute material (usually in the vicinity of the adhesion plane) in the form of a number of cracks oriented along the direction of the external load. On the basis of the above observations and considering also the study of composite specimens with unequal portions of original and new material[6], it can be concluded that it is possible to affect the compatibility of various types of natural building stones (which appear to be more or less incompatible from the mechanical point of view) by a simple modification of the geometry of the configuration.

As a next step the variation of some critical "overall" (average) mechanical properties of the composite specimen was explored. For this purpose the modulus of elasticity, E, is plotted versus the inclination of the interface in Figure 7a. It should be emphasized at this point that in the discussion following the term modulus of elasticity designates the slope of the stress - strain curve of the composite specimen. In this context the stress is calculated as the ratio of the externally applied load over the total cross sectional area of the specimen while the strain is obtained as the change of length of the specimen as it is measured from loading plate to loading plate and reduced overall the total length of the composite specimen.

As can be seen from Figure 7a the modulus of elasticity varies according to an almost sinusoidal law. Its value for $\theta=0^{o}$ is equal to that of the authentic stone while for $\theta=90^{o}$ it approaches the modulus of elasticity of the Alfopetra stone, although it appears to be slightly higher. The most astonishing observation, however, is the fact that the variation of the modulus of elasticity of the composite specimen is non-monotonous. A clear minimum corresponds to $\theta\approx15^{o}$ with a value equal to about 1.5 GPa which is slightly lower even compared to the respective value of the original stone. On the contrary, for the complementary inclination angle, $\theta=75^{o}$, a clear maximum of about 3.5 GPa is observed, which exceeds even the value of the modulus of elasticity of the Alfopetra stone.

The behaviour of the peak stress, i.e. the stress level corresponding to the abrupt stress drop observed at the end of the linear portion of the stress - strain curve, is of similar nature. However, it is to be noted that for $\theta=0^{o}$ the peak stress of the composite specimen reaches a value equal to about 6.7 MPa, exceeding significantly that of the original stone which is less than 4.5 MPa. Again for $\theta\approx15^{o}$ a clear minimum is observed equal to about 4 MPa, a value slightly lower from that of the Kenchreae stone. The maximum value corresponds to $\theta\approx60^{o}$ and its value is about 12 MPa well comparable to the respective value of the Alfopetra stone.

The last quantity studied is the slope of the strain energy density plotted versus the strain level. This quantity is studied because it has been concluded from a recent series of experiments with conchyliates shellstone specimens that all conventional mechanical constants depend more or less on the size of the specimen[7]. Thus, their use becomes rather inefficient both for practical applications as well as for purely scientific analysis. On the other hand, it is

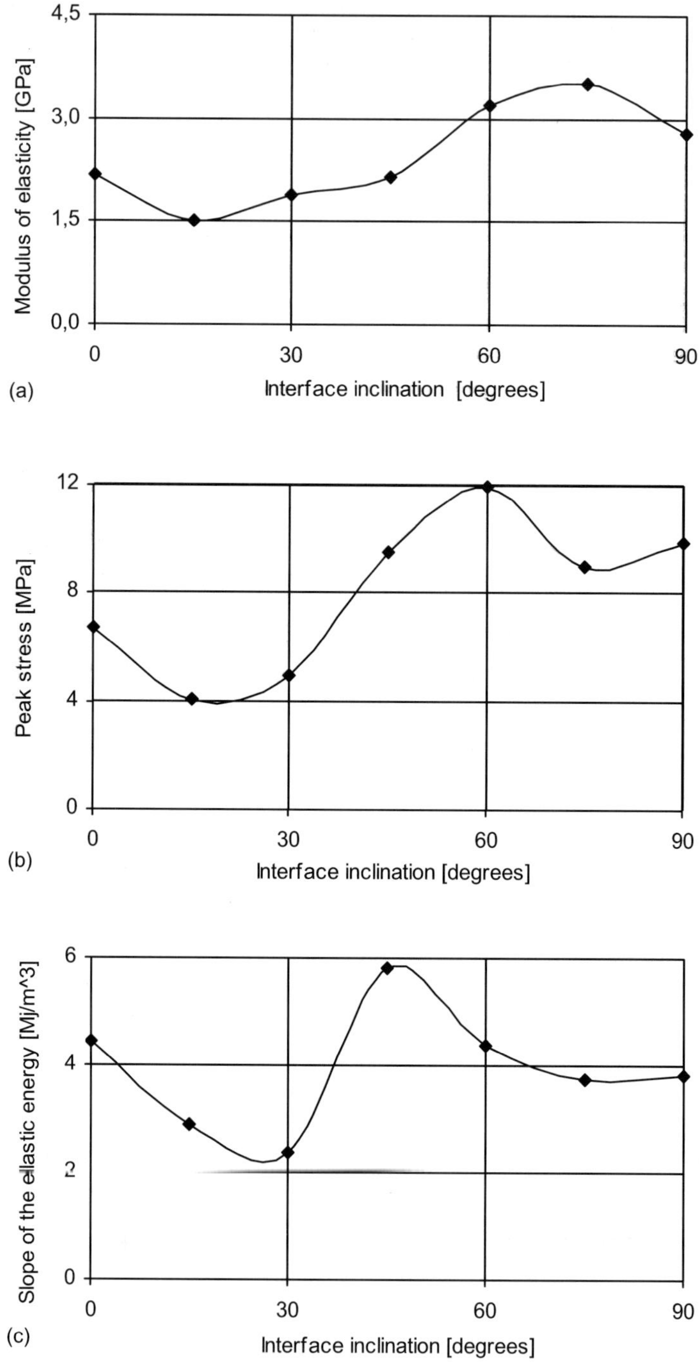

Figure 7. The variation of (a) the modulus of elasticity, (b) the peak stress and (c) the slope of the elastic strain energy density versus the strain level as a function of the inclination of the interface of the composite specimen.

to be accepted that the working load of such materials in practice is restricted to only a portion of the σ-ε graph, well below the peak stress. It appears thus reasonable that this portion of the axial stress - axial strain curve must be studied in a more detailed manner. To this end, the variation of the accumulated strain energy density versus the axial strain for the working-load portion of the axial stress - axial strain curve was studied and it appeared that its slope is almost insensitive to the changes of the size of the specimens[7,8].

Following the above discussion the dependence of the slope of the strain energy density on the interface inclination is plotted in Figure 7c and it can be seen that its behaviour is qualitatively similar to that of the modulus of elasticity and the peak stress. Indeed a clear minimum is observed for an inclination angle between 15° and 30° while a maximum appears for an angle θ equal to about 45°.

As a final step the variation of the strain field in the immediate vicinity of the interface of the original and the substitute stones is plotted in Figures 8 and 9 versus the applied stress, for two typical specimens with θ=15° and θ=60°, respectively. From Figure 8 it is seen that both the axial and the transverse strains appear to be incompatible to each other, and a strong strain discontinuity is observed in the vicinity of the interface. For the particular case with θ=15° the ratio of the axial strain in the new stone over the respective strain in the original stone is equal to about 5.6 while for the trans-

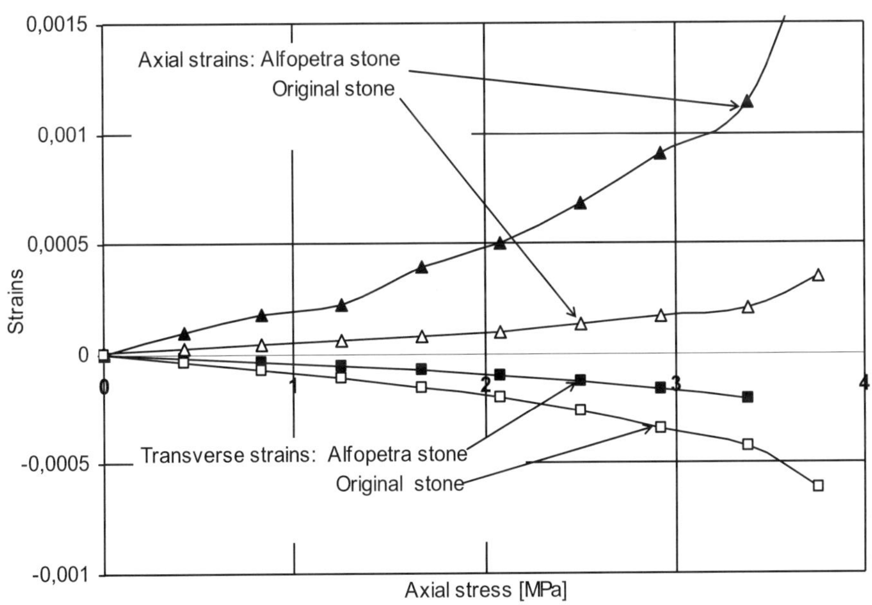

Figure 8. The variation of the axial (triangular symbols) and the transverse (square symbols) strains for the original (empty symbols) and the new stone (filled symbols) in the immediate vicinity of the interface for θ=15°.

verse strains the respective ratio is equal to about 2. It is clear that such a discontinuity could be responsible for local failures in the vicinity of the adhesion plane due to the mismatch of the strain field components. Therefore it can be concluded that the role of the adhesive material is paramount for the gradual transition from the intense strain field developed in the new material to the weak strain field in the original one. However, with increasing inclination angle the strain discontinuity tends to be eliminated, as can be seen, for example, for the angle θ=60° shown in Figure 9. It is seen from this figure that the strain mismatch is drastically reduced and the ratio of the strains developed in the original and the new stone does not exceed 1.5 for both the axial and the transverse strains. Once again it is concluded that the inclination of the interface affects drastically all the aspects of the mechanical compatibility between original and new stone.

4. CONCLUSIONS

The problem of combining mechanically incompatible stone materials in a single specimen subjected to unconfined compression was investigated in the present work. Composite specimens consisting of equidimensional parts

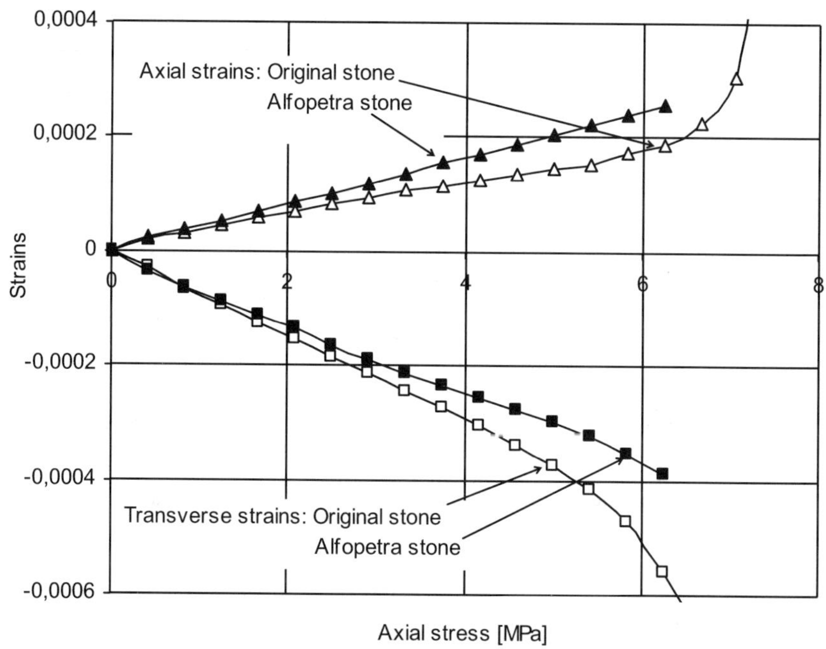

Figure 9. The variation of the axial (triangular symbols) and the transverse (square symbols) strains for the original (empty symbols) and the new stone (filled symbols) in the immediate vicinity of the interface for θ=60°.

of two different stones were used. The parameter studied in particular was the inclination of the connecting interface, which varied from 0° to 90°. Although the two stones judged on the basis of their independent mechanical properties are incompatible, it seems that for a range of inclinations from 30° to 90° there is satisfactory cooperation between the two materials, thus yielding an overall "apparent" behaviour representing an intermediate state between the two.

For inclinations between 0° and 90° the resulting behaviour approaches that of the weakest and the strongest material, respectively. The inclinations of 15° to 75°, however, appear to produce an absolute minimum and an absolute maximum of strength compared to the strength of the two stones.

In sum the variation of the "overall" (average) mechanical properties of the composite specimens appears to depend on the slope of the connecting interface according to a non-monotonous law. Absolute values of the various mechanical properties, however, varied between relatively broad limits exceeding in some cases those of both constituent materials.

The key to understanding the observed behaviour are the strong discontinuities of the strain field components (both axial and transverse) recorded close to the interface, since they are responsible for the cracking initiation in either of the two materials. The combination of the two strain components with the direction of loading (i.e. the interface inclination) controls the way cracking will start and develop.

The use of composite specimens could be a powerful means for studying the effect of different types of interface on better protecting the weaker of the two materials depending on the inclination of the sloping interface.

REFERENCES

1. N. L. Ninis and S. K. Kourkoulis, in: *Proceedings of the 6th National Congress on Mechanics*, edited by E. C. Aifantis and A. N. Kounadis, (Giahoudi & Giapouli Press, Thessaloniki, Hellas, 2001), **I**, pp. 348-356.
2. S. K. Kourkoulis, I. Vardoulakis and N. L. Ninis, in: *Proceedings of the International Congress on Geology and Geotechnics in the Preservation of Historical and Cultural Heritage*, edited by G. Lollino (CNDCI Publishing, Tourin, Italy, 2000), pp. 831-839.
3. S. K. Kourkoulis, N. L. Ninis, and V. Latsiou, in: *Proceedings of the International Conference on the Influence of Traditional Mathematics and Mechanics on Modern Science and Technology*, edited by G. C. Sih and C. P. Spyropoulos (Eptalofos S.A., Athens, Hellas, 2004), pp. 317-325.
4. A. Dresher and I. Vardoulakis, Geometric softening in triaxial test on granular materials, *Geotechnique* **32**, 291-303 (1982).
5. H. E. Read and G. A. Hegemier, Strain softening of rock, soil and concrete - A review article, *Mechanics of Materials* **3**, 271-294 (1984).
6. S. K. Kourkoulis and N. L. Ninis, in: *Proc. 34th Conf. on Solid Mechanics* (Institute of Fundamental Technological Research of the Polish Academy of Sciences, Warsaw, Poland 2002), pp.149-150.

7. S. K. Kourkoulis, N. L. Ninis and V. N. Bakolas, in *Proceedings of the 12th International Conference on Experimental Mechanics*, edited by C. Papalettere, (Mc Graw Hill, New York, 2004) pp. 12.3-222/1-8.
8. S. K. Kourkoulis, in *Proceedings of the 7th National Congress on Mechanics*, edited by A. Kounadis, C. Providakis, G. Exadaktylos, (HSTAM & Technical University of Crete, 2004) **II**, pp. 1-8.

Chapter 2.4

MECHANICAL CHARACTERISTICS OF ROMAN "OPUS CAEMENTICIUM"

Carlo Giavarini, Alessandro Samuelli Ferretti, Maria Laura Santarelli
CISTeC – Research Centre of Science and Technology for the Conservation of Cultural Heritage, University of Rome "La Sapienza", via Eudossiana, 18, 00184 Rome, Italy

Abstract: The masonry of many Roman buildings had a composite structure made of two external brick walls and a concrete nucleous; this inner nucleous was a concrete composed by a pozzolanic lime mortar containing large pieces of stones, marbles, tuff and bricks. The mechanical characteristics of the wall depended almost exclusively on the strength of the opus caementicium. As a consequence, it is essential to know the mechanical characteristics of the opus caementicium. An extended experimental project has been carried out in the University of Roma "La Sapienza" to define these characteristics and they are described in this paper.

Keywords: opus caementicium; roman concrete; mortar; pozzolana; lime; mechanical properties; stones; tuff.

1. THE ROMAN OPUS CAEMENTICIUM

It is a common belief that the extraordinary development of civil engineering and of the building industry in roman times was correlated to the quality of the building materials and to the attention reserved to their production and selection. The astonishing rapidity of the Romans in rising huge buildings such as the Maxentius Basilica (Basilica Nova) was possible thanks to the availability of great amounts of materials produced in industrial quantities (Giavarini, 2005). One of the most important and innovative techniques was based on the so called "opus caementicium", which in practice replaced the use of stones in the construction of bearing structures and, generally, of the whole building, including arches and vaults or domes.

The masonry of many roman buildings had a composite structure made of two external brick walls (opus latericium) and a concrete nucleous or core (Figure 1). The inner nucleous was a conglomerate composed by a pozzolanic

S.K. Kourkoulis (ed.), Fracture and Failure of Natural Building Stones, 107–120.

lime mortar containing large pieces of stones marbles, tuff and bricks. Sometimes the external walls were made of square pieces of tuff (opus reticulatum).

In fact, the term "caementum" was referred to the pieces of stones inserted into the mortar and tamped inside. Without the side bricks, the opus caementicium was used for the foundations, but it was normally included inside layers of bricks (opus latericium) or tuff (opus reticulatum). The bricks were triangular in shape disposed as indicated in Figure 1 (Samuelli, 2000, Giavarini, 2005).

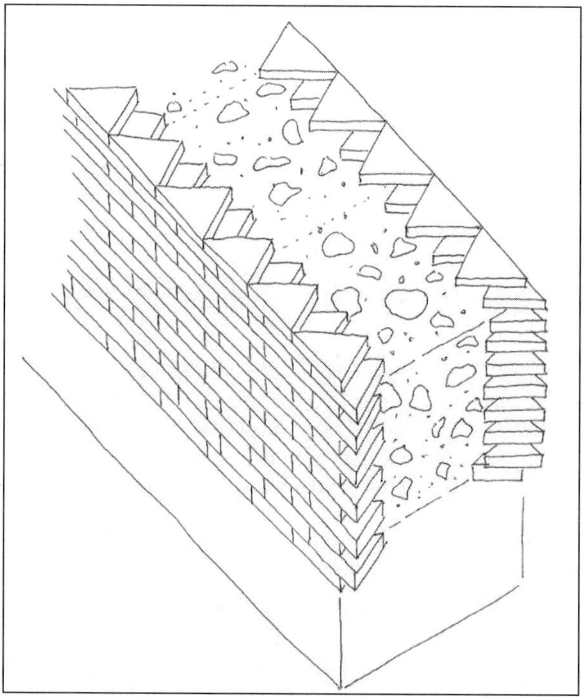

Figure 1. Opus caementicium contained within two facing walls of triangular bricks (opus latericium).

Considering the most common walls delimited by triangular bricks and obtained by diagonally cutting square bricks (typically 20 x 20 cm) the real bearing width of the masonry was only 7 cm, i.e. absolutely negligible when referred to walls with a thickness of some meters (e.g. 4-5 m for the Maxentium Basilica walls). Therefore, the mechanical characteristics of the wall depended almost exclusively on the resistance of the opus caementicium: in fact, 14 cm (7 + 7) out of 500 cm is only about 3%.

As a consequence, for the evaluation of the structural behaviour of a roman building, it is essential to know the mechanical characteristics of the opus caementicium. Published works on this subject are relatively scarce: some data are reported by Rondelet (1831); more information has been published by Lamprecht (1968) who gives, however, values which are not in agreement with Rondelet and with the present work; this is due to the fact that the samples were taken from waterproofing materials made with special care and with carefully selected small aggregates ("cocciopesto"). Some structural considerations are also due to DeLaine (1985).

Generally, the building materials used in ancient times are considered "rigid systems", compared to elastic materials such as steel and concrete, widely used in modern buildings. This is not completely true; in fact, materials such as the Roman opus caementicium show a "strange" elastic behaviour in a relatively large range of stresses in spite of the fact that it is more subjected to static deformation than modern materials.

An experimental research on the mechanical characteristics of "opus caementicium" can add useful information for the evaluation of the safety margins of the still standing roman monuments and, moreover, it can contribute to the understanding of the design criteria of the ancient architects.

2. THE EXPERIMENTAL WORK

An extended experimental work was carried out by CISteC with the help of the Materials & Structures Laboratory of the University of Roma "La Sapienza". Portions of masonry as well as the constituent materials (bricks, mortar, opus caementicium, different types of stones) were tested in the laboratory, both using original ancient samples (when available) and new samples prepared in the lab following the ancient recipes and procedures.

The resistance to axial stresses, mainly compression, was considered most important. The elastic modulus (ratio between stresses and corresponding strains) and the Poisson's coefficient (ratio between axial and transverse deformations) were determined from the linear elastic portion of the deformation curves. During the testing it was important to evaluate how the deformations evolved after the peak stress. Important physical properties included were the density, the porosity, and the ductility. A number of non-elastic deformations were also important, as well as the dimensional changes due to temperature and to the absorption of humidity.

2.1 Bricks

The components of the opus caementicium first studied were bricks. Bricks were used both as "side walls" of the opus caementicium and inside it as fragments (recycled from other uses).

For the tests, prismatic samples were used. Their base had dimensions 15x15 mm while their height was 30 mm. The square base was perpendicular to the applied load; the samples were suitable for compression, direct traction and elastic modulus tests. Four strain gauges were applied to the rectangular faces during the compression tests, while the traction tests were carried out by applying two metal plates each connected to a steel wire, in order to ensure a good aligning of the load. A number of tests carried out on modern bricks of similar strength allowed the definition of a "shape coefficient" necessary to correlate samples with different dimensions. A "rigid" test equipment was used with controlled deformation, suitable for tests beyond the peak stress.

Ductility was considered to be the ratio between the peak axial deformation (during the compression test) and the deformation after the sample collapsed and reduced its resistance down to 50% of the original one: $\sigma = \varepsilon_{50\%} / \varepsilon_{ult}$. The results of the tests carried out on about 30 samples taken from different roman bricks (II-IV century a. D.) are shown in Table 1 and Figures 2 and 3. The stress - strain curves are characteristic of a fragile material. Considering the different origin of the bricks and, therefore, the different production materials and procedures, the results are not homogeneous. For the compressive resistance the values vary from about 10 to almost 40 N/mm^2.

A study was also carried out on 30 bricks (hand-made in the laboratory) to evaluate the influence of the specimen shape on the compression strength. The shape factor of the first series of specimens (15x15x30 mm) was 1.00; the shape factor of the second series (cubic specimens, 30x30x30 mm) was 1.14; finally, the third series (125x125x30 mm) had a shape factor of 1.26.

2.2 Stones

A number of pieces (average dimensions 5-25 cm) of bricks and of various stones were inserted in the opus caementicium mortar: mostly tuff, but also marble, travertine, basalt, etc. Tuff square pieces were sometimes used as external components ("walls") of the opus caementicium, as well as the characteristic tuff truncated pyramids of "opus reticulatum". Therefore, it was important to evaluate their contribution to the total strength of the opus. Tuff stones were particularly important in the vaults and as roofing concrete components due to their lower density.

Table 1. Characteristics of roman bricks (29 specimens)

		Average	Standard dev.	Variation coeff.
Compression strength, f_c	N/mm^2	17.02	5.90	0.35
Tensile strength, f_t	N/mm^2	3.33	1.25	0.30
Young's modulus	N/mm^2	13400	4748	0.35
Ductility ratio	-	2.26	0.35	0.16
Apparent volumic mass	kN/mm^3	24.00	1.02	0.04
Dry volumic mass	kN/mm^3	16.03	0.91	0.06
Real volumic mass	kN/mm^3	27.70	0.46	0.02
Porosity	%	13.9	4.37	0.31

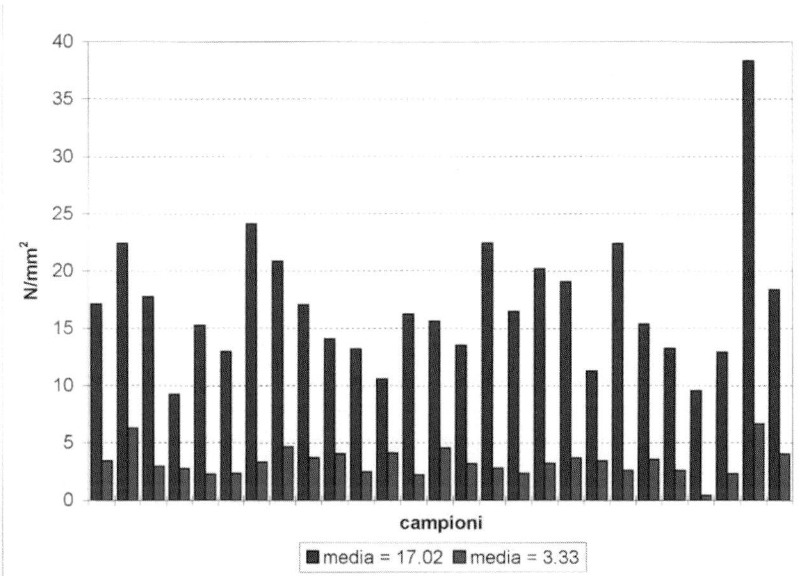

Figure 2. Histogram of the compression (long bars) and the traction (short bars) test on brick specimens.

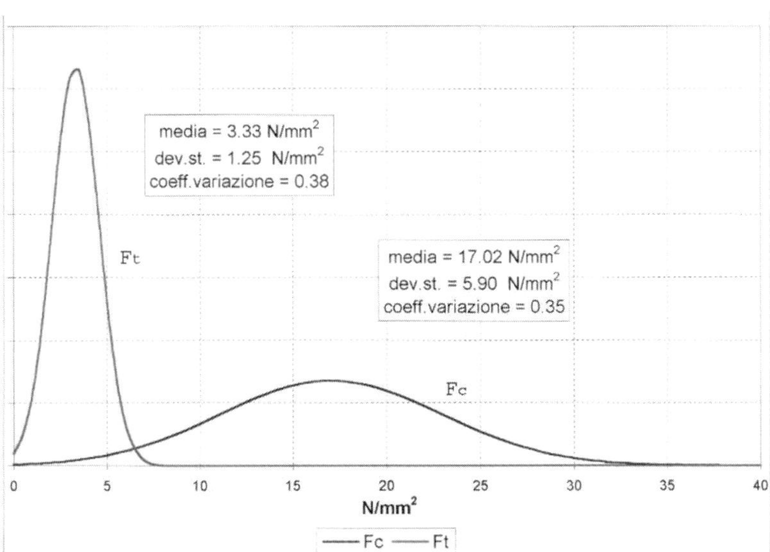

Figure 3. Gauss distribution of the compression (Fc) and traction (Ft) test on bricks.

The stones considered in our experimental work were different types of tuff and travertine. The elastic modulus and the results of the compression tests performed on parallelepiped samples are shown in Table 2.

Table 2. Compression strength and elastic modulus on tuffs and travertine samples

Sample	Volume mass kN/m^3	Compressive strength, f_c N/mm^2	Elastic modulus E N/m^2
Soft tuff	11.63	1.96	--
	12.78	3.66	--
Medium tuff	14.21	9.88	--
Peperino tuff	18.15	36.03	--
	18.85	50.16	16950
Travertine	24.88	44.37	--
	24.65	38.95	64540

2.3 Mortars

Ancient mortars cannot be tested because of the practical difficulty to have suitable test samples. Therefore, the samples (40x40x160 mm) were prepared in the laboratory by using industrial lime ("calce fiore") and pozzolanic material from Pomezia ($\phi \leq 2$ mm). The recipe, taken from Vitruvius (1997), used one part by volume of lime, three parts of pozzolana and ~ 0.7 parts of water (1.39 liters per kg of lime); the weights of lime and pozzolana were about the same. The exact amount of total water was determined by the slump test, using a small scale (1/3) Abram's cone. It was found that a 5 mm slump, referred to the original height of 100 mm, could provide both sufficient workability and good consistency.

The term "harena" used by Vitruvius was supposedly referred to the coarse pozzolana from the quarries near Rome, while the term "pulvis puteolana" indicated the finer material from Pozzuoli. The samples were subjected to the traction test following the Italian standards. The flexural strength was deduced from a three point bending test; the two fragments obtained were then inserted in the compression equipment. For each curing period (7, 28, 90, 180, 360 days) 15 samples were prepared in order to have 15 values of the traction resistance and 30 values for the compression strength. The average values of the compression, traction and elastic modulus at various curing periods are given in Table 3.

Table 3. Mechanical characteristics of the mortars

Curing time [days]	f_c, compression strength [N/mm^2]			f_t, traction strength [N/mm^2]			Elastic modulus [N/mm^2]		
	Ave.	St. dev.	St.var. %	Ave.	St. dev.	St.var. %	Ave.	St. dev.	St.var. %
7	5.92	0.15	2.6	0.85	0.09	10.5	969	40	4.3
28	9.68	0.67	6.9	1.31	0.12	9.1	3429	127	9.1
90	13.32	1.24	10.1	1.35	0.14	10.1	2960	131	4.4
180	13.04	0.77	6.0	1.09	0.06	5.0	3244	103	3.0
360	12.07	1.02	8.5	0.95	0.10	10.2	3077	110	2.9

The compression strengths were high, especially when related to the Italian standards for the lime/pozzolana mortars (2.5 N/mm^2 at 28 days curing); values in this range are usually characteristic of cement - sand mortars. The unusual decrease of the strengths after 180 days has not been explained.

An important conclusion can, also, be drawn by considering the stress-strain curves (not reported here): the large deformations recorded after the peak stress (in the seven days tests) indicated that the mortar can tolerate quite large settings even when significant hardening is reached.

2.4 Ancient roman concrete

2.4.1 Mechanical tests

The availability of original samples is very restricted for obvious reasons (the responsibles do not like any destruction of the ancient monuments!); moreover, the ancient samples are usually far from perfection, due to the presence of various defects (cracking, missing parts, etc.).

Thanks to the interest of the Soprintendenze responsible for the Italian monuments, a number of samples were taken from:

- Warehouses of Nerva in the ancient harbour of Emperor Claudius in Ostia.
- Villa Adriana in Tivoli (three exedras hall).
- Maxentius' Basilica in Rome.

The composition of roman opus caementicium has been already described, however in the actual study the following were observed:

- the mortar content in the opus caementicium varied between about 40% and 60%.
- the quality of the pozzolana was better in the more stressed structures.
- the aggregate (bricks and stones, mostly tuff) were roughly placed in horizontal layers.
- in the barrel vaults the content of light materials as aggregates was significantly higher; as an example, the vaults of the Basilica of Maxentius showed higher pumice contents.

The results of the tests, as well as the size and form of the samples, are shown in the Table 4. The data of the table and additional experimental results suggest that the compressive strengths of a "good" opus caementicium can be estimated to be averagely 5-6 N/mm^2.

2.4.2 Wetting-drying cycles

The purpose of this part of the experimental work was the evaluation of the effect of wetting and drying cycles on a number of ancient cores taken from ancient roman monument.

Table 4. Mechanical tests on ancient opus caementicium samples

	Origin	Shape	Vol. mass kN/m^3	f_c N/mm^2	f_t N/mm^2	Elast. mod. N/mm^2	Poisson's ratio
1	Nerva's warehouses	Cylinder	15.9	4.29	-	2550	-
2	"	"	15.3	2.22	-	2500	-
3	"	"	14.6	0.98	-	800	-
4	Villa Adriana	Prism	17.7	5.87	0.88	9170	-
5	"	"	17.4	6.7	-	3000	0.18-0.19
6	"	"	17.7	4.5	0.77	5740	-
7	Slope fill. Terrace	Cylinder	13.5	2.35	-	-	-
8	"	"	13.5	2.43	-	-	-
9	Caem.. Second vault	"	13.5	6.07	-	-	-
10	"	"	15.0	4.97	-	-	-
11	"	"	14.8	5.82	-	-	-
12	"	"	-	4.84	-	-	-
13	First wall	Prism	-	6.0	-	2800	-
14	Filling N-E side	"	-	4.2	-	1750	-
16	S-W foundation	"	-	5.3	-	3500	-
17	"Ladrone" arch	"	16.5	6.16	-	-	-

Small dimensional variations, when referred to huge walls, could give interesting results and, possibly, explain a number of phenomena (cracks, failures, variations of wall heights) previously attributed exclusively to ground movements.

The consolidation project of Domus Tiberiana included the application of a number of tie bars and the drilling of the walls to allow the bars to be put in place. The most suitable cores obtained from the drilling process were used for the present study; their characteristics and the coarse aggregates (>10mm) are listed in Table 5. Observed by stereoscopic microscope, the thin sections of all mortars showed the presence of a fine red pozzolana and tuff (aggregate <4mm). In sample 2 the tuff was a typical roman yellow tuff, while in the other samples the tuff was a red tuff (*lionato* tuff). An aggregate with dimension >4mm, made of a grey and a grey/red pozzolana, was present in the mortars of the samples 1 and 4, respectively.

Table 5. Main characteristics of the core samples

Sample	Diameter (cm)	Length (cm)	Pozzolana type	Coarse aggregates
1	10	40	Red	Red tuff (~ 33%)
2	9	37	Red	Yellow tuff (~14%) Brick (~18%)
3	9	52	Red	Red tuff (~33%)
4	9	43	Red	Red tuff (~29%) Marble (~2%)
5	8	33	Dark red	Brick (~30%) Yellow tuff (~12%)

Three couples of metallic datum-points were applied on all cores in various points in order to obtain three different measures in each sample (Figure 4). Each sample was longitudinally immersed in tap water covering only half of its diameter to damp gradually the total core by capillary rise and to allow the entrapped air to escape. A series of measurements were taken at regular periods for a month or more. Afterwards, the wet samples were left in open air in the laboratory for a month or more until they were dry. Six complete wetting-drying cycles were carried out on each sample.

The results of the wetting - drying cycles are given in Figure 5. The data show that, on average, the concrete's behaviour is similar for all cycles, i.e. the length variation (ΔL/L) of samples of roman concrete subjected to a number of wetting/drying cycles is quite repeatable. For the purposes of this research it was assumed that, on average, the length variation of the samples is about 3-4‰, or 3-4 mm for each meter. Such a variation is quite important for a material normally considered dimensionally stable.

Figure 4. The core samples used for wetting-drying cycles, showing large pieces of aggregates, typical of the roman opus caementicium.

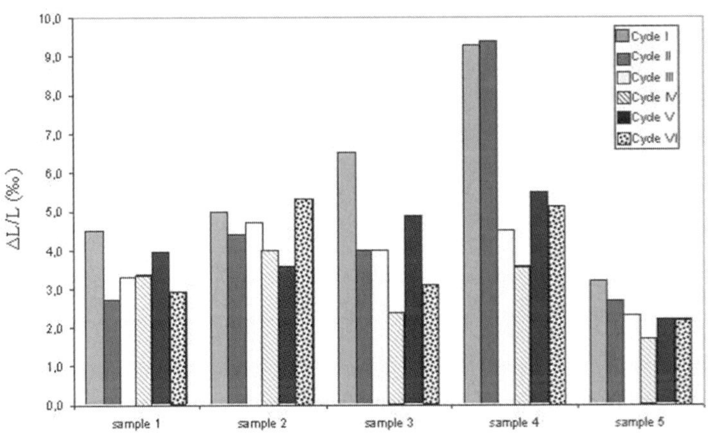

Figure 5. Visualization of the dimensional variations during wetting/drying cycles on five samples of opus caementicium.

2.5 Opus caementicium prepared in the laboratory

Prismatic elements of opus caementicium (300x300x600 mm) were prepared in the laboratory. Several samples were then taken from the prismatic elements by means of a core-drilling procedure. The samples were tested by the equipment shown in Figure 6. The sample rests on two semi-circular steel cradles, lined by a layer of teflon. A central steel cradle with similar teflon strips applies the shearing load. The gap between the loading and the supporting devices must be accurately defined, because it must allow a "guillotine" behaviour, without introducing any disturbing effect. The two supports rest on rollers, in order to avoid interference with the applied axial load. The test is performed by applying a given value of compressive stress σ, which is kept constant by means of an automatic device, and then increasing the shearing action until collapse is reached. The results are shown in Figure 7.

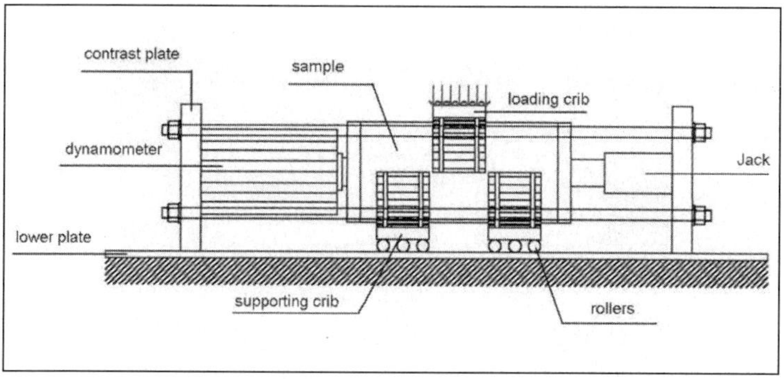

Figure 6. Compression-shear equipment for testing the laboratory cylindrical specimens.

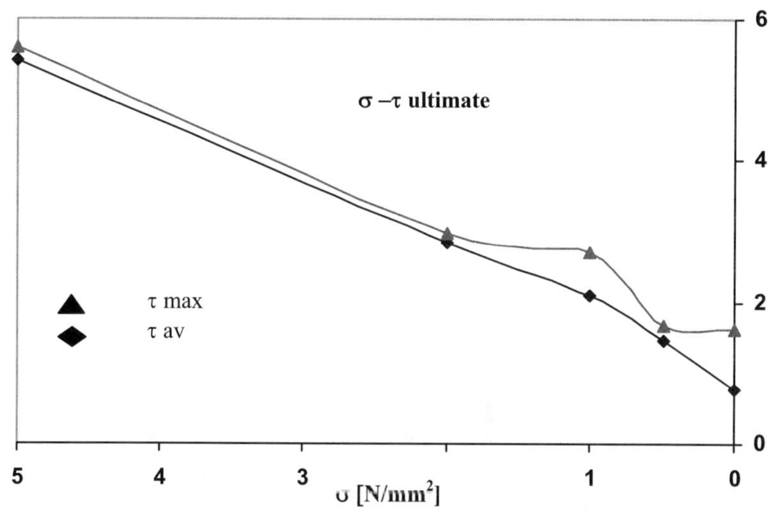

Figure 7. Results of the shear-compression ultimate tests on opus caementicium samples.

2.6 Roman-type wall made in laboratory

The purpose of the last phase of the work was to test panels representing portions of the whole Roman wall, composed of two external layers of bricks containing the core of opus caementicium (Figure 8). The triangular bricks were handmade, in order to obtain a strength (average 19.3 N/mm^2) similar to that of the original bricks. The filling of the core followed the procedure suggested by a number of authors (Adam 1984, Giuliani 1990). The wall samples were built on steel plates in order to move them from the building site to the testing machines. Two of the samples were subjected to axial compression. The most significant results are the following:

- the ultimate average strength is 6.3 N/mm^2.
- the wall samples follow good linear behaviour up to around 5 N/mm^2.
- the collapse is preceded (at around 4.5 N/mm^2) by vertical (columnar) cracking. The Poisson's ratio in the elastic region increases from its initial value of 0.18 to more than 1 near collapse.

Figure 8. An example of roman – type wall during the preparation in the laboratory (Engineering Faculty, University of Rome "La Sapienza").

The remaining eight wall samples were subjected to combined shear compression, by using two different pieces of equipment, especially made for the test. The first one (Figure 9), which was employed in case the ratio τ/σ was larger than one, was composed of two diagonal steel bands contrasting an actuator (jack and dynamometer); in this way equal components of compression and shear were imposed: $\sigma=\tau$. The second equipment was of the guillotine type as it is shown in Figure 10. The results are shown in Figure 11. In the low compression region the structure behaves as a cohesive material, while for higher values of σ a friction mechanism prevails. The friction coefficient, in this case, is considerably lower (0.2 instead of 0.8). This is presumably due to the sliding of the lateral brick layers.

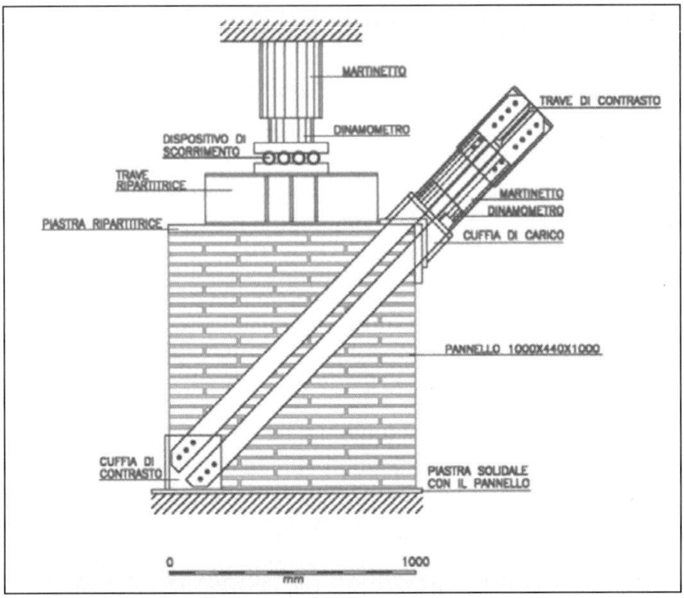

Figure 9 . Diagonal-type equipment for testing the roman wall.

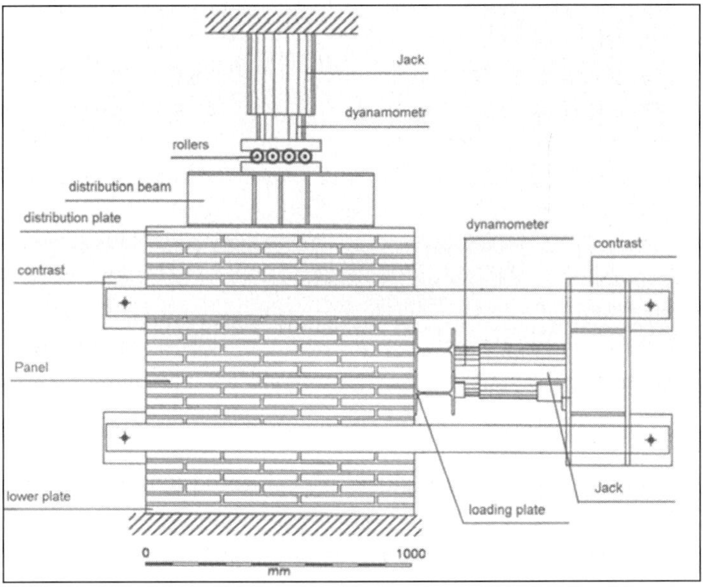

Figure 10. Guillotine-type equipment for testing the roman wall.

3. CONCLUSIONS

Some important mechanical characteristics of the construction materials used by ancient Romans have been measured both on authentic archaeological

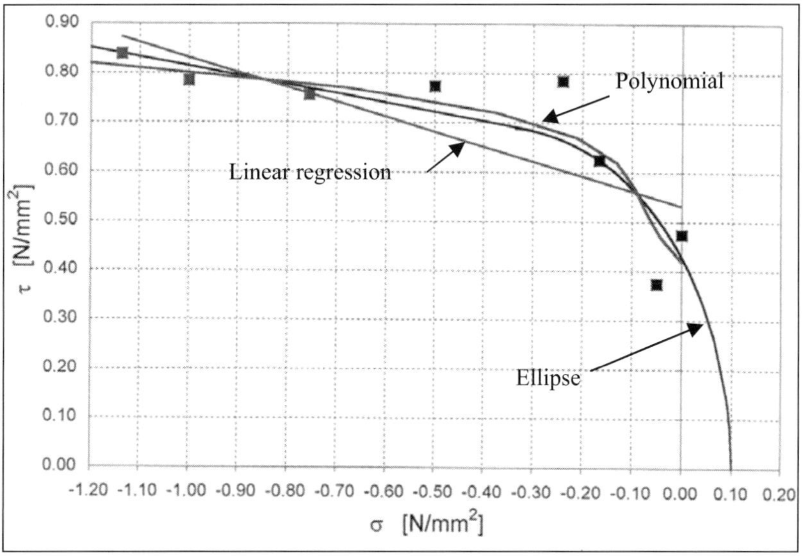

Figure 11. Shear-compression tests on roman walls. Experimental results and interpolations.

samples as well as on new samples prepared following the roman receipts. Mortars, concrete (opus caementicium) and pieces of masonry have been tested. The mechanical characteristics of roman walls depend almost exclusively on the resistance of the opus caementicium. For compositions in the range of the samples used in this work, it can be safely assumed that the compression strength is equal to 5 N/mm^2. The shear stresses that can be foreseen are usually "safe" with respect to the corresponding strength.

One of the more significant results of this study is that a linear-elastic modelling of masonry buildings is possible. As a matter of fact the linearity of the opus caementicium extends for more than two thirds of the total range, up to collapse. The dimensions of the stones composing the opus caementicium seem sufficiently small to admit the use of Finite Elements analysis.

The results provide useful information for solving problems of structural analysis related to roman monuments. The behaviour of roman masonry is influenced by the humidity contained in the opus caementicium. In fact, dimensional variation of 3-4‰ was measured during wetting - drying cycles: referred to huge walls, these changes could be responsible of cracks and failures.

REFERENCES

Adam, J.P., L'arte di Costruire presso i Romani. Materiali e Tecniche, Milano, Longanesi, 1984.

DeLaine J., An Engineering Approach to Roman Building Techniques. The Baths of Caracalla, in Rome. British Archaelogical Reports, 1985.

Giavarini, C. (Ed.), La Basilica di Massenzio, L'Erma di Bretschneider, Roma, 2005.
Giuliani, F.C., Edilizia nell'antichità, Roma, INS, 1990.
Lamprecht, H. O, Opus caementicium, Dusseldorf, 1968.
Rondelet, G., Trattato teorico e pratico dell'arte di edificare, Mantova, 1831.
Samuelli Ferretti A,. Opus Caementicium, The Roman Concrete. Domus Aurea and Basilica of Maxentius: Stress Analysis and Safety, in *Proceedings of the 12th International Brick/Block Masonry Conference*, Madrid 2000.
Vitruvius, De Architectura, Torino, Einaudi, 1997.

Chapter 3: Masonry

Chapter 3.1

CONTINUUM MODELLING OF MASONRY STRUCTURES UNDER STATIC AND DYNAMIC LOADING

I. Stefanou[1], J. Sulem[2] and I. Vardoulakis[1]

[1]Dept. of Applied Mechanics and Physics, National Technical University of Athens, Greece, istefanou@mechan.ntua.gr, I.Vardoulakis@mechan.ntua.gr; [2]CERMES, Ecole Nationale des Ponts et Chaussées/LCPC, Institut Navier, Paris, France, sulem@cermes.enpc.fr

Abstract: Ancient masonry structures can be seen as a set of rigid blocks where the deformation is taking place at the interface between the blocks. Rocking, twisting and sliding between them are possible mechanisms that actually take place under static or dynamic loading. A continuum model for regular block structures is derived by replacing the quotients of the discrete equations by corresponding differential quotients. The homogenisation procedure leads to an anisotropic Cosserat Continuum. For elastic block interactions the dispersion relations of the discrete and the continuous models are derived and compared.

Key words: masonry; block structure; 3D Cosserat; out of plane; in plane; dynamics; elasticity; homogenisation.

1. INTRODUCTION

The numerical analysis of discontinuous blocky structures can be dealt using discrete and finite element codes. In the latter case, special interface elements are needed in order to account for the unilateral kinematics of the rock joints. The interest of developing continuous models for discrete structures is that discrete type analyses are very computer time intensive and, at least for periodic structures, one might argue that a homogenised continuum model would allow a much more elegant and efficient solution. One could list the practical relevance of the development of continuum models: (a) it is extremely flexible when used with numerical methods, since no interface elements are needed and since the topology of the finite element is independent of block size and geometry (one mesh can be used to study several different structures); (b) quite a number of analytical solutions can be provided that

S.K. Kourkoulis (ed.), Fracture and Failure of Natural Building Stones, 123–136.

can be used as benchmarks for discrete codes; (c) unconditionally stable integration through implicit algorithms can be used, unlike discrete models where conditionally stable explicit integration schemes are used.

However, the computational efficiency comes at a price. Continuum models are usually based on micro-mechanical mechanisms, which govern the material behaviour in the medium to large wavelength range. When the characteristic length of the macroscopic deformation pattern is smaller than a certain multiple of the characteristic fabric length of the material, then the applicability of the continuum model reaches its limit. Another important limitation of the homogenisation of layered or blocky structures with classical continuum theories is that they cannot account for elementary bending due to inter-layer or inter-block slip and may thus considerably overestimate the deformation. In order to overcome these limitations and to expand the domain of validity of the continuum approach one has to consider the salient features of the discontinuum within the frame of continuum theories with microstructure. The 2D-Cosserat theory has been used with some success in the recent years for analysing blocky and laminated systems[1-3]. The enriched kinematics of the Cosserat continuum allows to model microelement systems undergoing rotations, which are different from the local rotations of the continuum. Various failure modes such as inter-block slip and block tilting can then be described. In these previous studies developed in the frame of two-dimensional Cosserat theory, only one rotational mode of the block has been considered. This work is extended in this paper by considering rock twisting in addition to rock sliding. This rotational mode is frequently observed on masonry structures under dynamic loading.

The problem of a three-dimensional regular block structure is addressed here. Starting with elastic behaviour of the joints, the dynamic differential equations for the discrete 3D structure are derived. A continuum model is obtained by replacing the difference quotients of the discrete equations by corresponding differential quotients. The homogenisation procedure leads to an anisotropic 3D-Cosserat continuum. For elastic block interactions the dispersion relations of the discrete and the continuous models are derived and compared. The domain of validity of the continuous approach is discussed by comparing the dispersion function of the discrete and the continuous system.

2. THE DISCRETE MODEL FOR 3D STRUCTURE AND INFINITESIMAL DEFORMATION

An idealized model of a masonry wall is considered here (Figure 1). Six others surround each block. The main concern is the accuracy with which the continuum model reflects the domain of rigidity set by the size of the blocks. The elasticity of the blocks and the joints' elasticity are lumped at the block faces for simplicity. Fully elastic joint behaviour is assumed. Also, it is assumed that the interaction between the block is concentrated in six points of the faces as shown in Figure 1.

Moments and normal and shear forces are written as:

$$\begin{aligned} Q_{kl} &= c_Q \Delta x^i_{kl} \\ N_{kl} &= c_N \Delta x^i_{kl} \\ M_{kl} &= c_M \Delta \phi^i_{kl} \quad or \quad M_{kl} = c_T \Delta \phi^i_{kl} \end{aligned} \tag{1}$$

where c_Q, c_N, c_M and c_T are the elastic shear, normal, bending and torsional stiffness respectively and Δx and $\Delta \phi$ are the relative translations and rotations of block k with block l at various contact points.

Using d'Alembert's principle, the motion equations of block (i,j) of the three-dimensional assembly of blocks are obtained as following:

$$\begin{aligned} & c_N\left(u_{i+2,j}-2u_{i,j}+u_{i-2,j}\right)+c_Q\left(u_{i+1,j+1}+u_{i-1,j+1}+u_{i-1,j-1}+u_{i+1,j-1}-4u_{i,j}\right)+ \\ & \qquad +\frac{b}{2}c_Q\left(\phi_{i+1,j+1}-\phi_{i+1,j-1}+\phi_{i-1,j+1}-\phi_{i-1,j-1}\right)=m\ddot{u}_{i,j} \end{aligned} \tag{2}$$

$$\begin{aligned} & c_Q\left(v_{i+2,j}-2v_{i,j}+v_{i-2,j}\right)+c_N\left(v_{i+1,j+1}+v_{i-1,j+1}+v_{i+1,j-1}+v_{i-1,j-1}-4v_{i,j}\right)- \\ & \qquad -c_Q a\left(\phi_{i+2,j}-\phi_{i-2,j}\right)+c_N\frac{a}{2}\left(\phi_{i-1,j+1}-\phi_{i+1,j-1}+\phi_{i-1,j-1}-\phi_{i+1,j+1}\right)=m\ddot{v}_{i,j} \end{aligned} \tag{3}$$

$$\begin{aligned} & ac_Q\left(v_{i+2,j}-v_{i-2,j}\right)+\frac{a}{2}c_N\left(v_{i+1,j+1}-v_{i-1,j+1}+v_{i+1,j-1}-v_{i-1,j-1}\right)- \\ & -\frac{b}{2}c_Q\left(u_{i+1,j+1}-u_{i-1,j-1}+u_{i-1,j+1}-u_{i+1,j-1}\right)-a^2c_Q\left(\phi_{i+2,j}+\phi_{i-2,j}+2\phi_{i,j}\right)- \\ & -\frac{1}{4}\left(b^2c_Q+a^2c_N\right)\left(\phi_{i+1,j+1}+\phi_{i-1,j+1}+\phi_{i-1,j-1}+\phi_{i+1,j-1}+4\phi_{i,j}\right)+ \\ & +c_M\left(\phi_{i+2,j}+\phi_{i+1,j+1}+\phi_{i-1,j+1}+\phi_{i-2,j}+\phi_{i-1,j-1}+\phi_{i+1,j-1}-6\phi_{i,j}\right)=J_3\ddot{\phi}_{i,j} \end{aligned} \tag{4}$$

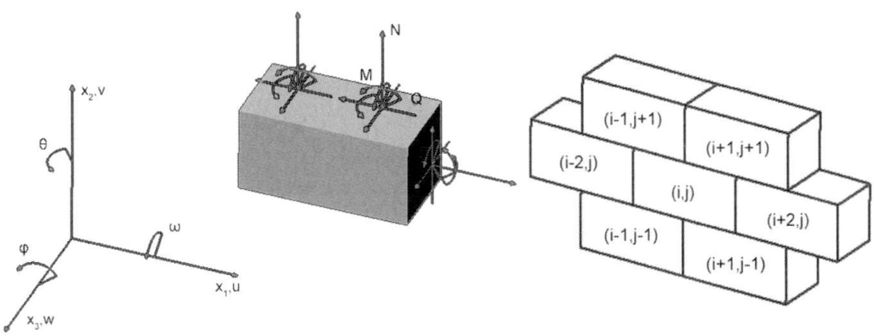

Figure 1. Statical and geometrical configuration. Each block is 2a long, b wide and b high.

$$\left(w_{i+2,j}+w_{i+1,j+1}+w_{i-1,j+1}+w_{i-2,j}+w_{i+1,j-1}+w_{i-1,j-1}-6w_{i,j}\right)+a\left(\theta_{i+2,j}-\theta_{i-2,j}\right)+ \\ +\frac{a}{2}\left(\theta_{i+1,j+1}-\theta_{i-1,j+1}+\theta_{i+1,j-1}-\theta_{i-1,j-1}\right)+\frac{b}{2}\left(\omega_{i+1,j-1}-\omega_{i-1,j+1}+\omega_{i-1,j-1}-\omega_{i+1,j+1}\right)=\frac{m}{c_Q}\ddot{w}_{i,j} \tag{5}$$

$$\frac{a}{2}c_Q\left(w_{i-1,j+1}+w_{i-1,j-1}-w_{i+1,j+1}-w_{i+1,j-1}\right)- \\ -\frac{1}{4}a^2c_Q\left(\theta_{i-1,j+1}+\theta_{i-1,j-1}+\theta_{i+1,j-1}+\theta_{i+1,j+1}+4\theta_{i,j}\right)+ac_Q\left(w_{i-2,j}-w_{i+2,j}\right)+ \\ \frac{1}{4}bac_Q\left(\omega_{i-1,j-1}-\omega_{i-1,j+1}+\omega_{i+1,j+1}-\omega_{i+1,j-1}\right)-a^2c_Q\left(\theta_{i-2,j}+\theta_{i+2,j}+2\theta_{i,j}\right)+ \\ +c_M\left(\theta_{i+2,j}+\theta_{i-2,j}-2\theta_{i,j}\right)+c_T\left(\theta_{i+1,j+1}+\theta_{i-1,j+1}+\theta_{i-1,j-1}+\theta_{i+1,j-1}-4\theta_{i,j}\right)=J_2\ddot{\theta}_{i,j} \tag{6}$$

$$\frac{1}{2}bc_Q\left(w_{i+1,j+1}+w_{i-1,j+1}-w_{i-1,j-1}-w_{i+1,j-1}\right)+ \\ +\frac{1}{4}bac_Q\left(\theta_{i+1,j+1}-\theta_{i-1,j+1}+\theta_{i-1,j-1}-\theta_{i+1,j-1}\right)+c_T\left(\omega_{i+2,j}+\omega_{i-2,j}-2\omega_{i,j}\right)- \\ -\frac{1}{4}b^2c_Q\left(\omega_{i+1,j+1}+\omega_{i-1,j+1}+\omega_{i-1,j-1}+\omega_{i+1,j-1}+4\omega_{i,j}\right)+ \\ +c_M\left(\omega_{i+1,j+1}+\omega_{i-1,j+1}+\omega_{i-1,j-1}+\omega_{i+1,j-1}-4\omega_{i,j}\right)=J_1\ddot{\omega}_{i,j} \tag{7}$$

where m and J_i the mass and the moments of inertia of each block.

The above equations describe the dynamic behaviour of the block assembly depicted at Figure 1. Looking carefully at these equations we distinguish two uncoupled sets of equations, i.e. Eqs.(2-4) and Eqs.(5-7), which correspond to two different modes of motion-deformation of the blocky structure. These modes describe the in and the out of plane deformation of the structure. Therefore the initial problem is finally separated in two independent problems, which is a common approach in the theory of plates. The in plane mode of deformation is examined at the paper of Sulem and Mühlhaus[2].

3. THE CONTINUOUS MODEL

In the above system of equations (Eqs.(2–7)) the discrete coordinates (i,j) are replaced by the continuous ones (x_1,x_2) and instead of $u_{i\pm1,j\pm1}$, $v_{i\pm1,j\pm1}$, $w_{i\pm1,j\pm1}$, $\varphi_{i\pm1,j\pm1}$, $\theta_{i\pm1,j\pm1}$ and $\omega_{i\pm1,j\pm1}$ one writes $u(x_1\pm a, x_2\pm b)$, $v(x_1\pm a, x_2\pm b)$, $w(x_1\pm a, x_2\pm b)$, $\phi(x_1\pm a, x_2\pm b)$, $\theta(x_1\pm a, x_2\pm b)$ and $\omega(x_1\pm a, x_2\pm b)$. Then the u, v, w, ϕ, θ and ω degrees of freedom are developed into a Taylor series up to the second order:

$$f\left(x_1\pm a,x_2\pm b\right)=f\left(x_1,x_2\right)\pm a\frac{\partial f}{\partial x_1}\pm b\frac{\partial f}{\partial x_2}+a^2\frac{\partial^2 f}{\partial x_1^2}+b^2\frac{\partial^2 f}{\partial x_2^2}\pm ab\frac{\partial^2 f}{\partial x_1 x_2}+O^3 \tag{8}$$

Applying the aforementioned procedure to Eqs.(2-7) one obtains the following system of second order differential equations:

$$2a^2\left(2c_N+c_Q\right)\frac{\partial^2 u}{\partial x_1^2}+c_Q 2b^2\frac{\partial^2 u}{\partial x_2^2}+c_Q 2b^2\frac{\partial \phi}{\partial x_2}=m\frac{\partial^2 u}{\partial t^2} \tag{9}$$

$$2a^2\left(c_N+2c_Q\right)\frac{\partial^2 v}{\partial x_1^2}+2c_N b^2\frac{\partial^2 v}{\partial x_2^2}-2a^2\left(c_N+2c_Q\right)\frac{\partial \phi}{\partial x_1}=m\frac{\partial^2 v}{\partial t^2} \tag{10}$$

$$\begin{aligned}2a^2\left(c_N+2c_Q\right)\frac{\partial v}{\partial x_1}-2b^2c_Q\frac{\partial u}{\partial x_2}-2\left(2a^2c_Q+\frac{1}{4}b^2c_Q+\frac{1}{4}a^2c_N-3c_M\right)a^2\frac{\partial^2\phi}{\partial x_1^2}-\\-2\left(\frac{1}{4}b^2c_Q+\frac{1}{4}a^2c_N-c_M\right)b^2\frac{\partial^2\phi}{\partial x_2^2}-2\left(b^2c_Q+2a^2c_Q+a^2c_N\right)\phi=J_3\frac{\partial^2\phi}{\partial t^2}\end{aligned} \tag{11}$$

$$6a^2\frac{\partial^2 w}{\partial x_1^2}+2b^2\frac{\partial^2 w}{\partial x_2^2}+6a^2\frac{\partial \theta}{\partial x_1}-2b^2\frac{\partial \omega}{\partial x_2}=\frac{m}{c_Q}\frac{\partial^2 w}{\partial t^2} \tag{12}$$

$$\begin{aligned}-6a^2c_Q\frac{\partial w}{\partial x_1}+a^2b^2c_Q\frac{\partial^2\omega}{\partial x_1\partial x_2}+2\left(2c_M-\frac{9}{4}a^2c_Q+c_T\right)a^2\frac{\partial^2\theta}{\partial x_1^2}+\\+2\left(c_T-\frac{1}{4}a^2c_Q\right)b^2\frac{\partial^2\theta}{\partial x_2^2}-6a^2c_Q\theta=J_2\frac{\partial^2\theta}{\partial t^2}\end{aligned} \tag{13}$$

$$\begin{aligned}2c_Qb^2\frac{\partial w}{\partial x_2}+c_Qa^2b^2\frac{\partial^2\theta}{\partial x_1\partial x_2}+\left(c_M+2c_T-\frac{1}{4}c_Qb^2\right)2a^2\frac{\partial^2\omega}{\partial x_1^2}+\\+\left(c_M-\frac{1}{4}c_Qb^2\right)2b^2\frac{\partial^2\omega}{\partial x_2^2}-2c_Qb^2\omega=J_1\frac{\partial^2\omega}{\partial t^2}\end{aligned} \tag{14}$$

Examining again Eqs.(9)-(14) one distinguishes two uncoupled sets of equations, i.e. Eqs.(9)-(11) and Eqs.(12)-(14), which correspond respectively to the in and the out of plane deformation of the structure.

4. IDENTIFICATION OF A COSSERAT ANISOTROPIC CONTINUUM

In a three-dimensional Cosserat continuum each material point has three translational degrees of freedom (u; v; w) and three rotational degrees of freedom $\omega^c{}_i$. The index c is used to distinguish the Cosserat rotation from the rotation:

$$\omega_{ij}=\frac{1}{2}\left(u_{i,j}-u_{j,i}\right);\ (.)_{,i}=\frac{\partial(.)}{\partial x_i}\ i=1,2,3 \tag{15}$$

For the formulation of the constitutive relationships one needs deformation measures, which are invariant to rigid body motion, i.e. the conventional strain tensor:

$$\varepsilon_{ij} = \frac{1}{2}\left(u_{i,j} + u_{j,i}\right) \tag{16}$$

The relative rotation of each material point is given by:

$$\omega_i^{rel} = \omega_{ji} n_j - \omega_i^c \tag{17}$$

and the gradient of the Cosserat rotation, which is called the curvature of the deformation, is given by:

$$\kappa_{ij} = \frac{\partial \omega_j^c}{\partial x_i} \tag{18}$$

Eqs.(16,17) are combined to give the following deformation quantities for the 3D-Cosserat continuum:

$$\begin{array}{l|l|l|l}
\gamma_{11} = \dfrac{\partial u_1}{\partial x_1} & \gamma_{12} = \dfrac{\partial u_1}{\partial x_2} + \omega_3^c & \gamma_{13} = \dfrac{\partial u_1}{\partial x_3} - \omega_2^c & \gamma_{23} = \dfrac{\partial u_2}{\partial x_3} + \omega_1^c \\
\gamma_{22} = \dfrac{\partial u_2}{\partial x_2} & \gamma_{21} = \dfrac{\partial u_2}{\partial x_1} - \omega_3^c & \gamma_{31} = \dfrac{\partial u_3}{\partial x_1} + \omega_2^c & \gamma_{32} = \dfrac{\partial u_3}{\partial x_2} - \omega_1^c \\
\gamma_{33} = \dfrac{\partial u_3}{\partial x_3} & & &
\end{array} \tag{19}$$

The twelve deformation quantities (Eqs.(18, 19)) are conjugate in energy with twelve stress quantities. First, one has the nine components of the non-symmetric stress tensor σ_{ij}, which are conjugate to the non-symmetric deformation tensor γ_{ij}, and second one has nine moment stresses (moment per unit area), which are conjugate with the nine components of the deformation curvature tensor κ_{ij}. Force and moment equilibrium at the element (dx_1, dx_2, dx_3) lead to (Figure 2):

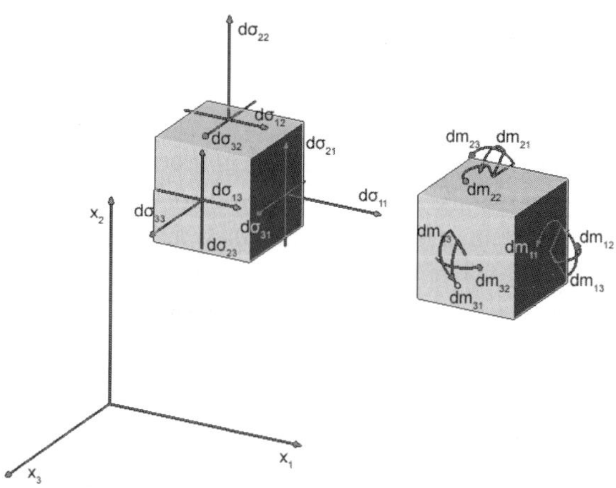

Figure 2. Stresses on element (dx_1, dx_2, dx_3).

$$\begin{aligned}
&\sigma_{11,1}+\sigma_{12,2}+\sigma_{13,3}-\rho\ddot{u}_1=0\\
&\sigma_{21,1}+\sigma_{22,2}+\sigma_{23,3}-\rho\ddot{u}_2=0\\
&\sigma_{31,1}+\sigma_{32,2}+\sigma_{33,3}-\rho\ddot{u}_3=0\\
&m_{11,1}+m_{21,2}+m_{31,3}+\sigma_{32}-\sigma_{23}-J_1\ddot{\omega}_1^c=0\\
&m_{12,1}+m_{22,2}+m_{32,3}+\sigma_{13}-\sigma_{31}-J_2\ddot{\omega}_2^c=0\\
&m_{13,1}+m_{23,2}+m_{33,3}+\sigma_{21}-\sigma_{12}-J_3\ddot{\omega}_3^c=0
\end{aligned}\tag{20}$$

Following the hereunder constitutive relations an equivalent Cosserat continuum is identified that describes the studied blocky structure.

$$\begin{array}{l|l}
\sigma_{11}=\dfrac{a}{b^2}\left(c_Q+2c_N\right)\cdot\gamma_{11} & m_{11}=\dfrac{a}{b^2}\left(c_M+2c_T-\dfrac{1}{4}b^2c_Q\right)\cdot\kappa_{11}\\
\sigma_{22}=\dfrac{1}{a}c_N\cdot\gamma_{22} & m_{22}=\dfrac{1}{a}\left(c_T-\dfrac{1}{4}a^2c_Q\right)\cdot\kappa_{22}\\
\sigma_{33}=0 & m_{33}=0\\
\sigma_{12}=\dfrac{1}{a}c_Q\cdot\gamma_{12} & m_{12}=\dfrac{a}{b^2}\left(2c_M+c_T-\dfrac{9}{4}a^2c_Q\right)\cdot\kappa_{12}+\dfrac{1}{2}ac_Q\cdot\kappa_{21}\\
\sigma_{21}=\dfrac{a}{b^2}\left(c_N+2c_Q\right)\cdot\gamma_{21} & m_{21}=\dfrac{1}{2}ac_Q\cdot\kappa_{12}+\dfrac{1}{a}\left(c_M-\dfrac{1}{4}b^2c_Q\right)\cdot\kappa_{21}\\
\sigma_{23}=0 & m_{23}=-\dfrac{1}{a}\left(\dfrac{1}{4}a^2c_N+\dfrac{1}{4}b^2c_Q-c_M\right)\cdot\kappa_{23}\\
\sigma_{32}=\dfrac{1}{a}c_Q\cdot\gamma_{32} & m_{32}=0\\
\sigma_{13}=0 & m_{13}=-\dfrac{a}{b^2}\left(2a^2c_Q+\dfrac{1}{4}b^2c_Q+\dfrac{1}{4}a^2c_N-3c_M\right)\cdot\kappa_{13}\\
\sigma_{31}=\dfrac{3a}{b^2}c_Q\cdot\gamma_{31} & m_{31}=0
\end{array}\tag{21}$$

The geometric configuration of the considered structure is anisotropic even for square blocks as each block has four neighbours in the x_2 direction and only two neighbours in the x_1 direction (Figure 1).

5. THE DISPERSION FUNCTION

The domain of validity of the above representation of a blocky structure by a Cosserat continuum is evaluated by comparing the dynamic response of the discrete and the homogenized structures. The dynamic response of a structure is characterized by its dispersion function, which relates the wave propagation velocity to the wavelength of the input signal. For elastic behaviour it is possible to derive analytical solutions for the dispersion function of the discrete and the continuous systems by using discrete and continuous Fourier transforms as presented below. Here, one is interested in the out of plane dynamic response of the structure. For the in plane comparison of the blocky structure the reader is referred to the previous work of Sulem and Mühlhaus[2], where a similar analysis is followed showing that the Cosserat model gives a good approximation for wavelengths bigger than five times the size of the block. First we will derive the expression of the dispersion function for the continuous system.

The Fourier transform of a function is:

$$G(\theta_1,\theta_2,\nu)=\mathcal{F}\left\{g(x_1,x_2,t)\right\}=\int_{-\infty}^{+\infty}\int_{-\infty}^{+\infty}\int_{-\infty}^{+\infty} g(x_1,x_2,t)e^{2\pi i(\theta_1 ax+\theta_2 by+\nu t)}dx_1dx_2dt \tag{22}$$

where $i=\sqrt{-1}$, θ_1 and θ_2 the wave numbers at x_1 and x_2 direction respectively and ν the frequency.

The inverse transform is:

$$g(x_1,x_2,t)=\mathcal{F}^{-1}\left\{G(p,q,\nu)\right\}=\int_{-\infty}^{+\infty}\int_{-\infty}^{+\infty}\int_{-\infty}^{+\infty} G(p,q,\nu)e^{-2\pi i(\frac{\theta_1}{a}x_1+\frac{\theta_2}{b}x_2+\nu t)}d\theta_1 d\theta_2 d\nu \tag{23}$$

The Fourier transform will be denoted as:

$$g(x_1,x_2,t)\rightleftharpoons G(\theta_1,\theta_2,\nu) \tag{24}$$

In general the wave speed V depends on the wave numbers θ_1 and θ_2. The dispersion function $V(\theta_1, \theta_2)$ is determined by insertion of the transformed unknown functions w, ω, and θ and their transformed derivatives into Eqs. (12)-(14). This leads to the following homogeneous system of equations:

$$\begin{pmatrix} a_1+Y & ia_2 & ia_3 \\ -ia_2 & b_1+b_3Y & b_2 \\ -ia_3 & b_2 & c_1+c_2Y \end{pmatrix}\begin{pmatrix} W(\theta_1,\theta_2,\nu) \\ \Theta(\theta_1,\theta_2,\nu) \\ \Omega(\theta_1,\theta_2,\nu) \end{pmatrix}=\mathbf{0} \tag{25}$$

This system possesses non-trivial solutions when its determinant vanishes, that is:

$$\det\begin{pmatrix} a_1+Y & ia_2 & ia_3 \\ -ia_2 & b_1+b_3Y & b_2 \\ -ia_3 & b_2 & c_1+c_2Y \end{pmatrix}=0 \tag{26}$$

where

$$\begin{aligned} a_1 &= -6c_Q\theta_1^2-2c_Q\theta_2^2 \\ a_2 &= -6c_Q a\theta_1 \\ a_3 &= 2c_Q b\theta_2 \\ b_1 &= -2\left(2c_M-\frac{9}{4}a^2c_Q+c_T\right)\theta_1^2-2\left(c_T-\frac{1}{4}a^2c_Q\right)\theta_2^2-6a^2c_Q \end{aligned} \tag{27}$$

$$b_2 = c_Q ab\theta_1\theta_2$$
$$b_3 = {J_y}/{m} \qquad \text{(27 cont'd)}$$
$$c_1 = -2\left(c_M + 2c_T - \frac{1}{4}c_Q b^2\right)\theta_1^2 - 2\left(c_M - \frac{1}{4}c_Q b^2\right)\theta_2^2 - 2b^2 c_Q$$
$$c_2 = {J_x}/{m}$$

and

$$Y = m\left(\frac{\theta_1 V_1}{a} + \frac{\theta_2 V_2}{b}\right)^2 \qquad (28)$$

Y is the solution of the following characteristic polynomial equation:

$$AY^3 + BY^2 + CY + D = 0 \qquad (29)$$

where the coefficients are:

$$\begin{aligned} A &= b_3 c_2 \\ B &= a_1 b_3 c_2 + b_3 c_1 + b_1 c_2 \\ C &= a_1 b_1 c_2 + a_1 b_3 c_1 - a_2^2 c_2 + b_1 c_1 - a_3^2 b_3 - b_2^2 \\ D &= a_1 b_1 c_1 - a_3^2 b_1 + 2a_2 b_2 a_3 - a_2^2 c_1 - a_1 b_2^2 \end{aligned} \qquad (30)$$

Concerning the discrete system a function $g_{i,j}(t)$ can be written as:

$$g_{i,j}(t) = g(x_{1i}, x_{2j}, t) = III_{a,b} f(x_1, x_2, t) \qquad (31)$$

where:

$$x_{1j} = j \cdot a, \quad x_{2j} = j \cdot b,$$

$$III_{a,b} = \sum_{n,k=-\infty}^{n,k=+\infty} \delta(x_1 - na)\,\delta(x_2 - kb)$$

(Dirac comb) and δ is the Dirac function. The discrete Fourier transform of function $g_{i,j}(t)$ is denoted as:

$$g_{i,j}(t) \rightleftharpoons G(\theta_1, \theta_2, \nu) \qquad (32)$$

and consequently:

$$g_{i\pm1, j\pm1}(t) \rightleftharpoons e^{\mp i\theta_1 \mp i\theta_2} G(\theta_1, \theta_2, \nu) \qquad (33)$$

Similarly to the continuous case, the dispersion function is determined by applying the Fourier transform to the discrete system of Eqs. (5)-(7). This leads to the following homogeneous system of equations:

$$\begin{pmatrix} a_1+Y & \mathrm{i}a_2 & \mathrm{i}a_3 \\ -\mathrm{i}a_2 & b_1+b_3Y & b_2 \\ -\mathrm{i}a_3 & b_2 & c_1+c_2Y \end{pmatrix} \begin{pmatrix} W(\theta_1,\theta_2,\nu) \\ \Theta(\theta_1,\theta_2,\nu) \\ \Omega(\theta_1,\theta_2,\nu) \end{pmatrix} = \mathbf{0} \tag{34}$$

This system has the same form as the one derived from the continuum model and it similarly possesses non-trivial solutions when its determinant vanishes:

$$\det \begin{pmatrix} a_1+Y & \mathrm{i}a_2 & \mathrm{i}a_3 \\ -\mathrm{i}a_2 & b_1+b_3Y & b_2 \\ -\mathrm{i}a_3 & b_2 & c_1+c_2Y \end{pmatrix} = 0 \tag{35}$$

where

$$\begin{aligned} a_1 &= 4c_Q\cos\theta_1\left(\cos\theta_1+\cos\theta_2\right)-8c_Q \\ a_2 &= -2c_Q a\sin\theta_1\left(2\cos\theta_1+\cos\theta_2\right) \\ a_3 &= 2c_Q b\cos\theta_1\sin\theta_2 \\ b_1 &= \left(4c_T-a^2c_Q\right)\cos\theta_1\cos\theta_2-4a^2c_Q\cos^2\theta_1-4c_M\sin^2\theta_1-a^2c_Q-4c_T \\ b_2 &= bac_Q\sin\theta_1\sin\theta_2 \\ b_3 &= J_y/m \\ c_1 &= \left(4c_M-b^2c_Q\right)\cos\theta_1\cos\theta_2-4c_T\sin^2\theta_1-b^2c_Q-4c_M \\ c_2 &= J_x/m \end{aligned} \tag{36}$$

and

$$Y = m\left(\frac{\theta_1 V_1}{a}+\frac{\theta_2 V_2}{b}\right)^2 \tag{37}$$

Y is the solution of the following characteristic polynomial equation:

$$AY^3+BY^2+CY+D=0 \tag{38}$$

where the coefficients are:

$$\begin{aligned} A &= b_3c_2 \\ B &= a_1b_3c_2+b_3c_1+b_1c_2 \\ C &= a_1b_1c_2+a_1b_3c_1-a_2^2c_2+b_1c_1-a_3^2b_3-b_2^2 \\ D &= a_1b_1c_1-a_3^2b_1+2a_2b_2a_3-a_2^2c_1-a_1b_2^2 \end{aligned} \tag{39}$$

Eq.(37) is formally the same as Eq.(29).

6. VALIDATION OF THE COSSERAT CONTINUUM - HOMOGENISED vs. DISCRETE MODEL

In order to compare the dynamic response of the discrete and the continuous systems, Eq.(29) and Eq.(37) are solved. The following dimensionless quantities are defined:

dimensionless wave velocity: $$V = \sqrt{\frac{1}{c_Q}\frac{Y}{\theta_1^2+\theta_2^2}} \tag{40}$$

phase velocity: $$\omega = \sqrt{\frac{Y}{m}} \tag{41}$$

dimensionless wave lengths: $$w_1 = \frac{2\pi}{\theta_1},\; w_2 = \frac{\pi}{\theta_2} \tag{42}$$

The wave propagation is analysed successively in the x_1 (θ_2=0) and the x_2 direction (θ_1=0). The dimensionless wave shear velocity versus the dimensionless wavelength for the Discrete and the Cosserat models are presented in Figures 3 and 4. In Figures 5 and 6 the phase velocity for the three different modes corresponding to the three real roots of the corresponding characteristic Eqs.(29) and (38) are depicted. It is observed that the Cosserat model gives a good approximation for wavelengths bigger than five times the size of the block ($\theta_2 < 0.6$).

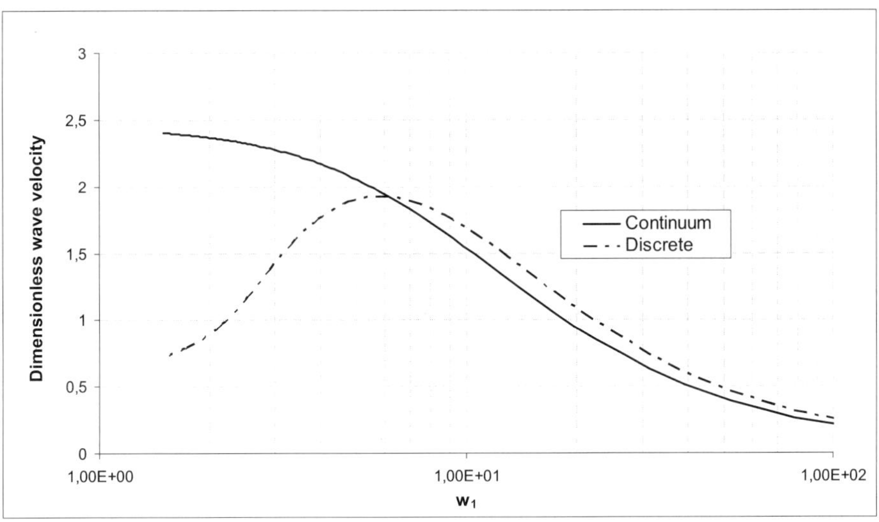

Figure 3. Dimensionless wave velocity for the Discrete and the Cosserat models (c_N=5c_Q, b=a) for wave propagation in x_1-direction.

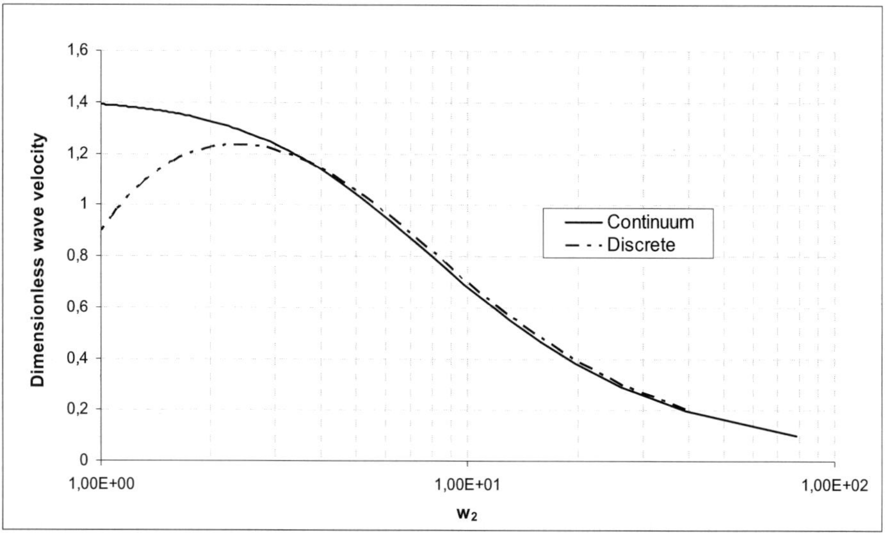

Figure 4. Dimensionless wave velocity for the Discrete and the Cosserat models (c_N=5c_Q, b=a) for wave propagation in x_2-direction.

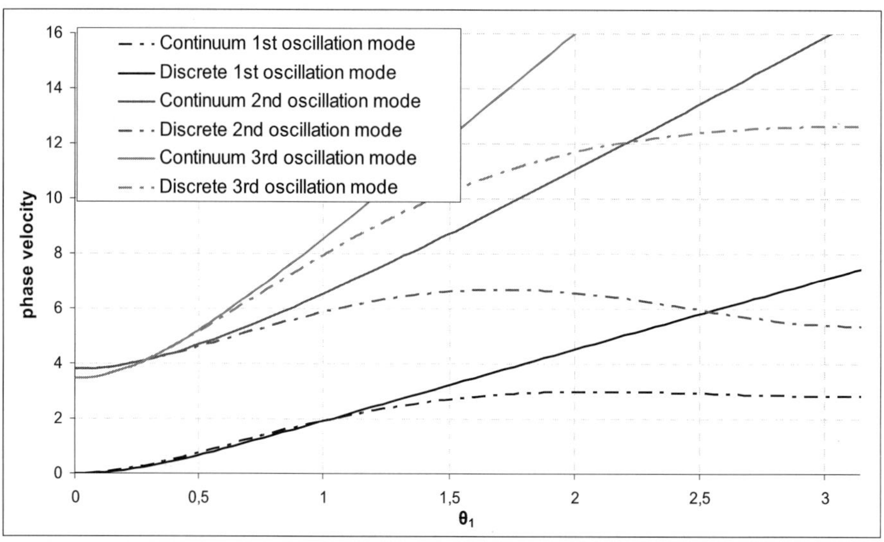

Figure 5. Phase velocity for the Discrete and the Cosserat models (c_N=5c_Q, b=a) for wave propagation in x_1-direction.

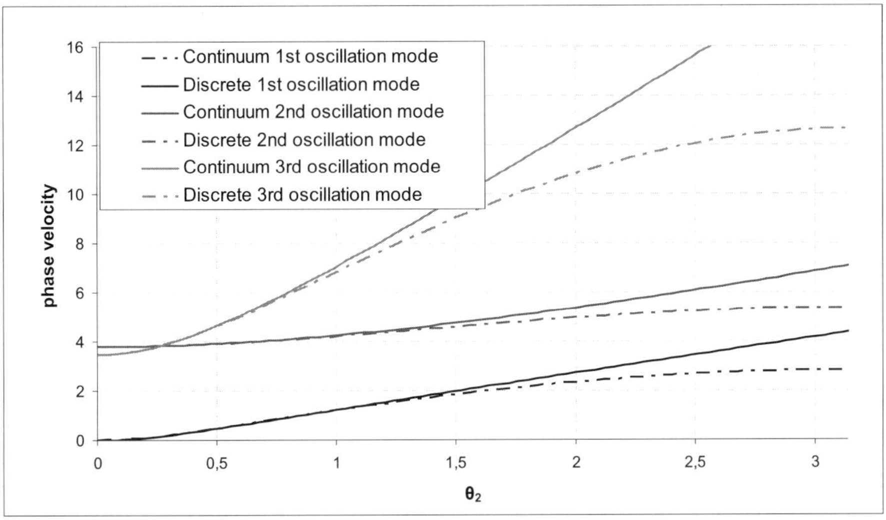

Figure 6. Phase velocity for the Discrete and the Cosserat model (c_N=5c_Q, b=a) for wave propagation in x_2-direction.

7. CONCLUSIONS

When dealing with blocky or layered structures or more generally with any structure where inhomogeneities are visible, one can address the question of modelling the behaviour of such a structure either by considering each inhomogeneity individually and solving the problem as in the distinct or discrete-element methods, or by considering the salient features of the discontinuum within the framework of generalised continuum theory.

The three-dimensional dynamic behaviour of a wall was studied here. Both the discrete and the 3D Cosserat model of the aforementioned structure are described, similarly to plate theory, by two uncoupled sets of equations, which correspond to the in- and to the out-of-plane motion of the blocky structure. The in plane motion of the wall was previously studied by Sulem and Mühlhaus[2]. Concerning the out of plane wall behaviour, the homogenisation procedure leads to similar conclusions with the in-plane analysis of the structure. The Discrete and the 3D Cosserat approaches coincide for wavelengths five times bigger the size of the block. However, the Cosserat model becomes increasingly inaccurate for smaller wavelengths. Generally one could assert that the Cosserat model appears to be the natural starting point for the development of continuum models for blocky structures. In the future, the derived continuous model is going to be extended in the plasticity domain. Multi-yield plasticity criteria will be developed and applied to express and simulate the interblock sliding, tilting and twisting failure modes.

ACKNOWLEDGMENTS

This research is sponsored by the General Secretariat for Research and Technology in Greece and the French Ministry of Foreign Affairs in the frame of the bilateral S & T cooperation between the French and Hellenic Republic (2005-2007): "Nouvelles méthodes d'analyse numérique du comportement mécanique des monuments anciens - Application à l'Acropole".

REFERENCES

1. H.-B. Mühlhaus, Continuum models for layered for layered and blocky rock, *Comprehensive Rock Eng. Pergamon Press* **2**, 209-230 (1993).
2. J. Sulem and H.B. Mühlhaus, A continuum model for periodic two-dimensional block structures, *Mechanics of cohesive-frictional materials* **2**, 31-46 (1997).
3. M. Cerrolaza, J. Sulem and A. El Bied, A Cosserat non-linear finite element analysis software for blocky structures, *Int. J. of Advances in Eng. Soft* **30**, 69-83 (1999).
4. J. Sulem, H.B. Mühlhaus and P. Untereiner, Discrete and continuous models for dry masonry columns, *Journal of engineering mechanics* (1997).
5. I. Vardoulakis and J. Sulem, Bifurcation Analysis in Geomechanics (Blackie Academic and Professional, Glascow, 1995).
6. R.D. Mindlin, Micro structure in Linear Elasticity, *Arch. Ration. Mech. Anal.* **16**, 51-78 (1964).
7. A.-S.-J. Suiker, A.-V. Metrikine and De Borst R., Comparison of wave propagation characteristics of the Cosserat continuum model and corresponding discrete lattice models, *International Journal of Solids and Structures* **38**, (2001).
8. E. Pasternak and H.B. Mühlhaus, Generalised homogenisation procedures for granular materials, *Journal of Engineering Mathematics* **52**, 199–229 (2005).

Chapter 3.2

MECHANICAL BEHAVIOUR OF MASONRY STRUCTURES STRENGTHENED WITH DIFFERENT IMPROVEMENT TECHNIQUES

Maria Rosa Valluzzi[1] and Claudio Modena[2]
[1]*Department of Architecture, Urbanism and Survey, University of Padova, via Marzolo 9, 35131 Padova, Italy;* [2]*Department of Construction and Transportation Eng, University of Padova, via Marzolo 9, 35131 Padova, Italy*

Abstract: The paper describes the main repair and strengthening techniques proposed to preserve historic masonry buildings subject to severe loading conditions. Based on experimental and modeling studies and on real cases applications, design and feasibility problems, execution and efficiency aspects of the interventions are discussed.

Key words: historic masonry; seismic improvement; strengthening and repair; injections; bed joint reinforcement technique; compatibility; FRP.

1. INTRODUCTION

A large number of materials and application techniques are available, and new ones are continuously being developed to strengthen and repair historic masonry buildings. The more recent developments mainly focus on finding appropriate solutions for satisfying at the same time conservation requirements, structural safety prescriptions and the specific, very variable structural problems of every single historic building. As far as this last issue is involved it must be pointed out that the same material and the same technique can give completely different results in terms of efficiency and type of structural behaviour. In such a content, the main factors involved are: the type of masonry (e.g. regular brick or rubble stone), the structural component (wall or vault), the combination of structural solutions (e.g. well connected or loose walls, floors made of masonry vaults or timber beams) and of actions (particularly in seismic or not seismic situations).

S.K. Kourkoulis (ed.), Fracture and Failure of Natural Building Stones, 137–156.

As a result, continuous effort is required to assess the real feasibility and efficiency of each technique in different application conditions and to give simple but reliable design rules.

The present paper briefly describes the researches that have been conducted at the University of Padova (based both on in situ and laboratory tests) on most of these topics regarding both traditional and innovative materials and solutions, under the basic assumption that all can give useful contributions if appropriately designed and applied.

Both stone and brick masonry structures are considered, with particular emphasis on the most severe loading conditions which they can be subjected to during their life (e.g. seismic actions, mainly for stone buildings and creep phenomena for heavy brick constructions). Case studies are also presented showing the major steps of the design process and the practical application of the research results.

2. STRENGTHENING OF MASONRY BUILDINGS IN SEISMIC AREAS

Stone masonry buildings are very common in many Mediterranean and North-European historical centres, often in seismic areas, and represent a typology particularly sensitive to horizontal load effects. In fact, they are often characterized by the use of poor materials and the construction of an irregular morphology (often revealing multi-leaf sections), affected by the presence of voids; moreover, they frequently show a general scarce collaboration among structural components (walls and floors or roof), which can led to particularly brittle collapse mechanisms (detachment of the layers and out-of-plane expulsions of material) (Giuffrè, 1993; Doglioni et al., 1994; Giuffrè and Carocci, 1999). Under these conditions, stone masonry constructions are particularly vulnerable under seismic loads, which basically represent the most severe actions for them.

From a structural point of view, it is well known that the proper behaviour under seismic loads is performed by constructions able to activate the so-called "box-behaviour". It concerns the presence of effective connections between bearing walls, between walls and horizontal components (floors and roof), and the contribution of "sufficiently" stiff floors or roof (Tomazevic, 1999).

At a global level, the main structural problems of masonry buildings under seismic actions are due to:

- the lack of connections among structural components (wall to wall, walls to floors and roof),
- the presence of horizontal structures (floors and roof) having scarce in-plane stiffness,

- the irregular morphology, due to continuous modification, stratification and superposition that have occurred during time.

At the local level, additional vulnerability is due to the peculiarity of the multi-leaf stone walls and, in particular, mainly to:

- the weakness of the internal core,
- the lack of interlocking among the layers.

The limits in the knowledge of the mechanical behaviour of existing masonry structures under seismic actions were clearly evidenced by the effects of the last seismic events that occurred in Italy, in particular the 1997 Umbria-Marche earthquake, which drew many experts from several countries. Despite the not particularly high magnitude (5.6 R), a severe damage scenario was observed: brittle out-of-plane mechanisms of collapses (partial or global overturning of façades or at the corners), local expulsions and losses of material were largely detected in the buildings (Figure 1) (Binda et al., 1999; Penazzi et al., 2001). A severe damage was also detected in buildings retrofitted with common upgrading interventions (as substitutions of timber floors and roofs with reinforced concrete slabs and hollow tiles, jacketing, etc…), systematic-cally applied after a previous earthquake incident in that area in 1979 (Figure 2). Such effects were due to the "hybrid" behaviour activated from the combination of new and old structures, unpredictable by the common available assessment procedures, based on the "a priori" assumption of the effecttive activation of the "box" behaviour. Moreover, many defects in the application of the consolidation techniques on walls (nets not sufficiently connected for the jacketing, unsuccessful injections of cement grouts not supported by proper investigations, etc.) were detected (Figure 3).

(a) (b) (c)

Figure 1. Out-of-plane mechanisms detected in the unretrofitted buildings: (a) overturning and collapse of the façade due to the lack of connection between floors and wall and among walls; (b) external layer expulsion due to the lack of connection of layers in the thickness of the wall; (c) collapse of the corner due to the excessive nearness of the openings.

Figure 2. Brittle mechanisms detected in the upgraded buildings: (a) collapse of the upper floor of a building due to the substitution of the timber roof with a heavier r.c. and hollow tiles mixed one, supported by poor masonry walls; (b) rear of the same building where a large overturning of the façade occurred, due to the high percentage of openings, combined with the presence of a heavy r.c. floor supported by a single layer of the poorly connected double-layer masonry; (c) detail of the eccentric loading on the wall.

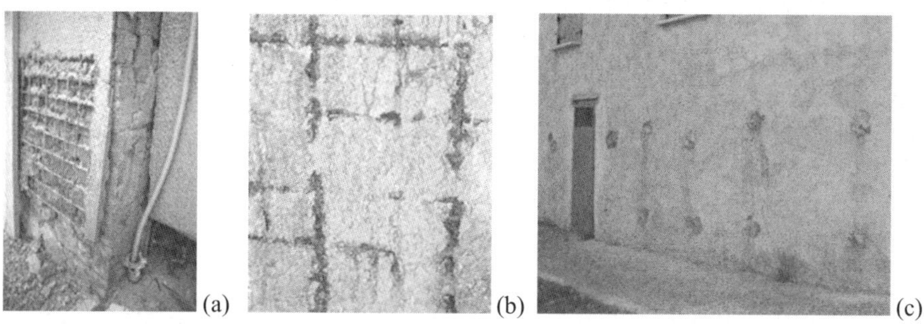

Figure 3. Effects of uncorrect design and execution of consolidation techniques in seismic zone: (a) loss of effectiveness of a wall strengthened with jacketing due to the scarce connection of the steel net through the thickness and the lack of overlapping of the reinforcement at the corners; (b) corrosion of the steel net; (c) leackages of grouts due to the uninjectability of a wall.

After this experience the need of suitable procedures for seismic assessment of masonry building and even whole centers was clear. Based on the extensive in situ observations a specific survey form and a matrix of the collapse mechanisms detectable in various types of buildings (single, row or complex aggregates) were set up (Binda et al. 2004). In this context, computational procedures based on the macro-modelling approach were calibrated and updated, in order to be able to interpret the real behavior of old buildings (Valluzzi et al., 2004; 2005) and to perform simulation of possible intervention. Finally, the rehabilitation of traditional techniques was mainly recognized as the most cautious way of intervention in order to avoid possible incompatibility with original material and structures.

2.1 Improvement of the structural global behavior

First, techniques able to improve the box behavior of the buildings should be preferred, as the local strengthening of the masonry is practically ineffective (or even worsening) when global out-of-plane collapse can be activated (Penazzi et al., 2001). Tying is one of the most effective interventions improving the collaboration among walls and avoid the trigger of out-of-plane mechanisms, even maintaining lighter timber floors and roof (Tomazevic, 1999) (Figure 4). In comparison with heavier substitutions with r.c. horizontal structures and the connected concrete tie-beams, or reinforced injections, metal ties are also easy to be removed and allow improving the seismic response without altering the original structural function (Figure 5).

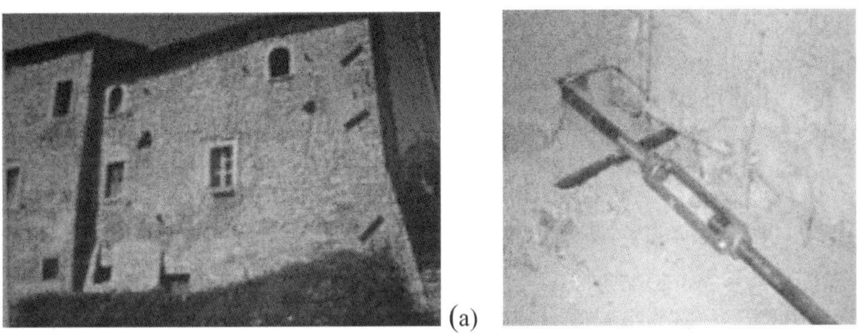
(a) (b)

Figure 4. Application of metal ties in old masonry buildings: (a) the presence of traditional ties and their regular distribution garanteed the optimal behavior during the 1997 Umbria-Marche earthquake of this building in its original condition; (b) application of a modern metal tie in an existing building.

(a) (b) (c)

Figure 5. Effects of concrete tie-beam and reinforced injections in stone masonry buildings: the contiguity of two different materials (original masonry and r.c.) provokes a damage in the continuity of the wall and espose the wall to brittle out-of-plane collapses (a) (b); reinforced injections ineffective to restrain the main façade of a church (c).

As for innovative materials, FRP (Fiber Reinforced Polymer), either in the form of laminates or rods, are recently considered also in restoration problems of monuments, due to their good mechanical properties and their ability to wrap structural components and even whole buildings (used as confinement). Nevertheless, regular surfaces (e.g. as in brick rather then stone masonry) are needed for the correct exploitation of the techniques, and durability aspects (resins are used at the interface with the substrate) should be carefully taken into account for real application on historic buildings or monuments. Mechanical performances related to the application of such materials will be shortly described in the following.

To contribute to the correct distribution of horizontal loads to the shear walls of the building, the in-plane stiffness of existing timber floors can be rehabilitated by using homogeneous materials together with the original ones. A strengthening technique has been recently experimentally set up at the University of Padova; it is performed by the placing of planks above the main beams of the floor and of one or more layers of boardings, positioned in different directions above the original extrados, connected by hardwood dowels, which has been proved to be of high efficiency with minimum obtrusiveness (Modena et al., 1998; 2004). The intervention is executed from the extrados of the floor, thus preserving possible precious intradoses, and enable the placing of systems (electrical, water, etc.) without impair walls at their base (Figure 6).

After stiffening, the resulting beam has a T-shaped section, whose web (original beam) and flange (new plank) are wooden made and a deformable connection between the flange and the web. The existing boarding, having a thickness of about 2-2.5 cm, which separates web and flange, is usually not considered as part of the compound section. The simplified design method for mechanically jointed beams with deformable fasteners was developed by

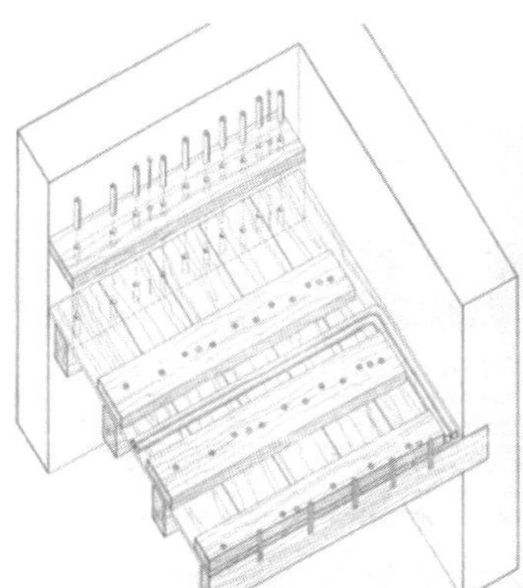

(a)

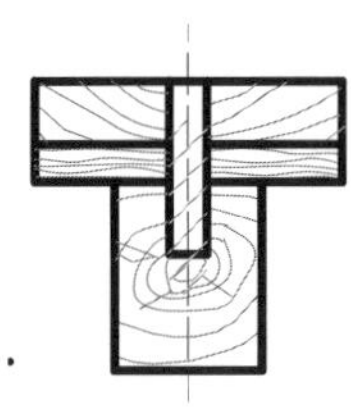

$$(EJ)_0 = E_1J_1 + E_2J_2$$
$$(\mathrm{EA})_\mathrm{S} = \mathrm{E_1A_1 + E_2A_2}$$
$$(\mathrm{EA})_\mathrm{P} = \mathrm{E_1A_1 \cdot E_2A_2}$$

$$(EJ)_{eff} = (EJ)_0 + \gamma \frac{(EA)_P}{(EA)_S} \cdot d^2$$

$$\gamma = \left(1 + \pi^2 \cdot \frac{(EA)_P}{(EA)_S} \cdot \frac{s_i}{K_i \cdot l^2} \right)^{-1}$$

(b)

Figure 6. Drawing of the stiffening intervention by using dry hardwood pins to connect planks and/or boardings to the original timber floors (a) and equations for calculation of the compound T-section (b).

by Möhler (1956) and assumed by the Eurocode 5 (2003). The compound section is characterized by the effective bending stiffness *(EJ)*$_{eff}$ as in the equations of Figure 6, where subscripts 1 and 2 refer to the stiffening plank and the existing beam respectively, d is the distance between the centroids of plank and beam, γ is the coefficient of connection efficiency ($0<\gamma<1$, 0 and 1 refer to infinitely deformable and rigid connection, respectively), l is the beam length, s_i is the dowels' spacing and K_i is the value of instantaneous slip modulus. This last parameter is conventionally estimated for the serviceability limit state (K_s) and for the ultimate limit state (K_u) from the experimental load-displacement curve obtained by push-out tests (UNI-EN 26891) (Figure 7).

Several experimental campaigns were carried out (1997-2004), testing various types of connection (wood or steel dowels, screws or rivets, also in combination among them), having different length, and without or in presence of different types of glue (vinyl, melamine, etc.) (Modena et al, 1997; 2004). Results showed that the best performance is given by dry wooden dowels having medium diameters (22-26 mm) and length of 130 mm, to which screws and a very thin layer of vinyl glue can give further support during the assembly phase (Figure 8). It is worth to remark that dry connections are more easily removable and therefore they represent the most suitable solution to the conservation criteria.

After experimental validation, the proposed strengthening technique was applied in the framework of some conservation work on the fifteenth-century complex Cà Duodo (Monselice, Padova, Italy), a relevant example of Venetian late gothic style (Modena et al., 1998). The building has three floors, having a relatively valuable intrados with tempera paintings, thus only the extrados was basically accessible (Figure 9). The strengthening intervention was carried out on all floors (first and second), for a total surface of 553 m^2. The 240 red pine beams presented a slender section (about 11x24 cm) and a

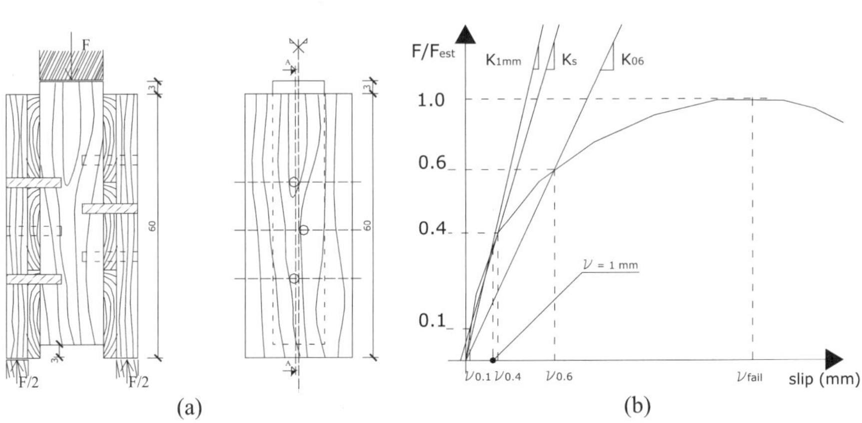

Figure 7. Push-out tests on compound sections: samples (a) and experimental curves for the definition of the slip moduli (b).

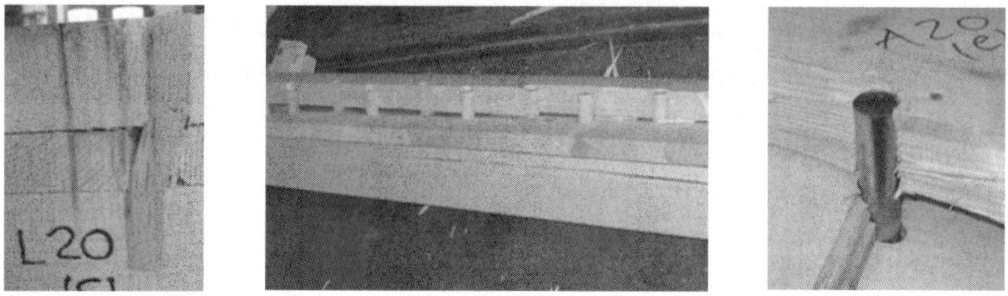

(a) (b) (c)

Figure 8. Results on application of various connections: rupture of dowel due to too small diameter (20 mm) (a); optimized diameter (22-26 mm) allow a combined shear-flexural behavior, avoiding both rupture of pin and damage of plank and beam, as checked after a bendng test (b); greatest wooden diameter (29 mm) or metal pins (even for small diameters, e.g. 12 mm) reveal to be too stiff and provoke damage of both existing and reinforcing materials (c).

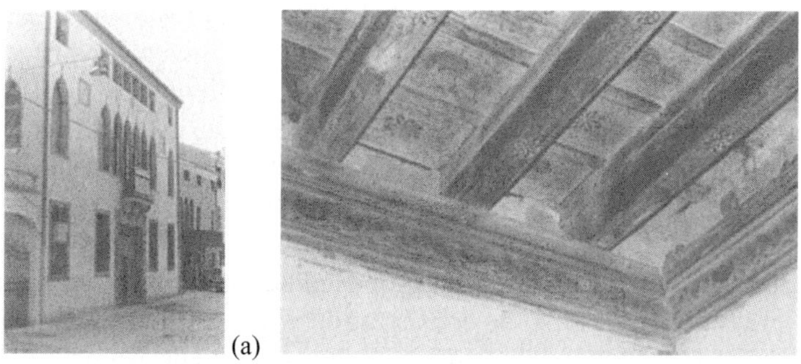

(a) (b)

Figure 9. View of palazzo Ca' Duodo in Monselice (Padova) (a) and of the intrados of the floors (b).

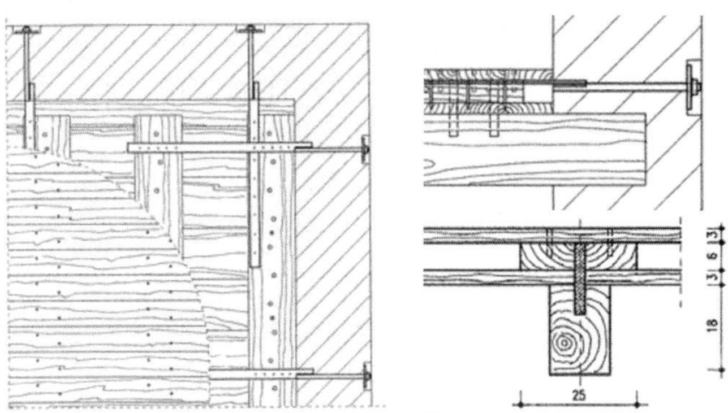

Figure 10. Sketches of the intervention executed in Ca' Cuodo: Drawing of the stiffening intervention by using dry hardwood pins to connect planks and/or boardings to the original timber floors, and details of the floor-to-walls connection (quotes in cm).

length of an average of 480 cm, spaced of about 56 cm. Preliminary investigations confirmed that the wood was in good condition, also in proximity to the bearing points on the walls.

The connections between beams and planks were made by hard-wood (beech-tree) dowels with diameter of 22 mm coupled with screws to facilitate the assembly phase. The longest beams (560 cm of span) were strengthened with 8x25 cm planks and 30 beech pins spaced from 11 to 30 cm, whereas for the shortest ones (430 cm of span) 5x25 cm planks 24 dowels were used (see details in Figure 10).

The stiffening planks were sawn before being placed on the boarding and their intrados were leveled in order to adhere perfectly to the existing surface. Planks were preliminarily fixed to the existing boarding with screws, in order to facilitate the following intervention phases; in case of no propping, they can also improve the adhesion between the straight planks and the often permanently deflected beams of existing floor. The position of the dowels is staggered of about half diameter from the longitudinal beam axis and they are not aligned in order to avoid longitudinal splitting. The pre-bored holes are about 1-2 mm smaller than the diameter of the pins and are cleaned with compressed air before forcing the dowels using a simple hammer (Figures 11(a,b)). It is also possible to use planks shorter than the floor span, avoiding their insertion into the load bearing walls. Such solution, still being statically admissible, allows placing all the technical installations in the floor thickness (Figure 11c) avoiding the opening of dangerous horizontal chases in the walls. The placement of installation and/or thermal and sound insulating panels in the empty space between adjacent planks can complete the intervention, if required.

2.2 Strengthening techniques for walls

Once the global behavior has been improved and the possibility of brittle out-of-plane effects has been eliminated or minimized, it is possible to increase the mechanical strength of shear walls, in order to raise the ultimate load. Among several solutions, injections are particularly able to consolidate the internal core of rubble multi-leaf stone masonry walls and improve the connection among the leaves (Valluzzi et al., 2004), with more mechanical

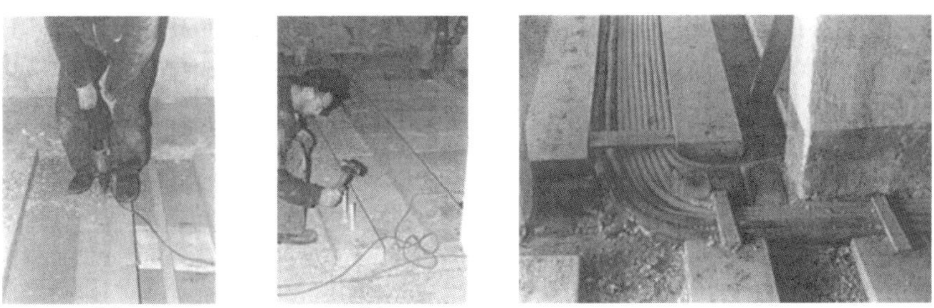

(a) (b) (c)

Figure 11. The phases of the intervention: fixing of the dowels on the strengthening planks (a) and detail of the technical installations placed between two planks (b).

compatibility than jacketing (Modena et al., 1998a). Most specific problems related to the injections are connected to the uncertainties of the masonry (features and composition of the section, size and distribution of voids, absorption of the mortar, etc.), especially in the case of rubble stone multi-leaf walls. Therefore, the correct application of the intervention requires: ((i) the investigation of the wall section in order to check its injectability, (ii) the selection of grouts with appropriate properties and adequate compatibility (chemical, physical and mechanical) with the original masonry, (iii) the proper procedure of injection itself (choice of the number and distribution of the injection holes, injection pressure, times of execution), (iv) the control of the effectiveness of the consolidation by using specific investigation techniques before, during and after the intervention (Berra et al., 1992; Binda et al., 1993; Valluzzi et al., 2003; da Porto et al., 2003). A comprehensive research conducted at the University of Padova demonstrated that low-strength grouts (e.g. natural hydraulic-lime based) can provide a full compatibility (chemical, physical and mechanical) with the original materials; moreover, they allow making more reliable analytical formulations for the prediction of the mechanical strength of the injected walls (Valluzzi et al., 2004) (Figure 12).

A significant result was that the use of high strength grouts (compared to the original strength of the walls) can influence the strength of the injected infill, but it is not able to increase with the same contribution the final strength of the wall (see Figure 12, where symbols are as follows: $f_{wc,s}$ and $f_{wc,0}$ are the compressive strength of the original and of the injected wall respectively, f_{gr} and f_{cyl} are the compressive strength of the grout and of the injected cylinders, which is assumed comparable to the strength of the consolidated infill; V_{inf} is the volume of the core). For real cases, prediction of strength can be approximated by assuming $f_{wc,0}$ from a double flat jack test and by simply measuring dimensions of walls and the compressive strength of the grout (taken as reference strength for the injection) by a common laboratory test.

Finally, to verify the efficiency of the techniques, the use of non destructive methods, as sonic pulse velocity and radar tests, can be successful, especially if in combination with inspections and endoscopies. An example of application of the sonic tomography on a vertical section of the curtain walls of Cittadella (Padova, Italy) before and after injections is given in Figure 13.

To improve the strength of the walls, deteriorated mortars can be replaced by selected proper materials with the deep repointing technique (involving at least 6-7 cm of the thickness) as it is shown in Figure 14. Combination of different techniques may be also adopted aiming to increase the performance of walls suffering various structural problems (Valluzzi et al. 2004).

Application of FRP laminates can improve the shear strength of the walls or consolidate arches and vaults (Figure 15) (Valluzzi et al., 2002; 2001). It requires regular surfaces (in order to reduce thickness of substrate at the interface) thus it is more suitable for brick walls. Some experimental results carried out at the University of Padova demonstrated the high effectiveness of

(a) (b)

Figure 12. Experimental wall ready to be injected and preliminary check of injectability (a); comparison between cylinders and wall increment of compressive strength, and analytical model for strength prediction after injection (b). The relation is reliable for low-strength grouts, e.g. for ratio $f_{gr}/f_{wc,0}$ not exceeding 4.

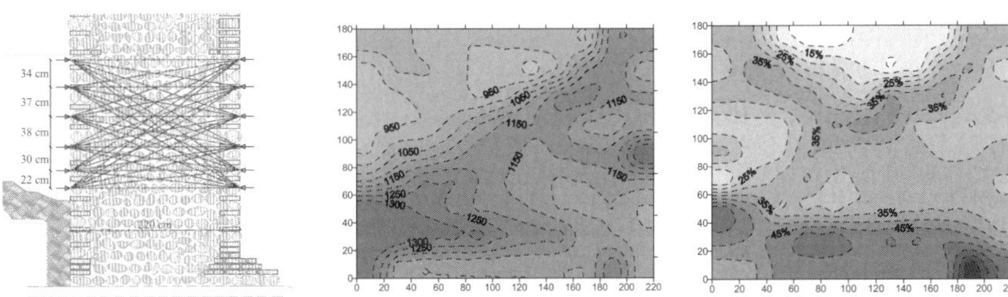

(a) (b) (c)

Figure 13. Results of the sonic tomography: (a) drawing of the wall section, map of velocity after injections (b) and their variation compared to the original condition (c) (darker colours corresponds to higher velocities).

(a) (b)

Figure 14. Deep repointing executed in an experimental wall: mechanical excavation of the joints (a) and phase of filling of the joints with new mortar (b) (Valluzzi et al. 2004).

medium-strength (and less expensive) products (e.g. glass fibers, instead of carbon). In particular, the use of three types of fibers (carbon, glass and polyvinyl-alcohol, labelled as CFRP, GFRP and PVA in the following) applied in different configurations (at one side or both sides) and geometry (as a squared grid or diagonally) on the faces of a total of 24 walls was considered (Valluzzi et al. 2002). Failure of the plain panels was due to brittle splitting (ultimate load of 101 kN) (Figure 15a). Splitting failure with a clear diagonal crack was also obtained for all single-side reinforced panels, whereas the ultimate load was in many cases lower than the reference one; their average value are: 91 kN for CFRP, 96 kN for GFRP and 101 kN for PVA for the grid arrangement, 116 kN for CFRP and 112 kN for GFRP for the diagonal one. The samples exhibited a clear bending deformation during the loading phases along the unstrengthened diagonal, due to the eccentricity of the reinforcement (Figures 15(b,c)). Among the on-side reinforced specimens, the diagonal strengthening configuration always revealed a higher effectiveness in comparison with the corresponding squared grid set up.

In the case of double-side strengthening the failure mechanism changed to the loss of collaboration between the reinforcement and the substrate, due to the detachment (peeling) of the superficial layer of the masonry (see Figures 15 (d,e)), particularly serious for the CFRP-reinforced panels, or the rupture of the FRP strips along the diagonal itself. Despite the lower strength, the advantages of grid reinforcement distribution was confirmed by the more spread crack patterns (see Figure 15f), whereas a clear splitting crack appeared in all the diagonally reinforced panels. The ultimate strength increase was anyway noticeable in almost all cases: 104 kN for CFRP, 115 kN for GFRP and 148 kN for PVA for the grid arrangement, 146 kN for CFRP and 175 kN for GFRP for the diagonal one.

The abovementioned behavior is also in agreement with numerical simulations, as shown in Figure 16. GFRP strengthening applied on both sides gave the best behavior, both in terms of strength and failure mode: the crack spreading provided sufficient signals of crisis far before collapse, indicating a less brittle global behavior. It is worth to remark that the use of those innovative materials should be particularly cautious in historical heritage, as incompatibility and durability problems can not be disregarded in such cases. Experimental validations are therefore fundamental in order to prevent possible alterations or damage.

3. STRENGTHENING OF MASSIVE MASONRY STRUCTURES SUBJECTED TO CREEP

Brick-masonry load bearing elements of heavy historic structures as towers, heavily loaded pillars, and large masonry walls (i.e. curtains), frequently exhibit very typical mechanical deterioration phenomena like:

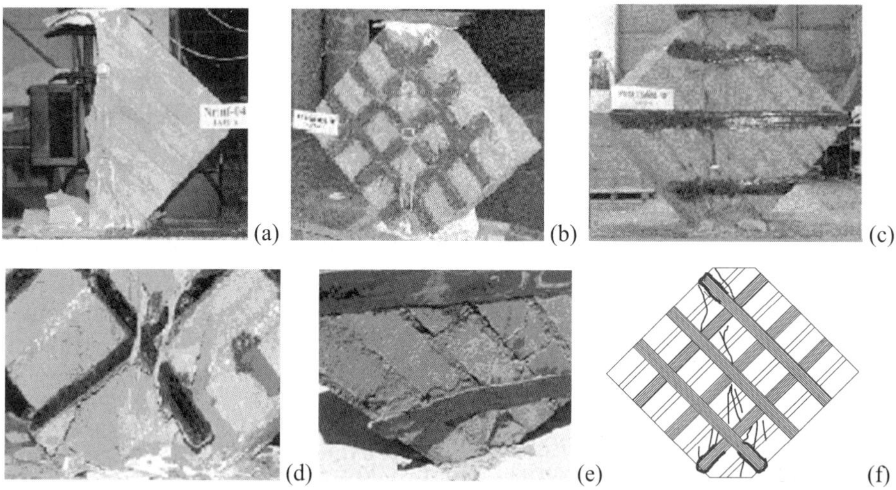

Figure 15. Mode of failure of the panels under diagonal loads: (a) reference panel (unstrengthened); (b) (c) one-side strengthened panels: (b) grid pattern, (c) diagonal pattern; double-side strengthened panels: detail of the peeling in the anchoring zone on grid (d) and diagonal (e) pattern; (f) crack distribution on grid pattern.

Figure 16. FEM simulation of distribution of stresses in the tested panels: (a) unreinforced sample (highest level of splitting stresses are concentrated in the core); (b) diagonal pattern reinforcement (tensile stresses spread from the core); (c) grid pattern reinforcement (stress peaks are shifted to anchoring areas).

- formation of vertical or sub-vertical, thin but very diffused cracks,
- b) more or less local detachment of the outer leaf in multiple leaf walls (Binda and Anzani, 1993; Anzani et al., 2000).

Such a particular crack pattern is often not attributable to common causes of damage like seismic events, foundation settlements, instantaneous increase of external loads (e.g. for added storey or building changing) or to chemical, physical and mechanical degradation of the basic materials. On the contrary, it is due to the prevalent effect of the dead load and the related time dependent phenomena, often combined with cyclic loads, like wind actions, thermal and hygroscopic excursions, or bell ring oscillations (in bell-towers) (Binda et al., 1991; 1992). This damage, generally disregarded in those structures, in comparison with other more evident critical conditions (large cracks, out-of-plumb, relevant deformations, etc.), can induce sudden and unexpected brittle collapse, as observed in several cases (e.g. the Civic Tower of Pavia, the Bell Tower of San Marco in Venice and the Cathedral of Noto in Italy; the Bell Tower of St. Magdalena in Goch in Germany) (Figure 17) (Binda et al., 1992; 2001), even for stress values lower (40% to 60%) than the strength of the masonry under short term static loads.

In the last years, several experimental campaigns jointly performed by the University of Padova and the Polytechnic of Milan, demonstrated that the creep damage can be effectively counteracted by minimal reinforcement of the superficial layer of the masonry, properly inserted into the bed joints. The "bed joint reinforcement" technique is executed by removing the mortar in the bed joint for few centimeters (about 5-7), by placing the reinforcement (small bars or plates) and by the final filling of the cut with a proper embedding material (Figure 18) (Binda et al., 1999a; Valluzzi et al., 2004a). Both-sides application is more effective, but also in case of single-face accessibility or large-section walls, the insertion of transverse pins in holes sealed with grouts are recommended. Moreover, the use of more compatible mortars (hydraulic lime-based) as embedding material can prevent incompatibility with original materials. The technique does not require particular skills and

Figure 17. Collapse of the Civic Tower of Pavia (1989) (a) and of the Cathedral of Noto (Siracusa, 1996) (b), Italy.

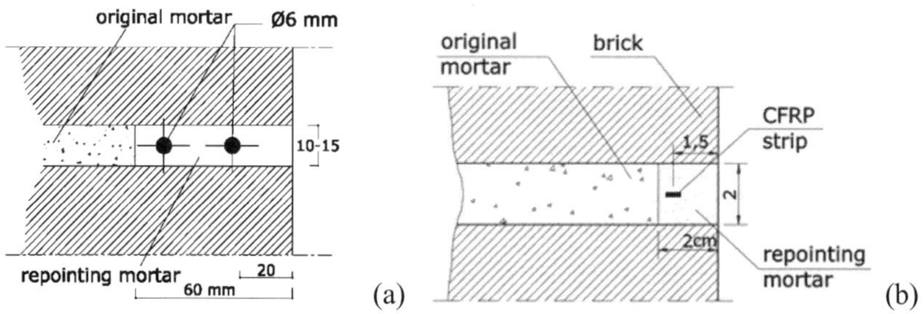

Figure 18. Sketch of the bed reinforcement technique with steel bars (a) and thin FRP strips (b).

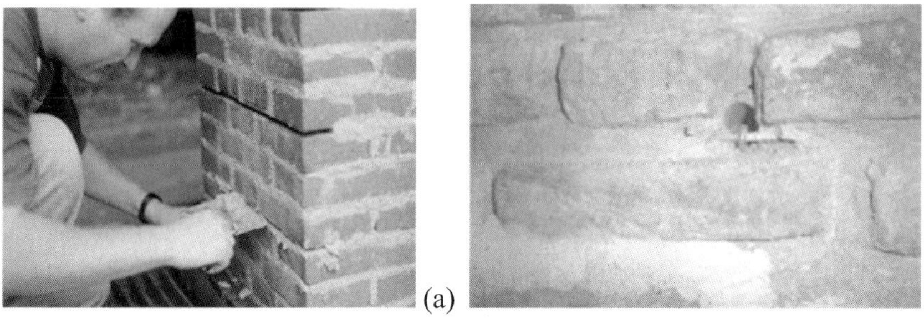
(a) (b)

Figure 19. Bed joint reinforcement technique: (a) detail of embedding phase in strengthened joint; (b) arrangement of bar and transverse tie into an excavated joint before final sealing of hole.

skills and tools during application; some care is required in some operative phases (cutting of the bed joint, cleaning, repointing) but it can be performed quite easily and quickly (Figure 19).

First experimental investigations (Modena et al., 2002; Valluzzi et al., 2004a) dealing with applications of low diameter (5 mm or 6 mm) stainless steel reinforced bars, showed the efficiency of the intervention in reducing the transverse dilation due to high compression stresses, as the bars bear the tensile stresses otherwise directed to the bricks. Therefore, even if a signify-cant increment in strength is not detectable, the technique is able to act direct-ly on the limitation of the cracks development, thus improving the perform-ance and the safety of the structure.

Further investigations were focused on identifying the possible application of FRP materials, aimed at checking, by laboratory tests, both advantages and disadvantages in their use, before any possible in-situ application. Monotonic, cyclic and creep simulating tests were performed first on panels strengthened with circular section carbon bars (CFRP) (5 mm in diameter) (Valluzzi et al., 2003) and then, recently, with thin rectangular section CFRP strips (1.5x5 mm) (Figure 21) (Saisi et al., 2004; Valluzzi et al., 2005a).

The advantages of the use of carbon fibers instead of steel reinforcement are mainly related to their complete corrosion immunity, but many aspects still need to be deeply investigated. Despite their high strength, FRP are very brittle and inductile to bending and folding (e.g. for anchorage); they are sensitive to high temperatures and constitutive laws able to describe the interface behaviour among reinforcement-mortar-brick are not comprehensively known in masonry yet.

Thin strips have more advantages in comparison with circular bars, due to their higher flexibility to the unevenness of the joints and the better behaviour against splitting failures, which lead to possible more superficial and less obtrusive interventions (see Figures 18 and 20).

Figure 21 shows the vertical and the horizontal deformation vs. stress diagrams of the creep tests on experimental panels. It can be noticed that the reinforcement has not great influence on the strength and the vertical deform-

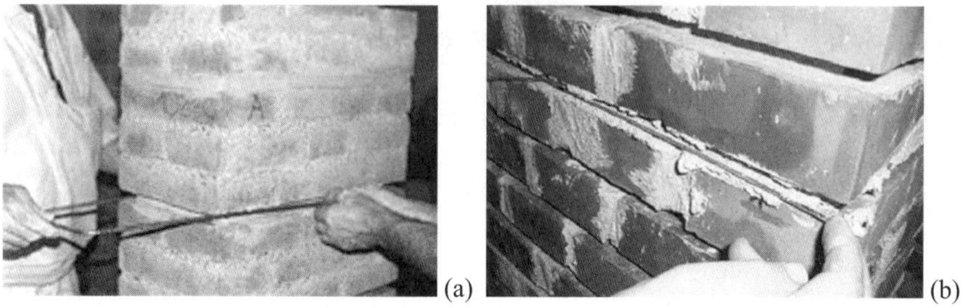

Figure 20. Example of panels strengthened with different reinforcements: (a) steel bars; (b) thin CFRP strips.

ations, whereas a great role is attributed to the control of the horizontal deformations. In particular, the reinforcement was able to reach the tertiary creep conditions at deformations around the 70% of the original case; moreover, in such ultimate phase, the prisms reinforced by the technique D, showed also an increment of about the 25% in the strength.

It is worth to say that, despite a number of researches and efforts devoted to FRP materials, until now real applications have been only performed with stainless steel bars in combination with lime-based repointing mortars, as they can guarantee effectiveness and durability, even in exceptional conditions not easy to predict.

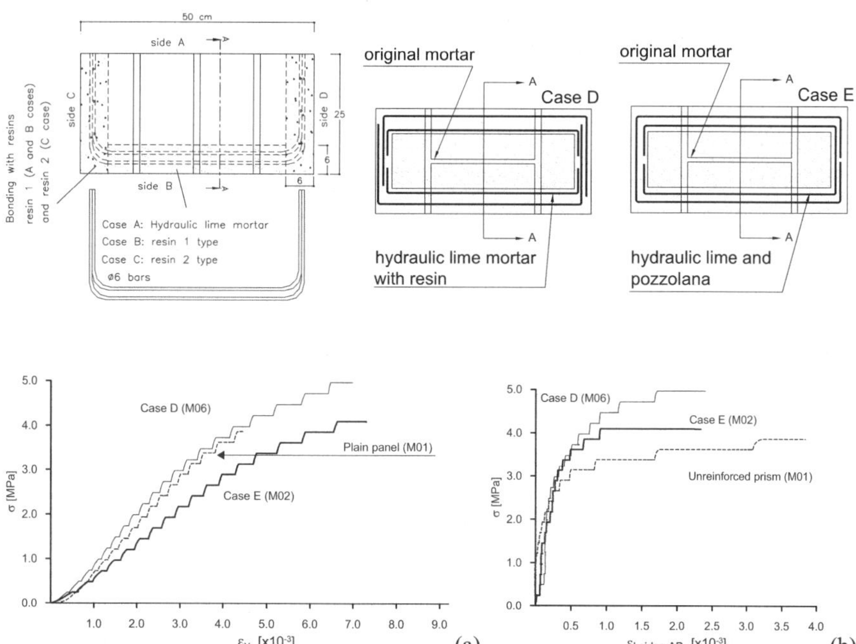

Figure 21. Draws of the laboratory samples (sections) and creep test results: vertical (a) and horizontal (b) deformations of the panels.

3.1 Interventions on case studies

The experimental validation of the performance of the bed reinforcement technique allowed proposing it for the consolidation of several historic constructions and monuments at risk in Italy, in order to counteract the dilation under high compressive stresses by improving the material toughness. It is worth to remark that, in real cases, more damage typologies can coexist, thus the proper selection and combination of different intervention techniques should be considered. In many cases, both injections and partial and limited rebuilding (the traditional "scuci-cuci") are used, to reduce the stress concentration and to replace the most damaged resistant parts, respectively. These techniques act locally by improving the mechanical behaviour of the material, which contributes to the rehabilitation of the proper load bearing capacity of the structure. In the following, some representative case studies of towers and structural components of churches in Italy are depicted (see Figures 22-24).

As suggested by the laboratory experience, both in repair and strengthening conditions, the bed joints reinforcement technique is easily and quickly performed; moreover, aesthetic aspects of the façade can be improved or maintained, depending on the specific conservation requirements.

4. CONCLUSIONS

Different studies on the application of repair and strengthening techniques on historic masonry structures in severe loading conditions have been presented. In seismic areas, the rehabilitation of masonry building has to be performed first at global level, by favouring the collaboration of the structural components (walls, floors and roof). Tying and the use of compatible tech-

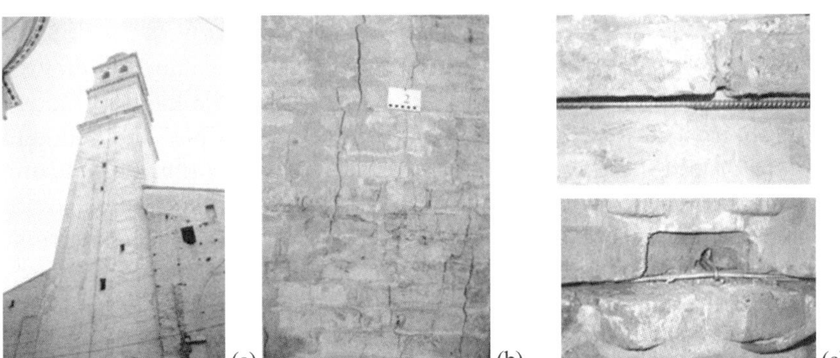

Figure 22. St. Giustina bell tower (Padova): general view (a), typical creep crack (b), application of the technique (c).

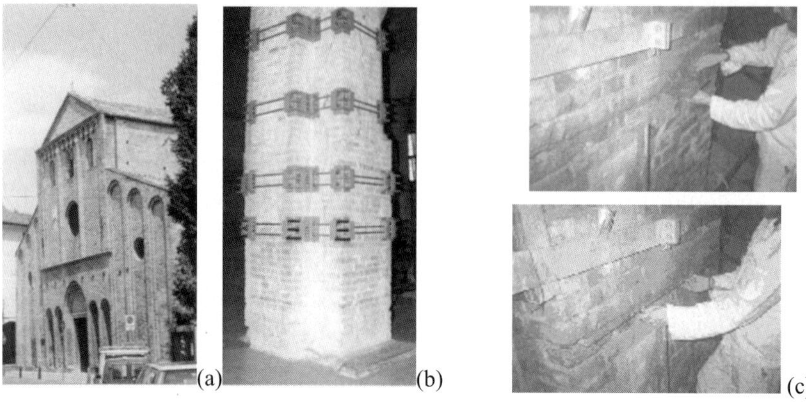

Figure 23. St. Sofia church pillars (Padova): general view of the church (a), provisional measures on a cracked pillar (b), application of the technique (c).

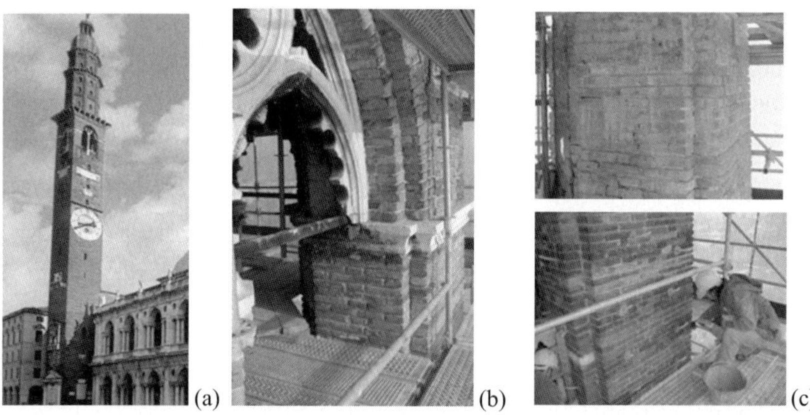

Figure 24. Civic tower of Vicenza: general view (a), cracked pillar of belfry (b), joints strengthened by structural repointing and combination with injections (c).

niques for the stiffening of floors and roof can represent a solution of minimum intervention, in order to preserve as much as possible the historical identity of cultural heritage. Strengthening of walls by injections and repointing requires a proper design of grouts, with particular attention to compatible binders. In particular, for grout injections, injectability on site and laboratory tests are recommended. The importance of the design and of the site control of the intervention has been pointed out for all the mentioned techniques, as well as the necessity of using appropriate preliminary diagnostic procedures for the effectiveness of the intervention.

The progressive damaging of masonry structures subjected to high long term compressive loads can be effectively counteracted by the insertion of small diameter bars into the bed mortar joints. Laboratory experiments simulating both monotonic and creep loads showed a significant reduction in the horizontal dilation and a consequent minor diffusion of the cracks.

FRP materials should be considered with caution in such ambit, but have a great potential in improving the mechanical strength of masonry, provided that experimental validation is available before real applications.

REFERENCES

Berra, M., Binda, L., Anti, L., Fatticcioni A., 1992, Non destructive evaluation of the efficacy of masonry strengthening by grouting techniques, Proc. of the International Workshop, Milan, Italy, pp. 63-70.

Binda, L., Modena, C., Saisi, A., Tongini Folli, R., Valluzzi, M.R., 2001, Bed joints structural repointing of historic masonry structures, 9th Canadian Masonry Symposium 'Spanning the centuries', New Brunswick, Canada, (on CD-ROM).

Binda, L., Modena, C., Valluzzi, M.R., Zago, R., 1999a, Mechanical effects of bed joint steel reinforcement in historic brick masonry structures, 8th SF&R International Conference, London, U.K., (on CD-ROM).

Binda, L., Anzani A.. 1993, The time-dependent behaviour of masonry prisms: an interpretation, The Masonry Society Journal, 11(2):17-34.

Binda, L., Anzani, A., Gioda, G., 1991, An analysis of the time-dependent behaviour of masonry walls, 9th International Brick/Block Masonry Conference, Berlin, Germany, vol. 2, pp. 1058-1067.

Binda, L., Cardani, G., Saisi, A., Modena, C., Valluzzi, M.R., Marchetti, L., 2004, Guidelines for restoration and improvement of historical centers in seismic regions: the Umbria experience, Proc. of SAHC2004: IV Int. Seminar on Structural Analysis of Historical Constructions - possibilities of experimental and numerical techniques, Padova, Italy, pp. 1061-1068.

Binda, L., Gambarotta, L., Lagomarsino, S. & Modena, C., 1999, A multilevel approach to the damage assessment and seismic improvement of masonry buildings in Italy, in: Seismic Damage to Masonry Buildings, A. Bernardini Ed., Balkema, Rotterdam.

Binda, L., Gatti, G., Mangano, G., Poggi, C., Sacchi Landriani, G., 1992, The collapse of the Civic Tower of Pavia: a survey of the materials and structure, Masonry International, 6(1):11-20.

Binda, L., Modena, C., Baronio, G., 1993, Strengthening of masonries by injection technique, 6th North American Masonry Conf., Vol. I, Philadelphia, U.S.A., pp. 1-14.

da Porto, F., Valluzzi, M.R., Modena, C., 2003, Performance assessment of different consolidation techniques for multi-leaf stone masonry walls, Proc. of International Conference on the Performance of Construction Materials in the New Millennium, Il Cairo, Egypt, pp. 403-412.

Doglioni, F., Moretti, A., Petrini, V., 1994, Le chiese e il terremoto, LINT, Trieste, Italy.

Eurocode 5, 1995, Design of timber structures.

Giuffrè, A., 1993, Sicurezza e conservazione dei centri storici. Il caso Ortigia, Laterza, Bari, Italy.

Giuffrè, A., Carrocci, C., 1999, Raccomandazioni per la sicurezza e la conservazione del centro storico di Palermo, Laterza, Bari, Italy.

ICOMOS/ISCARSAH, 2001, Reccomandations for the analysis, conservation and structural restoration of Architectural heritage, Paris, Sept. 2001.

Modena, C., Tempesta, F., Tempesta, P., 1997, Una tecnica a secco basata sull'uso del legno per l'adeguamento statico di solai in legno, L'edilizia 11/12: 23-32.

Modena, C., Tempesta, F., Tempesta, P., 1998, Una tecnica a secco basata sull'uso del legno per l'adeguamento statico di solai in legno, L'edilizia 3/4: 38-45.

Modena, C., Valluzzi, M.R., Garbin, E., Da Porto, F., 2004, A strengthening technique for timber floors using traditional materials', Proc. of SAHC2004: IV Int. Seminar on Structural Analysis of Historical Constructions - possibilities of experimental and numerical techniques, Padova, Italy, pp. 911-921.

Modena, C., Valluzzi, M.R., Tongini Folli, R., Binda, L. 2002, Design choices and intervention techniques for repairing and strengthening of the Monza cathedral bell-tower, Construction and Building Materials, Special Issue, Elsevier Science Ltd., 16(7):385-395.

Modena, C., Zavarise, G., Valluzzi M.R., 1998a, Modelling of stone masonry walls strengthened by r.c. jackets, 4th International Symposium on Computer Methods in Structural Masonry, Pratolino, Florence (Italy), vol. 4, pp. 285-292.

Möhler, K., 1956, Über das Tragverhalten von Biegeträgem und Druckstäben mit zusammengesetztem Querschnitt und nachgiebigen Verbindungsmitteln, Karlsruhean, Germany.

Penazzi, D., Valluzzi, M.R., Saisi, A., Binda, L., Modena, C., 2001, Repair and strengthening of historic masonry buildings in seismic areas, Int. Congr. More than Two Thousand Years in the History of Architecture Safeguarding the Structure of our Architectural Heritage, Bethlehem, (Palestine), Vol. 2, Section 5.

Saisi, A., Valluzzi, M.R., Binda, L., Modena, C., 2004, Creep behavior of brick masonry panels strengthened by the bed joints reinforcement technique using CFRP thin strips, Proc. of SAHC2004: IV Int. Seminar on Structural Analysis of Historical Constructions - possibilities of experimental and numerical techniques, Padova, Italy, pp. 837-846.

Tomazevic, M.., 1999, Earthquake-Resistant Design of Masonry Buildings (Series on Innovations in Structures and Construction), Vol 1, Imperial College, London, U.K.

Valluzzi, M.R., Bernardini, A., Modena, C., 2005, Seismic vulnerability assessment and structural improvement proposals for the building typologies of the historic centers of Vittorio Veneto (Italy)', Proc. of STREMAH 2005, Malta, pp. 323-332.

Valluzzi, M.R., Binda, L., Modena, C., 2004a, Mechanical behavior of historic masonry structures strengthened by bed joints structural repointing, Construction and Building Materials, Elsevier Science, 19(1):63-73.

Valluzzi, M.R., Cardani, G., Binda, L., Modena, C., 2004, Seismic vulnerability methods for masonry buildings in historical centres: validation and application for prediction analyses and intervention proposals, Proc. of the 13th World Conference on Earthquake Engineering, Vancouver, B.C., Canada (on CD-ROM).

Valluzzi, M.R., da Porto, F., Modena, C., 2003, Grout requirements for the injection of stone masonry walls, Proc. of International Conference on the Performance of Construction Materials in the New Millennium, Il Cairo, Egypt, pp. 393-402.

Valluzzi, M.R., da Porto, F., Modena, C., 2004a, Behavior and modeling of strengthened three-leaf stone masonry walls, RILEM Materials and Structures, MS 267 37:184-192.

Valluzzi, M.R., Disarò, M., Modena, C., 2003, Bed joints reinforcement of masonry panels with cfrp bars', Proc. of the International Conference on Composites in Construction, Rende (CS), Italy, September 2003 (Bruno-Spadea-Swamy Ed.), pp. 427-432.

Valluzzi, M.R., Tinazzi, D., Modena, C., 2005a, Strengthening of masonry structures under compressive loads by frp strips: local-global mechanical behaviour, Science and Engineering of Composite Materials, Special Issue, Freund Publishing House, Tel Aviv, Israel, 12(3):203-218.

Valluzzi, M.R., Tinazzi, D., Modena, C., 2002, Shear behavior of masonry panels strengthened by frp laminates, Construction and Building Materials, Elsevier Science, Special Issue, 16(7):409-416.

Valluzzi, M.R., Valdemarca, M., Modena C., 2001, Behavior of brick masonry vaults strengthened by frp laminates, ASCE Journal of Composites for Construction 5(3):163-169.

Chapter 3.3

STRESS-FAILURE ANALYSIS OF MASONRY STRUCTURES UNDER EARTHQUAKE LOADING

C. A. Syrmakezis, P. G. Asteris, A. K. Antonopoulos, O. A. Mavrouli
Institute of Structural Analysis and Aseismic Research, National Technical University of Athens, 9, Iroon Polytechniou, 157 73 Zografou Campus, Athens, Greece

Abstract: During the structural simulation of historical structures and monuments made of masonry, high demands for reliable model formulation, often, cannot be satisfied by the use of two-dimensional finite elements. Especially in the case of earthquake loading vertically to the plane of the structure, the use of three-dimensional solid finite elements is strongly recommended in order to obtain accurate results. In this paper, a methodology for the stress computation and failure evaluation of masonry structures, using this type of finite elements, is presented. Given the stress analysis results, the three-dimensional stress state is transformed into an equivalent biaxial and failure analysis is performed implementing failure criteria especially adapted to masonry structures. Demonstration of the methodology is performed on a case study, a masonry arch bridge. The efficiency of the methodology is discussed and conclusions are derived.

Key words: 3D finite elements; solid finite elements; masonry structures; stress analysis; failure analysis.

1. INTRODUCTION

In the case of masonry structures, difficulty in determining their structural response and failure mechanisms arises from the variability of their composite materials, mainly blocks and mortar, as well as from the diversity of the structure layout. Some important factors affecting the behaviour of the overall non-homogenous system are the building techniques applied, the inherent material mechanical properties, masonry shapes, the dimensions and the construction quality. Concerning the evaluation of composite materials' mechanical properties, many uncertainties are introduced due to the large dispersion of data and the scarce of the experimental data. As a result, in-

S.K. Kourkoulis (ed.), Fracture and Failure of Natural Building Stones, 157–166.

complete assessment of the masonry's structural response is achieved, especially in the case of a seismic event, where further uncertainties occur due to the phenomenon' s random character.

In order to perform structural computer aided stress analysis, mathematical models simulating real structures are formulated, comprising simulation of materials, geometry, loading and boundary conditions. For masonry structures, the finite element method is highly recommended due to its flexibility in modeling three-dimensional structures and the accuracy of the results obtained. Distribution of masonry's self-weight along its height is efficiently taken into account by attributing to each finite element, a mass.

The finite element (FE) method of analysis implements various types of finite elements, with a main distinction between one- dimensional (bar), two-dimensional (shell, plate, membrane) and three-dimensional (solid) elements. In order to describe composite material properties and their influence on structural performance, micro- or macro-analysis is used, depending on the accuracy desired. During micro-analysis, blocks, mortar and their interface are simulated separately, while for macro-analysis, discretization of the model is based on the assumption of a homogenous material, representing masonry. Its properties are defined by taking into account the quantified occurrence of composite materials. Combinations of micro- and macro-analysis can be used, depending on the aim of the analysis.

The Young's Modulus of Elasticity, E, is a parameter that determinatively influences the masonry's response. Either in the case of 2-D or in the case of 3-D elements implementation, the masonry material is considered isotropic or orthotropic, depending on whether distinct directional properties are exhibited. For micro-analysis purposes, materials are assigned with a value that can be determined from experimental, destructive or non destructive testing. In the case of macro-analysis, the Young' Modulus value that corresponds to the homogenized masonry is selected either using proper equations from world literature or by analytical evaluation for a prism consisting of two blocks and a mortar joint.

2. STRESS AND FAILURE ANALYSIS OF STRUCTURES

The finite element method is used to obtain values of stresses developed within the structure, on the nodes of each FE. Evaluation of failure strongly depends on the criterion chosen to determine the conditions under which failure occurs. The methodology used in this paper, proposes two failure criteria, in terms of normal and shear stresses, applied on finite elements nodes, for biaxial elasticity assumptions. As a consequence, the output of the triaxial stress state analysis has to be converted into an equivalent plane state one, through transformation equations. Reduced two-dimensional stress state, in its principal form, is afterwards processed using the failure criteria and conclusions concerning failure are drawn.

3. APPLICATION OF THE FE METHOD: THE SOLID FINITE ELEMENT

The prevalent use of two-dimensional FE for the stress analysis of masonry structures is based on the, often made, assumption that masonry's geometrical characteristics and loading conditions permit it to be accurately enough simulated by plane structural members. For historical buildings and monuments' analysis, particularities occurring, often impede such reduction assumptions, especially when standards for high analysis results are imposed.

The accuracy of the stress output is strongly related to the FE meshing and the density of the realized grid. In areas of openings and members' intersections, variations of stress distribution often appears to be more intense than in the rest of the wall surface, thus leading to the use of smaller two-dimensional FE, for the acquisition of detailed results. Consecutively, FE thickness, usually representing the wall width, results to be considerably great in proportion to their area dimensions and shells' (FE activating six degrees of freedom) final geometry, contradicts their two-dimensional character.

Inadequate model formulation is also observed in the case of architectural particularities, e.g. walls with width variations. Assumptions required to effect dimensionality reduction in order to use two-dimensional elements for the analysis may be proved misleading with inaccurate stress analysis results.

In order to overcome the problem, implementation of three-dimensional "solid" FE is recommended. In Figure 1, discretization of a model representing a dam, using "solid" FE, is illustrated. When dynamic loading vertically to the wall plane is imposed, such as in case of a dam subjected to hydrodynamic pressure or earthquake, implementation of "solid" elements permit the acquirement of data related to stress variation along the width of the structure.

Standard types of "solid" FE are: the 4-node tetrahedron pyramid, the 6-node pentahedron wedge and the 8-node hexahedron brick. Refinement is achieved using mid-side nodes. For the formulation of the strain-displacement matrix, calculation and inversion of the Jacobian matrix is needed[1]. Given the stiffness matrix of the FE, the displacements of the nodes are obtained.

4. MASONRY SEISMIC ANALYSIS USING THE SOLID FINITE ELEMENT

The procedure followed for the analysis of masonry structures subjected to earthquake excitation, is presented through a case study.

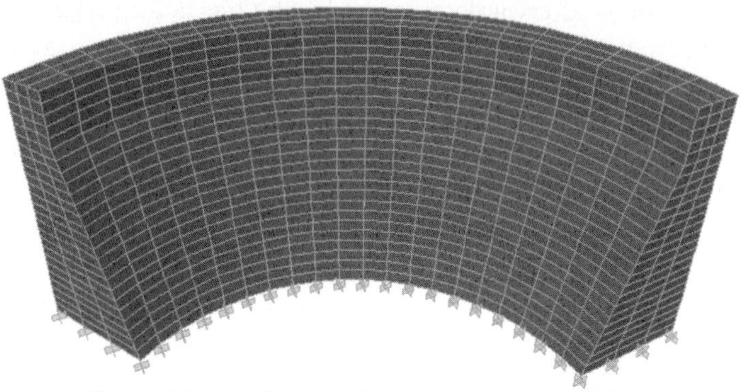

Figure 1. Model discretization of a dam using "solid" FE.

Analysis comprises three stages: model simulation, stress calculation and failure evaluation.

The selected structure (Figure 2) is a curved masonry arch bridge, located near a village of Peloponese named Manari. The bridge was built in the late 19th century, by French engineers and it consists part of the secondary railway network of Greece[2]. It is made of ashlar masonry, of optimum construction quality. Evaluation of its present state and possible rehabilitation for new type of trains, have rendered necessary the analysis of the structure.

5. STRUCTURE SIMULATION

In order to perform the analysis of the masonry bridge a model was formulated using solid FE elements, so as to acquire detailed out-of-plane stress and failure results in case of an earthquake event. Eight and six isoparametric

Figure 2. Masonry arch railway bridge in Manari, Greece.

solid elements were implemented, activating three translational degrees of freedom. Masonry was considered isotropic. The macro-analysis model is shown in Figure 3.

The software used was SAP2000 V.9. Earthquake simulation was made according to EAK2000 (actual Greek Seismic Code) and a quasi-static earthquake loading was considered for peak ground acceleration equal to A=0.24 g.

For actions due to railway operation, loadings were considered as suggested by the Eurocode 1.

6. STRESS EVALUTION FOR FAILURE ANALYSIS

Stress analysis data were obtained for the three principal stresses on each FE node. According to the procedure presented in this paper, failure analysis implied adaptation of the 3D problem of determination of mechanical failure to 2D elasticity assumptions, through substitution of the third principal stress by its effect on the other two[3]. The proposed process had to be repeated three times in order to determine stress and failure state for each plane. Each time a principal stress σ_1, σ_2 or σ_3 was eliminated and the set consisting of the two remaining principal stresses, (σ_2,σ_3), (σ_1,σ_3) or (σ_1,σ_2) respectively, was assumed to be applied on the nodes.

Calculation of the effect of the eliminated principal stress on the remaining two was made using the Hooke's law. In Figure 4 the principal stresses are shown for a triaxial stress state, while in Figure 5 its equivalent biaxial stress state is presented after elimination of the principal stress σ_3^T (the index T denotes the triaxial state while B denotes the respective biaxial one).

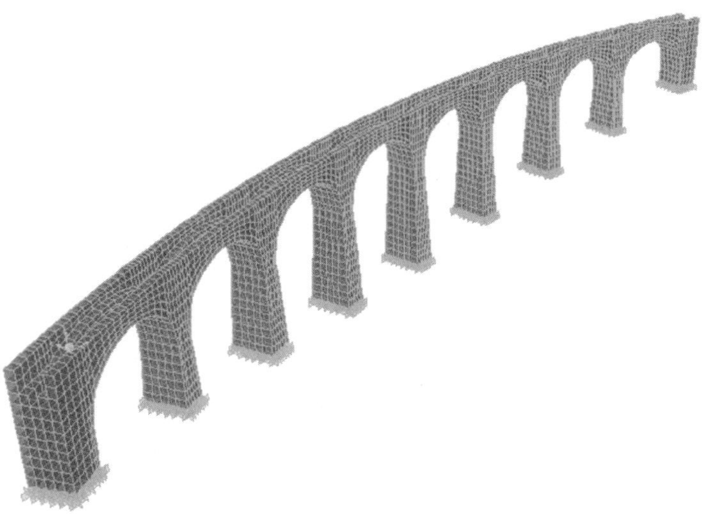

Figure 3. Model discretization of the bridge using "solid" FE.

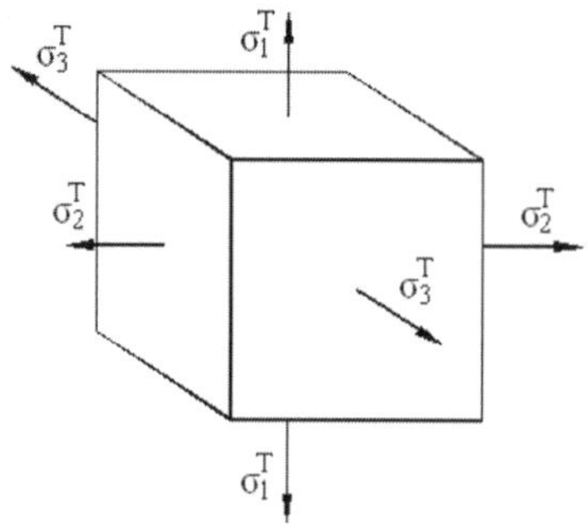

Figure 4. Three-dimensional stress state.

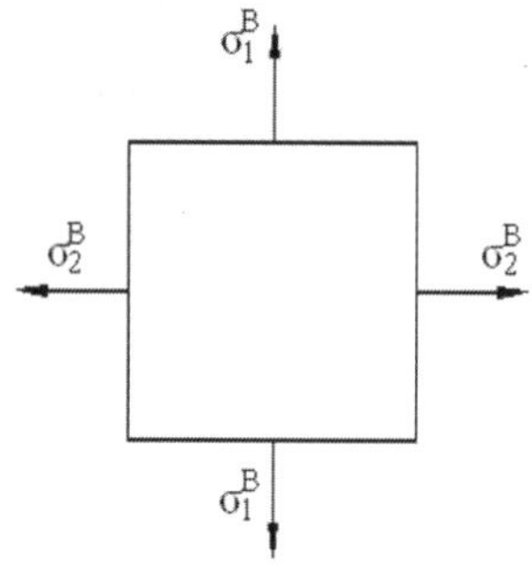

Figure 5. Equivalent two-dimensional stress state.

For a three-dimensional stress state, the strains on plane (1,2) are equal to:

$$\varepsilon_1^T = \left[\sigma_1^T - \nu \cdot \left(\sigma_2^T + \sigma_3^T\right)\right]/E \qquad \varepsilon_2^T = \left[\sigma_2^T - \nu \cdot \left(\sigma_1^T + \sigma_3^T\right)\right]/E \tag{1}$$

while for a biaxial state:

$$\varepsilon_1^B = \left[\sigma_1^B - \nu \cdot \sigma_2^B\right]/E \qquad \varepsilon_2^B = \left[\sigma_2^B - \nu \cdot \sigma_1^B\right]/E \tag{2}$$

Equivalence of the two stress states presumes equivalence of strains. In order to reduce the three-dimensional stress state ($\sigma_1^T, \sigma_2^T, \sigma_3^T$) to the biaxial ($\sigma_1^B, \sigma_2^B$) one, the following equations have to be satisfied:

$$\sigma_1^B = \sigma_1^T - \nu \cdot \sigma_3^T/(1-\nu) \qquad \sigma_2^B = \sigma_2^T - \nu \cdot \sigma_3^T/(1-\nu) \tag{3}$$

Respectively, for the plane (2,3) the transformation equations are:

$$\sigma_2^B = \sigma_2^T - \nu \cdot \sigma_1^T/(1-\nu) \qquad \sigma_3^B = \sigma_3^T - \nu \cdot \sigma_1^T/(1-\nu) \tag{4}$$

Finally for plane (1,3) it holds:

$$\sigma_1^B = \sigma_1^T - \nu \cdot \sigma_2^T/(1-\nu) \qquad \sigma_3^B = \sigma_3^T - \nu \cdot \sigma_2^T/(1-\nu) \tag{5}$$

7. FAILURE ANALYSIS

Following a recently introduced methodology[4,5] and described shortly in the present paper, two failure criteria were established for the elaboration of stress analysis data. The failure theories for isotropic materials are not applicable for masonry under biaxial stresses because they are derived on the basis of the invariant state of stress concept, where the stress orientation has no effect on the strength. In this section, failure criteria are proposed in a generalized form for masonry under biaxial stresses, taking also into consideration its anisotropic (orthotropic) nature since it is a composite material.

Adaptation of the initial Von Mises criterion to masonry structures has been made[4]. The modified criterion is expressed in a three-dimensional system, where the third axis, vertically to the plane of direct stresses represents applied shear stress on the node, resulting in the semi-empirical failure surface, presented in Figure 6, for zero shear stress level. The four gray-shaded areas are defined as follows:

$$S_1 : \sigma_{xx}^2 + \sigma_{yy}^2 - \sigma_{xx}\sigma_{yy} + 3\,\tau^2 - f_{wc}^2 = 0, \text{ for } \sigma_{xx} \leq 0 \text{ and } \sigma_{yy} \leq 0 \tag{6}$$

$$S_2 : \sigma_{yy} + (1 - \sigma_{xx} / a)\sqrt{f_{wc}^2 - 3\tau^2} = 0, \quad \text{for } \sigma_{xx} \geq 0 \text{ and } \sigma_{yy} \leq 0 \tag{7}$$

$$S_3 : \sigma_{xx} + \sigma_{yy} - a = 0, \quad \text{for } \sigma_{xx} \geq 0 \text{ and and } \sigma_{yy} \geq 0 \tag{8}$$

S_4: symmetrical to S_2 with respect to the bi-sectional level of the first quadrant.

Concerning a it holds that:

$$a = \left(f_{wt} / f_{wc}\right)\sqrt{f_{wc}^2 - 3\tau^2} \tag{9}$$

where f_{wc}, f_{wt} denote the strength of masonry in compression and tension, respectively.

An improved criterion for masonry structures can also be used for the evaluation of failed areas[5]. The mortar bed joints, due to their continuous nature, can be assumed as dividing the media into layers of equal thickness and thus give to the masonry the appearance of a laminated composite material. For the expression of an analytical failure model of masonry, therefore, a polynomial that is available already for composite materials is proposed. This failure surface in the stress space can be described by Eq. (10):

$$\begin{aligned} f\left(\sigma_x\ \sigma_y\ \tau\right) &= F_1\sigma_x + F_2\sigma_y + F_{11}\sigma_x^2 + F_{22}\sigma_y^2 + F_{66}\tau^2 + 2F_{12}\sigma_x\sigma_y + 3F_{112}\sigma_x^2\sigma_y + \\ &+ 3F_{122}\sigma_x\sigma_y^2 + 3F_{166}\sigma_x\tau^2 + 3F_{266}\sigma_y\tau^2 - 1 = 0 \end{aligned} \tag{10}$$

In order to determine the coefficients of the proposed cubic tensor polynomial (Eq. 10), an evaluation of the mechanical characteristics of masonry is performed using the experimental data of Page[6], through a least squares approach. The above mentioned methodology[4,5] is followed here and the failure surface can be defined as:

$$\begin{aligned} &2\ 27\sigma_x + 9\ 87\sigma_y + 0\ 573\sigma_x^2 + 1\ 32\sigma_y^2 + 6\ 25_\tau^2 - 0\ 30\sigma_x\sigma_y + 0\ 009585\sigma_x^2\sigma_y + \\ &+ 0\ 003135\sigma_x\sigma_y^2 + 0\ 28398\sigma_x\tau^2 + 0\ 4689\sigma_y\tau^2 = 1 \end{aligned} \tag{11}$$

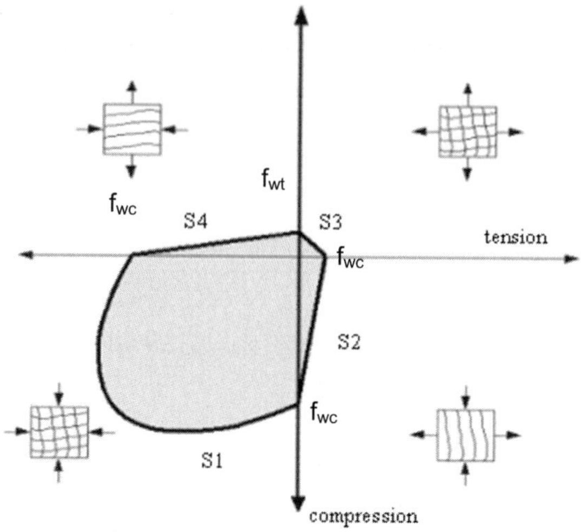

Figure 6. The modified Von-Mises Failure criterion.

The validity of the failure criterion is demonstrated by comparing the analytically derived failure surface of Eq.(11) with the existing experimental results of Page[6]. About 70 experimental data are depicted in Figure 7, together with the analytic curves in principal stress terms for the failure surface of Eq.(11). The satisfactory agreement between the analytical and experimental data is apparent for this general failure surface with a non-symmetric curve.

Processing the stress data, so as to obtain failure information, can be made using the FAILURE software, using both the above decribed approaches[4,5]. Input data are comprised by information with respect to model definition, transformed biaxial stress data for each node and values of masonry's strength

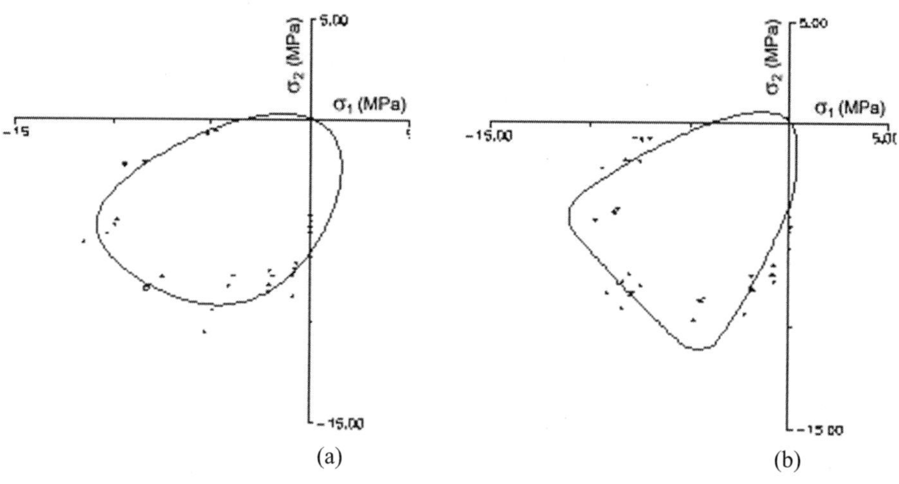

Figure 7. Masonry failure criterion in principal stress terms: a) q=22.5°, b) q=45.0°.

in compression and tension. Outputs of FAILURE software include graphical failure results with a distinction between the four different ways of failure, that is failure under biaxial compression, failure under biaxial tension, failure under compression parallel to one principal axis and tension to the other, and vice versa.

8. FAILURE RESULTS

The proposed procedure was applied on the case study. In Figure 8 graphical failure results are illustrated for a longitudinal section on the mid-level of the curved masonry bridge. Failure in this case is due to σ_1^D, σ_2^D (elimination of σ_3^T). For the alternative considerations of elimination of σ_1^T or σ_2^T, the results are presented in Figures 9 and 10, respectively. Superposition of all three failure cases can provide the most unfavorable results, vertically to the plane of the bridge.

Red colored areas (darker areas in black and white printing) represent failure under biaxial tension, while green colored areas illustrate failure under tension/compression. Blue areas have not been affected by earthquake. Colored areas are indicative of the location of failure on the wall surface and not of the plane on which failure occurs.

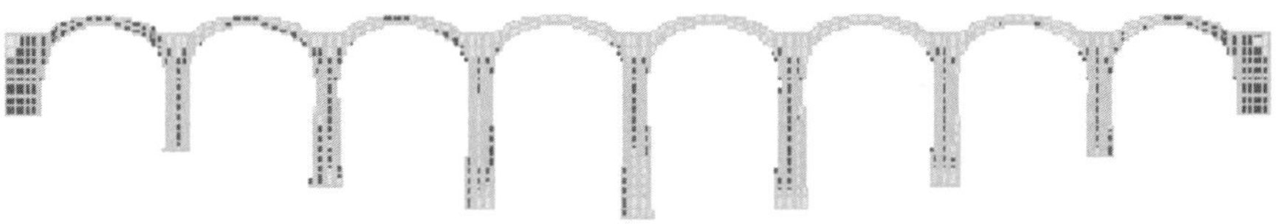

Figure 8. Failure results of the masonry bridge due to σ_1^D, σ_2^D.

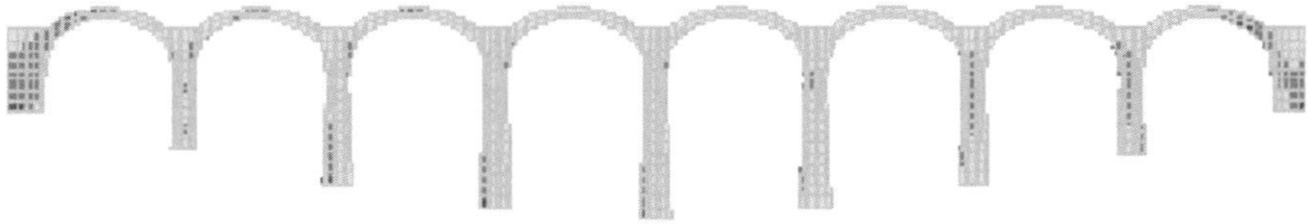

Figure 9. Failure results of the masonry bridge due to σ_1^D, σ_3^D.

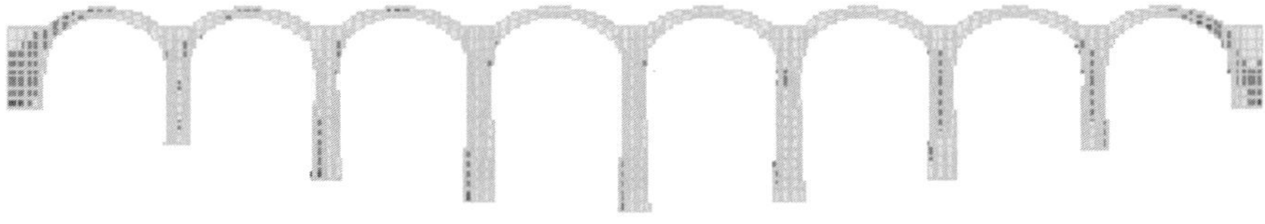

Figure 10. Failure results of the masonry bridge due to σ_2^D, σ_3^D.

9. CONCLUSIONS

In this paper, a methodology for a computer aided evaluation of stress distribution and resultant failure in masonry structures has been presented. The motivation for the implementation of three-dimensional solid elements has been discussed, proving that such finite elements can be used in order to overcome the problem that arises from shell elements reduction assumptions.

The use of solid elements has been demonstrated through a case study, an existing masonry arch bridge. The 3D model has been formulated and stress analysis has been performed. Triaxial stress state has been transformed into an equivalent biaxial one, and transformation equations have been extracted. The procedure has been repeated three times and results have been obtained for all three sets of stress states (σ_1^D,σ_2^D), (σ_2^D,σ_3^D), (σ_1^D,σ_3^D). Reduction of the third stress has permitted the use of the 2D modified Von Mises failure criterion, especially adapted to masonry structures.

Graphical failure results have been obtained for a longitudinal section on the mid-level of the curved masonry bridge, each time for a stress state corresponding to one of the three planes of the structure (1,2), (1,3) and (2,3), providing important additional information, related to failure resulting from out of plane stresses.

ACKNOWLEDGMENTS

This paper is presented within the frame of the European project WIND-CHIME: Wide-Range Non-intrusive devices toward Conservation of Historical Monuments in the Mediterranean Area.

REFERENCES

1. O. C. Zienkiewicz and R. L. Taylor, *The Finite Element Method, Basic Formulation and Linear Problems*, p. 153 (1967).
2. S. E. Sourtzi, Parametric Investigation of Static and Aseismic Behaviour of Arch Masonry Bridges, Master Thesis, National Technical University of Athens (Athens, Greece, 2005).
3. C. A. Syrmakezis, A. K. Antonopoulos and O. A. Mavrouli, Analysis of Historical Structures Using Three Dimensional Solid Elements, in *Proceedings of the 10th Int. Conf. Civil, Structural and Environmental Engineering Computing*, Rome (2005).
4. C. A. Syrmakezis, P. G. Asteris, Masonry Failure Criterion under Biaxial Stress State, *Journal of Materials in Civil Engineering, ASCE*, **13**(1), 58-64, 2001.
5. P. G. Asteris, Analysis of Anisotropic Nonlinear Masonry, PhD thesis, Dept. of Civil Engineering, National Technical University of Athens (Athens, Greece, 1999).
6. A. W. Page, The biaxial compressive strength of brick masonry, *Proc. Instn. Civ. Engrs., Part 2*, **71**, 893-906 (1981).

Chapter 3.4

COMPATIBILITY OF MATERIALS USED FOR REPAIR OF MASONRY BUILDINGS: RESEARCH AND APPLICATIONS

Luigia Binda, Antonella Saisi and Cristina Tedeschi
DIS – Dept. of Structural Engineering, Politecnico of Milan, Piazza Leonardo da Vinci 32, 20133 Milan, Italy

Abstract: The research emphasizes the necessity of a deep knowledge of the structure morphology, the materials, and their characteristics and eventual the state of damage and its causes when approaching the repair of an historic masonry building,. The relevant damages surveyed in stone-masonry buildings after the Umbria-Marche earthquake (1997-98), together with the contributions of several theoretical and experimental studies carried out in the '90s has confirmed the need to improve the knowledge of the seismic response of old masonry buildings and of the reliability of retrofitting techniques. Retrofitting or repair of this damage is a very difficult task. In many cases grout injection or wall jacketing fail due to incompatibility with the construction technique of the walls. The present paper describes and critically analyses the main repair techniques applied to the repair of stone masonry and their effectiveness.

Key words: stone masonry; earthquake; strengthening; repair; grout injections; jacketing.

1. INTRODUCTION

The 1997 earthquake which hit Umbria and Marche in Italy, clearly showed that several repair and strengthening work carried out in recent times (after the previous earthquake of 1979), were not of real effectiveness. On the contrary some interventions performed on historic stone-masonry buildings caused damages to the original structure (Penazzi et al., 2000).

In fact, after the 1976 Friuli earthquake and following, the collapse of old masonry dwellings made with poor masonry walls and timber floors, convinced structural engineers and authorities that in order to resist the seismic

S.K. Kourkoulis (ed.), Fracture and Failure of Natural Building Stones, 167–182.

actions the masonry buildings should be treated from the structural point of view as structures having "box" behaviour. This meant stiff connections between the bearing and shear walls, stiff connection between walls and floors and floors stiff in their plane.

These assumptions caused a systematic substitution of timber floors and roofs, with concrete-clay floors, stiff reinforced concrete tie beams at every floor, reinforced injections through the walls, jacketing of very poor walls. The same assumptions were also supported by the Italian Seismic Code.

The last earthquakes showed that these techniques hardly apply to the stone-masonry buildings when walls are made by multiple leaf stone masonry with poor connection between the leaves, weak mortar and irregular stones. Collapses of heavy roofs, hammering of adjacent buildings by the ones which were heavily repaired, out of plane collapses, ineffectiveness of the grout injections, failure of jacketing, are some of the phenomena shown by the 1997 earthquake; nevertheless the techniques themselves can be successful in the case of other types of masonry (e.g. brick masonry or regular stone masonry in new buildings) (Binda et al., 1999).

The ineffectiveness of these techniques in the case of the above mentioned masonry buildings are mostly due to the incompatibility in terms of stiffness between the original and the new structural elements, to the poor workmanship, but mainly to the lack of knowledge on the material and structural behaviour of historic buildings (Penazzi et al., 2000; Penazzi et al., 2001) .

The authors have proposed within Italian Research Contracts supported by the Italian Department of Civil Protection a multilevel approach based on site and laboratory investigation at the level of historic centres with the aim of studying the vulnerability of stone-masonry buildings made with irregular texture and multiple leaf walls and to calibrate appropriate mathematical models (Binda et al., 2004a and 2004b).

In the meantime the authors continued their research on the effectiveness of two techniques: grout injection and deep repointing, showing that a good knowledge of the morphology and behaviour of masonry contributes to successful application of these techniques particularly to irregular stone masonry.

2. MATERIALS AND MASONRY CONSTRUCTION TECHNOLOGY

The structural performance of a masonry can be understood provided the following factors are known: (i) its geometry; (ii) the characteristics of its texture (single or multiple leaf walls, connection between the leaves); (iii) the physical, chemical and mechanical characteristics of the components (brick, stone, mortar); (iv) the characteristics of masonry as a composite material (Binda et al., 2003).

The worst defect of a masonry wall is to be not monolithic in the lateral direction, and this can happen for instance when the wall is made by small pebbles or by two external layers even well ordered but not mutually connected, containing a rubble infill (Giuffrè, 1993; Binda et al., 2003). This causes the wall to become more brittle particularly when external forces act in the horizontal direction (Figure 1). The same problem can happen under vertical loads if they act eccentrically.

Given the great number of existing cross sections and the great influence of the construction technique, a systematic study on the mechanical behaviour of stonework masonry should in fact begin from an extensive investigation of the different geometry and building techniques which takes into account the different layers constituting the wall and the kind of constraints which may or may not be present between the layers themselves (Binda et al., 2003).

3. LACK OF KNOWLEDGE AND FAILED REPAIRS

The last earthquake effects have shown that: a) for some building typologies and masonry morphologies the adopted structural models need to be adjusted to their real behaviour, b) the retrofitting techniques applied after the previous earthquake of 1979 still need improvement.

The interventions, carried out according to the code suggestions were made in order to retrofit all the existing buildings (damaged and undamaged) assuming the safety criteria applied to new buildings designed and constructed according to the current seismic code. Masonry and historic masonry structures were considered too weak to bear future earthquakes, therefore they needed invasive interventions to respect the imposed safety coefficients.

The code suggested that these criteria can be attained for all the masonry buildings, especially for the weakest ones by:

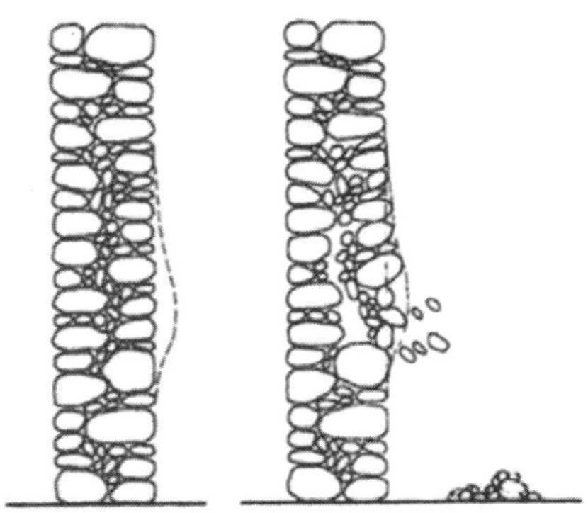

Figure 1. Deformation and failure of a two leaves wall (Giuffrè, 1993).

a) substituting the original timber floors and roofs with reinforced concrete ones.
b) constructing r.c. tie beams in the wall thickness at every floor level and under the roof.
c) jacketing and/or injecting the walls in order to improve their shear strength.

As it is well known, the first two types of intervention are intended to improve the structural response of the building: type a) by ensuring "rigid floor" action, type b) by connecting loadbearing and shear walls in order to prevent out of plane failures. A type b) intervention of course can improve the overall strength of shear walls, as the strength of the existing masonry spandrel beams can be increased and the equilibrium even after shear cracks appear in the masonry piers and spandrel beams is insured. Type c) interventions can obviously increase both the in-plane shear strength and the out of plane flexural strength of masonry walls. The term "can" is used intentionally, as the effectiveness of these retrofitting techniques highly depends on the type of masonry and masonry structure they are applied to.

Even if experimental and analytical research has been carried out in the past decades on these techniques, nevertheless the effectiveness was always checked in terms of strength increase rather than on chemical, physical and mechanical compatibility with the original masonry (Modena et al., 1997a; Binda et al., 1997). Few research was carried out in this direction on the effectiveness of grout injections (Tomazevic and Turnsek, 1982; Tomazevic, 1992; Binda et al., 1993; Binda et al., 1994; Modena and Bettio, 1994; Bettio et al., 1996; Laefer et al., 1996; Valluzzi et al., 2004) and of jacketing (Modena and Bettio, 1994; Modena et al., 1997b). The conclusions recommended a careful approach and suggested a previous knowledge of the masonry wall morphology and of the masonry characteristics, since some types of walls could be not injectable. Similar conclusions were reached for jacketing; where connectors through the wall of the opposite two reinforcing nets (especially for multiple leaf stone masonry) could not be realised, the system was failing (Figure 2).

The failures due to inadequate application of repair interventions are the most difficult to interpret. The questions to be answered are: what actually

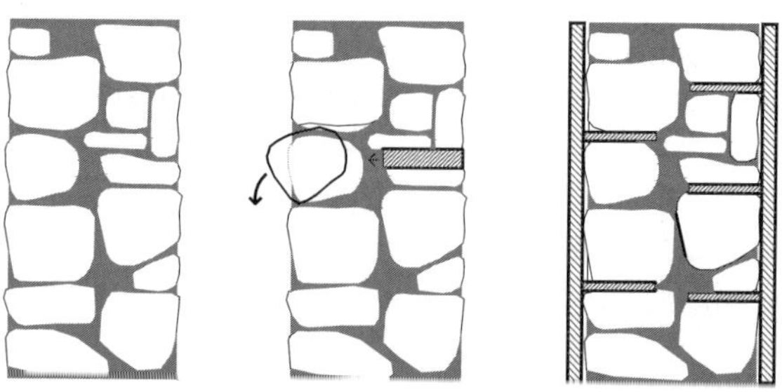

Figure 2. Difficulty in the application of jacketing to multiple leaf stone masonry.

caused the failures and how did they develop and how serious were their consequences to the overall response of the building.

In this context a further problem has been defined, that is the modality of intervention on buildings repaired in the past with unsuccessful techniques. Up to now, the question was often solved by the complete demolition of the building and the volume reconstruction in concrete (Figure 3).

In the following, a critical review is made of some retrofitting techniques.

3.1 Concrete ties and roof and floor substitution

Concrete ties are usually inserted where timber floors and roofs are substituted by mixed concrete and clay block structures. In these cases a concrete tie is built at every floor. The tie is positioned along the four sides of the structure as a connection floor to walls. In an existing building while the roof concrete tie can be positioned on the whole thickness of the top wall, the ties at each floor can only be inserted in part of the section after partial demolition of it. In this case it is very difficult to realise a stiff connection to the existing wall. In general this connection is very difficult when the wall is made of a multiple leaf irregular stone masonry.

The damages observed more frequently were the following: (i) partial eccentric loading of the walls (Figure 4), (ii) lack or poor connection of the tie beam to the walls (Figure 5). The seismic events, then, showed that these elements cannot transmit the horizontal actions to the walls and neither can connect the two masonry leaves, of which one remains free and can rotate freely and overturn (Figure 6).

The collapse mechanism of the masonry is not for in plane shear as expected after the floor substitution, but a partial overturning mechanism of the external leaf of the wall which starts for lower values of the expected collapse coefficient. Some details visible in the upper part of the Figures 4 and 6 suggest that the intervention contributed to reduce the already weak connection between the leaves to the very critical section where the walls are connected to the floor. In fact, in those connections the confining actions of the floors are applied and most probably less uniformly distributed. Even the contribution of the new internal wall (Figure 6), perpendicular to the collapsed facade was completely missing in the collapse, possibly due to the restoration interventions.

In this technique it is important to realise an effective connection between the tie beam and the masonry. In the case of the tie beam at the roof level, which can rest on the whole section of the wall, the connection is difficult because it should be realised by vertical metal connectors inserted in the wall from the top. Once again, this connection is seldom possible in a multiple leaf stone masonry. Furthermore the stiffness of the concrete roof can be too high compared to the one of the existing wall and the roof can hammer the wall and cause a partial collapse (Figure 7).

Figure 3. The building repaired in the 80s and partially failed, has been demolished and an r.c. structure rebuilt.

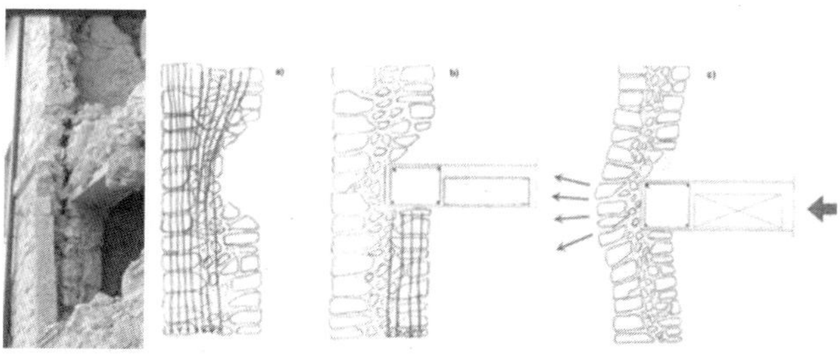

Figure 4. Effect of eccentric loading due to r.c. tie beam positioning and failure of the tie beam insertion at each floor under vertical and horizontal actions (Borri and De Maria, 2004)

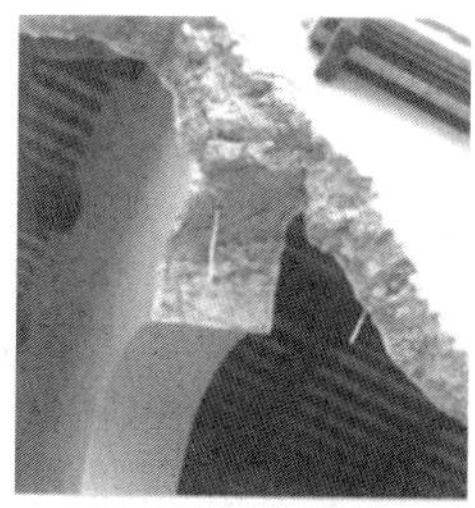

Figure 5. Difficult connection between the roof tie beam and the wall.

Figure 6. Out-of-plane collapse of a wall with r.c. tie beams.

Figure 7. Roof hammering the masonry walls.

3.2 Wall and pier jacketing

The aim of the technique is to better connect the different leaves of a wall in damaged conditions producing a new section constituted by the old one increased by the two jacketed reinforced parts. The idea behind it is to have a thicker section, to increase compressive, tensile and shear strength and ductility (Modena et al., 1997b). The same technique has also been applied to connect load-bearing and shear walls and also large cracks, as well. The technique consists in positioning a reinforcing net (ϕ= 6 to 8mm) on both faces of a wall, connecting the two nets with frequent steel connectors and applying on the two faces a cement mortar based rendering, which constitutes a sort of slab. The masonry panel, then, acquires high strength and stiffness, which is not always a positive point when considering the overall behaviour of the building.

This technique was extensively applied particularly to irregular multiple leaf stone-walls in Italy and it is recommended by the Italian Code. Nevertheless, its execution on site is not very easy due to the inhomogeneity of the walls, to the cost and difficulty of connecting the two faces of the wall (Figure 2). In fact, it is possible to observe frequently local failures of jacketed walls, almost always very clearly connected to poor detailing. Examples are shown in Figures 8(a,b), representing failures respectively due to insufficient steel.

The most widespread mistakes made on site are described in the following together with the consequent damages: (i) lack of connection between the nets in orthogonal walls and in correspondence to the floors; they cause discontinuities between the walls (Figure 8a), (ii) lack of overlapping between two different sheets of the net (Figure 8c), (iii) absence of steel transversal connectors (Figure 8d), (iv) use of too short connectors (Figure 9), (v) lack of uniformity of distribution of the repaired areas in the structure; this can cause torsion stresses due to non uniform distribution of the stiffness.

Furthermore, a low durability of the technique was often observed, due to insufficient thickness of the steel cover with consequent steel corrosion (Figures 9b,c). This problem is of a great importance in building where capillary rise and diffuse moisture are widely recognisable.

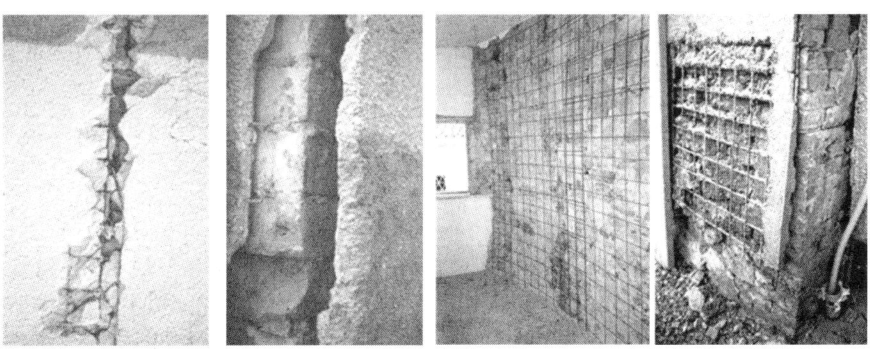

(a) (b) (c) (d)

Figure 8. (a) Failure due to insufficient steel mesh overlapping and (b) insufficient transversal ties confining action; (c) lack of connection between nets; (d) absence of connectors.

(a) (b) (c)

Figure 9. (a) Too short connectors; (b), (c) corrosion of the steel net in a jacketed wall.

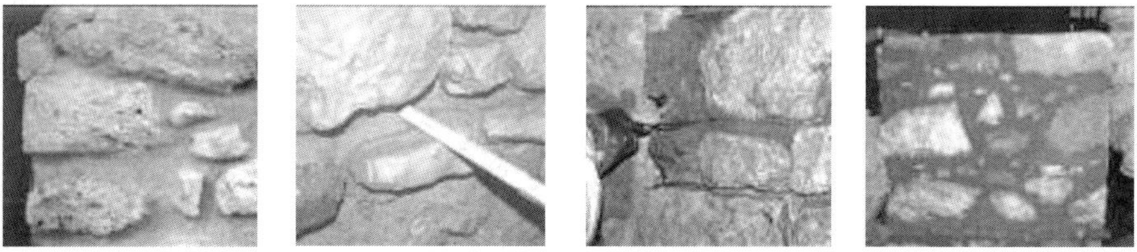

(a) (b) (c) (d)

Figure 10. (a) Joint after cleaning, *(b)* detail of the joint depth, *(c)* first layer of repointing, *(d)* after intervention.

A proposal for a confining technique for the walls was made by Binda et al. (2005), (Corradi et al. 2005), based on deep re-pointing of the wall carried out on both sides (Figure 10). Tests were carried out on site and gave encouraging results, particularly when the technique is applied together with grout injection.

3.3 Grout injection

Repair and retrofitting of masonry is extensively performed by grout injection, which for years have been regarded as a suitable technique to restore the homogeneity, uniformity of strength and continuity of masonry walls. Research has been carried out in these last years on the effectiveness of the technique. Testing on small-scale models under horizontal loads has also been performed in order to study the response of masonry repaired by injection of grouts, and compare it to other techniques (Tomazevic and Turnsek, 1982; Tomazevic, 1992).

In general, the aims of the technique are: (i) to fill large and small voids and cracks increasing the continuity of the masonry and hence its strength, (ii) to fill the gaps between two or more leaves of a wall, when they are badly connected. The aim can be fulfilled only knowing with good precision the morphology of the wall section, the materials constituting the wall and their composition in order to avoid chemical and physical incompatibility with the grout, the crack distribution, the size, percentage and distribution of voids (Binda et al., 1997; Binda et al., 1993; Binda et al., 1994; Laefer et al., 1996; Valluzzi et al., 2004).

As known, the success of thc injcction tcchnique can be limited when badly applied, due to the masonry morphology, to the desegregation and sedimentation of the grouts, to the mix characteristics (grain size distribution) and to the operative technique.

Thc main problems related to grout injection can be summarized as follows: a) lack of knowledge on the size distribution of voids in the wall, b) the difficulty of the grout to penetrate into thin cracks (2-3 mm), even if microfine binders are used, c) the presence in the wall, of fine and large size voids, which make difficult choosing the most suitable grain size of the grout (injecting large size voids with a fine grained mix can in fact induce segregation), d) the segregation and shrinkage of the grout due to the high rate of absorption of the material to be consolidated, e) the difficulty of grout penetration, especially in the presence of silty or clayey materials, f) the need for sufficiently low injection pressure to avoid either air being trapped within the cracks and fine voids or even wall disruption.

Therefore, the effectiveness of a repair by grout injection depends not only on the characteristic of the mix used, but also on its mechanical properties and on the injection technique adopted and once again on the knowledge of the wall type. The injectability of the grout is influenced also by its compatibility with the masonry to be repaired. The technical improvements of the last years have lead to the development of new grouts with specific properties, such as a low salt content and an ultra fine size of aggregate, and have also shown how to optimise the injection methodology, such as the injection pressure or the distance between the injectors, in relation to the masonry characteristics. Multiple leaf walls can be made with very poor mortars and stones but have very low percentage of voids (less than 4% of voids are not injectable) and have internal filling with loose material, which is not injectable. Figures 11 and 12 show two of the cases where injection was clearly proved to be very poor.

Injectability tests proposed (Binda, et al., 1993) can be carried out in the laboratory on materials sampled from the internal part of walls. The sampled material can be inserted in cylinders and be injected in laboratory (Figure 13). Compressive and splitting tests on the injected cylinders in laboratory can be carried out on the cylinders after the time necessary to reach the hardening of the grout.

The methodology used for testing the injectability of the material is presented in Figure14. The sequence of operations in laboratory is shown in Figure 15 while Figure 16 presents a comparison between cylinders filled with different materials and injected with different grouts.

Figure 11. Poor results of applied injecttion.

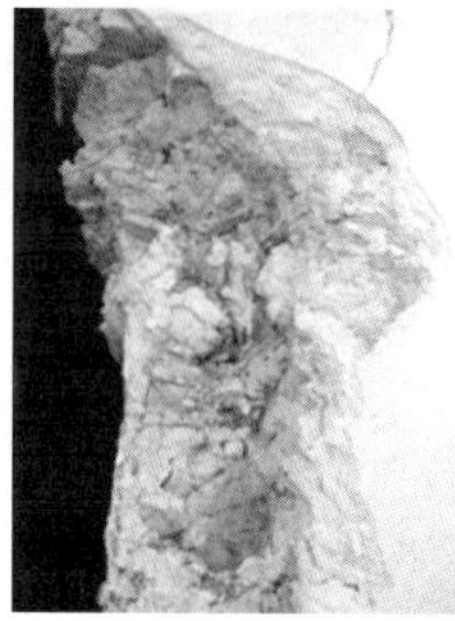

Figure 12. Only some spots were injected in the case of this wall with very low percentage of voids.

Figure 13. Cylinders can be filled with the materials sampled on site in order to evaluate the masonry injectability.

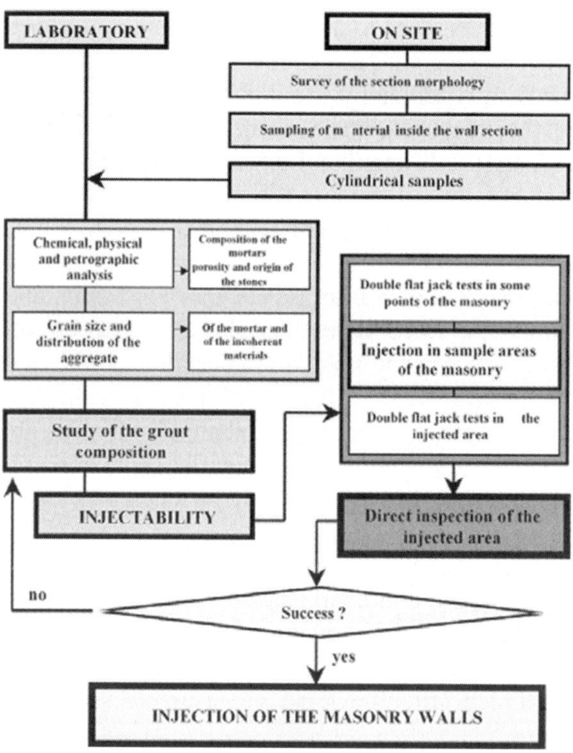

Figure 14. Methodology used for testing the injectability of the masonry

Figure 15. Phases of the injection in laboratory.

Injectability tests can be carried out directly on site on sampled area, locally dismantling the masonry. Non destructive tests as sonic tests can also be carried out on site, before and after injection in order to detect the penetration and diffusion of the grout.

4. OLD AND TRADITIONAL RETROFITTING AND REPAIR TECHNIQUES

Even if the scientific knowledge reached much lower levels as today, the mitigation of the earthquake effects by improving the structural behaviour and studying better details was certainly tried. Special techniques were suggested since the past centuries for repairing the damaged buildings and retrofit the structures. In the following a brief description of the techniques is reported referred to the different structural elements.

4.1 Foundations

The suggested way for mitigating the seismic effects was, since Plinius in the "Historia naturalis", to enlarge the dimension of the contact between the foundation and the soil. The same suggestions were given by Scamozzi (Scamozzi, 1615). Geometrical rules (plan dimension of the foundation equal to 1/6 of the height of the building) were given in 1783 by Sarti. Underpinning with brick or stone-masonry in order to reach lower strongest layers of the soil was also used in the past centuries. Leonardo da Vinci also mentioned reversed arch foundation and enlargement of the foundations. Actually this technique is still used today, were concrete beams are used. In 1909 M. Viscardini (Barucci, 1990) was proposing something similar to the base isolator

Figure 16. Comparison among cylinders filled with different materials and injected with different grouts in a laboratory.

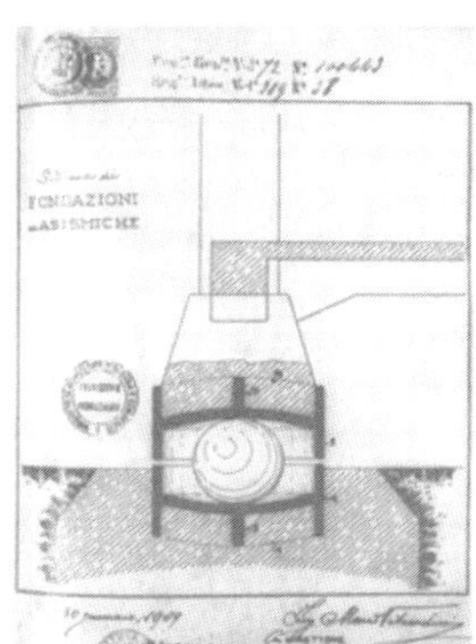

Figure 17. Design of foundations in seismic areas (1909).

(Figure 17). Similar proposals were made by C. Pesenti and others, as to insert between the foundations and the soil a layer of highly elastic material obtained by injection of wood or lead.

4.2 Walls

The use of steel or timber tie rods to connect wall to wall and wall to floor was known since the Byzantine times (Saint Demetrius in Thessaloniki, Aghia Sophia in Istanbul, 4th and 5th century). Their use continued in the Gothic architecture not only for seismic protection but also for collect the thrust of arches and vaults. In the 15th century, systems of rods were applied in seismic areas for restoring verticality of out of plane walls. Tie rods in seismic areas were suggested systematically through the 17th, 18th and 19th century (Milizia, 1554; Rondelet, 1832). This technique was also applied in Calabria in 1878 after an earthquake. In Umbria it was prescribed by the Recommendations published by the municipality of Norcia and in Sicily at the beginning of the 20th century (Archivio Storico, 1861) (Figure 18).

Another system largely adopted in Umbria was the use of buttresses against the existing walls (Figure 19).

Repointing and reconstruction of partially collapsed walls was also frequently adopted in the past; the same applies for the technique of adding a new leaf to increase the thickness of the wall.

A way of retrofitting the whole structure was the use of shear frames against the load-bearing and the shear walls (Figure 20) (Barucci, 1990). At the beginning of the 20th century reinforced masonry was also introduced in seismic areas (Genovese, 1915). In order to avoid hammering between two adjacent buildings a separation joint was realised by demolishing and reconstructing the end wall (Figure 21).

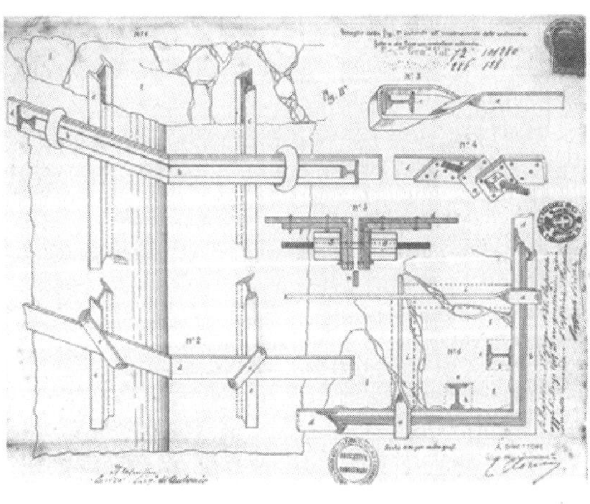

Figure 18. Connection wall to wall by tie rods (1909)

Figure 19. Addition of buttresses

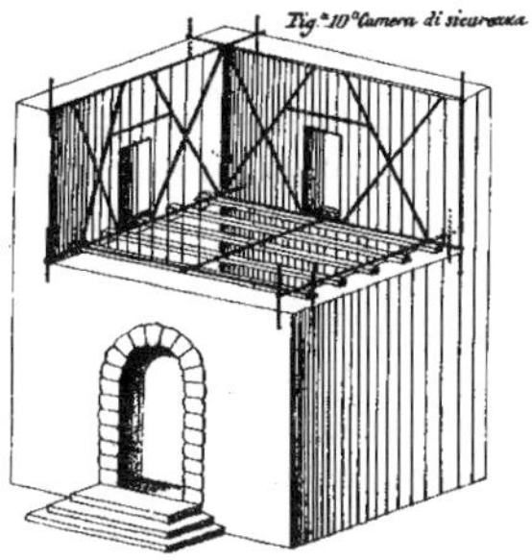

Figure 20. Timber framed walls (1876).

Figure 21. Joint cut between two houses.

4.3 Floors and roofs

These two important elements were continuously repaired against the earthquake, sometimes by substituting the highly damaged ones with stronger ones (steel beams and depressed vaults instead of timber); nevertheless a great deal of repair of the timber structures was also carried out.

Special care was given to the construction or the repair of timber roofs, with the description of all the most important details. It was considered safer to connect the roof beams to the walls and the use of trusses in order to avoid large thrusts. Some of the mentioned interventions have failed due to repeated seismic events, so it is impossible to criticise their effectiveness. Most timber ties and repair of timber floors and roofs simply failed for lack of maintenance; the same was for repointing and thick renderings when poorly made but also due to incompatibility with the substrate. The addition of a third leaf to existing walls usually failed by separation of the new leaf especially when bad connections were realised between the new and the other leaves. Steel ties, when appropriately applied, were successful everywhere; so were timber and floor roofs provided neat good connection existed between the walls and the timber elements.

5. MULTILEVEL APPROACH TO THE ANALYSIS OF THE VULNERABILITY OF STONE - MASONRY BUILDINGS

The research was supported by the Italian Department of Civil Protection, involving Universities and Cultural Property Regional Offices.

The aim of the research was to set up, for historic centres, systematic data-bases storing the information useful in order to prepare rescue plans and to design interventions for the preservation of the cultural heritage. Such information deals with: i) the technological and constructive characteristics of the surveyed buildings, ii) the material and structure properties (with particular reference to the constructive techniques and to the materials used for load-bearing masonry), iii) the materials and the techniques used for restoration before the earthquake, iv) the collapse mechanisms of the buildings and structures due to the earthquake, considering also the ones already retrofitted.

The object of the aforementioned research was not the single building, but the whole historic centre (even if small). Therefore, the strategic aim was also, besides collecting information on the effectiveness of the repair techniques adopted in the distant and recent past, to define a methodology for the analysis of the vulnerability of a building patrimony previously considered as minor, but with meaningful testimonies of cultural heritage. Hence there was the necessity of defining a "minimal" investigation program, eventually carried out by the Municipality or by the Province or Region, in order to support the designers in choosing the right analytical models for the safety definition and the appropriate intervention techniques for their projects (Binda et al., 1999; Binda et al., 2004a and 2004b).

It is possible to state that the seismic vulnerability assessment of historical buildings should consist of an articulated procedure which first of all takes advantage of two sources of information: indirect (as archives and bibliographic information, collected for reconstructing the evolution of the building from origin and its load history, also through the study of the earthquakes occurred in the past) and direct (as geometrical and photographic survey; typological analysis of the building, aimed at understanding the rules of behaviour in the process of formation and growing of the built types; stratigraphic survey, when possible, for gaining chronological information; survey of the masonry section and surface texture; survey of the crack pattern; analysis of the main structural elements including load-bearing walls, roofs, floors and vaults, staircases, and of their connections, damages, and effectiveness of past repair; laboratory characterisation of material samples; on site tests).

6. CONCLUSIONS

The recent frequent earthquakes in Italy provided an opportunity to learn from the failure mechanisms that occurred to both non retrofitted and retrofitted buildings. As in the case of conservation of monumental buildings, compatible repair techniques and materials have to be applied even to a simple dwelling. Knowledge of the structure typology and masonry morphology, of the material's chemical, physical and mechanical properties is necessary through an onsite and laboratory investigation carried out on each building.

When using new techniques and materials, experimental research has to be carried out before, not only on the mechanical behaviour but also on the physical and chemical compatibility with the existing structure and materials.

The collected data have to be documented for future interventions, possibly in the form of a data base extended to the whole historic centre.

The repair and retrofitting techniques have to be properly chosen according to the structure and material characteristics. There is not a single technique for every masonry or for every structural element, but the most appropriate for every case.

Guidelines should be available for both designers and end users describing the way of choosing the most reliable intervention.

ACNOWLEDGMENTS

The research has been partially supported within the GNDT - Framework Program 2000-2004 "Simulation of earthquakes and damage scenarios in urban areas. Vulnerability of historic centres and cultural heritage" supported by the Civil Protection Agency, Italy and by MIUR - Cofin, 2002.

REFERENCES

Archivio Storico Comunale di Norcia, 1861, Consigli e Riformanze reg. 151, cc 65-70. *Consigli dal 1858 al 22 gennaio 1861.*

Barucci C., 1990, *La casa antisismica: prototipi e brevetti*, Roma.

Bettio, C., Modena, C. and Riva, G., 1996, The efficacy of consolidating historical masonry by means of injections, *7th North American Masonry Conf.*, Notre Dame, USA, pp. 458-471.

Binda, L., Modena, C. and Baronio, G., 1993, Strengthening of masonries by injection technique, *6th North American Masonry Conf.*, Vol. I, Philadelphia, pp. 1-14.

Binda, L., Modena, C., Baronio, G. and Gelmi, A., 1994, Experimental qualification of injection admixtures used for repair and strengthening of stone masonry walls, *10th Int. Brick/Block Masonry Conf.*, Calgary, Vol.2, pp. 539-548.

Binda, L., Modena, C., Baronio, G., Abbaneo, S., 1997. Repair and investigation techniques for stone masonry walls, *Construction and Building Materials*, Vol. 11, N.3, pp. 133-142.

Binda, L., Baronio, G., Gambarotta, L., Lagomarsino, S. and Modena, C., 1999, Masonry constructions in seismic areas of central Italy: a multi-level approach to conservation, *8th North American Masonry Conference,* Austin, USA, CD-ROM.

Binda, L., Penazzi, D., and Saisi, A., 2003, Historic masonry buildings: necessity of a classification of structures and masonries for the adequate choice of analytical models, *6th Int. Symp. Computer Methods in Structural Masonry (STRUMAS VI)*, Ed. T.G.Hughes & G.N. Pande, Computers & Geotechnics Ltd, Roma 22-24/09/2003, pp. 168-173.

Binda, L., Cardani, G., Saisi A., Modena, C. and Valluzzi, M.R., 2004a, Multilevel approach to the analysis of the historical buildings: application to four centers in seismic area finalised to the evaluation of the repair and strengthening techniques, *13th Int. Brick/Block Masonry Conf.*, Rai, Amsterdam, CDROM.

Binda, L., Cardani, G., Saisi, A., Modena, C., Valluzzi, M.R. and Marchetti, L., 2004b, Guidelines for restoration and improvement of historical centers in seismic regions: the Umbria experience, *IV Int. Seminar on Structural Analysis of Historical Constructions,* Padova, pp. 1061-1068.

Binda L., Borri A., Corradi M., Tedeschi C., 2005, Experimental evaluation of shear and compression strength of masonry wall before and after reinforcement: deep re-pointing, *1st Canadian Conference on Effectiveness Design of Structures*, Hamilton, Ontario 10-13/07/2005, CD-ROM, pp. 293-304.

Borri, A. and De Maria, A., 2004, Alcune considerazioni in materia di analisi e di interventi sugli edifici in muratura in zona sismica, *XI Congresso Nazionale L'ingegneria Sismica in Italia*, Genova.

Corradi, M., Tedeschi, C., Binda, L., Baronio, G., 2005, Riparazione strutturale mediante ristilatura profonda dei giunti, *Proc. Workshop: Dalla conoscenza e dalla caratterizzazione dei materiali e degli elementi dell'edilizia storica in muratura ai provvedimenti compatibili di consolidamento*, 16-17/12/2004, DIS – Politecnico di Milano, Par. 2, pp. 75-84.

Genovese, M., 1915. *Trattato di costruzioni antisismiche.*

Giuffrè, A., 1993, *Sicurezza e conservazione dei centri storici: Il caso Ortigia*, Editrice Laterza, Bari.

Laefer, D., Baronio, G., Anzani, A., and Binda, L., 1996, Measurement of grout injection efficacy for stone masonry walls, *7th North American Masonry Conf.*, Notre Dame, USA, Vol. 1, pp. 484-496.

Milizia, 1554. *Principij di architettura*, Venezia.

Modena, C. and Bettio, C., 1994, Experimental characterization and modeling of injected and jacketed masonry walls, *Proc. Italian-French Symposium Strengthening and Repair of Structures in Seismic Area*, Nizza, pp. 273-282.

Modena, C., Binda, L. and Anzani, A., 1997a, Investigation for the Design and Control of the Repair Intervention on Historical Stone Masonry Wall, *7th Int. Conf. and Exhibition, Structural Faults and Repair 97,* Edinburgh, Vol. 3, pp. 233-242.

Modena, C., Zavarise, G. and Valluzzi, M.R., 1997b, Modelling of stone masonry walls strengthened by r.c. jackets, *Proc. STRUMAS – 4th Int. Symp. On Computer Methods in Structural Masonry*, Florence, Italy, pp. 285-292.

Penazzi, D., Valluzzi, M.R., Cardani, G., Binda, L., Baronio, G. and Modena, C., 2000, Behaviour of historic masonry buildings in seisimic areas: lessons learned from the Umbria-Marche earthquake, *12th Int. Brick/Block Masonry Conf.,* Madrid, pp. 217-235.

Penazzi, D., Valluzzi, M.R., Saisi, A., Binda, L. and Modena, C., 2001, Repair and strengthening of historic masonry buildings in seismic areas, *Int. Congr. More than Two Thousand Years in the History of Architecture Safeguarding the Structure of our Architectural Heritage*, Bethlehem, (Palestine), Vol. 2, Section 5.

Rondelet G., 1832. *Trattato teorico e pratico dell'arte di edificare*, Libro VII.

Scamozzi, 1615. *Dell'idea dell'architettura universale*, Venezia.

Tomazevic, M. and Turnsek, V., 1982, Verification of the seismic resistance of masonry buildings, *Proc. of the British Ceramic Society: Loadbearing Brickwork*, n. 30, pp. 360-369.

Tomazevic, M., 1992, Laboratory and in situ tests of the efficacy of grouting and tying of stone masonry walls. *Int. Workshop on Effectiveness of Injection Techniques and Retrofitting of Stone and Brick Masonry Walls in Seismic Area*, ed. by L. Binda, Milan, pp. 95-116.

Valluzzi, M.R., da Porto, F., Modena, C., 2004, Behaviour and modeling of strengthened three-leaf stone masonry walls, *Materials and Structures*, MS 267, Vol. 37, April 2004, pp. 184-192.

Chapter 4: Conservation Approaches, Applications, Case Studies

Chapter 4.1

THE DIFFICULT CHOICE OF MATERIALS FOR THE RECONSTRUCTION OF THE CATHEDRAL OF NOTO

Luigia Binda, Giuliana Cardani, Claudia Tiraboschi
DIS - Politecnico di Milano, P.za L. da Vinci, 32, 20133, Milan, Italy

Abstract: In 1996, the Cathedral of Noto (Sicily), suddenly collapsed partially, after being damaged by the 1990 earthquake, which hit the eastern part of Sicily. As the continuous propagation of cracks and damages in the pillars (made with external leaves in limestone and filled internally with rubble) due to the weak materials used and the poor construction adopted was considered as the main cause of the collapse, it was decided to rebuilt the collapsed and part of the damaged pillars adopting the traditional techniques and the same materials used in the past, with improvements of the pillars technique of construction. A difficult task was the choice of the stones for the reconstruction, first of all the choice of a suitable quarry and then the control of the stone properties at each delivery.

Key words: limestone; Noto cathedral; experimental investigation.

1. INTRODUCTION

The Cathedral of Noto (Sicily, Italy) dedicated to S.Nicolò was built after an earthquake which hit the eastern part of Sicily in 1693. On March 13, 1996, the Cathedral of Noto suddenly collapsed partially, after being damaged by the 1990 earthquake. The four left pillars of the central nave, the vault, the domes of the right lateral nave, the transept roof and vaults and three quarters of the central dome were lost. Following destruction a choice was made to reconstruct the collapsed part with local traditional materials. The Cathedral had been built in different phases beginning in 1764. In 1780, the dome collapsed and the church was reopened in 1818. In 1848, the dome collapsed again due to an earthquake, it was rebuilt, and the church reopened again in

S.K. Kourkoulis (ed.), Fracture and Failure of Natural Building Stones, 185–200.

1862 although the dome was not completely finished until 1872. In 1950, the Cathedral was restored with new plasters and paintings, and the timber roof was substituted by a concrete structure. The work continued until 1959.

An extensive investigation was carried out by D.I.S. Politecnico di Milano laboratory, responsible L. Binda, on the materials and structures, in order to: (i) find the causes of the collapse, (ii) characterise the materials, (iii) identify the structural damages on the remaining parts of the church, (iv) choose the materials for repair and reconstruction.

The extensive experimental and numerical investigation carried out after the removal of the ruins, showed that the collapse started from the pillars (one or more), due to the damages they had already accumulated before the earthquake. The pillars showed a state of damage due to compression with vertical cracks covered. The poor construction technique and the use of the weak limestone were probably the cause of the damages to the pillar of the Catherdral, even if a clear crack pattern was reported to have appeared only after the 1990 earthquake.

A unanimous decision was made by the authorities and the Noto citizens to reconstruct the collapsed parts of the Cathedral, as they were before the collapse, with the same type of materials used in the past, improving the structural elements, which were indicated by the investigation as being the weakest. A difficult task was the choice of the stones for the reconstruction. First of all it was necessary to find the right quarry, since several quarries were closed in order to protect the environment. The choice was made by testing physically and mechanically the stones which were coming from different quarries. Once the quarry was chosen continuous experimental studies were made on the cut stones statistically sampled in order to control their quality. A rather difficult task was the quality control on site. Rather complicated was also the preparation of the stones in order to reduce the influence of their rather high absorption on the mortar hardening and on the bond between stones and mortar.

The methodology used for the investigation and the results will be reported and discussed and some guidelines will be suggested for the future intervention, also in case of small reconstructions in seismic areas, after an earthquake damage.

2. CHARACTERIZATION OF THE ORIGINAL STONE

The losses caused by the collapse were the following: 4 pillars of the right part of the central nave and one of the 4 pillars supporting the main dome and the transept, the complete roof and vault of the central nave, three quarters of the drum and dome with the lantern, the roof and vault of the right part of the transept and part of the small domes of the right nave (Figures 1, 2). The right pillars were named A, B, C, D, E; the left ones were named A′, B′, C′, D′, E′ starting from the internal part of the façade (Figure 1).

The investigation carried out by the authors, together with the designers of the reconstruction, showed that the collapse certainly developed starting from one or more of the right pillars of the central nave. Being the support for the dome, these pillars consisted of a multiple leaf structure in which the external leaf made with regular stones confined a central rubble masonry core made with calcareous stones of different dimension and shape. The external leaf (except for the base of the piers) was made with regularly cut blocks from the local "travertine" also called calcareous tuff. This material came from sedimentary carbonate deposition in the presence of turbulent waters, and it is rich in voids of various shape and dimensions, which previously contained vegetarian and organic parts later on dissolved. The height of the blocks varies from 24 to 26 cm and the thickness, which is very small

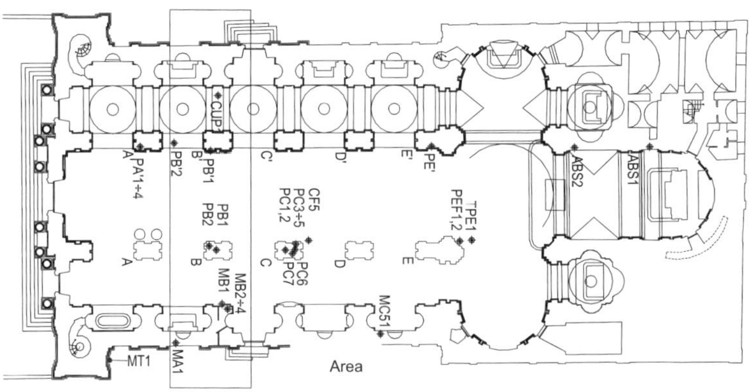

Figure 1. The Noto Cathedral: plan of the remaining parts and samples locations.

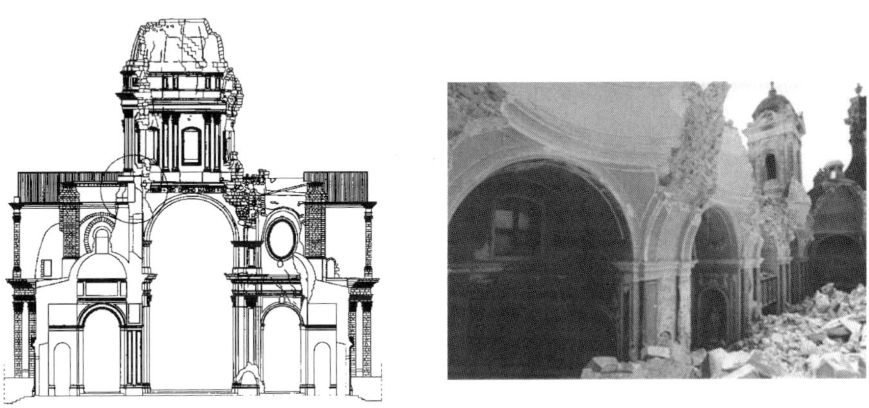

(a) (b)

Figure 2. (a) The transversal section and (b) the collapsed Noto Cathedral.

compared to the pillar dimensions, varies from 25 to 30 cm. No effective connection was realized between this external leaf and the core (Figures 3, 4). The external part of the base is made with regular blocks of limestone (calcarenite) which have a greater thickness and a better strength than the travertine (see next sections).

The inner part of the pillars represents 55% of the entire section, while it represents 58% in the piers supporting the dome. This part, with irregular by-cut stones up to the half of the total height, was made with large round river pebbles (Figure 3). Nevertheless, every 50 cm a course, made with small stones and mortar, was inserted in order to obtain certain horizontality (Figure 4). Scaffolding holes were left everywhere, some crossing the whole section. The mortar made with lime and a high fraction of very small calcareous aggregates appeared to be very weak. Also, the bond between the mortar and the stones was very weak; in fact, it was possible to sample stones and pebbles from the interior of the pillars without any difficulty and with the stones completely clean of mortar.

The left pillars, still covered with a thick plaster, showed vertical cracks at the bottom (Figure 5a), but the concern that the damage could be inside and perhaps even present before the 1990 earthquake, suggested that the plaster made in the fifties should be removed and a series of large vertical cracks were found, some of which were filled with the gypsum mortar used for the

Figure 3. Detail and horizontal reconstruction of a pier section.

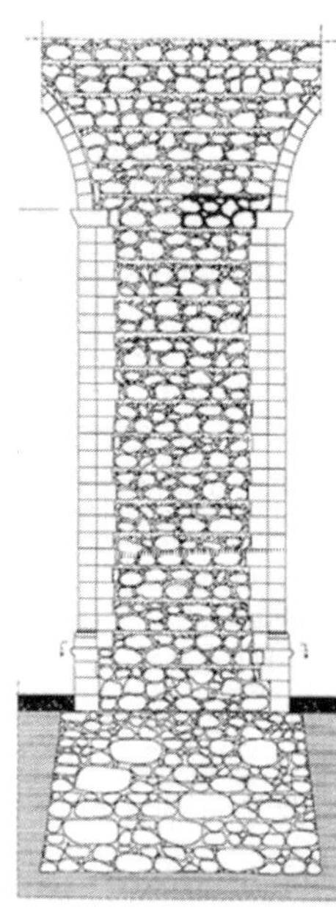

Figure 4. Reconstruction of a vertical section.

(a) (b)

Figure 5. a) Large crack found in a pillar and b) Example of a crack filled with gypsum mortar in the sixties

plaster (Figure 5b). This finding was the proof that the damage was already present before the last earthquake. The pre-existing crack pattern showed clearly damage from compressive stresses, indicating long term damage dating probably even before the rendering. This damage would have probably progressed even without the earthquake, which only accelerated the collapse[1].

Four main types of stones were used in the construction of the Cathedral:

1. limestone (Noto stone, a calcarenite), as regularly cut blocks in the external leaf of pillars and built-in columns, but only for the base, and also used for the internal leaves as sharply cut pebbles,
2. "travertine", as the limestone, but in much larger quantities,
3. "*giuggiolena*", a sort of compact "travertine" for arches and dome voussoirs and
4. round (boulder) river stones used in the internal part of the pillars.

For each type of stone some blocks were sampled in order to carry out chemical, petrographic, physical, and mechanical tests[1]:

1. The Noto limestone (a calcarenite) comes from a calcareous rock with fine grains, finely porous, and of light yellow colour. Observed with polarized light in a thin section, the stone has the characteristics of a fine grain limestone with small pores rich in foraminifer calcareous fossils. Furthermore small black and ochraceous masses were present.
2. The "travertine" is a carbonate rock of light yellowish color with high percentage of voids probably from the same quarries as the calcarenite. In thin section, the carbonate nature is confirmed; numerous irregular voids appear and the material is mainly calcite with very fine grain size which are clean or impregnated with a fine brown material probably clay.
3. The "giuggiolena" is composed by carbonatic rock fragments with fine grain, friable and of brownish colour. In thin section the rock appears to be made by rather round elements with dimension around 0.1 mm; the calcitic cement is scarce and the material is very porous.

Table 1. Physical test on calcarenite and "Giuggiolena" specimen

SPECIMEN		Bulk density dry (kg/m^3)	Bulk density satured (kg/m^3)	Absorption coeff. (total immers. 24h) (%)	IRS (kg/m^2.min)
Calcarenite	CNP-C51.1	1726	1969	14	5.55
	CNP-C51.2	1746	1989	14	6.78
	CNP-C51.3	1689	1947	15	5.81
"*Giuggiolena*"	G.1	1585	1858	17	2.44
	G.2	1598	1881	18	2.50
	G.3	1536	1834	19	2.46

From each stone, 3 to 5 cylinders of 50 mm diameter and 100 mm height were cored for the physical and mechanical tests. It was impossible to have larger dimensions, perhaps more representative of the material, due to the small dimension of the cut stones. Chemical analyses were also carried out in order to find the eventual presence of sulfates.

The physical tests were only carried out on the calcarenite and on the "*giuggiolena*"; in fact, the "*travertine"* voids were filled with mortar or clay and therefore the results were not reliable. It can easily be seen that the stone absorbs a very high quantity of water when saturated, and that the IRS coefficient is very high. This fact usually means low strength and low durability.

Compression and indirect tensile (splitting) tests were carried out on dry and saturated specimens. The results and discussion are reported in the following. Due to the peculiar use of the results and to the dimension of the specimens, the tests did not refer to a specific code of standard[2].

Uni-axial compression tests: no material was put between the specimen and the machine platen due to the dimension of the specimens which already assures pure compression in the central part. For the vertical displacements two LVDT's were applied between the plates and three extensometers (DD1) were directly applied to the specimen. For the lateral displacements a ring clip gauge was used. A hydraulic press with a load cell of 30 to 50 KN was used for the tests, and the tests were carried out in displacement-control at a speed of 4 μm/s. The specimens were tested under two different conditions: dried to constant mass and saturated up to constant mass.

For the detection of the tensile strength, the authors have chosen the splitting test which is easy to perform on cylinders and gives reliable results. The test was carried out according to the Italian code specification (UNI 6135, 1972). Also in this case, the specimens were tested dry and saturated.

The calcarenite used for the external leaf of the pillars is characterized by a fairly good strength when tested dry up to constant mass (18.0 average N/mm^2 in compression, 2.2 N/mm^2 in tension for cylinders of 50 mm diameter and 100 mm height) and by a much lower strength when tested saturated up to constant mass (11.6 N/mm^2 in compression and 1.3 N/mm^2 in tension), as showed in Figure 6. This means a reduction of 35.5% in the compressive strength and of 39% in the tensile one. The "*travertine*" used in the core of the pillars and in the external leaf above the base has a much lower strength than the calcarenite (5.2 N/mm^2 in compression as an average value measured on dry prisms of 100x100x200 mm dimension). The different mechanical characteristics of thc two stones were also detected by sonic tests in laboratory and on site[3]. The other type of stone, very light, called "*giuggiolena*" was used for the construction of the arches and dome. This stone also shows a different behaviour when tested dry or saturated: 5.3 N/mm^2 in compression and 1.0 N/mm^2 in tension, when dry, 5.05 N/mm^2 in compression and 0.8 N/mm^2 in tension, when saturated.

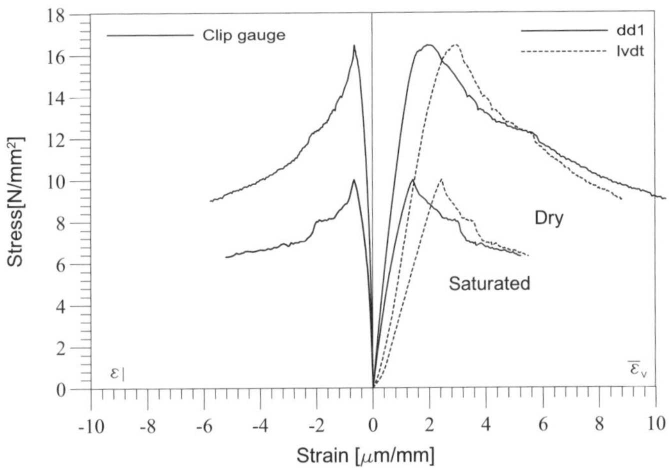

Figure 6. Stress-strain plots of compression tests on saturated and dry calcarenite specimens.

3. CHOICE OF THE NEW STONE

The extensive experimental and numerical investigation carried out after the removal of the ruins showed that the collapse started from the pillars (one or more), due to the damages they had already accumulated before the earthquake. The pillar showed a state of damage with vertical compression cracks covered by the plaster.

A unanimous decision was made by the authorities and by the Noto citizens to reconstruct the collapsed parts of the Cathedral as they were before the collapse, with the same materials used in the past, improving the structural elements which were indicated after the investigation as being the weakest. It was decided to rebuild not only the collapsed pillars of the central nave, but also to demolish and rebuild the survived ones. This was a possible alternative, taking into account the necessity of symmetry in the structure due to the fact that Noto is situated in an active seismic area.

3.1 Search for the quarries

Following experimental investigation, it was concluded that the new stones should be taken from the original or similar quarries. It was necessary to find the right quarry, since several quarries were closed in order to respect the environment. In order to choose a limestone adequate for the reconstruction, samples were taken from different quarries. Even if it was impossible to use the same limestone as the original one, care had to be taken in order to use new stone with similar characteristics to the ones used for the construction of the Cathedral for several reasons. The new stones should have: a) chemical, physical and mechanical compatibility with the original materials, b) good mechanical and physical characteristics and c) minimum salt content for a better durability.

Six quarries situated in the area of Noto were visited by the authors and by the designers (Figure 7). The stones sampled from the quarries were sent to the DIS-Laboratory in Milan for characterization. As to the quarries, the stones were named E, N, D, Dg, S, B, I (D and Dg come from the same quarry). Cylinders were cored from the stones, of 50 mm diameter and 100 mm height. The dimensions are not always representative of the stone structure, but they were chosen for uniformity allowing comparing the results of the tests to those of the original stones. Physical and mechanical tests were carried out after mineralogical identification of the material[1,4].

Figure 7. Limestone quarries.

3.2 Test on the sampled new stones

Physical tests: apparent bulk density in dry and saturated state is very important because the limestone behaves differently in the two states and this property certainly influences the mechanical strength and the long-term behaviour of the stone. The absorption is very important for durability and the initial rate of suction defines the retentivity of the mortar in order to obtain a good bond between mortar and stone.

The results of the tests are reported in Table 2. It has been confirmed that the density shows a high increase in the saturated state for all the stones. The lowest absorption belongs to the stone I followed by Dg, D, S and B. The best rate of suction (for the bond with mortar) belongs to D, Dg and B.

Mechanical tests: uniaxial compression and splitting tests were carried out on the seven types of stone. The procedures were the same as for the original stones. These results were compared in order to choose the appropriate quarries and were also compared to the ones obtained from the tests carried out on the stones recovered from the collapsed pillars of the Cathedral[4,5].

Table 2. Results of physical tests

SPECIMEN	Bulk density dry (kg/m^3)	Bulk density satured (kg/m^3)	Absorption coeff. (total immers. 24h) (%)	IRS ($kg/m^2 \cdot min$)
E	1679	1956	16	2.70
N	1744	2026	16	3.16
D	1864	2103	13	1.68
Dg	1853	2079	12	1.53
S	1770	2033	15	2.37
B	1818	2060	14	1.53
I	2070	2221	7	1.97

The mechanical properties: Compressive and tensile strength, F_c and F_t, respectively, Young's secant modulus, E_{sec} and Poisson's coefficient, $\Delta\varepsilon_l/\Delta\varepsilon_v$ are reported in Table 3. The secant modulus was calculated in an interval between 30% and 60% of the peak stress: in this interval the behaviour of the material could be considered as linear. The test was carried out on specimens dry and saturated to constant mass and the specimen dimensions were: 50 mm diameter and 100 mm height.

Table 3. Compressive and indirect tensile test results

SPECIMENS (50x100 mm)	Dry				Saturated			
	F_c [N/mm^2]	E_{sec} (30-60%) [N/mm^2]	$\Delta\varepsilon l/\Delta\varepsilon v$ (30-60%)	F_t [N/mm^2]	F_c [N/mm^2]	E_{sec} (30-60%) [N/mm^2]	$\Delta\varepsilon l/\Delta\varepsilon v$ (30-60%)	F_t [N/mm^2]
D	23.8	7487	0.25	1.7	10.9	4707	-	1.1
Dg	21.1	6530	0.18	2.0	10.0	4187	0.30	1.1
S	17.0	5617	0.15	1.5	7.8	3450	0.29	1.0
N	15.7	6137	0.19	1.3	11.6	5327	0.30	1.2
E	12.3	5627	0.23	1.4	10.2	4230	0.34	1.2
B	20.7	8690	0.17	2.0	11.6	6197	0.43	1.2
I	14.6	8720	0.09	2.4	10.5	4863	0.07	1.4

3.3 Choice of the quarry

In order to verify the performance of the stones and choose the right quarry, the values of bulk density, absorption by total immersion, Young's modulus and tensile strength were reported against the compressive strength of dry and saturated specimens (Figure 8(a,b,c,d)). Even though the number of specimens was low and statistically not representative, it was possible to find some linear correlation.

It is also possible to conclude from Tables 2 and 3 that the two stones of the quarry D and the one of the quarry B gave similar values and seem to have the best behaviour; furthermore the stone B tends to have a higher Young's modulus that can indicate also a better hardness.

The capillary rise test on the stone D and B gave respectively capillary rise coefficient values of 0.0105 and 0.0114 ($kg/m^2min^{1/2}$), which are very similar. In order to better distinguish the durability of the two stones Amsler tribometer tests were carried out (according to the code R.D. 16/11/1939) which allow to calculate the surface loss under a turning carborundum brush. The stone D with a loss of 27.91 mm after 1,000 m running showed to be softer than the stone B which lost only 17.51 mm.

Therefore the two types of stones coming from the quarries D and B were chosen as suitable for the reconstruction.

4. CHARACTERIZATION OF THE STONES AT EACH DELIVERY

Following the study carried out both on site and in laboratory and the decision to also demolish and reconstruct the remaining left pillars, the design for the new elements was to be decided. The design should respect the geometry of the original pillars, and the technique of construction would provide a section with a certain amount of filling, but with much better connections. The choice of good mortars was important since the limestone used is a very soft material and sensible to the moisture content. Furthermore, the mortars had to be compatible with the stones and be free from sulphates.

Once the materials had been chosen, a constant quality control from the site was required by the designers. A difficult task was the quality control on site. Rather complicated was also the preparation of the stones in order to reduce the influence of their high absorption on the mortar hardening and on the bond between stones and mortar.

Minimum values for the compressive and tensile strength were specified for the contractor involved in the reconstruction, respectively, as 16.5 N/mm^2 dry, 9.5 N/mm^2 saturated in compression and 1.45 N/mm^2 dry, 0.95 N/mm^2 saturated in tension on cylinders cored from the stone blocks with dimensions:

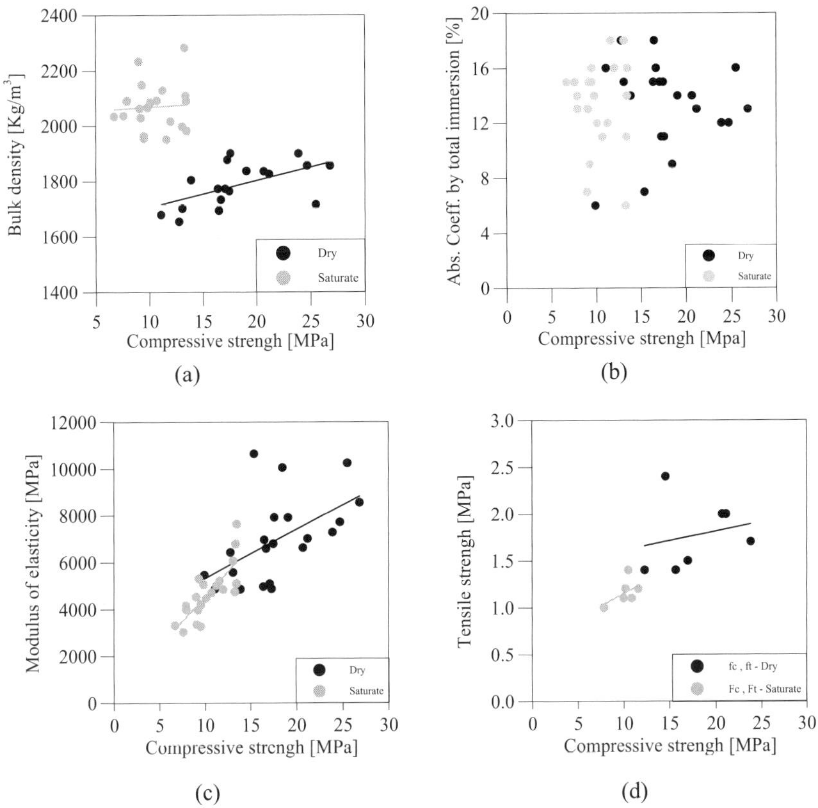

Figure 8. Tentative correlations between compressive strength (a) and, respectively bulk density, absorption coefficient for total immersion (b), Young's modulus (c) and tensile strength (d) for dry and saturated specimens.

80 mm diameter and 160 mm height. These new dimensions were chosen due to the fact that the blocks could be cut with dimensions of 200x200x400 mm.

Unfortunately, after the design was approved, due to the difficulty of finding good quarries, the contractor could only provide the type D stone for the foundations and the type B for the base of the new pillars. For the remaining parts of the pillars two types of stones called NTB and BS were found; but their strength was rather low (even below the minimum values, particularly in tension). Nevertheless, the designers accepted this material, taking into account that the minimum values incorporated large safety factors.

4.1 Experimental results

In laboratory, a continuous control of the stone quality was made every time a new supply came from the quarry to the site[6]. A similar control was made on the mortar sampled on site during the visit of L. Binda and/or G. Baronio who were consultants for the Prefect of Syracuse[3], responsible for the work of reconstruction made with the public support of the Province of Syracuse.

The cylinders were cored from each block in two directions A (rift plane) and B (perpendicular to rift plane) in order to characterize the eventual anisotropy of (Figs.9a and 10a). In Figures 9(b,c) the results of the compressive tests and in Figures 10(b,c) the results of the tensile strength carried out on stone cylinders in the state dry and saturated during subsequent testing campaigns are presented, compared to the strength limits required by the contractors; they show the variability in strength of the different types of stones and how the quality of the stones tended to decrease at each delivery.

The difference between the two directions A and B is not so evident, indicating low anisotropy. Subsequent deliveries of stone blocks show a decrease of compressive strength values below the requested one, while the indirect tensile strength varies according to the stone type. Large difference in strength shown by dry and saturated specimens was also clear. Saturation of the stones always induces a reduction in compressive strength of about 40%.

The values obtained for the original limestone seemed to be more homogeneous around an average value of 17.89 N/mm^2. If the designers wanted to

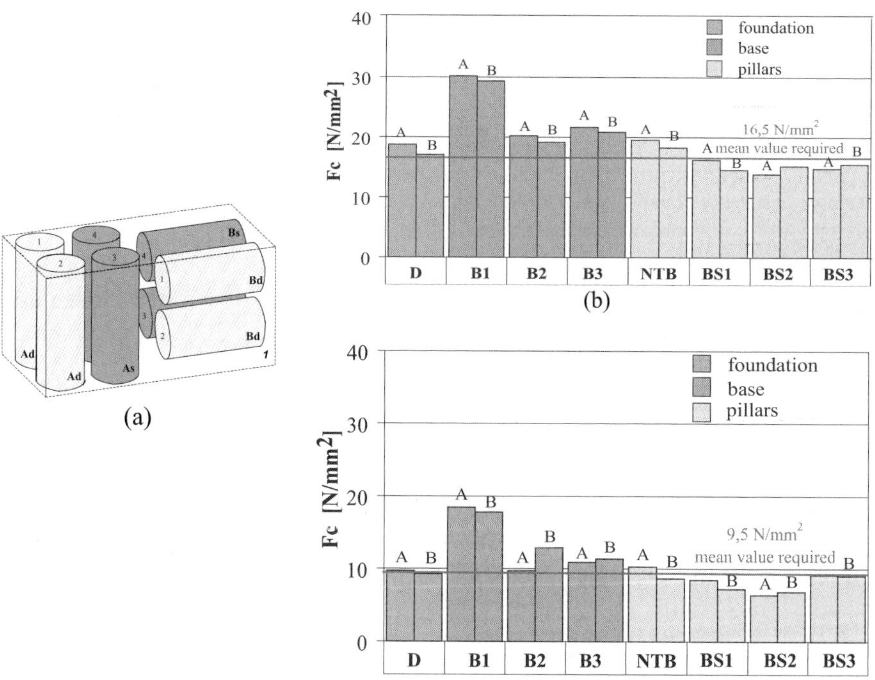

Figure 9. (a) Scheme of the stone specimen coring; (b) compressive strength in two directions of dry cylinder type D, B, NTB, BS at various times (numbers after letters indicate different supplies) and (c) compressive strength in two directions of saturated cylinder type D, B, NTB, BS at various times.

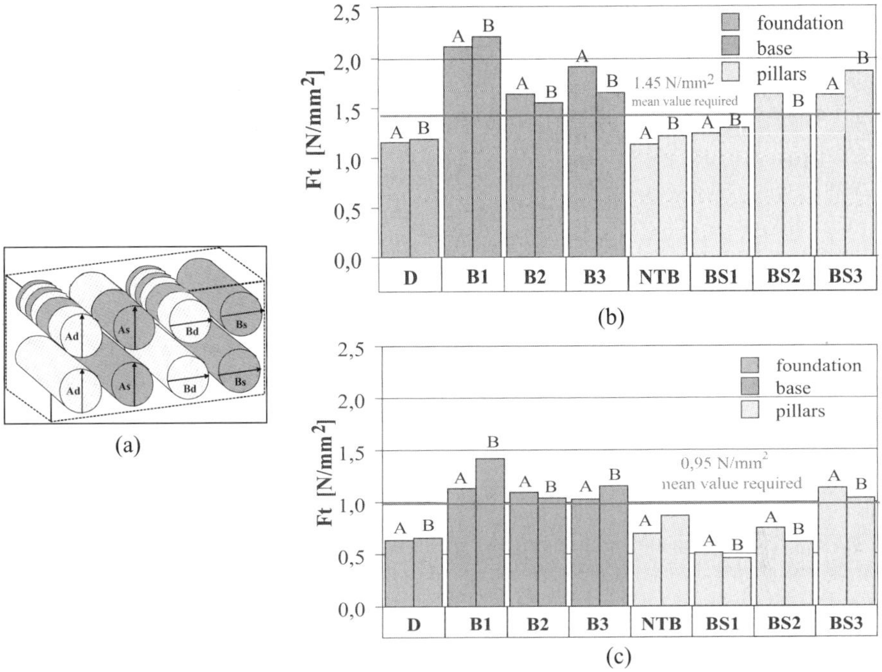

Figure 10. (a) Scheme of the stone specimen coring. (b) tensile strength in two directions of dry cylinder type D, B, NTB, BS at various deliveries (numbers after letters indicate different supplies) and (c) tensile strength in two directions of saturated cylinder type D, B, NTB, BS at various deliveries.

use a stone with similar strength as the original one then the S stone should be used. As a slightly higher strength was needed, stones B or Dg were employed. It is in any case better to use a quarry, which gives stones with less scattering in strength.

Figure 11 shows the elastic modulus against the compressive strength of the stones from quarry B, achieved by two different strain measurement devices. There is a good correlation between the values, in particular with the measurements made with the instrument LVDT positioned between plates, while measurements made with DD1 directly applied on the specimens are more scattered. The lower strength values are referred to saturated specimens and the higher to the dry specimens. This behaviour is quite similar for all the analysed stones as shown in Figure 12.

The consultant requested to saturate the stones on site. This request was based on the observed high absorption of the stones which could influence the behaviour of the mortar used for the joints. The presence of swelling clay inclusions in the stones (due to the characteristics of the quarry) caused the fracture of some percentage of stones during the immersion in water (Figure 13). This phenomenon suggested using the operation also as a quality control useful to eliminate the defective stones.

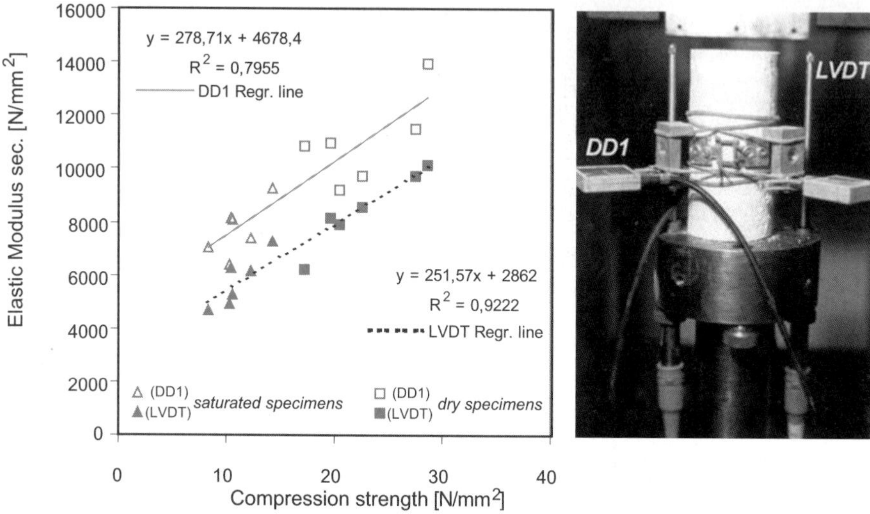

Figure 11. Correlation among the compressive strength values and the elastic secant modulus for the calcarenite B with different measurement devices.

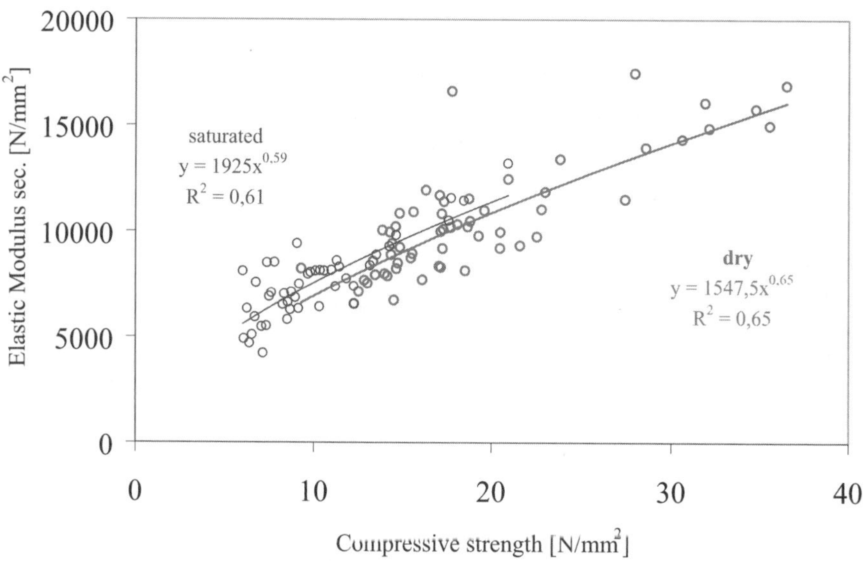

Figure 12. Compressive strength against elastic secant modulus of all the different types of characterised stones distinguished between saturated and dry specimens.

5. CONCLUSIONS

The high level of difficulty of choosing and controlling the materials for the reconstruction of partially collapsed historic buildings has been shown.

Figure 13. Creaked blocks in quarry.

It was decided to rebuild not only the collapsed pillars of the central nave, but also to demolish and rebuild the lowest part of the survived ones. This was a possible alternative, taking into account the necessity of symmetry in the structure due to the fact that Noto is situated in an active seismic area.

The efforts made and the frustrations encountered during reconstruction show the difficulties in applying the approach to the following steps: i) finding a better technique for the reconstruction of the collapsed pillars due to the original use of materials that were too weak, ii) choosing the quarries and the most appropriate type of stones, to replicate the original ones, iii) choosing the most appropriate and compatible mortars following the decision that the traditional ones could not be remade, iv) studying the right thickness of the joints and the most appropriate grain size distribution of the mortar aggregates, v) controlling the supply of materials and the workmanship.

A rather difficult task, together with many others, was the quality control on site. Complicated was also the preparation of the stones in order to reduce the influence of their high absorption on the mortar hardening and on the bond between stones and mortar.

Determining the stone properties at each delivery was important in order to control the quality of material. In fact, it was possible to detect the variation in strength due to different quarry. This action allowed in fact keeping under control the minimum values also, even if they were accepted by the designer.

REFERENCES

1. G. Baronio, L. Binda, C. Tedeschi, C. Tiraboschi, Characterization of the materials used in the construction of the Noto Cathedral, in *Construction & Building Materials*, Special Issue, Elsevier, vol.**17**, N. 8, pp. 557-571 (2003).

2. L. Binda, C. Tiraboschi., G. Mirabella Roberti, G. Baronio, G. Cardani, Measuring masonry material properties: detailed results from an extensive experimental research, Part I 'Tests on masonry components', Report 5.1, CD-ROM, April 1996, pp. 1-263 (Milano, 1996).
3. L. Binda, A. Saisi, C. Tiraboschi, S. Valle, C. Colla, M. Forde. Application of Sonic and Radar Tests on the Piers and Walls of the Cathedral of Noto, Int. Symposium. on: Cattedrale di Noto e Frauenkirche di Dresda: Indagini e Studi per il Progetto di Ricostruzione, 10-12/02/01, pp. 91-106, (Noto, 2002).
4. L. Binda, G. Baronio, C. Gavarini, R. De Benedictis, S. Tringali. Investigation on Materials and Structures for the Reconstruction of the Partially Collapsed Cathedral of Noto (Sicily), *6° Int. Conf. Structural Studies, Repairs and Maintenance of Historical Buildings*, STREMAH 99, Germany, p.323-332 (Dresden, 1999).
5. L. Binda, Baronio G., Contratto di Consulenza: Studio sullo stato di conservazione dei materiali e delle strutture e sulla possibilità di un loro recupero in base ai risultati delle prove e dei rilievi di laboratorio e in situ, Final Report, Contratto Prefettura di Siracusa, (Milano, 1998).
6. G. Baronio, L. Binda, G. Cardani, C. Tedeschi, The Difficult Choice of Traditional Materials for the Reconstruction of a Partially Collapsed Historic Building: the Cathedral of Noto, 9NAMC, *9th Int. North American Masonry Conf.*, 1-4/6/2003, South Carolina, USA, CD-ROM, pp. 942-953, (Clemson, 2003).

Chapter 4.2

TECHNOLOGICAL AND CONSERVATION ASPECTS VERSUS URBAN APPEARANCE IN A STONE-BUILT ENVIRONMENT: AN EVALUATION APPROACH

A. Lobovikov-Katz

Centre for Conservation and Western Galilee College, Akko; Faculty of Architecture and Town Planning, Technion - Israel Institute of Technology, Technion City, Haifa, 32000, Israel; anna@technion.ac.il; lokatz@netvision.net.il

Abstract: An approach for the evaluation of the architectural impact of stone features and the deterioration and conservational interventions on historic buildings is proposed. The approach involves analysis of the geometric and material characteristics of the building material (stone) with a view to provide an evaluation tool (ETC) for the correlation between the stone unit and its built environment.

Key words: ETC analysis; correlation; hierarchy levels; complexity grade; building material (stone).

1. INTRODUCTION

The ETC analysis helps reconcile the often contradictory aspects present in almost every conservation effort: how to balance between preservation of the architectural/aesthetic aspects on the one hand - and the needs of material conservation on the other.

The two main groups of factors[1,2] influencing the visual image of the built environment are considered by means of ETC: stone technology and material factors, and architectural and urban factors. The first group deals with the parameters of the stone units – from the original ones through their deterioration to conservation; the second group deals with the "macro geometry" of the built environment and its elements. The information thus acquired and analyzed would contribute to planning the conservational intervention and the maintenance mode.

S.K. Kourkoulis (ed.), Fracture and Failure of Natural Building Stones, 201–214.

ETC analysis is multi-level and was initially developed with particular focus on stone, but is applicable to other selected characteristics of historic buildings. It is suitable both in preliminary field studies and in processing their results.

2. DESCRIPTION OF ETC

The ETC analysis is implemented in two parallel directions: "A" and "B" by means of similarly designated tables.

Direction "A" deals with the spatial/visual characteristics of building and its elements on different levels.

Direction "B" deals with the technological characteristics of the built environment, including the conservation aspects. Implementation of the two directions permits the subsequent synthesis.

ETC considers a building as a whole in both its material and aesthetic aspects and aims at clarifying and defining the role of stone in this combination of art and technology. The "architectural geometry" analysis in "A" "breaks down" the building to the level of single stones, with a view to better understanding the whole system and provision of a compatible conservation mode and maintenance strategy.

2.1 "A" Tables

Tables of this type help to analyze the "architectural geometry" - the spatial/visual characteristics of the building and its elements on different levels (see examples of the analysis/documentation Tables: A1; A2; A3).

2.1.1 Architectural geometry characteristics

In the present context, the architectural shape of a building/structure under consideration may be studied through a number of its characteristics:

- complexity degree (grade);
- spatial spreading directions of the structure as a whole and its parts as related to the nearest environment;
- physical size, overall and that of the parts;
- "transparency" (openness) level of the structure's shell: "open" / "closed" ratio – relative visual "weight" of openings (windows, doors, etc.) versus the solid built surfaces;
- size and proportion of the built volume versus the adjoining open space (street, square, etc).

2.1.2 Hierarchy levels

The analysis is also related to different hierarchy levels in the building:

- level 1 – the building as a whole;
- level 2 – the parts of the building (facades, bulks, wings);
- level 3 – the elements of building - structural, functional, decorative (door and window openings, niches, corners, pilasters and other protruding and recessed elements);
- level 4 – architectural and structural details.

2.1.3 "A" Table mapping (schematic illustrative description of spatial/visual characteristics).

On the basis of both the architectural geometry characteristics and the hierarchy levels it is possible to illustrate the spatial/visual characteristics.

2.1.3.1 **Spatial axes system (column "r" of "A" Table)** (Scheme 1)

- **"x"** – (in horizontal plane) parallel to spreading of the building front along the street (or other urban/rural space).
- **"y"** – (in horizontal plane) normal to the axis of the main façade/street (or other urban/rural space).
- **"z"** – (normal to horizontal plane) along the height of the building and its different parts and elements.

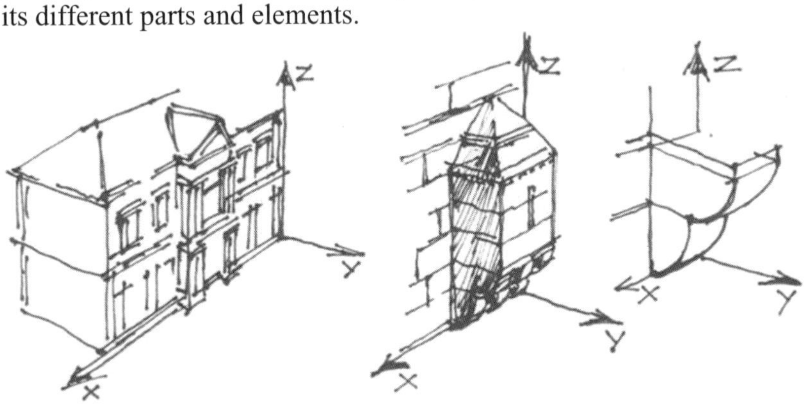

Scheme 1

Spatial axes system and its application to the analysis of the building and its components

2.1.3.2 Complexity grade (column "p" of "A" Table).

Grade 1 – building consisting of a single architectural volume.

Grade 2 – building composed of a very small number of primary architectural volumes.

Grade 3 – complex composed of multiple primary architectural volumes. (See Scheme 2).

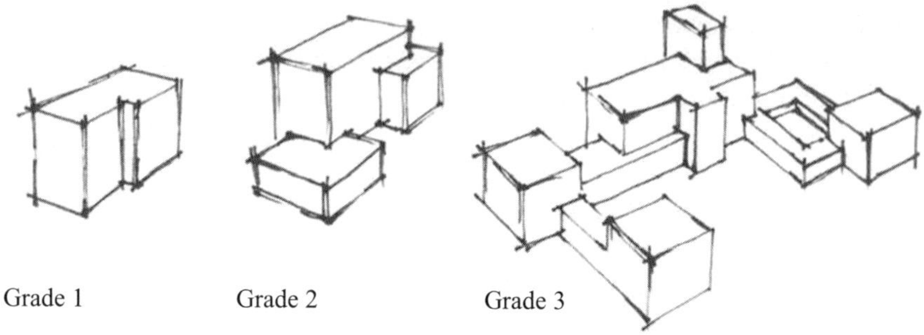

Scheme 2

Complexity grades

2.1.3.3 General geometry and "inner" complexity study of building and its parts (columns "a" to "i" of "A" Table).

This section of the "A" Table uses the same axes as in paragraph 2.1.3.1 (spatial environmental context): "x" – width, "y" – horizontal depth, "z" – height, and provides 3D information on the building design: dominant direction (if any), straight lines and/or curves of different types, etc.

The approach used in the complexity study of the building as a whole (see §2.1.3.2) is applicable for an "inner" complexity study. In this case each "A" Table (A1 to A4) will refer to a specific hierarchy level as in paragraph 2.1.2 (see Scheme 2).

2.1.3.4 "Transparency" level of building shell (column "q" of "A" Table).

This section concerns the "open"/"closed" ratio - the relative visual "weight" of openings-windows, doors, arcades, - vs. the built mass of facades/ bulks. The value varies from 1 (most solid) to 5 (most open) (Scheme 3).

Scheme 3

"Transparency" level

2.1.3.5 Vertical variation of design character (columns "j" to "m" of "A" Table).

This section concerns the vertical variations (if any) of the character of the building (number of the storey types). The separate treatment of the height aspect is due to its special role in the architectural image. It is studied in terms of the architectural linkage of the building to its grounds on the one hand, and to the skyline on the other.

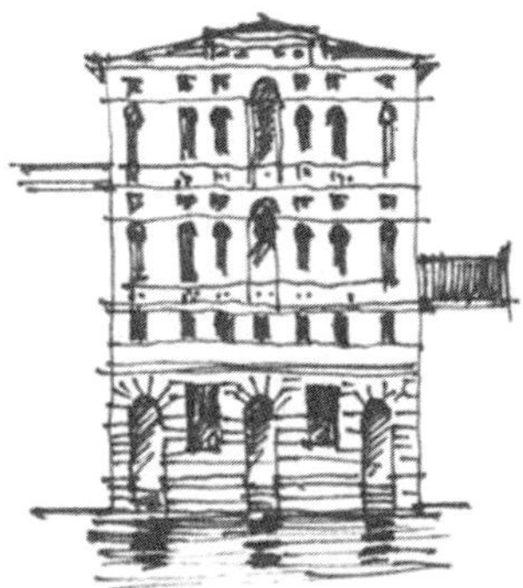

Scheme 4

Design metamorphoses over the height of the building

2.1.3.6 Spatial environmental context (columns "n, o" of "A" Table).
A 3D analysis of the relationship between the building and the adjoining urban/rural space and their proportion ratio (height of the building façade versus width of space). The spaces may be both urban and rural, i.e. the building may border a street or a valley, etc.

Corresponds to three axes: x, y, z.

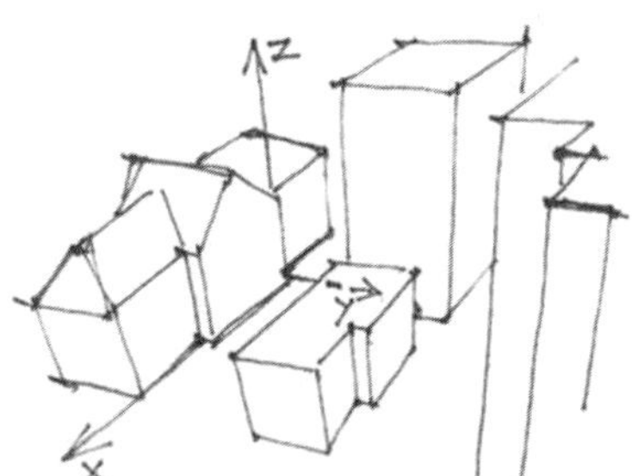

Scheme 5

Building – urban context interaction via spatial axes

2.1.4 "A" Table: special features

A building on the overall level will be studied through a single "A" Table A1, where "1" denotes hierarchy level 1 (see §2.1.2). Table A2 refers to hierarchy level 2 – parts of the building. The scope and the number of the Tables A3 (elements) and A4 (details) will depend on the complexity grade as well as on that of the parts and elements. This may be illustrated by study of the height of the building. The height (column "k") is defined by a number of storeys. The "A" table provides an option to specify the maximum number of storeys in the building as a whole (in A1) and for each of its parts (in A2).

2.1.5 "A" Tables: scope of object types

The "A" Tables are also applicable to the study of other types of structures, e.g. bridges, street walls, sculptures, stone benches, fountains, pavements, stairs, etc. These may be examined in accordance with their type, size and location. A sculpture, for example, may be considered as an element (hierarchy level 3), and a fountain as a part of a building (hierarchy level 2, dominant geometrical dimension e.g. "y").

2.2 "B" Tables

The "B" Tables" concentrate on the analysis of the technological characteristics of the built environment, including conservation aspects. As in the "A" Tables, the columns are related to characteristics and the rows – to components.

The "B" Tables deal with the building via its stone material, and bring out the role and input of the material in the artistic image of the building.

The list of components presents different types of stone units and their location in the building shell.

The different features of stone derived from geological data and from application of specific technologies in its quarrying and dressing, as related to function and design - are described as "characteristics".

The principal characteristics may be presented as follows:

2.2.1 Geometrical characteristics (columns "a" to "d" of "B" Table).

The geometrical characteristics are linked to aspects of both architectural design and building technology. They describe the actual size and proportions of a stone unit in building and testify to the history of its construction and the technologies applied.

2.2.2 Characteristics of material (columns "e" to "g" of "B" Table).

Listing of the type(s) of stone used in the building.

2.2.3 Material deterioration (columns "h" to "i" of "B" Table).

The deterioration aspect is touched upon briefly with a view to a complete scope of the correlation between architectural and technological factors. 2D transformations (color changes) and 3D transformations (shape, texture) are listed, both graded 1 to 5.

2.2.4 Technological characteristics (columns "j" to "p" of "B" Table).

Technological characteristics refer to the dressing mode, the technology involved (building in stone, stone facing), and the mortars. "B" Table clarifies the connection between location, size, proportions, stone type, dressing mode of the stone unit and its deterioration.

2.3 Synthesis

The data analysis via the "A" and "B" Tables brings out and clarifies the interaction of different design and deterioration factors influencing the building/structure under discussion. This synthesis contributes to planned conservational intervention by estimating the relative weight of technical (material, technological) and aesthetical (architectural, artistic) aspects.

3. ETC IMPLEMENTATION EXAMPLE

The ETC implementation is illustrated here by a case of the historic "Assicurazioni Generali" building (Arch. M. Piacenini) in Jerusalem, erected during the British Mandate (1935). The main purpose of this example is to introduce ETC – not to present an extended study of the building itself.

3.1 "A" Tables and "B" Tables

On level 1 both the analysis and synthesis in this example are related to the data in brief. On levels 2 and 3 the analysis deals in brief with selected parts and elements of the building. On level 3 elements relevant to part No. 4 (of the list in Table A2) are considered.

3.2 Synthesis

Building No. 1: "Assicurazioni Generali", arch. M. Piacenini, 1935.

This office building has preserved its public function to the present time. Its architectural image is clearly dictated by its stone material: the stone, through its assorted parameters, influenced the architectural design and played an active role in the spatial aesthetic result.

The complexity grade of the building is 1 – it is a kind of a box, with a small eastern part of it having an addition of one storey. However, application of contrasting types of stone dressing mode enhances this otherwise slight difference and totally changes the design result. The eastern part has a coarse stone[3] dressing throughout its height with a narrow vertical interval

Table 1. "A" Table. Spatial/visual characteristics of building and its components – example: hierarchy level 1.

Table type: **Spatial/visual characteristics**		Table code
Building: Assicurazioni Generali	Building No. 1	
Hierarchy level: building	Level No. 1	**A1**
Location on building:		
Architect, period of construction: Marcello Piacenini, 1935		
Address: 1, Shlomzion Hamalka st., Jerusalem		
Function: public		

	Characteristics	General geometry									Height				Urban		Complexity grade (1-3)	Transparency (1-5)	Spatial directions
		Dominant direction			Spatial character of arch. lines						Storey types	Storeys number	Ground storey special	Top storey special	Distance from street/road (m)	Street/ building ratio			
					Straight lines			Curves											
		X			X			X											X
			Y			Y			Y										Y
	Components			Z			Z			Z									Z
N		a	b	c	d	e	f	g	h	i	j	k	l	m	n	o	p	q	r
1											4	5					1	3	
2																			
3																			
4																			
5																			
6																			
7																			

Table 2. "A" Table. Spatial/visual characteristics of building and its components – example: hierarchy level 2 – parts of the building.

Table type: **Spatial/visual characteristics**		Table code
Building: Assicurazioni Generali	Building No. 1	**A2**
Hierarchy level: building parts	Level No. 2	
Location on building: facades Jaffa Rd., Bar Kochba sq.	Part No. 1, 2, 3, 4	
Architect, period of construction: Marcello Piacenini, 1935		
Address: 1, Shlomzion Hamalka st., Jerusalem		
Function: public		

Characteristics / Components		General geometry									Height				Urban		Complexity grade (1-3)	Transparency (1-5)	Spatial directions
		Dominant direction			Spatial character of arch. lines						Storey types	Storeys number	Ground storey special	Top storey special	Distance from street/road (m)	Street/ building ratio			
					Straight lines			Curves											
		X			X			X											X
			Y			Y			Y										Y
				Z			Z			Z									Z
N		a	b	c	d	e	f	g	h	i	j	k	l	m	n	o	p	q	r
1	Ground storey level (all facades)										1	1			0		1	3	
2	Eastern part: except ground storey										2	4			0		1	2	
3	Western part: except ground storey										1	3			0		1	3	
4	Main entrance façade (Bar Kochba sq.)										3	4			0		2	2	
5																			
6																			
7																			

Table 3. "A" Table. Spatial/visual characteristics of building and its components – example: hierarchy level 3 – elements of the part No. 4 of the building (see *Table 2*).

Table type: **Spatial/visual characteristics**		Table code
Building: Assicurazioni Generali	Building No. 1	**A3**
Hierarchy level: building elements	Level No. 3	
Location on building: main entrance façade	Part No. 4	
Architect, period of construction: Marcello Piacenini, 1935		
Address: 1, Shlomzion Hamalka st., Jerusalem		
Function: public		

Components / Characteristics		General geometry									Height				Urban		Complexity grade (1-3)	Transparency (1-5)	Spatial directions
		Dominant direction			Spatial character of arch. lines						Storey types	Storeys number	Ground storey special	Top storey special	Distance from street/road (m)	Street/ building ratio			
					Straight lines			Curves											
		X			X			X											X
			Y			Y			Y										Y
				Z			Z			Z									Z
N		a	b	c	d	e	f	g	h	i	j	k	l	m	n	o	p	q	r
1	Niche of entrance opening, of the windows of the storeys 2,3)										2	3			0		1	2	
2	Ground storey														0		2	3	
3	Balconies (storeys 2,3)										1						1		
4	Windows and balcony doors' openings (storeys 2,3,4)										2						1		
5	Sculpture of the lion and inscriptions of the upper storey																		
6																			
7																			

Table 4. "A" Table. Spatial/visual characteristics of building and its components – example: technological characteristics – building material (stone).

Table type: **Technological characteristics – building material**		Table code
Building: Assicurazioni Generali	Building No. 1	**B**
Topic: building material	Stone	
Parts of building: façade: Jaffa st., main entrance	Part No.	
Architect, period of construction: Marcello Piacenini, 1935		
Address: 1, Shlomzion Hamalka st., Jerusalem		
Function: public		

	Characteristics	Geometry				Material (stone types)			Deterioration		Technology							
											Dressing modes				Building technique		Mortars	Location (storey No.)	Spatial directions
		Dominant direction		Size (cm)		Dolomitic limestone			2D transformation (color) 1 to 5	3D transformation (texture) 1 to 5	Very smooth	Coarse			Building	Facing	Pale		
		X		X															X
																			Y
	Components		Z		Z														Z
N	Location of stone on building	a	b	c	d	e	f	g	h	i	j	k	l	m	n	o	p	q	r
1	Building (general): minimum				25				0	0									
2	Building (general): maximum				35				1	1								1	
3	Special/ decoration stones								1										
4	Main entrance façade – ground storey, central part				25-35				1									1	
5	Ground storey (No 1) except main entrance portion				25-35				1									1	
6	Upper storeys				25-														
7																			

on the Jaffa Road façade, and the same coarse dressing mode of the ground storey, the main western structure with its smoothly dressed stone becomes a contrasted enclosure. The different stone dressing modes correspond to different design languages: coarse stone for traditional-looking elements (arched windows of the ground storey and the top storey of the eastern part) - versus the smooth modern three upper storeys of the western part, typical of the Italian architecture of the Mussolini period.

The deterioration grade is very low, mostly confined to color changes, most evident in the areas of coarse dressing. Thus the most important conclusions should relate to the cleaning and maintenance mode. As regards the

Figure 1. View of Generali building from Jaffa Road.

latter, special attention should be given to preserving the perfect condition of the smoothly dressed surfaces, especially of their monochromatic character. As for the coarse dressing of the ground storey, the special attention should be paid to chemical monitoring of decay processes at almost zero distance from the heavy traffic of that city area. From the historical point of view, change of color of that part of the building would not be a disadvantage, by virtue of its association with the architecture of the past.

The architectural image of the building may be defined as complex to contradictory. Even with its modern outline, it creates different impressions from different points of observation. The main entrance façade exemplified the dominance of modern architecture, while the two vertical strips of coarse dressing along both sides of the niche with the enclosed entrance appear as a random quotation from some other epoch.

Another aspect which should not be underestimated – is the close distance, typical of Jerusalem, between the observer and the building on eyelevel. If the observer moves in the east-west direction along the north façade of the building, he will see only coarse dressing of massive stones and large arched windows. Unless he looks back to the main entrance façade, he would probably never guess that the building was erected in 1935. In this context it should be mentioned that due to the official policy of exclusive stone building in Jerusalem[4], the Generali, like others of that period, is a composite concrete-stone construction, with thickness of the surface stone layer up to 30 cm. Thus it can be named "stone/concrete building" rather than "stone facing". This combination produced visual results similar to those of the traditional architecture in stone.

4. CONCLUSION

An example of a historic building study by means of ETC shows how the role of stone may be read through both geometrical and technological / material analysis. The "A" and "B" Tables clarify and bring out the role of stone, with its aesthetic and physical problems as well as its contribution, thereby permitting the development of a compatible conservation strategy and maintenance mode, thus preserving the authentic image and character of a historic monument, on the basis of full attention to the stone material.

REFERENCES

1. A. Lobovikov-Katz, Architectural Aspects of Preservation of Stone Building in the Context of a Historic City (Case Study of the Stone Regulation in Jerusalem), *Proceedings of the 5th International Symposium on the Conservation of Monuments in the Mediterranean, Sevilla, Spain, 5-8 April 2000, pp. 635-641, A.A. Balkema Publishers, Netherlands, 2002*

2. A. Lobovikov-Katz, “The architectural aspects of deterioration of building material (stone) in the historic monuments”, *Proceedings of the 6th International Symposium on the Conservation of Monuments in the Mediterranean Basin, Lisbon, April 7-20, 2004*
3. A. Shadmon, “The development of an integrated dimension-stone industry in Israel”, *Geological Survey of Israel, Jerusalem,1988*
4. A. Lobovikov-Katz, “The "Stone regulation" and stone building conservation in Jerusalem (the role of building material – stone – in urban conservation”, *Research Thesis (Doctor of Philosophy), Supervisor: Wachman, A., Turner, M., Technion, 2001*

Chapter 4.3

STRUCTURAL STABILITY OF HISTORIC UNDERGROUND OPENINGS IN ROCK

Two Case Studies from Israel

Yossef H. Hatzor[1], Michael Tsesarsky[2] and Roni C. Eimermacher[3]
[1] *Dept. of Geological and Environmental Sciences, Ben-Gurion University of the Negev, Beer-Sheva, 84105, Israel;* [2]*Faculty of Civil and Environmental Engineering, The Technion – Israel Institute of Technology, Haifa 3200, Israel;* [3]*Dept. of Maritime Civilizations, Haifa University, Haifa 31905, Israel*

Abstract: In this paper the structural stability of the roof in two historic monuments excavated underground in different dimensions and rock types is discussed. An emphasis is made on stability analysis of roofs excavated in bedded and jointed rocks, referred to as laminated Voussoir beams. It is shown that the significance of joint friction, spacing and orientation can not be overlooked because these factors dictate roof stability. For accurate modeling of such a problem numerical methods must be employed. In this work the power of numerical discontinuous deformation analysis (DDA) method is demonstrated.

Key words: tunneling; rock mechanics; voussoir; DDA; old monuments.

1. THE TEL BEER SHEVA SITE

1.1 Introduction

The 3000 year old underground water storage system, excavated at Tel Beer-Sheva in southern Israel, is discussed first. The rock-mass into which the reservoir is excavated is a horizontally bedded and vertically jointed, low strength, upper Cretaceous marly chalk, of the Gareb formation. In such a rock mass, prismatic blocks are expected to slide out of the loosened zone in the roof, up to a level where arching stresses are sufficiently high to interlock any further block displacement. The relevant stability issue for this monument is therefore the analysis of the arching mechanism in the immediate roof, which may be modeled as a layered and jointed beam, referred to as

S.K. Kourkoulis (ed.), Fracture and Failure of Natural Building Stones, 215–237.

laminated Voussoir[1]. Specifically, the height of the loosening zone and the magnitude of maximum compressive stress must be evaluated. Accurate determination of these factors enables determination of factors of safety against failure by crushing, shear along abutments, or buckling. Consequently, the required length and capacity of active support elements (rock bolts) can be designed. The problem is first solved using a semi-analytical solution for a *Voussoir* beam[2], and then it is solved numerically using the Discontinuous Deformation Analysis (DDA) method[3]. It is found that the joint spacing (Sj) and friction angle (ϕ), while ignored in the standard solution procedure[2] including later modifications[4-6], are of paramount importance in the correct modeling of the roof response to excavation. It is shown that the function ϕ_{req} (required joint friction angle for equilibrium) vs. Sj presents a minimum at which the arching mechanism is most effective. With joint spacing smaller or larger than the optimum, the arching mechanism becomes less effective, and failure by shear along the abutments prevails. Since the mean joint spacing in the site, $Sj_{,mean}$=25 cm, is much smaller than the optimal joint spacing value, it is concluded that the arching mechanism was not sufficiently effective during construction, as confirmed by the collapse of the entire center of the roof in historic times.

1.2 Structural setting

In the archeological site of Tel Beer Sheva, an ancient city dated back to the Late Iron stage 1,200 - 700 B.C. was explored (Figure 1). Modern excavation of the underground water system revealed that the roof of the main reservoir chamber had collapsed, probably during time of construction, and that the ancient engineers have erected a massive support pillar in the center of the chamber in order to support the remaining roof. The same plaster coating which was explored on the opening side walls at ground level was also discovered higher, above the level of the original roof, indicating the proximity of the failure episode to the original time of excavation.

The reservoir was excavated in horizontally bedded chalk with three vertical joint sets. The two most abundant sets are orthogonal with mean spacing of Sj=20 to 25 cm, and the mean bed thickness t = 50 cm. The intersection of these three joint sets creates a dense network of cubic blocks that form the roof of the excavation.

The roof collapsed into the shape of a three dimensional dome with three distinguishable structural zones[7]: zone 1 - the original roof level delimited by a vertical step with t = 50 to 125 cm; zone 2 - a sub horizontal plane parallel to an existing bedding plane delimited by a vertical step similar to boundary between zones 1 and 2; and zone 3 - the uppermost failure level which like zone 2 is parallel to an existing bedding plane. A structural map of the roof is shown in Figure 2. It can be seen that the center of the roof is comprised of zone 3 with a circular boundary, and that the external sections of the roof are comprised of zone 1 - the original roof level. The support pillar was erected directly below zone 3, and extensions were built in order to support the unstable transitions from zones 3 to 2.

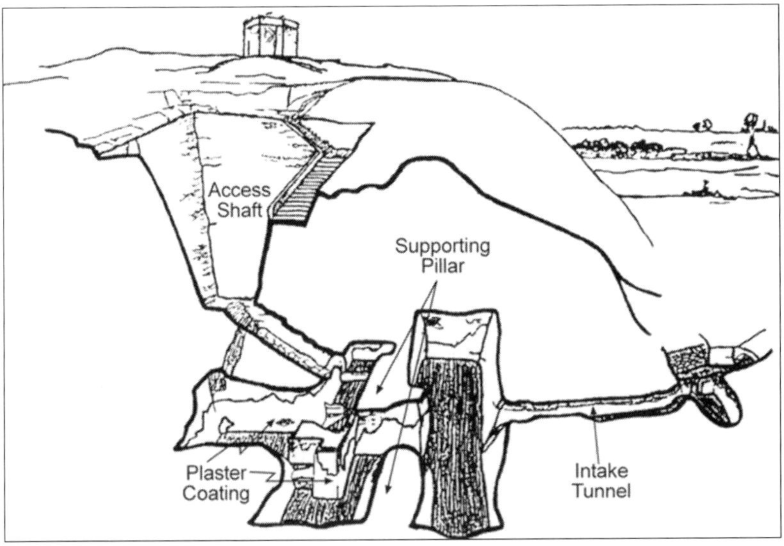

Figure 1. Artist conception of underground water reservoir in Tel Beer Sheva.

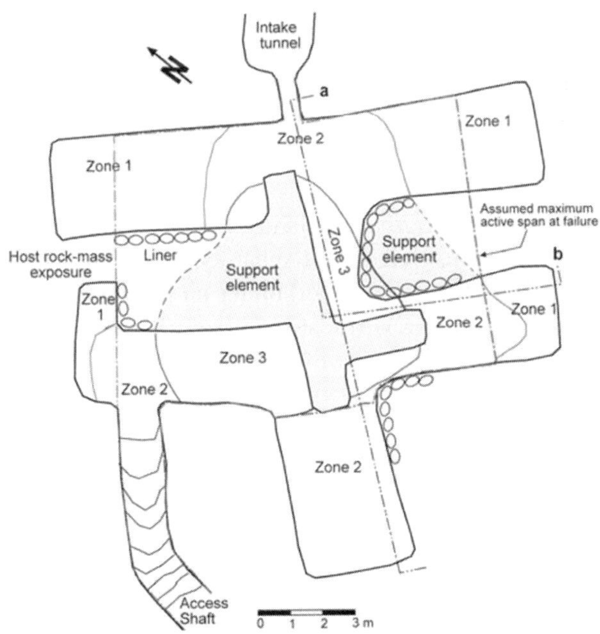

Figure 2. Structural map of roof showing dome structure after the historic collapse (modified after[7]).

The mapped roof is considered here as failed laminated Voussoir beam. The failed beam arrived at a new equilibrium after the collapse, with the aid of support measures taken by the ancient engineers, primarily in the form of the central support pillar (Figure 2). The spherical dome shape of the current roof indicates that the problem, although axis-symmetric, is really three dimensional. The stability analysis presented below however is limited to two dimensional solutions, and its validity for the three dimensional case must be verified.

1.3 Mechanical properties of rock

The chalk exhibits homogeneous porosity, between n = 27 to 31% and unit weight between γ = 18.1 to 20.1 kN/m^3. Atterberg limits of interbedded marl layers indicate relatively low plasticity and low swelling potential. The mechanical behaviour of the chalk was determined from 4 triaxail and 5 uniaxial compression tests, performed under a constant strain rate of 10^{-5} s^{-1}, in the stiff, hydraulic, servo controlled triaxail load frame at Ben-Gurion University. The confining pressures values used were 2, 4, 6, and 10 MPa. A linear Coulomb-Mohr failure envelope fitted to the peak strength values yielded a cohesion of 3.1 MPa and internal friction angle of 32°.

The chalks exhibit transverse isotropy, with bedding planes being planes of isotropy in the material. The uniaxial compressive strength values vary from 7 to 34 MPa, for bed-normal and bed-parallel compression respectively (Figure 3). Similarly, Young's modulus varies from 1.9 to 31 GPa for bed-normal and bed-parallel compression respectively. Finally, Poisson's ratio ranges from 0.05 to 0.19 again for bed-normal and bed-parallel compression.

Rock discontinuities are persistent, clean and tight exhibiting relatively planar surfaces. In order to evaluate the influence of surface roughness on joint shear strength direct shear tests were performed using the stiff, hydraulic, servo-controlled, direct shear system at BGU rock mechanics laboratory. The tests were performed under an imposed constant normal stress condition. Sample TBS-1 was tested in two consecutive segment shear tests under a constant shear velocity of 0.00127 mm/sec. Results are shown in Figure 4 where the influence of normal stress on friction is also presented. The peak friction angle is ϕ_p=47° while the residual one is ϕ_r=24°. The decrease in friction with increasing normal stress represents the degradation of asperities during shearing cycles in the relatively weak joint surface material.

Polished joint surfaces were studied further, by cyclic direct shear tests performed at a constant cyclic shear displacement rate of ± 0.0254 mm/sec. Shear displacement–shear stress curves for sample TBS–4 are shown in Figure 5 together with the resulting shear stiffness and friction angle values. The shear stiffness values increase with increasing normal stress, as expected. The friction angle exhibits anisotropy: during forward shear the friction angle reduces from ϕ=47° at σ_n=500 kPa to ϕ=38° at σ_n=2000 kPa, and during backward shear from ϕ=30° at σ_n=500 kPa to ϕ=35° at σ_n=2000 kPa.

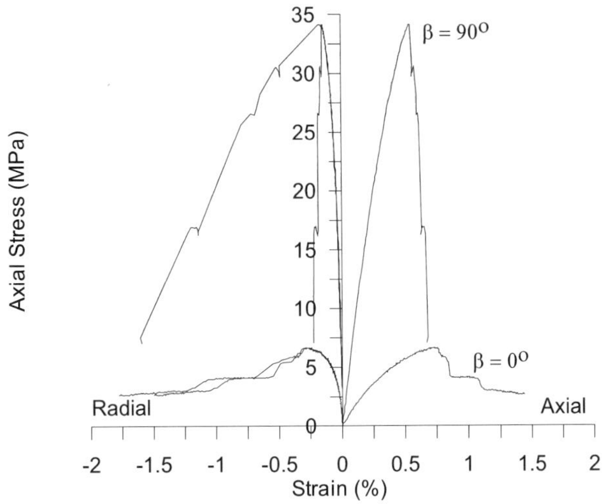

Figure 3. Comparison between bed parallel (β = 90°) and bed normal (β = 0°) uniaxial compression tests of Tel Beer Sheva marly chalk (Ghareb Fm.).

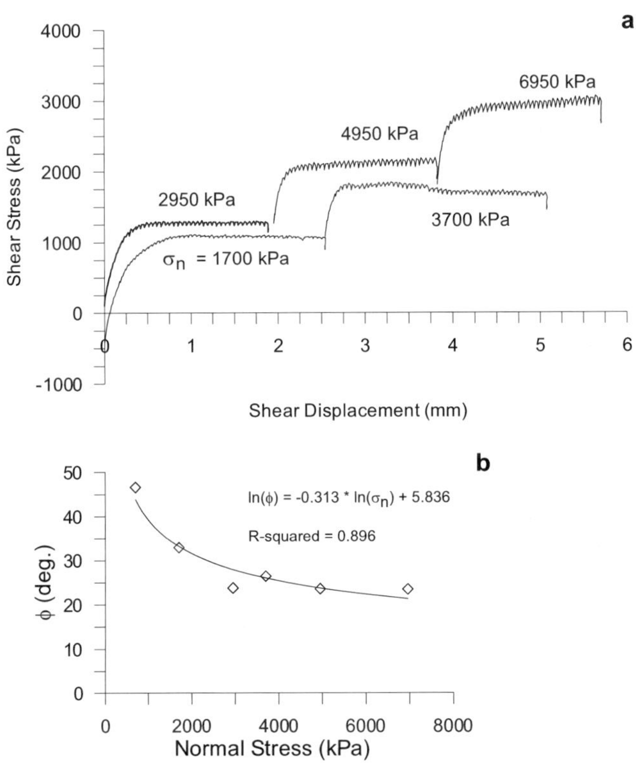

Figure 4. Direct shear test results for natural joints: (a) Segment direct shear test under constant normal stress, (b) Influence of normal stress on friction angle.

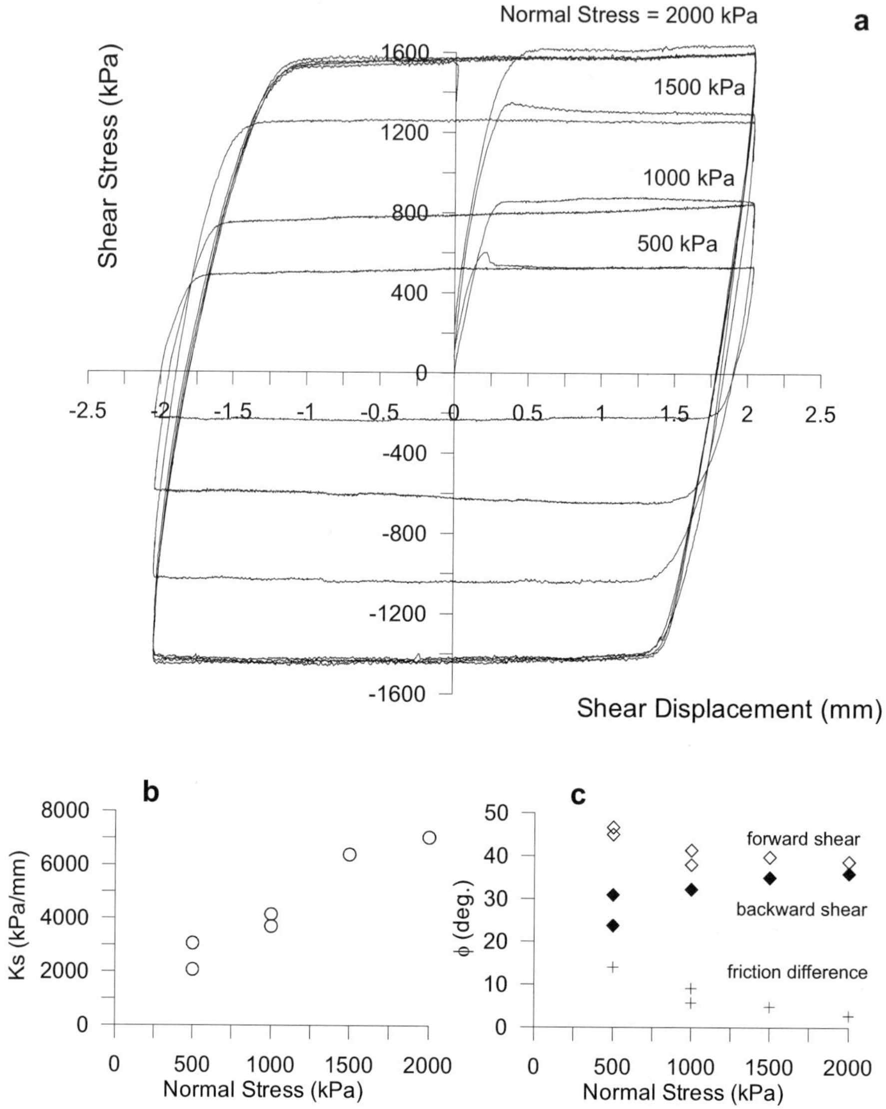

Figure 5. Results of cyclic shear tests on polished samples.

1.4 Stability analysis by the Voussoir beam analogue

The stability of the immediate roof at the site was first estimated using the iterative procedure for the static solution of a Voussoir beam, originally proposed by Evans[8] and later modified by Beer and Meek[2], Brady and Brown[4], Diederichs and Kaiser[5] and Sofianos[6]. The Voussoir model and the stress distribution assumed are shown in Figure 6 where S and t are the beam span and thickness respectively.

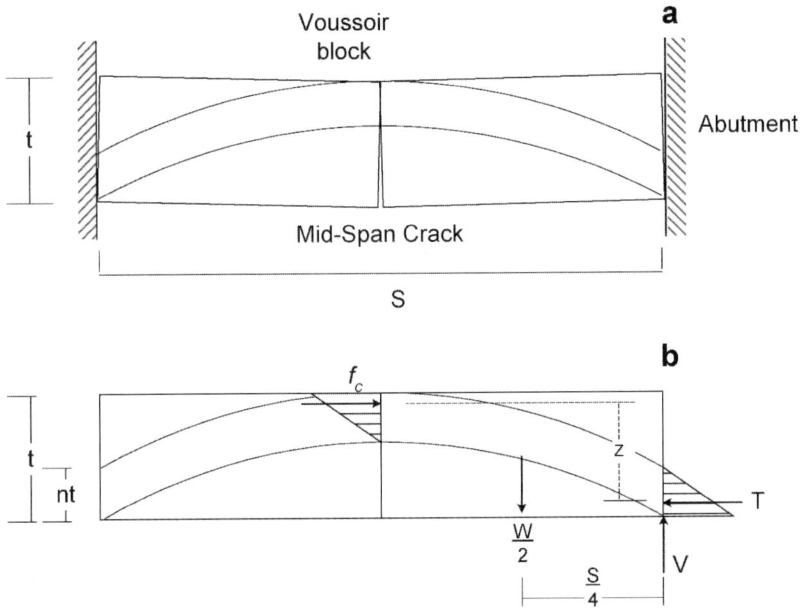

Figure 6. The Voussoir beam analogue (after Brady and Brown[4]).

The following mechanical parameters were used for the analysis: density ρ=1900 kg/m^3, elastic modulus E=7840 MPa, Poisson's ratio ν=0.17, uniaxial compressive strength UCS=27.6 MPa, joint cohesion c=0 MPa and joint friction angle ϕ=47°. The factors of safety against failure in compression (axial crushing) or shear (sliding of beam along vertical abutments) were calculated for a range of beam spans (S=5 to 16 m) and beam thicknesses (t=0.25 to 5 m) accounting for all possible geometries at time of failure, while assuming solid roof beams with no layers and only one vertical discontinuity at the centerline that are free to displace along the abutments. The results are shown in Figure 7. Assuming an active beam span of 8 m and beam thickness of 2.5 m at time of failure, the beam is perfectly stable against failure in compression. Similarly, with a friction angle of 47° the factor of safety against failure in shear approaches 2.0. These results are in contrast to field observations and therefore reflect the limitations of the Voussoir approach when analyzing layered and jointed beams.

1.5 Stability analysis by the numerical DDA method

The Voussoir analogue discussed above is limited in the sense that it does not allow for layered, and multiply jointed beams. Rather, it considers the beam as a continuous solid with only one discontinuity at the centerline, which is free to slide along the abutments. Furthermore, the Voussoir analogue can not handle multiple joints, and completely ignores joint friction.

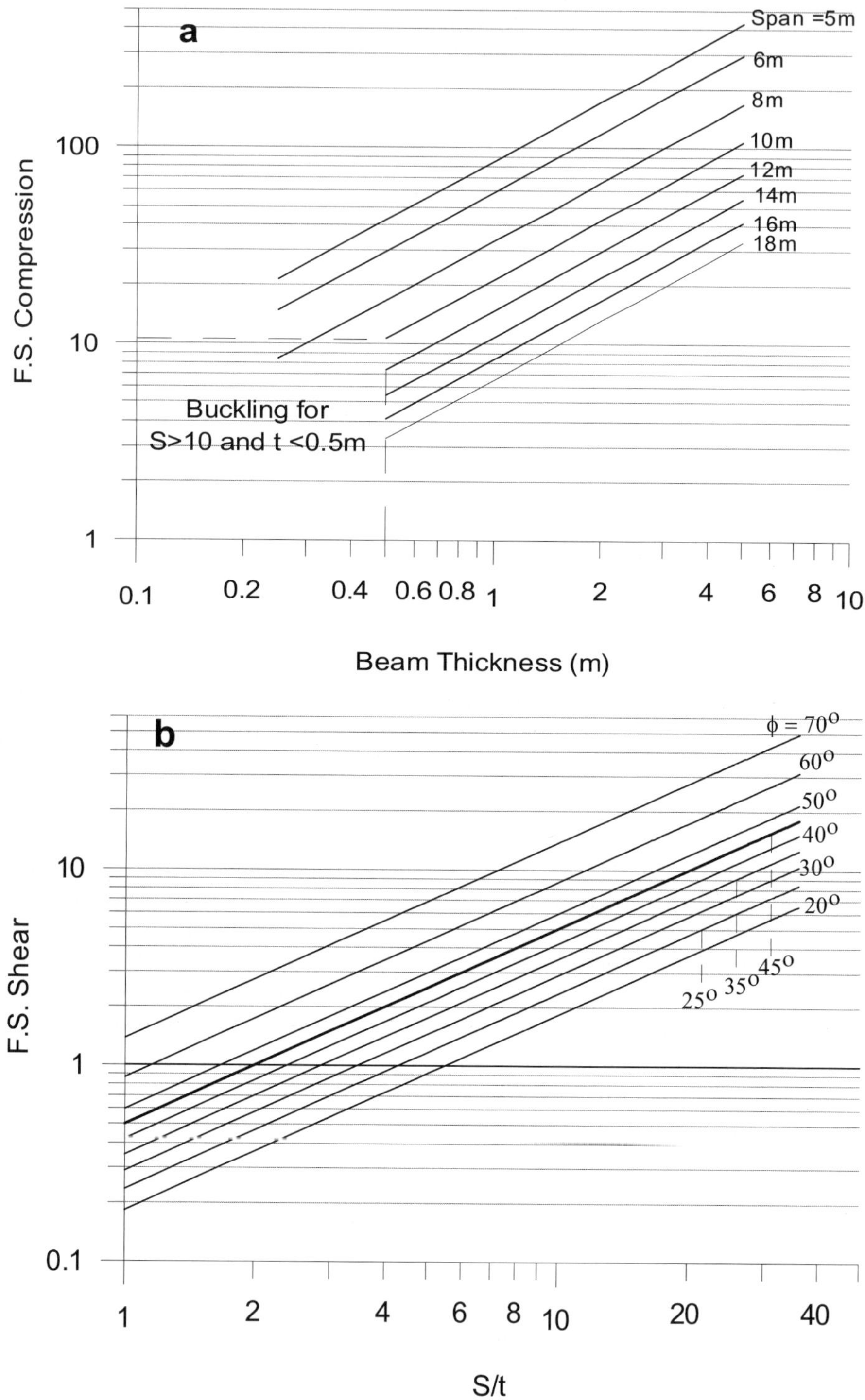

Figure 7. Results of Voussoir beam analogue for the Tel Beer-Sheva reservoir.

In order to account for more complex (and realistic) geometries numerical methods must be employed. In this paper the study is focused on the discontinuous deformation analysis (DDA) method of Shi[3]. Details on the DDA method are discussed by Jing[9], and a review of a decade of validations is presented by MacLaughlin and Doolin[10].

Both single and multiple layered beams are analysed with DDA, the geometry of which is shown in Figure 8.

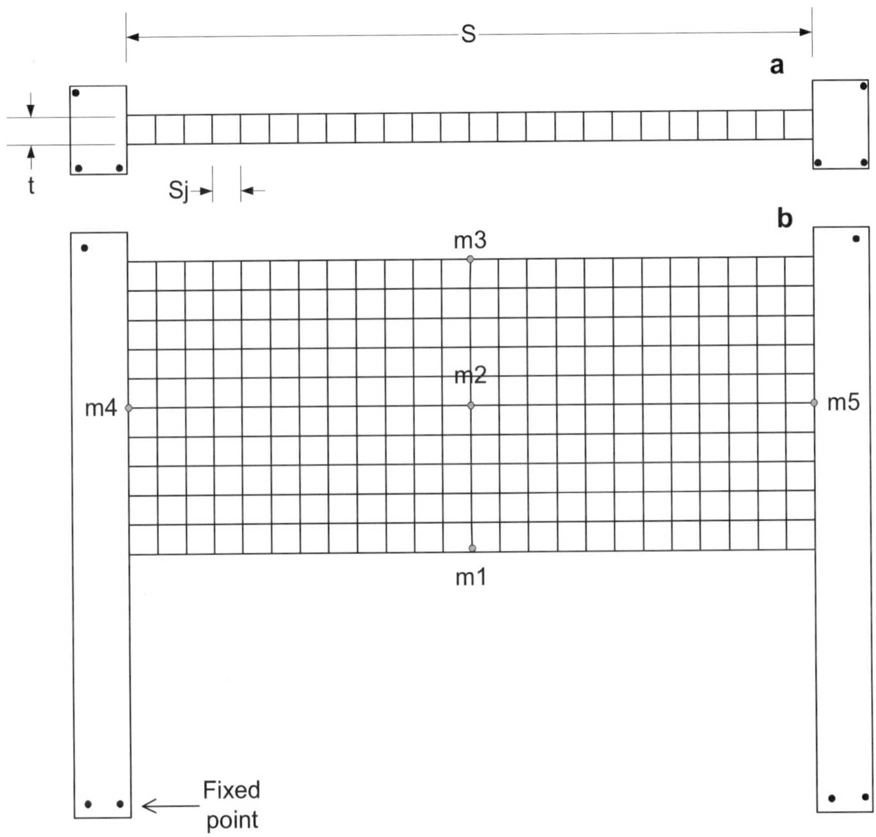

Figure 8. Single layered and multiple layered beam models studied in DDA.

Location of measurement points for analysis output are marked by circles in Figure 8 and are labelled mi (i = 1,2,...5). Two geometries are analyzed:

1) a single layer with t = 0.5 m modelling the immediate roof layer (Figure 8a);

2) a 5 m thick stack of horizontal layers each with t=0.5 m (Figure 8b), modeling the entire excavation roof. The active span in both configurations is S=8 m. The following input parameters are used in DDA (see Hatzor and Feintuch[11] for details): ρ=1900 kg/m^3, E=7840 MPa, ν=0.17, penalty stiffness

g(0)=1000 MN/m, time step size g(1)=0.00025 sec, penetration control parameter g(2) = 0.00025, and dynamic control parameter k01 = 1.

The effect of joint friction angle is demonstrated in Figure 9 for a single layered beam with thickness of t=0.5m and joint spacing of Sj=0.25m. The measured deformation variables (u, v, ω) of blocks in the single layered beam are shown in Figure 10 according to their position at the beam.

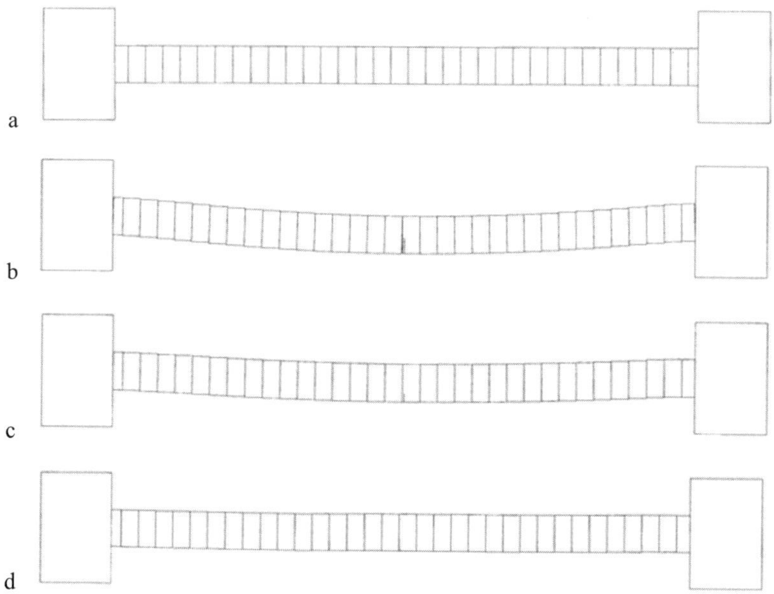

Figure 9. DDA graphic output of single layer deformation for different values of joint friction angle: (a) original geometry; (b) $\phi_{av} = 45°$; (c) $\phi_{av} = 75°$; (d) $\phi_{av} = 80°$.

The results of the DDA single layer model indicate that the required friction angle for stability, with immediate roof layer thickness of 0.5m and vertical joint spacing of 0.2 5m (as found in the field) is $\phi_{req} \geq 78°$. Since the available joint friction angle in the field is assumed to be $\phi_{av}=47°$, the DDA results explain the observed failure of the immediate roof in the field, delimited by the contour of zone 1 in the present-day structural map of the roof (Figure 2).

The response of a laminated Voussoir beam is shown in Figure 11. The laminated Voussoir is modelled by an 8 m span, 5 m stack of individual layers, each of 0.5 m thickness, simulating the general structure of the roof in the field. The computed deflections of measurement points m1, m2, and m3 (see Figure 8 for location) are plotted as a function of joint friction angle (open symbols in Figure 11) and of joint spacing (solid symbols in Figure 11). The influence of friction is analyzed for a case of constant joint thickness of Sj = 0.25 m, similar to the case in the field. The influence of joint spacing is analyzed for constant friction of $\phi = 47°$, as believed to be the case in the field.

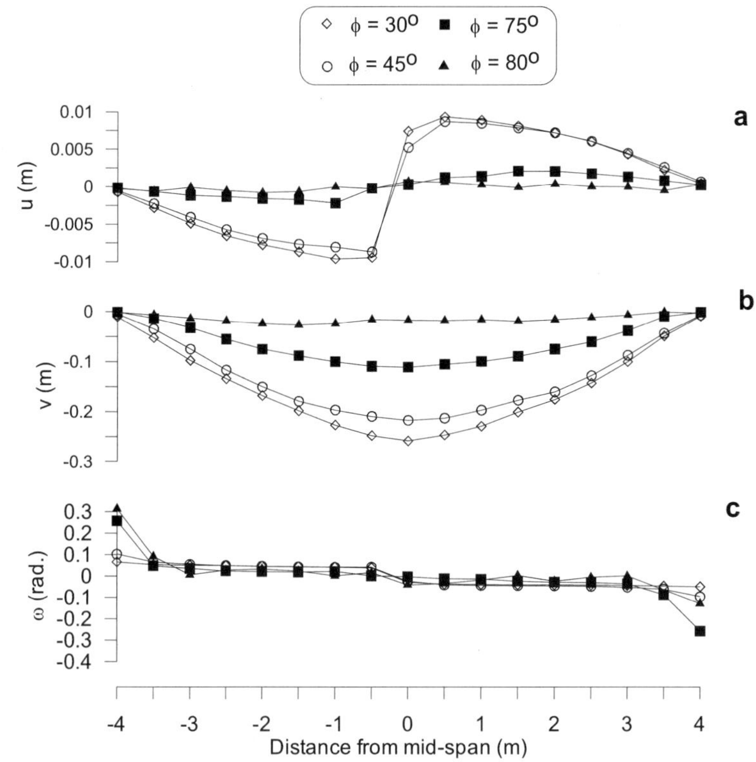

Figure 10. Deformation of the DDA single layer model, measured at the lowermost fiber of the beam: (a) horizontal displacement (u); (b) vertical displacement (v); and (c) rotation (ω).

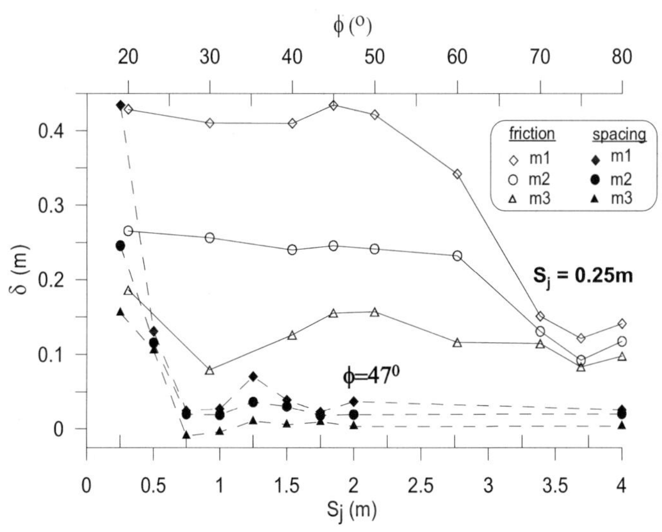

Figure 11. Response of a layered beam as computed by DDA (see text for details).

From Figure 11 it can be deduced that when friction angle is smaller than 70° vertical deflections in the immediate roof (m1) are excessive indicating instability, but with friction angles greater than 70° the composite beam stabilizes, and all measurement points deflect homogenously by about 10 cm.

The influence of spacing on layered beam stability is revealing as well. The layered beam, with a constant friction angle of $\phi = 47°$ seems to stabilize when joint spacing $Sj \geq 0.75$ m. For the analyzed friction angle, increase in joint spacing beyond Sj=0.75 m does not improve stability, indicating development of a stable arching mechanism as long as joint spacing is greater than the required value.

It would be intuitive to expect that the same optimal joint spacing value could be found for any value of interlayer friction. This analysis was conducted by Hatzor and Benary[1] with DDA for a slightly different beam configuration: Total beam thickness=2.5 m; *S*=7 m; t=0.5 m; γ=18.7 kN/m^3, E=2 GPa, $\nu = 0.2$. The results are shown in Figure 12.

The DDA results shown in Figure 12 clearly demonstrate that in a layered Voussoir beam of given geometrical and mechanical properties there is an optimal joint spacing which requires a minimal value of discontinuity shear strength in order to generate a stable arching mechanism. For the analyzed beam in Figure 12 the optimal joint spacing is Sj = 175 cm and the required friction angle for equilibrium is ϕ_{req}=24°. This result suggests that the optimal S/Sj ratio for equilibrium is $(S/Sj)_{eql}$=4.0.

The DDA results shown in Figure 12 also demonstrate that for every available discontinuity shear strength value, a corresponding joint spacing value necessary for equilibrium exists. For the analyzed case study with mean joint spacing of 0.25 m the required friction angle for stability is ϕ_{req}=80°, since ϕ_{av}=47° the layered roof could not remain stable once the opening was attempted. As shown in Figure 11 for a very similar beam configuration, the required joint spacing for equilibrium with ϕ_{av}=47° is 0.75 m, three times greater than the mean joint spacing in the field.

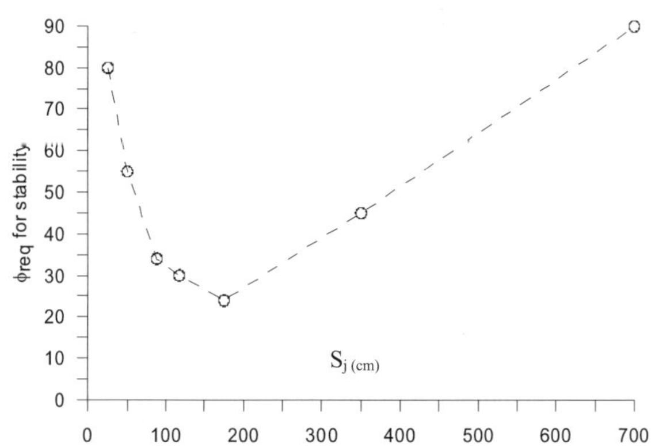

Figure 12. ϕ_{req} for stability vs. joint spacing, and optimal joint spacing for a given layered Voussoir beam[1].

2. THE FREEMASONS HALL AT ZEDEKIAH CAVE

2.1 Introduction

Zedekiah Cave has been used as an underground quarry below the city of Jerusalem from about 700 - 800 BC, and continuously until the end of the late Byzantine period, in order to extract high quality building stones for monumental constructions in Jerusalem and vicinity. The quarry is excavated underneath the old city of Jerusalem (Figure 13a) in a sub-horizontally bedded and moderately jointed, low strength, upper Cretaceous limestone of the Bina formation[14]. The underground quarry is 230 m long, with maximum width and height of 100 m and 15 m, respectively.

The most striking feature of the quarry is certainly the 30m span, unsupported central chamber, widely known as "Freemasons Hall", because of ritual ceremonies taken place at the chamber by Freemasons in recent times (Figure 13a). Site investigations revealed that large roof slabs in several side chambers have collapsed over the years, but that the roof of Freemasons hall remained intact.

In this section roof stability in Freemasons hall is analyzed and discussed. First a continuum mechanics approach is taken[12], followed by an attempt to apply the Voussoir beam analogue[2], and finally the numerical DDA [3] is used to study the interaction between joint friction, spacing and orientation.

2.2 Clamped beam model

The 30 m span, unsupported roof of Freemasons hall seems intact in field inspection and therefore calls for a preliminary analysis based on continuum mechanics principles. Obert and Duvall[12] review the elastic solution for a clamped beam which provides deflections, shear forces and bending moments across the beam. This solution may be applicable for the immediate roof of Freemasons hall provided that the limestone bed comprising the immediate roof material is completely continuous with no intersecting joints. An accurate cross section through Freemasons hall is shown in Figure 14c and its location in the underground monument is shown in Figure 13b (section A-A'). The concept of Obert and Duvall[12] for a clamped beam model as applied to underground openings excavated in rock masses containing planes of weakness parallel to the roof is illustrated in Figure 15 below. Using field mapping and laboratory tests the relevant input parameters for the case of Freemasons hall are listed in Table 1. The results of the clamped beam analysis for the roof of Freemasons hall are listed in Table 2.

Application of the analytical clamped beam model predicts that the axial stresses that will develop in the beam will be $\sigma = \pm$ 10.5 MPa. While the upper fiber of the beam is safe against failure by crushing, the lowermost fiber

(a)

(b)

Figure 13: (a) Layout of Zedekiah cave superimposed on the old city of Jerusalem, (b) plan of Zedekiah cave, "Freemasons hall" delimited by dashed square.

(a)  (b) 

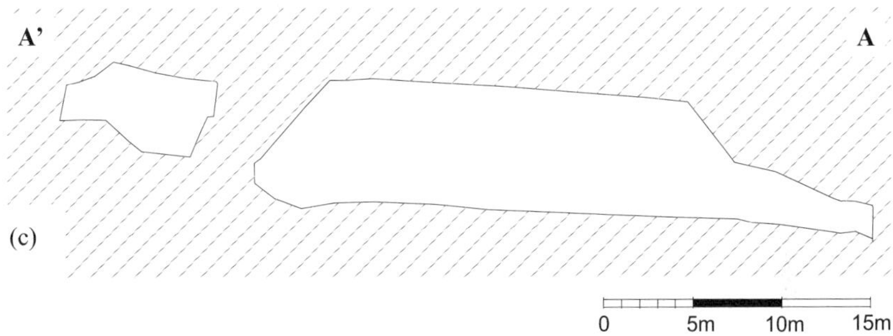

Figure 14. (a) The Bina formation outside Zedekiah cave, (b) A view towards Freemasons hall (Photograph by Y. Novitz [13]), (c) A cross section through Freemasons hall (for location see Figure 13b).

of the beam is unsafe against failure in tension; consequently, the clamped beam model predicts that a tensile fracture will initiate at the centerline and propagate upwards, disjointing the beam into two blocks, each of 15 m length. The stability of such a three – hinged beam can be estimated by application of the Voussoir beam analogue, discussed in section 1.4 above.

Table 1. Physical and mechanical properties of immediate roof in Freemasons hall.

Free span (L)	30 m
Beam thickness (t)	0.85 m
Unit weight (γ)	19.8 kN/m
Elastic modulus (E')	$8*10^3$ MPa/m^2
Uniaxial compressive strength (σ_c)	16.4 MPa (bedding parallel)
Tensile strength (σ_t)	2.8 MPa

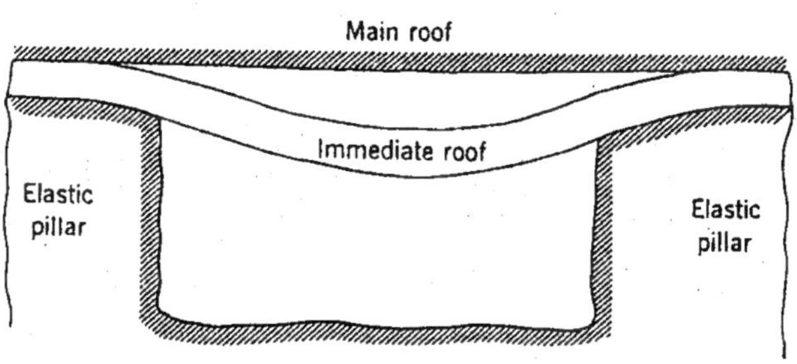

Figure 15. Deflection of a single layer on elastic pillars, after Obert and Duvall[12].

Table 2. Results of clamped beam analysis for immediate roof in Freemasons hall.

Maximum deflection at centerline (η)	8.67 cm
Maximum shear stress at abutments (τ_{max})	445.5 kPa
Maximum axial stress (σ)	10.5 MPa

2.3 Voussoir beam analogue

The Voussoir beam analogue can be applied to the immediate roof of Freemasons hall assuming that a tension crack formed at the centerline, as discussed in section 2.2 above. All required input parameters for such an analysis are listed in Table 1. Application of the iterative procedure[4] shows that the *n* value (see Figure 6) does not converge; after one iteration *n* becomes greater than 1.0. Using the modified approach suggested by Diederichs and Kaiser[5] by introducing incremental steps in *n*, reveals that when the system is supposed to attain equilibrium (when the extreme fiber stress is minimum, incidentally this always happens when $n = 0.75$[6]) the thickness of the compressive arch Z is negative. The meaning of this result is that under the given loads, geometry, and material properties, the beam will undergo buckling deformation leading to a "snap through" mechanism[4]. Indeed, in some roof sections in the site snap-through mechanism may be responsible for the observed failures (see Figure 13b). This is certainly not the case in the roof of Freemasons hall which still stands unsupported. Therefore, complete understanding of the mechanics of the roof in Freemasons hall requires a further, more robust, analysis which allows for interaction between blocks and incorporates friction laws for discontinuities. Such an analysis is presented in the next section using DDA.

2.4 Discontinuous deformation analysis (DDA)

The elastic solution for the roof predicts tensile fracture if modelled as a continuous beam. Application of the Voussoir analogue for the roof predicts a snap-through mechanism. Both scenarios did not materialize in Freemasons hall during its 2500 years history; field inspections suggest that the current immediate roof is the original one. These findings suggest that interactions between distinct blocks in the rock mass above the immediate roof may stabilize, rather than weaken, the roof. To explore this possibility the rock mass

Table 3. Rock mass structure at Zedekiah cave

Discontinuity Set	Genetic Type	Mean Orientation	Mean Spacing (m)	Joint friction angle
1	Bedding	08/091	0.85	41°
2	Shears	71/061	0.79	41°
3	Shears	67/231	1.48	41°
4	Joints	75/155	1.39	41°

around Freemasons hall is modeled as a mesh of distinct elements, using discontinuous deformation analysis[3]. The rock mass structure in Zedekiah cave consists of one set of sub-horizontal beds and three sets of inclined joints (Table 3).

The numerical analysis is limited to two dimensions because of the large number of blocks in the rock mass around Freemasons hall. Because of the inherent limitations of the two-dimensional solution, the complex rock mass structure is simplified and represented by two extreme end members: 1) A rock mass consisting of horizontal beds and vertical joints; 2) A rock mass consisting of horizontal beds and inclined joints. Furthermore, the actual spacing value for each joint set is doubled in the generation of the DDA mesh in order to reduce the total number of blocks. Even so, the total number of blocks in scenarios 1 and 2 are 368 and 735 respectively. Figure 17 displays these two end members as represented in a DDA mesh. The measurement point location for displacement and stress output are marked in Figure 17b for reference.

The two mesh configurations are modeled for response under gravitational loading for duration of 5 sec, an equivalent of 20,000 DDA time steps. A friction angle value of 41° is assumed for all discontinuities based on tilt tests of saw cut planes and measured surface profiles in the field. The deformation configuration is shown in Figure 18, where principal stress trajectories at the end of the computation are marked as well.

The response of the two mesh configurations is strikingly different. The vertical joint configuration (Figure 18a) exhibits some shear displacement along vertical joints that extend upwards from the abutments, but ultimately downward displacement is restrained, most probably due to the development of an effective arching mechanism in the roof. The inclined joints configuration in contrast exhibits ongoing downward displacement with no indications of stabilization. The downward displacement output data for the four measurement points are plotted in Figure 19. The onset of arching and displacement arrest is clearly demonstrated in the vertical joints configuration, where stabilization is indicated after 2000 time steps (Figure 19a). In the case of the inclined joints however all measurement points exhibit downward displacement at a constant velocity with the immediate roof undergoing downward displacement of 50 cm after 5 seconds of loading (Figure 19b). This result suggests that with an inclined joints configuration the development of an effective arching mechanism in the roof is hampered. This can also be detected by the erratic orientation of the principal stress trajectories in Figure 18b. The magnitude of the horizontal stress in the immediate roof (mp1) in the two structural configurations is shown in Figure 20. Clearly, the vertical joint configuration stabilizes under a constant horizontal stress of 650 kPa. The inclined joints configuration however displays an erratic stress behaviour that never attains equilibrium.

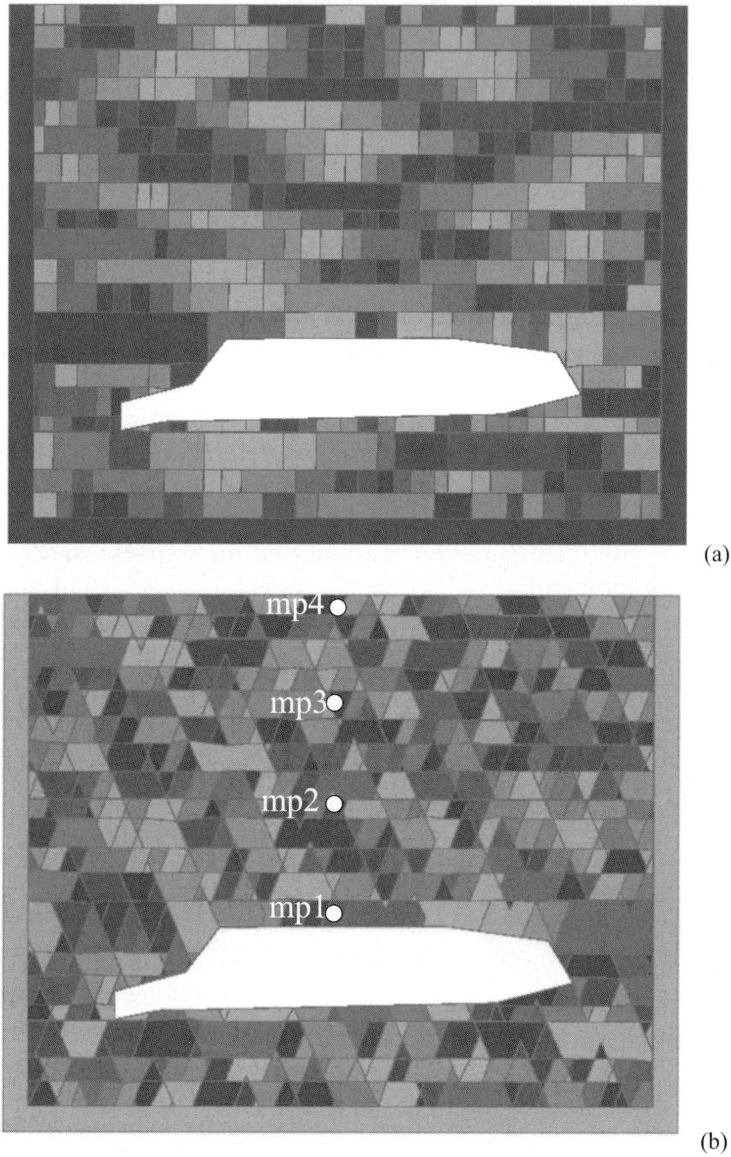

Figure 17. Two rock mass structures to be modelled in DDA: (a) horizontal beds with vertical joints, (b) horizontal beds with inclined joints.

3. SUMMARY AND CONCLUSIONS

Two historic monuments, excavated in bedded and jointed rock masses are discussed in this paper: 1) the 3000 year old water storage system at Tel Beer - Sheva, 2) the 2500 year old quarry at Zedekiah cave in Jerusalem. In Tel Beer-Sheva the immediate roof of the reservoir collapsed during construc-

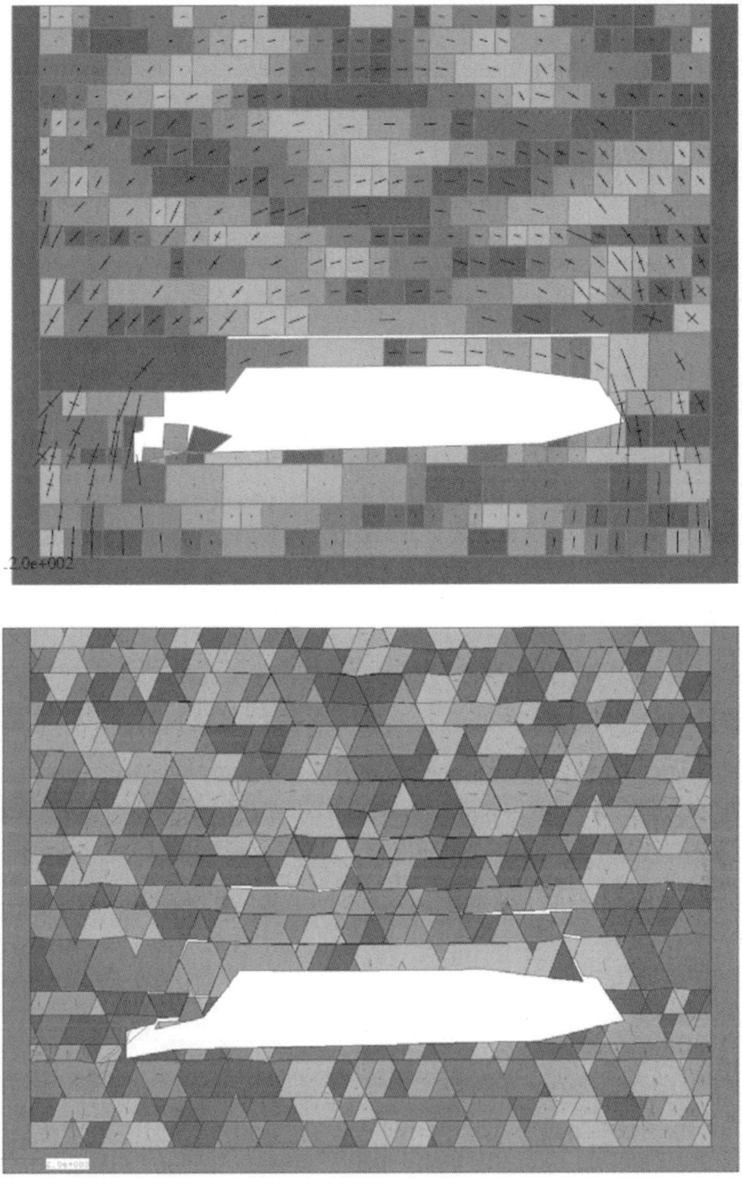

(a)

(b)

Figure 18. Deformed configurations after 5 seconds of gravitational loading: a) Vertical joints, b) Inclined joints.

struction while at Zedekiah cave the magnificent roof of the central 30 m span chamber, known as Freemasons hall, remained intact over the years. Both cases are used to test existing analytical methods (elastic theory and the Voussoir beam analogue) and to verify validity of numerical codes (the discontinuous deformation analysis method).

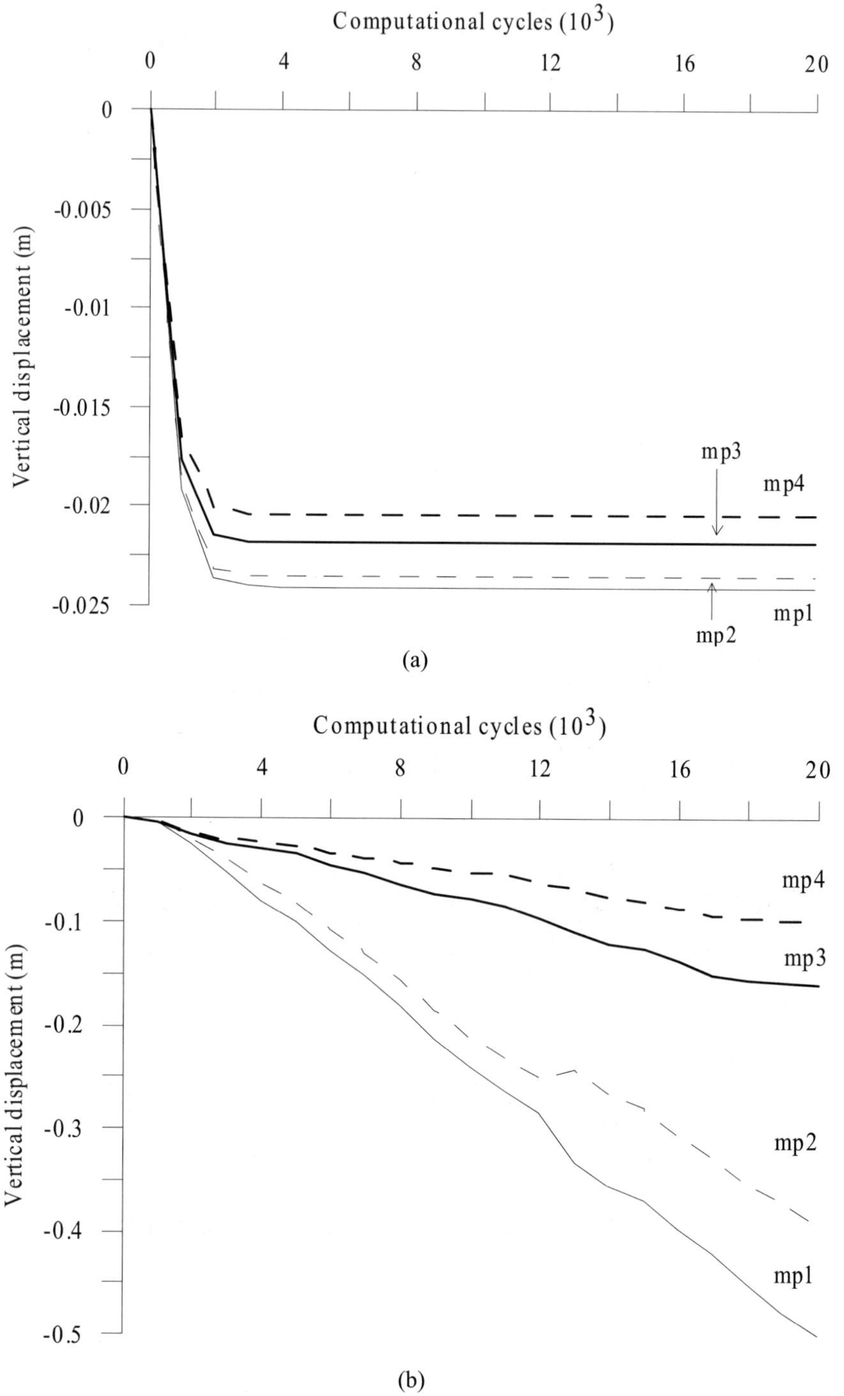

Figure 19. DDA measurement point output (for location see Figure 17b). (a) vertical joints configuration, (b) inclined joints configuration.

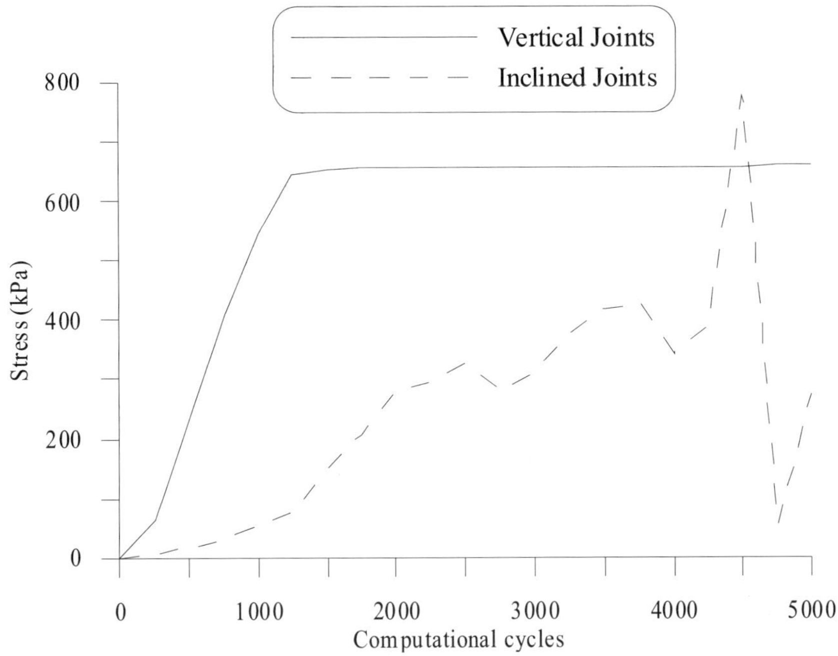

Figure 20. Horizontal stress development in vertical (solid line) and inclined (dashed line) joint configuration in the roof.

Back analysis of the immediate roof failure at Tel Beer-Sheva suggests that the Voussoir beam analogue may be un-conservative: it predicts stability against failure by shear or axial crushing, while in reality the roof has collapsed. Application of the DDA method which considers joint spacing and friction reveals that with the joint spacing value in the field the roof could not have remained stable against failure in shear along the abutments.

DDA further provides the following general conclusions:

- The ratio between free span (*S*) and joint spacing (Sj) is critical for immediate roof stability.
- Both joint spacing and friction must be considered in evaluation of immediate roof stability.
- For a given horizontal beam transected by vertical joints, there exists a unique function which relates beam equilibrium to joint spacing and friction. The unique function exhibits a minimum, which defines the optimal spacing-friction combination for the analyzed beam.

Back analysis of the immediate roof of Freemasons hall in Zedekiah cave suggests that application of the elastic solution for a clamped beam to the actual roof may be misleading. The predicted extreme fibre stresses in the immediate roof layer never materialize in reality because the material does not behave as a continuous solid but as a disjoint beam, the deflections in the field are therefore much smaller than predicted by either the elastic solution or the Voussoir beam analogue. Application of a distinct element numerical method, DDA, confirms this result and further helps explore the significance of a third parameter: joint orientation. It is shown that vertical joints are much more preferable for immediate roof stability than inclined joints. This could be because of better arching mechanism due to longer moment arms, and because of better interlocking between parallel joints due to inherent surface roughness.

ACKNOWLEDGEMENTS

The Tel Beer Sheva study was funded by Israel Nature and Parks Authority and by the US - Israel Bi-national Science Foundation (BSF) through contract # 9800399. The Zedekiah study was funded by Israel Ministry of National Infrastructure, the quarries rehabilitation fund. Drs. Y. Mimran and A. Sasson are thanked for critical review of the manuscript and for stimulating discussions.

REFERENCES

1. Y. H. Hatzor and R. Benary, The stability of a laminated Voussoir beam: Back analysis of a historic roof collapse using DDA, *Int. J. Rock Mech. Min. Sci.* **35**(2), 165-181. (1998)
2. M. Beer and J. L. Meek, Design curves for roofs and hanging-walls in bedded rock based on Voussoir beam and plate solutions, *Trans Instn Min. Mettal.* **91**, A18-22. (1982)
3. G. H. Shi, *Block System Modelling by Discontinuous Deformation Analysis*, Computational Mechanics Publications. (Southampton, UK, 1993).
4. B. H. G. Brady and E. T. Brown, *Rock Mechanics for Underground Mining,* (Kluwer Academic Publishers, Dordrecht, 2004).
5. M. S. Diederichs and P. K. Kaiser, Stability of large excavations in laminated hard rock masse: the Voussoir analogue revisited, *Int. J. Rock Mech. Min. Sci.* **36**(1), 97-117. (1999).
6. A. I. Sofianos, Analysis and design of an underground hard rock Voussoir beam roof, *Int. J. Rock Mech. Min. Sci,* **33**(2), 153-166.(1996).
7. R. Benary, *Stability of Underground Openings in Jointed Chalky Rock – A Case Study from Tel Beer-Sheva*, M.Sc. Thesis, Dept. of Geological and Environmental Sciences, Ben-Gurion University of the Negev (1996).
8. W. H. Evans, The strength of undermined strata. *Trans. Inst. Min. Metal.* **50**, 475-500. (1941)
9. L. Jing, Formulations of discontinuous deformation analysis for block systems, *Engineering Geology*. **49**, 371-381. (1998).
10. M. M. MacLaughlin, and D. M. Doolin, Review of validation of the Discontinuous Deformation Analysis (DDA) method. *Int. J. Numer. Anal. Meth. Geomech.* **29** (2005).
11. Y. H. Hatzor, and A. Feintuch, The validity of dynamic block displacement prediction using DDA. *Int. J. of Rock Mech. and Min. Sci.* **38**(4), 599 – 606. (2001).
12. L. Obert and W. I. Duvall. R*ock Mechanics and the Design of Structures in Rock* (John Wiley & Sons, New York, 1967).

13. Z. Safrai and A. Sasson, *Quarrying and quarries in the Land of Israel* (Hebrew, with English abstract, Ariel, Jerusalem, 2001).
14. A. Shadmaon. *The Bi'na limestone.* Geological Survey of Israel, Bulletin **24**, (Jerusalem, 1959).

Chapter 4.4

SEISMIC RESPONSE OF CLASSICAL MONUMENTS WITH FRACTURED STRUCTURAL ELEMENTS

Ioannis N. Psycharis
Laboratory for Earthquake Engineering, School of Civil Engineering, National Technical University of Athens, 9 Heroon Polytechniou Str, Polytechnic Campus, GR 157 80, Zografos, Greece, ipsych@central.ntua.gr

Abstract: Classical monuments, although made by stone or marble pieces placed one on top of the other without mortar, are stable against earthquakes. Their good seismic behaviour can be attributed to the sliding and rocking of the structural blocks during the strong ground shaking. Unfortunately, damages, which usually exist in such structures, significantly decrease this stability. Previous investigations on the dynamic response of classical columns showed that an initial inclination and/or corner cut-offs of drums may lead to collapse during a medium-size earthquake in spite of the fact that the structure has survived much stronger seismic events in the past. One type of damage, which is common in monuments, concerns fractures at the structural elements due to imperfections of the original material. During a strong earthquake, existing cracks open threatening the stability of the structure. In this paper, an investigation of the seismic response of monuments with fractured structural elements is presented. The distinct element method was used for the analysis and the model employed concerns a part of the Olympieion in Athens, Greece. The results show that the degree of the crack opening during an earthquake increases almost linearly with the peak velocity of the ground motion and the number of repetitions of the excitation. If significant shear and tensile strength exist at the crack interface, a stronger seismic excitation is required, in general, to cause failure. Cracks at column drums do not endanger the stability of the structure, unless they produce wedge-type pieces, which may slide during the earthquake.

Key words: earthquake response; classical monuments; cracks; fractures; distinct element method.

S.K. Kourkoulis (ed.), Fracture and Failure of Natural Building Stones, 239–255.

1. INTRODUCTION

1.1 Dynamic response of classical monuments

Classical monuments are made of carefully fitted stones (drums in the case of columns), which lie on top of each other without mortar. The dynamic response is dominated by the spinal form of the construction and is governed by the sliding and the rocking of the individual stones, independently or in groups. Therefore, it is quite different than the response of 'typical' structures.

The overall behaviour is nonlinear and sensitive. These characteristics are evident even in the simplest case of a rocking rigid block. The latter, in spite of its apparent simplicity, is a complicated problem, which attracted the attention of researchers since the end of the 19th century. The first attempt for the analytical treatment of its dynamic response was presented by Housner[1] in 1963. In the following years, many investigators examined the problem analytically or experimentally producing an impressive amount of research on this subject, which continues up to date.

In the contrary, relatively few investigations have been presented on the dynamic response of stacks of rigid bodies, as it is the case of classical monuments (for a list see Papantonopoulos et al[2]). This is mainly due to the growing complexity of the behaviour as the number of blocks increases. In this case, analytical solutions can be obtained only in simple cases, as for example for two-block assemblies (e.g. Psycharis[3]). If many blocks are involved, it seems that the response can be calculated only by numerical approaches. In the present analysis, the distinct (or discrete) element method was employed.

The complexity of the seismic behaviour of classical monuments originates from the fact that the structure continuously moves from one 'mode' of vibration to another; different joints are opened and different poles of rotation apply for each mode. The term 'mode' is used here to denote different patterns of the rocking response (for an example see Figure 1) and does not refer to the eigenmodes of the system, since spinal structures do not possess natural modes in the classical sense and the period of free vibrations is amplitude dependent. Note that the number of the possible modes of vibration increases exponentially with the number of the individual stone elements. Although the motion can be approximated by linear equations during each mode (for small rotations), the transition from one mode to another makes the overall response nonlinear. One of the consequences of the nonlinearity is that a column may collapse under a certain earthquake motion and remain stable under the same excitation magnified by a value greater than one.

Another interesting characteristic of the response is its sensitivity even to trivial changes of the parameters of the system or the excitation. This sensitivity is apparent in both experimental and analytical results. For example, experiments on the seismic behaviour of a marble model of a column of the Parthenon (Mouzakis et al[4]) showed that "identical" experiments might produce significantly different results (Figure 2), due to uncontrolled perturbations in the initial geometry of the column and/or the shaking table motion. Another effect of the response sensitivity is the significant out-of-plane displacements recorded for purely planar excitations; in some cases, the deformation in the direction normal to the plane of the excitation was of the same order of magnitude with the principal deformation[4].

During rocking, the pole of rotation of each block may move from one corner of the base to the other. This transition produces impact phenomena among adjacent structural elements and energy dissipation, causing a sudden decrease in the angular velocities. An equivalent coefficient of restitution can be determined, the value of which plays an important role to the response. Theoretical and experimental investigations (Aslam et al[5]) showed that the effect of this coefficient is not monotonic and that an increase in its value may decrease or increase the response in an unpredictable way. Note that in linear systems, increasing the coefficient of restitution always results in increasing the damping and decreasing the response.

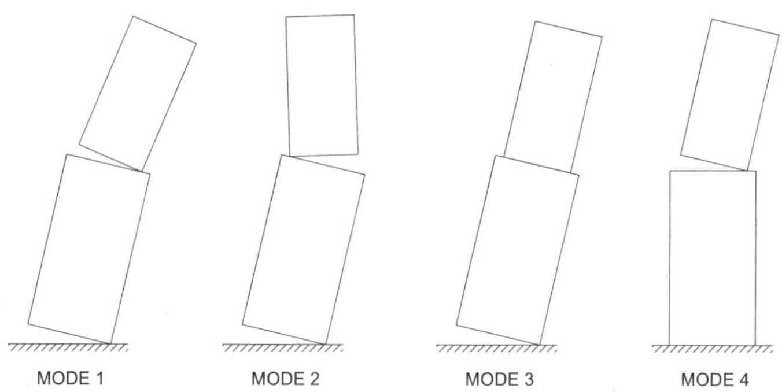

Figure 1. Modes of rocking for two-block assemblies[3].

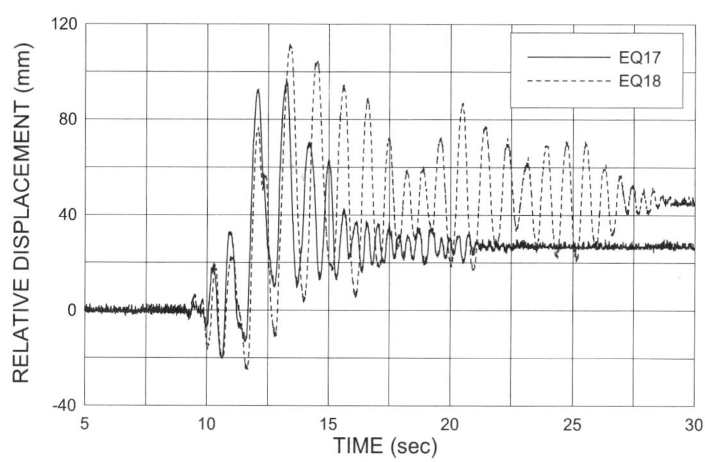

Figure 2. Top displacement of a model of a classical column for two "identical" experiments (shaking table results, Mouzakis et al[4]).

The vulnerability of classical monuments to earthquakes depends on two main parameters: the predominant period of the ground motion and the size of the structure. The former significantly affects the response and the possibility of collapse with low-frequency earthquakes being much more dangerous than high-frequency ones. In the first case, the response is characterised by intensive rocking; in the latter, significant sliding of the drums occurs, especially close to the upper part of the structure, while rocking is usually restricted to small values. This good seismic behaviour may be attributed to their large 'apparent' period, which increases with the amount of rocking. The size of the structure is another important parameter, with bulkier structures being much more stable than smaller ones of dimensions with the same aspect ratio.

1.2 Effect of existing damage

In spite of the lack of inter-connection among the stone elements, classical monuments in their intact condition are not, in general, vulnerable to 'usual' earthquake motions. As mentioned above, their large 'apparent' period and their large dimensions make them vulnerable only to long-period earthquakes. The energy dissipation, caused by rocking and sliding, has also a beneficial effect. This good seismic behaviour has been proved in practice, since many classical monuments are standing for more than 2000 years, although they are located in regions of extensive seismic activity, as Greece and Italy.

Unfortunately, imperfections are present in many monuments. They are caused by previous earthquakes, foundation failure, material deterioration and man interventions, as fire and vandalism. The most common imperfections are cut-off of drum corners, displaced drums, inclined columns and broken element stones. Previous analyses[6,7] have shown that such imperfections reduce significantly the stability and can lead to collapse even for middle-size earthquakes. An example of the significant reduction of the stability, produced by imperfections, is shown in Figure 3.

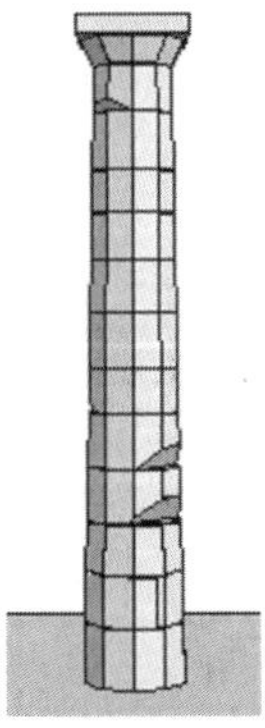

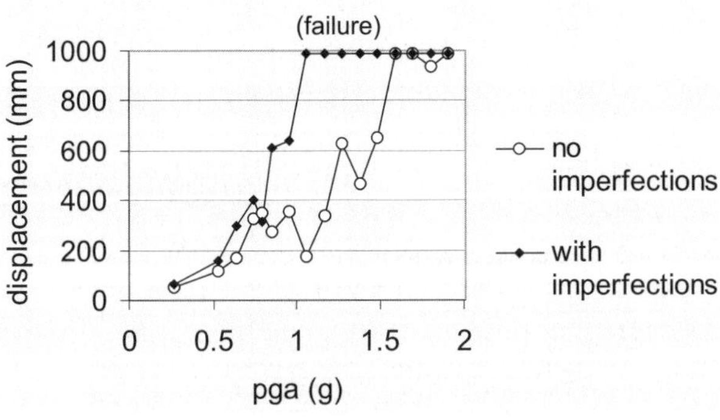

Figure 3. Top displacement of a model of the Parthenon Pronaos column with and without imperfections, for the Aigion, Greece, 1995 earthquake, scaled to several values of PGA[7].

In this paper, an investigation of the seismic response of a part of the Athens Olympieion is presented for several cases of fractures at the stone elements. Some of the cases examined are simplified representations of flaws that are displayed in the present state of the monument, while others are fictitious ones, aiming to the investigation of the effect of several parameters to the possibility of failure, as for example the position and the inclination of the crack.

2. NUMERICAL ANALYSIS

2.1 Model description

2.1.1 Geometrical data

In this paper, all the analyses were based on a numerical model of columns 7.5 and 7.6 of the SE corner of the Temple of Olympios Zeus (Olympieion) in Athens, Greece (see Figure 4a). The first digit of the column numbering refers to the row in the E-W direction, numbered from north to south, in which the column belonged in the original structure; the second one refers to the number of the column within the row, from east to west. The two columns considered here are linked with a three-beam marble architrave.

The total height of the columns is 16.81 m (one of the largest encountered in practice) including the base and the abacus. The main column is of varying diameter ranging from 1.92 m at the base to 1.57 m at the top. Under the column, a base drum of varying diameter from 2.51 to 1.92 m is placed and under it there is a square stone base. The capital is made from two pieces. The height of the architrave is 1.25 m, its width is 1.83 m and its length 5.50 m, equal to the axial distance of the columns.

The height of each drum is not constant, depending on the pieces of marble available at the site during the construction. Also, the number of drums varies from column to column. Thus, column 7.5 has 15 drums and column 7.6 has 14 drums. The drums are connected to each other by two steel dowels (randomly placed in the direction N-S or E-W) with a cross section varying from 9 to14 cm^2 and a length of about 12-14 cm (data supplied by M. Korres). Originally, connections also existed between the architrave beams which, in most cases, are missing today.

In the present condition of the monument, most drums of the columns are displaced by a few millimeters from their original position, especially close to the top. However, the most severe damage concerns a crack close to the middle of the span of the architrave and a clear almost vertical crack at drum #14 of column 7.6, right underneath the capital.

(a) (b)

Figure 4. (a) View of the south-east corner of Olympieion of Athens (Toelle–Kastenbein[8]). Numerical analyses concern the two leftmost columns; (b) numerical model.

2.1.2 Numerical model

The numerical model was based on the actual geometry of the structure (Figure 4b). The drums were represented by polyhedral pyramidal segments of 22-sided cross section and varying diameter according to the original structure. The number of sides considered is equal to the number of flutes of the real columns. All structural elements (blocks) were considered rigid. The marble density was taken equal to 2700 kg/m^3.

A Mohr-Coulomb constitutive model was adopted to describe the mechanical behaviour of the joints between adjacent structural elements. In the normal direction, the joint behaviour is governed by the normal stiffness coefficient, K_n, which relates the contact stress with the normal contact displacement. No tensile strength was considered, so this spring element is only active in compression. In the shear direction, an elasto-plastic stress-displacement law was assumed. The elastic range is characterised by the shear stiffness, K_s, while the shear strength is governed by the Coulomb friction coefficient, with no cohesive strength component.

The joint properties used are: K_n=5.0×10^9 Pa/m, K_s=1.0×10^9 Pa/m and friction angle φ=36.87° (equivalent friction coefficient, tanφ=0.75). The values of the stiffness coefficients were proposed by Papantonopoulos et al.[2], based on the numerical reproduction of the model column experiments[4]. The value of the friction angle is typical for marble.

The steel dowels connecting adjacent drums were considered by special springs (two at each joint) with elasto-plastic behaviour. Since real dowels do not offer any axial resistance and they practically act as shear connections, zero axial stiffness was assigned to these springs. The shear stiffness, K_s, at the elastic range was calculated assuming a cross section of 9 cm^2, shear modulus for steel G=77×10^6 kPa and an active length of 6 cm, which lead to a value of K_s=580000 kN/m for each dowel. The shear strength was considered equal to 220 kN, which corresponds to a yield stress of 240 MPa.

2.2 Method of analysis

2.2.1 Discrete element modelling

As it was mentioned above, the deformation and failure of classical temples is governed by the relative movement of the blocks. For such structures, discontinuous models, in which the structure is considered as a block assemblage and the joints are represented explicitly, should be used.

The distinct (or discrete) element method was proposed by Cundall in the '70's in the context of rock mechanics and later extended to 3D problems[9,10], leading to the code 3DEC[11] used in the present study. This method provides the means to apply the conceptual model of a masonry structure as a system of blocks, either rigid or deformable. Block deformability may be taken into account by internal discretisation of blocks into finite elements. However, in the present study only rigid blocks were used, as they were found to provide a sufficient approximation and reduce substantially the run times. The system deformation, and thus the non-linear material behaviour, is concentrated at the joints, where frictional sliding or complete separation may take place. As discussed in more detail by Papantonopoulos et al.[2], the discrete element method employs an explicit algorithm for the solution of the equations of motion of the blocks, taking into account large displacements and rotations.

The efficiency of the distinct element method and particularly of 3DEC to predict with satisfactory accuracy the seismic response of classical structures has been proved by comparison of numerical results with experimental data (Papantonopoulos et al.[2]). In that study, the experimental data were obtained from the shaking table response of a 1:3 scale model of a column of the Parthenon (Mouzakis et al.[4]). The experiments were reproduced numerically and it was proved that, in spite of the sensitivity of the phenomenon, the numerical analysis depicted with sufficient accuracy all the main features of the response, as the amplitude, the period and the residual displacements.

2.2.2 Damping

The comparison of the numerical results with the experimental data[2] showed that the introduction of damping in the numerical analysis reduces unreasonably the amplitude of the response during the strong shaking, leading to an underestimation of the deformation. On the other hand, zero damping leads to a better estimation of the response during the earthquake, but it does not attenuate the vibrations fast enough after the end of the seismic motion. For this reason, in the present analyses zero damping was applied during the first 12 sec of the earthquake motion, but 10% of critical mass-proportional damping at 0.3 Hz was added after that time. The damping was further increased to 20% of critical for t>30 sec in order to stop the free vibrations and obtain the residual deformation accurately. Mass-proportional damping was chosen instead of stiffness-proportional one, because the latter required a smaller time step of integration and longer run-time.

2.2.3 Seismic input

The seismic action was applied to the base of the numerical model by prescribing the 2 horizontal components of the motion. The records used were based on the two horizontal components of the Kalamata, Greece, 1986 earthquake, which were normalized to several levels of peak ground velocity (PGV), the same in both directions. The ground velocity was chosen as a means of the normalization, because it gives a better representation of the ground motion characteristics than the peak acceleration (PGA), for the response of the structures under consideration.

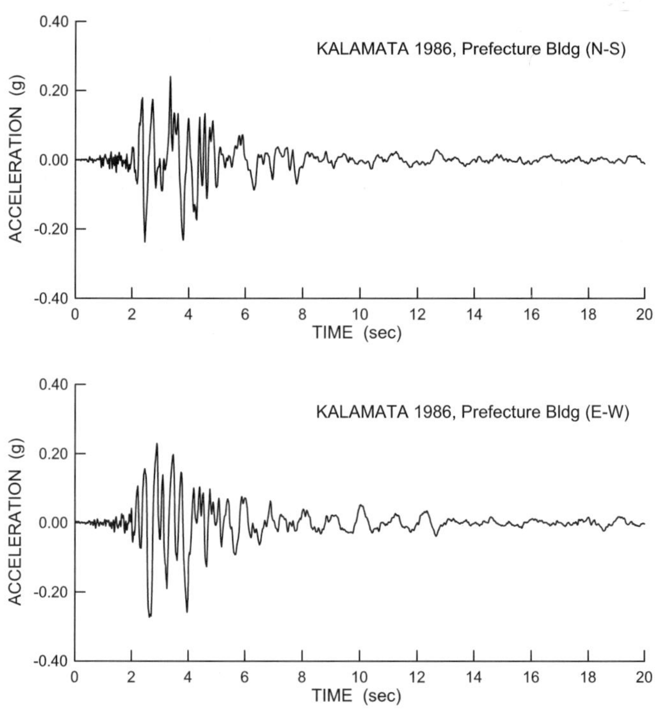

Figure 5. Horizontal components of the Kalamata, Greece, 1986 earthquake.

The Kalamata earthquake was recorded on hard ground at a distance of about 9 km from the epicenter and its magnitude was Ms=6.2. The record samples the near field strong motion that caused considerable damage to the buildings of the city of Kalamata. The duration of the strong motion is about 6 sec and the maximum accelerations are 0.24 g in the N-S direction and 0.27 g in the E-W direction. The corresponding peak velocities are 32.0 and 23.5 cm/s, respectively. Figure 5 shows the two horizontal components of the earthquake.

3. ANALYSES FOR THE CRACK AT THE ARCHITRAVE

As mentioned above, a crack exists at the southern and the middle beam of the architrave of columns 7.5 and 7.6. The simplified representation of this crack, shown in Figure 6, was considered in the analysis. No crack was assumed at the northern beam. For this geometry, gravity loads produce a vertical displacement of about 6 mm at the crack interface. In the following, the term 'relative displacement' is used to denote the additional dislocation at the crack faces, caused by the earthquake excitation, excluding the initial one due to gravity. In most cases, only friction was assumed at the crack interface; the same value of friction angle, φ=36.87°, which was used for the drum joints, was employed. In some cases, however, cohesion and tensile strength were also considered, in order to account for a semi-open crack.

An example of the results obtained is illustrated in Figure 7, in which the time-histories of the vertical displacement at the crack interface of the S beam, for the seismic action normalized to PGV=20 cm/s, are plotted. The value of the ground velocity considered corresponds to a medium-size earthquake and the results show permanent displacements equal to 169 mm for the left piece and 63 mm for the right. The variation of the residual displacements with PGV is shown in Figure 8. In general, the stronger the seismic motion the larger is the dislocation at the crack. As shown in Figure 8, initially, the displacements increase almost linearly with PGV but then they start increasing exponentially and eventually the architrave collapses.

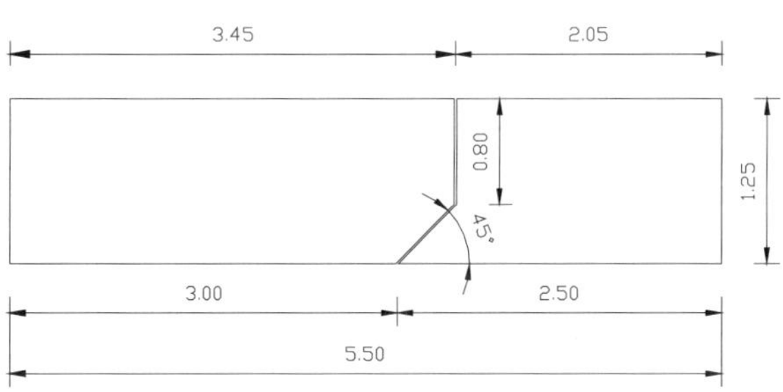

Figure 6. Geometry of the architrave crack considered in the analyses.

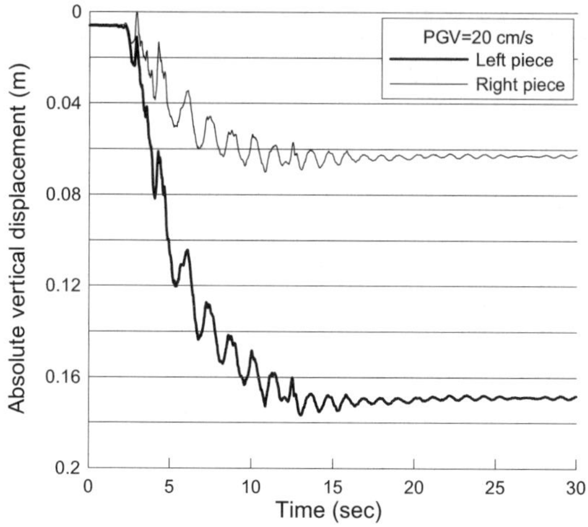

Figure 7. Absolute vertical displacement at the crack interface of the S beam of the architrave for the Kalamata earthquake normalized to PGV=20 cm/s.

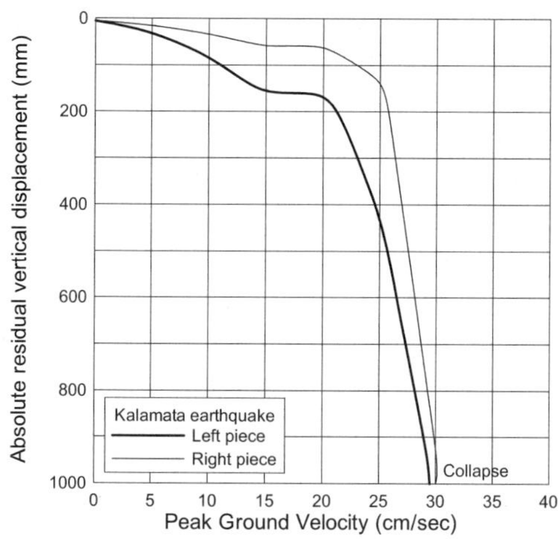

Figure 8. Absolute residual vertical displacement at the crack interface of the S beam of the architrave versus the peak velocity of the seismic excitation.

Fractures at structural elements of classical monuments are due to imperfections of the original material and, thus, the adjacent pieces are bonded initially. However, the crack strength weakens with the time, due to material deterioration. In order to examine this phenomenon, runs were performed including shear (cohesion) and tensile strength at the crack faces. In all cases, the tensile strength was equal to 80% of the cohesion while the value of the latter was varying. Representative results are shown in Figure 9. These results were obtained for the seismic input normalized to PGV=10 and 20 cm/s. It is interesting to note that, up to a certain value of c, the permanent displacements are practically independent of the cohesion and the tensile strength. In such cases, the strength of the joint was exceeded during the earthquake, the crack opened and the two pieces behaved like they were not bonded. In this sense, weak zones or partially cracked elements will behave as fully cracked ones, if the crack breaks during the earthquake.

For larger values of *c*, the internal forces are not capable to break the crack. As illustrated in Figure 9, the limit value of *c*, for which the crack does not break, increases with the intensity of the seismic motion. Since failure is associated with the opening of the crack, collapse occurs under weaker ground shaking for smaller values of the cohesion and the tensile strength. This is shown in Figure 10, in which the minimum value of PGV, required to cause failure, is plotted versus the cohesion, *c*. For small values of *c*, the architrave collapses when PGV becomes equal to 30 cm/s, independently of the exact value of the cohesion. For values of c>1000 kPa, however, the larger the value of *c* the stronger is the earthquake which is required to cause failure.

If a second earthquake hits the monument, the existing damage increases. Thus, the architrave may collapse even for small earthquakes, if they are repeated a few times. In Figure 11, the time-histories of the vertical displacement of the two pieces of the S beam are shown for the seismic motion repeated five times. A rather small earthquake with PGV=10 cm/s was considered in this case. The cumulative effect of the repetition of the ground shaking to the residual vertical displacements at the crack is shown in Figure 12. It is seen that initially the displacements increase almost linearly with the times of repetition of the earthquake, but, after a few events they start to grow exponentially and eventually the architrave collapses during the 6th earthquake. Actually, as it is seen in Figure 13a, the S beam was already close to failure

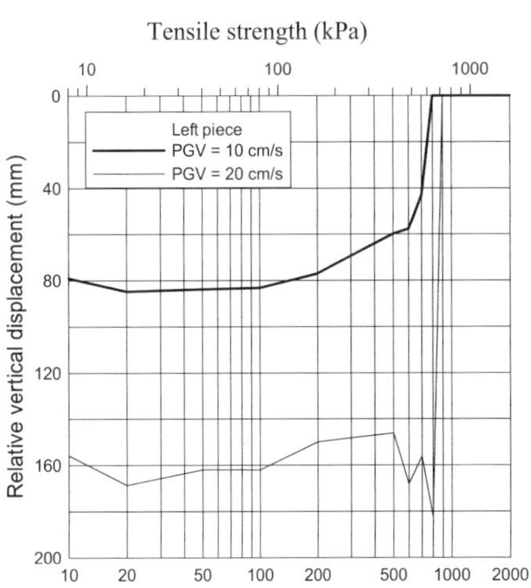

Figure 9. Relative residual vertical displacement of the left piece of the S beam of the architrave versus the cohesion considered at the crack interface.

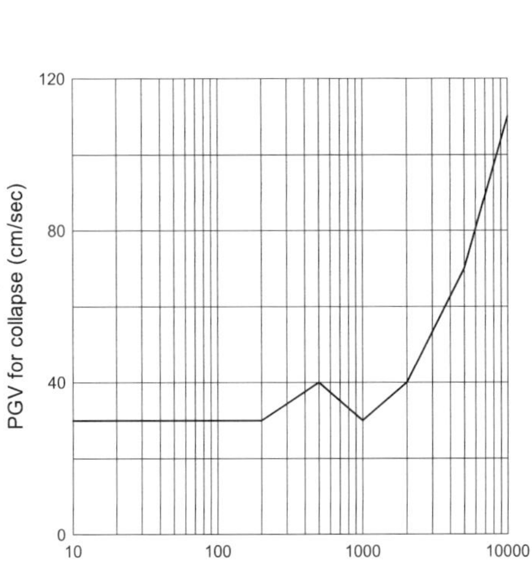

Figure 10. Variation of the minimum peak ground velocity, required to cause failure of the architrave, with the cohesion at the crack.

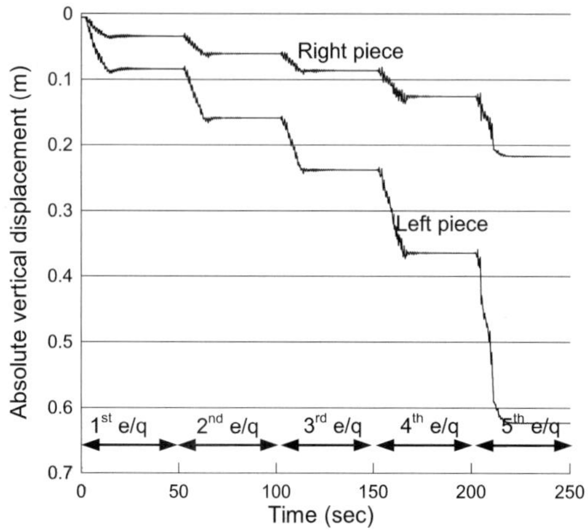

Figure 11. Absolute vertical displacement at the crack interface of the S beam of the architrave for the Kalamata earthquake normalized to PGV=10 cm/s, repeated 5 times.

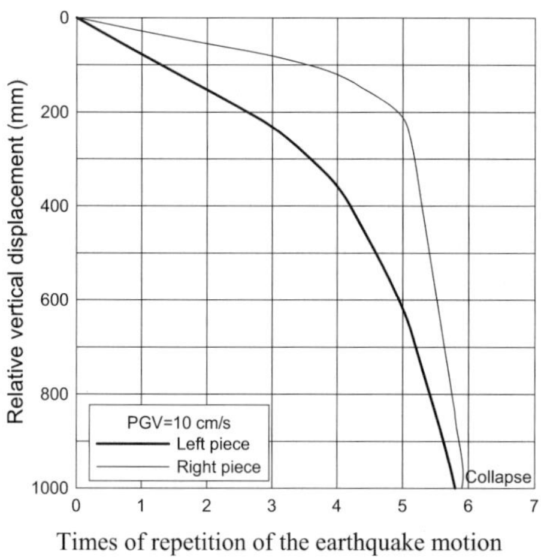

Figure 12. Relative residual vertical displacement at the crack interface of the S beam of the architrave versus the times of repetition of the seismic motion, for PGV=10 cm/s.

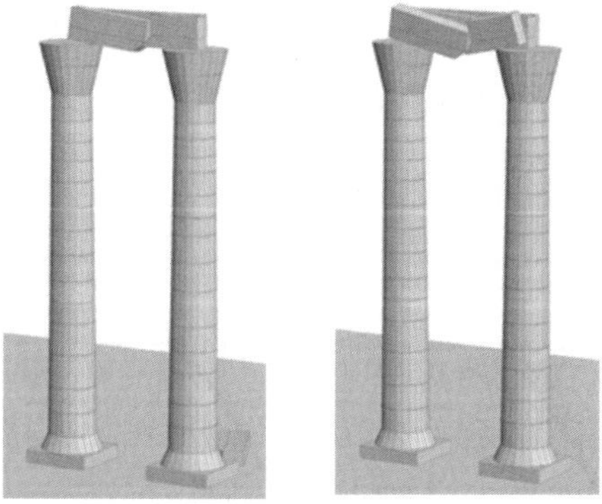
(a) (b)

Figure 13. Residual displacements of the architrave: (a) after 5 repetitions and (b) after 10 repetitions of the Kalamata earthquake with PGV=10 cm/s.

after the 5th event. It is interesting to note that the rest of the structure remained stable even after 10 repetitions of the base motion (Figure 13b), although the middle beam of the architrave had the same crack with the S beam.

4. ANALYSES FOR CRACKS AT THE DRUMS

4.1 System with two columns and an architrave

As mentioned above, drum #14 of column 7.6 is split in two pieces by an almost vertical crack. In the present condition of the monument, the two pieces are dislocated showing an opening of the crack of about 106 mm in the N-S direction (data supplied by M. Korres). The effect of this crack to the stability of the structure is examined here. In the numerical model, the crack was considered vertical through the centre of the drum and forming an angle of 30° with the longitudinal axis of the structure (E-W direction).

Figure 14 shows the residual crack opening, caused by the seismic motion normalized to several values of PGV. For comparison, the existing crack opening is shown with a dashed line. The crack opening increases almost linearly with the peak velocity of the seismic motion; however, the existence of the crack does not seem to affect significantly the collapse of the structure, which occurs for PGV=110 cm/s, i.e. for a very strong earthquake.

The repetition of the seismic excitation increases the opening of the crack almost linearly. This is shown in Figure 15, in which the results for PGV=20 and 30 cm/s are shown. For PGV=20 cm/s, the structure collapses during the 6th repetition of the earthquake. Failure starts from the architrave and extends to the E column while the W column with the broken drum remains standing. For PGV=30 cm/s, the structure remains stable even after 7 repetitions of the seismic action. This 'abnormal' behaviour is attributed to the non-linearity of the response, as discussed in the introduction.

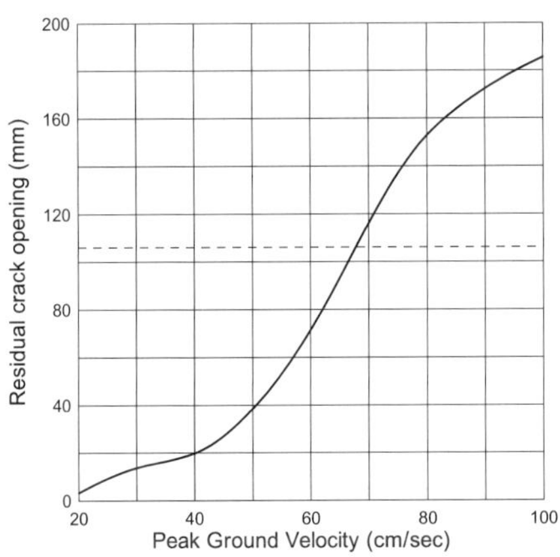

Figure 14. Residual opening of the crack at drum #14 of column 7.6. The dashed line corresponds to the existing opening of the crack.

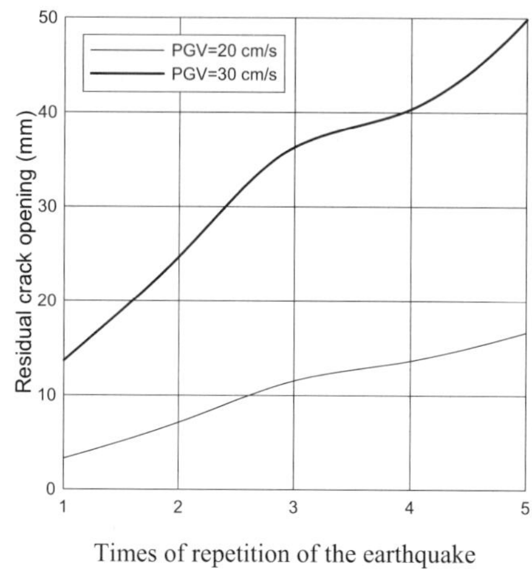

Figure 15. Residual opening of the crack at drum #14 of column 7.6 versus the number of repetition of the seismic motion.

4.2 Single column

The analysis presented above showed that the vulnerability of the structure to earthquakes is not affected significantly by the presence of the crack. This happens because a vertical crack through the centre of the drum does not produce instability, even if it opens a few centimeters. In order to investigate the effect of the crack orientation to the stability of the structure, analyses were performed with various types of cracks.

For these analyses, only column 7.6 (left column in Figure 4b) was considered and the crack was placed at the bottom drum #1 of the column (instead of drum #14), in order its effect to be more pronounced. Three types of cracks were examined, as shown in Figure 16: (A) vertical crack through the centre of the drum; (B) inclined crack by 45° with the cut-off piece not forming a wedge; (C) inclined crack as in type B, but with the upper piece forming a wedge. The variation of the crack opening with the peak value of the ground velocity is shown in Figure 17 for the three types of cracks. It is evident that vertical cracks (type A) do not open significantly, even for strong earthquake motions. In this sense, they do not seem to be dangerous for the stability of the structure. Inclined cracks of type B also do not seem to increase the probability of collapse, although they may open a few centimeters if the structure

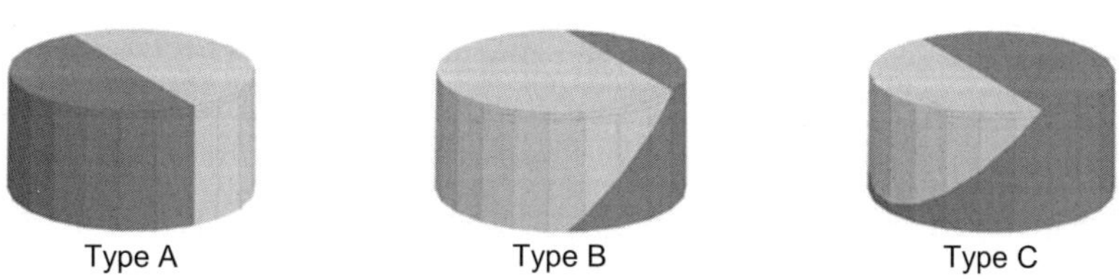

Figure 16. Types of cracks at drum #1 of the single column, considered in the analyses.

is exposed to strong seismic motions. It is interesting to note that the column does not collapse even for PGV=110 cm/s and shows a slightly better stability than crack type A, for which failure occurs for PGV=100 cm/s.

The danger of the crack increases exponentially if a wedge of type C is formed. As it is shown in Figure 17, in this case the column collapses at a much smaller earthquake with PGV=40 cm/s. It is obvious that the vulnerability increases with the size of the wedge, since the larger the sliding piece the larger is the loss of area of contact for the part of the column above the crack. In the present analysis, the loss of contact was almost equal to one half of the area of the drum (refer to Figure 16, type C).

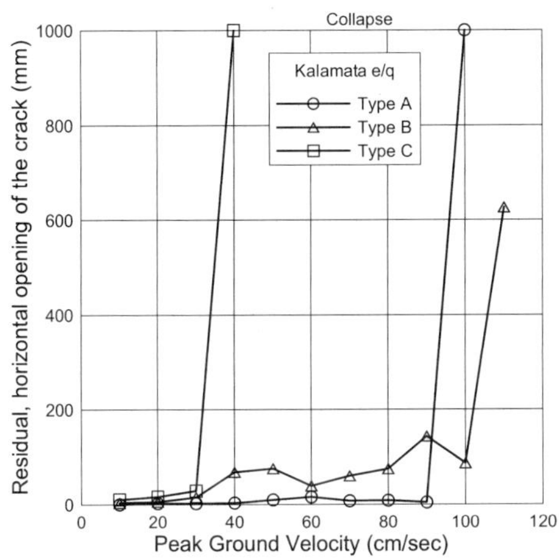

Figure 17. Residual opening of the crack at drum #1 of the single column versus the PGV of the seismic motion.

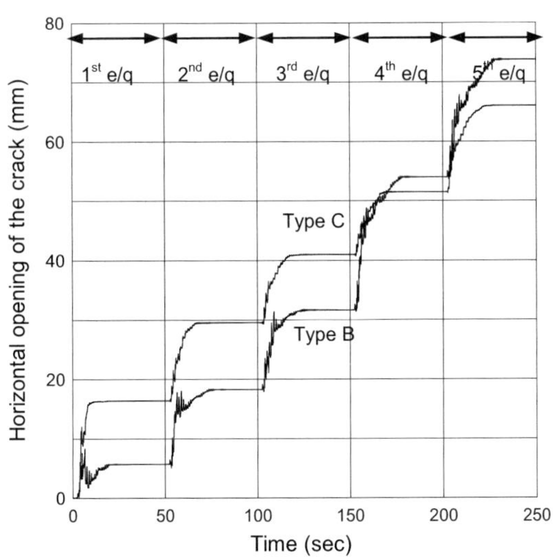

Figure 18. Time history of the opening of the crack at drum #1 of the single column for the Kalamata earthquake with PGV=20 cm/s.

As it was expected, the repetition of the seismic motion produces a cumulative effect on the residual opening of the crack, increasing the failure danger. This is shown in Figure 18, in which the time-history of the crack opening for a sequence of five earthquakes with PGV=20 cm/s is shown for the crack types B and C.

5. CONCLUSIONS

In this paper, an investigation of the seismic behaviour of classical monuments with cracked structural elements is presented. The analyses were performed for the model of columns 7.5 and 7.6 of the Olympieion of Athens, Greece, which are connected with an architrave, using the distinct element method. The conclusions drawn can be summarized as follows:

Cracks at the architrave are dangerous, first to the architrave itself, which may collapse, and then to the whole structure, because the architrave beams can cause damage to the columns, if they hit them during their downfall. In case that the crack is not initially fully open and represents a weak zone of the material, the damage, which an earthquake will cause to the structure, depends on whether the crack will break or not; if it breaks, the damage will be similar to the one that would occur for an open crack. In general, the risk of collapse increases as the strength at the crack interface decreases.

Cracks at column drums do not seem to comprise an immediate threatening to the stability of the structure, as long as they do not form wedges, the sliding of which produces loss of contact for the upper part of the column. In the latter case, critical is the size of the piece that is in danger to slide.

The analyses show that cracks are expected to open during earthquakes. The residual opening increases almost linearly with the intensity of the ground motion. The repetition of the seismic excitation further increases the dislocation of the adjacent pieces in an almost linear way. In this sense, even cracks at drums, which do not form wedges and at first place are not alarming, may become dangerous if not repaired, since they may open extensively after a number of earthquakes.

REFERENCES

1. G. W. Housner, The behavior of inverted pendulum structures during earthquakes, *Bul. Seismol. Soc. America* **53**(2), 404-417 (1963).
2. C. Papantonopoulos, I. N. Psycharis, D. Y. Papastamatiou, J. V. Lemos and H. Mouzakis, Numerical prediction of the earthquake response of classical columns using the Distinct Element Method, *Earthq. Eng. Struct. Dyn.* **31**, 1699-1717 (2002).
3. I. N. Psycharis, Dynamic behaviour of rocking two-block assemblies, *Earthq. Eng. Struct. Dyn.* **19**, 555-575 (1990).

4. H. P. Mouzakis, I. N. Psycharis, D. Y. Papastamatiou, P. G. Carydis, C. Papantonopoulos and C. Zambas, Experimental investigation of the earthquake response of a model of a marble classical column, *Earthq. Eng. Struct. Dyn.* **31**, 1681-1698 (2002).
5. M. Aslam, W. G. Godden and D. T. Scalise, Earthquake rocking response of rigid bodies, *J. Struct.Div. ASCE* **106**(ST2), 377-392 (1980).
6. I. N. Psycharis, D. Y. Papastamatiou and A. P. Alexandris, Parametric investigation of the stability of classical columns under harmonic and earthquake excitations, *Earthq. Eng. Struct. Dyn.* **29**, 1093-1109 (2000).
7. I. N. Psycharis, J. V. Lemos, D. Y. Papastamatiou, C. Zambas and C. Papantonopoulos, Numerical study of the seismic behaviour of a part of the Parthenon Pronaos, *Earthq. Eng. Struct. Dyn.* **32**, 2063-2084 (2003).
8. R. Toelle-Kastenbein, *Das Olympieion in Athen* (Boehlau Verlag, Köln, 1994).
9. P.A. Cundall, Formulation of a three-dimensional distinct element model - Part I: A scheme to detect and represent contacts in a system composed of many polyhedral blocks, *Int. J. Rock Mech. Min. Sci.* **25**, 107-116 (1988).
10. R.D. Hart, P.A. Cundall and J.V. Lemos, Formulation of a three-dimensional distinct element model - Part II: Mechanical calculations, *Int. J. Rock Mech. Min. Sci.* **25**, 117-125 (1988).
11. Itasca Consulting Group, 3DEC - Universal Distinct Element Code (Itasca, Minneapolis, USA, 1998).

Chapter 4.5

SPANNING INTERVALS: TOWARDS UNDERSTANDING THE ANCIENT GREEK OPTIMIZATION PROCEDURE FOR THE DESIGN OF HORIZONTAL BEAMS

C. Papantonopoulos
10 25th March, 162 33 Vyron, Attiki, Hellas, kl-pap@ath.forthnet.gr

Abstract: The paper is dealing with the effort made by the ancient Greek architects to design marble beams (coffer slabs, ceiling beams) using different techniques (stiffening/hollowing), in order to meet the relevant requirements, namely structural integrity and lightness. Comparative studies of various beams of different periods prove the consistency of the design in the ancient world. Interpreting the results of the investigation in terms of bending capacity and using the maximum bending stress as design criterion, it can be shown that for stress values greater than approximately 1.70-1.90 MPa a suitable stiffening ridge was employed. In order to facilitate the transportation of heavy marbles in the mountainous region of Apollo Epicurios at Bassae, instead of stiffening the porch beams, the architect decided to hollow them out reducing the weight by 50%. It can be shown that these U-shaped beams were hollowed out to such a degree, that the maximum bending stress remains less than of that of the initial block.

Key words: ancient Greek architecture; ceiling beam; coffer slab; temple of Apollo Epicurios.

1. INTRODUCTION

The structural system of ancient Greek monumental architecture is exceptionally simple: carefully dressed stone blocks closely fitted together in a post-and-lintel structure by means of clamps and dowels. It is characterized by a total absence of binding agents (mortars) and by the fact that the individual stone members achieve static equilibrium through their own weights. Eccentric loadings and lateral thrusts as a general rule are avoided. For centuries classical Greek construction was regarded as a simple pilling up of

S.K. Kourkoulis (ed.), Fracture and Failure of Natural Building Stones, 257–268.

stones and their superstructure as a simple pilling up of beams. However, a thoroughly investigation of ancient buildings techniques reveals the inventtiveness of architects in solving special construction problems.

The most remarkable example of the awareness of the ancient builders on the structural behavior of simply supported beams is the design of ceiling at Propylaea, in Athens. In the west porch of the central building two rows of Ionic columns support the marble ceiling, and the ceiling beams are arranged so that one comes over each column and one at the mid-span of each Ionic architrave. In order to avoid the concentrated load to be applied at the mid-span of architrave the architect adopted an ingenious solution[1]. He introduced an iron bar at the top of the architrave, as supporting element of the ceiling beam, which stopped about 0.90 m from each end (Figure 1). In this way Mnesicles transferred the load from the center of the archritrave to points near to the supporting columns and so reduced the strength demand to 33% and the deformation to 25%.

Apart from the simply supported beam, Greek architects also knew the principle of cantilever beam, where one end of the beam is built into a wall, the weight of which balances the load of the other end. Cantilever was used in military architecture; the stairways of Aghios Petros in Andros and Aghia Marina in Kea consist of blocks of stone cantilevered out from the walls. Another interesting example occurs in Parthenon, where iron bars as cantilever beams were built into the wall of the pediment to carry the main pediments statues, so relieving the cornice of their weight. Ancient builders used also the double cantilever beam, so arranged that the loads on the two arms balance.

2. THE DESIGN OF MARBLE CEILINGS

The tradition of monumental architecture imposed the design of a peripteral temple, with a box-within-a-cage appearance. A covered colonnade surrounded the cella; a rectangular hall representing the innernucleus of the

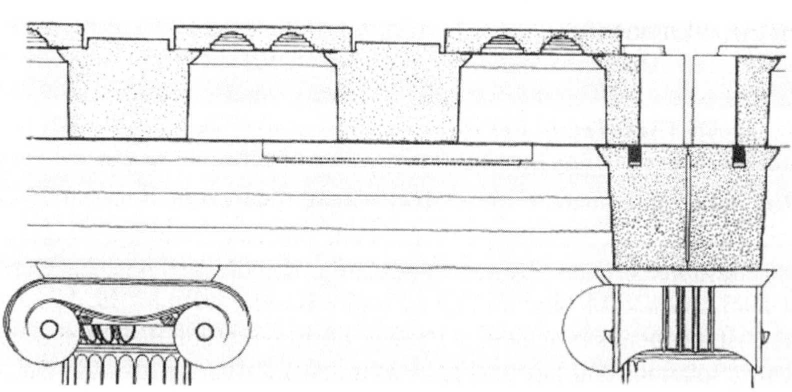

Figure 1. Propylaea at Athens: Ionic architrave showing insert iron beam[1].

building. Because of the short distance between the long colonnades and the cela walls, covering of the long flanks was a rather simple task. Usually, coffer slabs resting on the backer of the entablature and on the sidewalls, covered the long porticoes. On the other hand, the more spacious arrangement on the front (*pronaos*) and the rear (*opisthodomus*) porticoes of a temple, requested a more sophisticated solution. The end porches were exceptionally deep, and as a result long ceiling beams of rectangular cross section running from the inner face of the entablature of the façade to the outer face of that of *pronaos* (or *opisthodomus*) were used. These beams supported the coffer slabs. The first coffered ceilings were of timber, but when marble prevailed as the material used in monumental architecture, the already evolved wooden coffered ceilings began to give way to similar ceilings made of marble. These marble ceilings, which were true masterpieces of delicate carving (Figure 2), continued to be faithful to their timber models[2]. One of the finest examples of ceiling construction is the north porch of Erechtheion, where the marble beams were of the longest employed by the Periclean builders; 5.70 m in clear span.

Usually the coffer slab was carved out of one solid block. In some cases (Hephaisteion, Hieron Samothrace), in order to diminish the weight of the slab, the masons used thin slabs in which they cut either four or six lacunaria (*οπαία*) in one or two rows all the way through each slab to form the lower section of each coffer. After that the framed slabs were covered with extremely thin plates of marble (Figure 3), i.e the removable coffers themselves (*καλύμματα*).

Figure 2. Hieron Samothrace, coffer slabs. View from below[3].

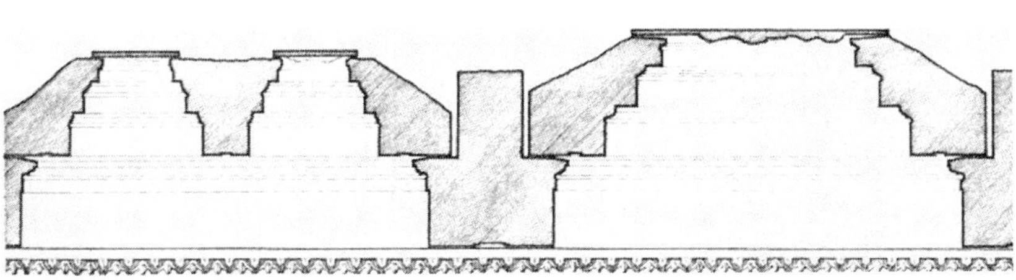

Figure 3. Hieron Samothrace. Cross section through the pronaos showing ceiling beams and coffer slabs[3].

It is well known that Greek architects worked in an environment of tradition and their beams were dimensioned primarily on aesthetics grounds (Figure 4). However, in an effort to strengthen those ceiling beams that carry heavy loads, the ancient builders allowed on the top of the beams a rib (ridge) of considerable height (Figure 3). Without altering the dimensions (height, breadth) of the visible part of the beam, they managed to improve its bending capacity with the addition of a stiffening ridge, which was invisible from below. This intentional ridge has to be distinguished from the flange of only a few centimeters high, which remains of the mantle normally allowed by the quarrymen all round the block (*απεργον*), and here left unworked because it came between the beds of coffer slabs. It is also worth noticing the so called relieving margin; a very slight depression of the edge of the block, only one or two millimeters high, so that flaking or spalling will not result from the pressure exerted by the coffer slab above (Figure 5).

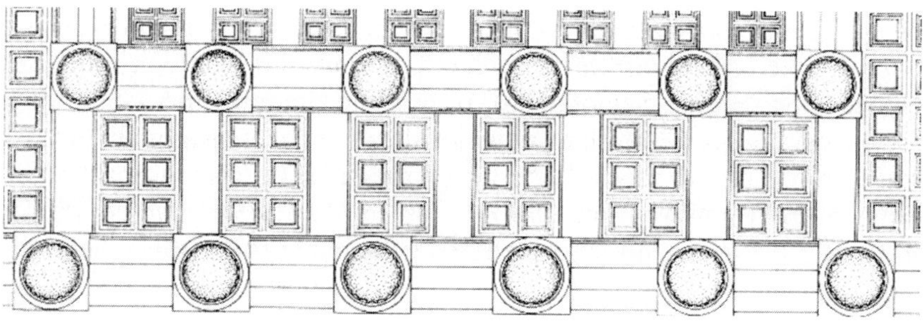

Figure 4. Parthenon west porch. View from below[4].

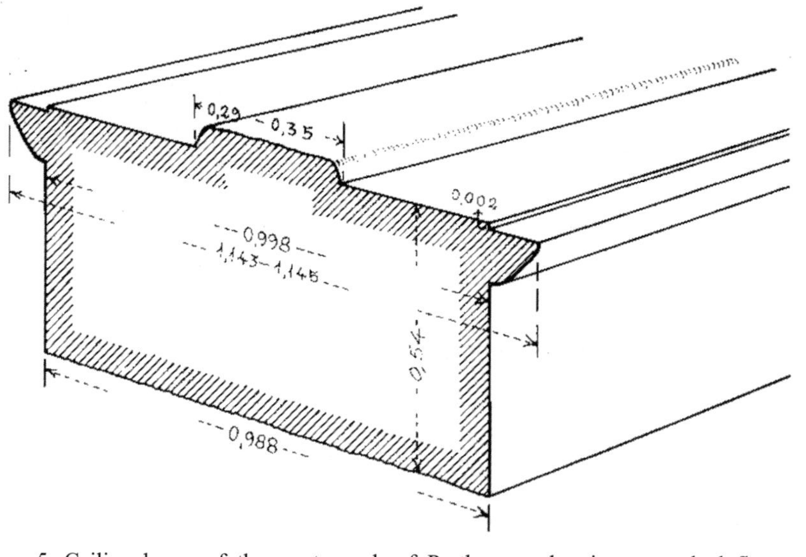

Figure 5. Ceiling beam of the west porch of Parthenon, showing unworked flange and relieving margins[4].

Table 1. Examined ceiling members.

Name	Place	Date (B.C.)	Ceiling member
Parthenon	Acropolis, Athens	447-432	Beam (western porch)
Propylaea	Acropolis, Athens	437-432	Beam (western hall)
Erechtheum	Acropolis, Athens	421-406	Coffer slab (Caryatid porch) Stiffened beam (north porch)
Hephaesteum	Agora, Athens	449-444	Beam (front porch)
Ares	Agora, Athens	440-436	Stiffened beam (front porch)
Nemessis	Rhamnus, Attica	436-432	Beam (front porch)
Apollo Epicurius	Phigalia Peloponessus	c. 450-425	U-shaped beam (north porch) Stiffened girder
Alea Athena	Tegea Peloponessus	c. 350	Coffer slab Beam (north porch) Stiffened girder
Zeus	Stratos, Acarnania	c. 321	Stiffened beam (front porch)
Hieron	Samothrace	c. 315	Beam (front porch) Stiffened beam (*pronaos*)
Apollo	Delos	460-454	Beam (front porch)

In order to clarify the design procedure, various beams of different parts of Greece, covering the Classical as well as the Hellenistic period, were investigated. A broad spectrum of several ceiling members (coffer slabs, beams, girders) with or without ridges was examined, as shown in Table 1 (the dates according to Dinsmoor[5]).

3. CEILING BEAMS – COFFER SLABS

3.1 Mechanical properties of natural building stones

Tests of compressive strength of Pentelic marble conducted in various laboratories using various types and dimensions of specimens and a variety of experimental techniques give a broad spectrum of results, ranging from 58.5 MPa to 116.0 MPa. Nevertheless, for the design of ceiling beams the maximum tensile strength due to bending is the appropriate design criterion. Diamantopoulou[6] gives the following values (Table 2) concerning the compressive/bending strength of different marbles in Greece.

In order to determine the bending strength normal and parallel to the foliation for the pentelic marble Zambas[7] reported three point bending tests, in two series of core specimens Φ3.5X150 mm. Papantonopoulos[8] reported similar tests in prismatic specimens 50X50X200 mm for the limestone of Apollo Epicurios. The abovementioned results are summarized in Table 3.

Regarding the Young's modulus of elasticity the mean values were specified equal to 23 GPa and 80 GPa for the marble in bending, and the limestone in compression, respectively. The Poisson's ratio was about 0.35 in both cases.

Table 2. Compressive/Bending strength of various Greek marbles (values in MPa).

Place	Compressive Strength	Bending Strength
Penteli	59.7	21.3
Paros	35.0	14.7
Volos	42.9	19.3
Aghia Marina	49.5	17.1
Aliveri	41.2	19.8
Skyros	44.2	22.4
Eretria	50.0	17.0
Tinos	82.7	11.0
Farsala	63.5	11.7

Table 3. Bending strength of Pentelic marble and Apollo Epicurios limestone (values in MPa).

Stone	Foliation	Number of tests	Mean Value	Standard Deviation
Pentelic marble	Parallel	44	19.4	3.1
	Normal	24	13.1	2.7
Epicurios limestone	Parallel	32	7.8	3.9
	Normal	5	5.8	-

3.2 Bending stresses of ceiling beams – coffer slabs

The geometrical characteristics (clear span, beam's dimensions, beam's interaxial) and the resulting bending stresses for seven different unreinforced beams (i.e. without ridge) are shown in Table 4. From Table 4 we notice that the maximum "allowable" bending stress is about 1.40-1.50 MPa.

The geometrical characteristics (clear span, beam's/ridge's dimensions, beam's interaxial) for six different reinforced beams (i.e. with ridge) are shown in Table 5. It should be noted that the ridge of the beam in Hieron (Samothrace) has a triangular shape in longitudinal direction (Figure 9). This rib, about 0.28 m thick, is about 0.50 m high in the center of the span, but tapers away to nothing at each end. In Table 6 a comparison is made between the bending stresses of the real reinforced beams (mentioned in Table 5) and the corresponding hypothetical unreinforced beams. Although the ancient design procedure is not known, the comparison approves the consistency of the design in the ancient Greek world.

Table 4. Unreinforced beams – Geometry - Bending stresses

Temple	Place	Clear Span (m)	Dimensions B/H (m)	Beam's Interaxial (m)	Bending Stress (MPa)
Parthenon	Western Porch	3.30	0.95/0.54	3.40	1.40
Propylaea	Western Hall	5.55	0.82/0.62	1.84	1.37
Hephaisteum[9]	Front Porch	4.00	0.54/0.25	1.40	1.25
Nemessis	Front Porch	2.95	0.36/0.25	0.92	1.31
Hieron	Front Porch	3.60	0.48/0.36	1.20	1.49
Athena Alea[10]	Front Porch	5.66	0.67/0.52	1.68	1.26
Apollo (Delos)	Front Porch	1.05	0.39/0.22	1.30	0.21

Table 5. Reinforced beams – Geometry

Temple	Place	Clear Span (m)	Beam's Dimensions B/H (m)	Ridge's Dimensions B/H (m)	Beam's Interaxial (m)
Erechtheum	North Porch	5.70	0.65/0.60	0.40/0.20	1.70
Ares	Front Porch	4.65	0.57/0.28	0.42/0.20	1.57
Stratos (1)	Front Porch	4.80	0.57/0.39	0.39/0.30	1.67
Stratos (2)	Girder	2.25	0.59/0.43	0.36/0.37	-
Athena Alea	Girder	1.05	0.87/0.40	0.60/0.25	-
Hieron	Pronaos	5.30	0.57/0.42	0.28/0.50	2.40

Using the maximum bending stress as design criterion, it is noticed from Table 6 that for stress values greater than approximately 1.70-1.90 MPa, a suitable stiffening ridge was employed. Speaking about the pentelic marble and taking into account the results presented in Table 3, the characteristic strength of the material (i.e. value with 5% probability to be exceeded) is equal to 14.4 MPa, which means that the maximum "allowable" bending stress (1.70-1.90 MPa) is less than 15% of the characteristic strength. The aforementioned results are also valid for the coffer slabs, where the bending stresses do not exceed the "permissible" value of 1.70 MPa (Table 7).

Table 6. Reinforced beams - Bending stresses (MPa)

Temple	Real Reinforced Beam	Hypothetical Unreinforced beam
Erechtheum	1.08	1.89
Ares	1.29	2.65
Stratos (1)	1.27	2.97
Stratos (2)	0.59	1.66
Athena Alea	0.92	2.04
Hieron	1.19	4.17

Table 7. Coffer slabs - Bending stresses

Temple	Place	Clear Span (m)	Dimensions B/H (m)	Bending Stress (MPa)
Erechtheum	Caryatid Porch	3.41	1.65/0.44	1.10
Athena Alea	Front Porch	5.66	1.00/0.55	1.64

3.3 The ceiling beams of the temple of Apollo Epicurios

The temple of Apollo Epicurios stands as one of the most remarkable and best-preserved monuments of classical architecture. The temple is near the site of ancient Arcadian town of Phigalia, 14 km from the village of Andritsaina; it stands on one of the natural plateaux (Bassae) of Mt. Kotylion at an altitude of 1130 m above sea level. The temple is only mentioned once in ancient literature, by Pausanias in 175 A.D. who praised the temple for its beauty and perfect jointing of the blocks and referred to the monument as one

Figure 6. Temple of Stratos[11]. View, showing the girder C-C'.

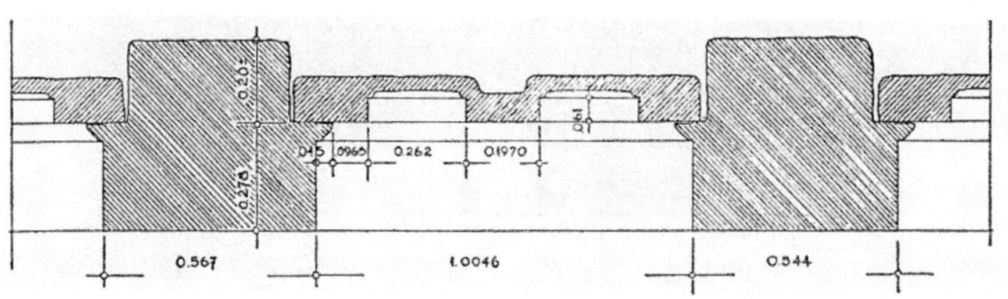

Figure 7. Temple of Ares. Cross section through the front porch, showing reinforced ceiling beams and exceptionally light coffer slabs[12].

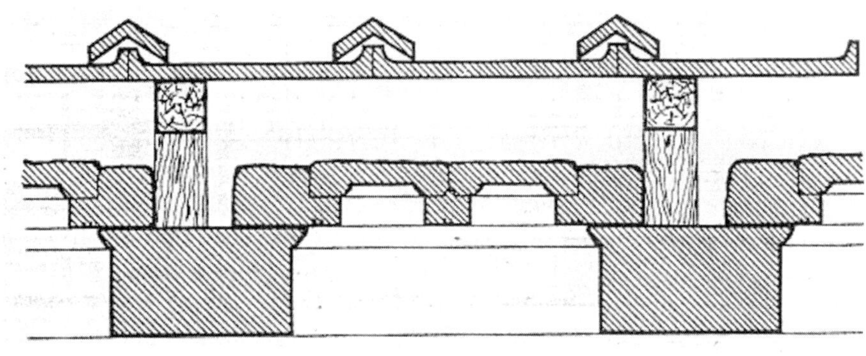

Figure 8. Temple of Nemessis. Cross section through the front porch, showing unreinforced ceiling beams and coffer slabs[13].

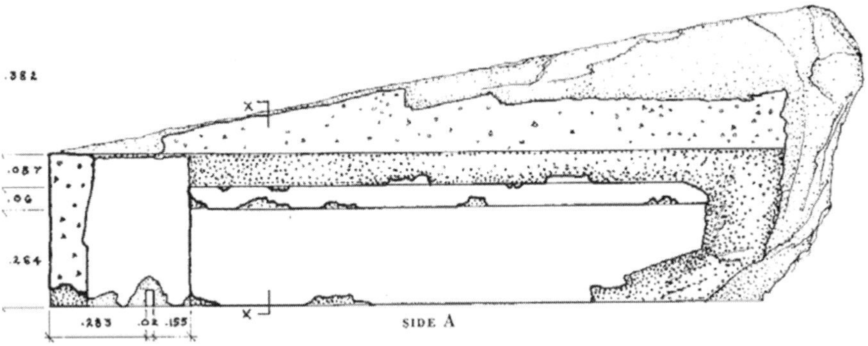

Figure 9. Temple of Hieron, Samothrace. Front view of a fragment, showing the triangular shape of the rib[3].

of Ictinos works. The main local building material is a light grey local limestone, which was very difficult to work, quarried in the vicinity. For this reason marble had to be imported for some of the finest architectural members: the ionic/corinthian capitals, the ceiling beams and coffers, the roof tiles, the frieze and the sculptured panels (*metopes*).

Considering the ceiling design, six different types of ceilings coffers were used. Local limestone was employed for the narrow peristyle ceilings on either flank; but over the great spans at front and rear, and also over the *pronaos* and *opisthodomus*, it was necessary to use the more compact marble. However, the U-shaped ceiling beams employed on the deep front porticos of the temple are rather unusual. Instead of retaining the full section of the marble, stiffening it by adding a ridge along the top, the ancient builders preferred to hollow them out (Figure 10). Dinsmoor[1] believed that the beams were hollowed out to take iron bars from which the beams were suspended. On the other hand, according to Coulton's interpretation[14], the architect reduced the beam's weight by about half at the quarry, in order to facilitate the long journey of the marbles from the island of Paros to the sanctuary (400 km by sea + 22 km overland). Nevertheless, calculating the bending stress of the full section of the beam a value of 1.557 MPa is specified, whereas the U-shaped section gives a maximum stress of 1.476 MPa. Although these values are below the acceptable level the ancient architect managed to find the

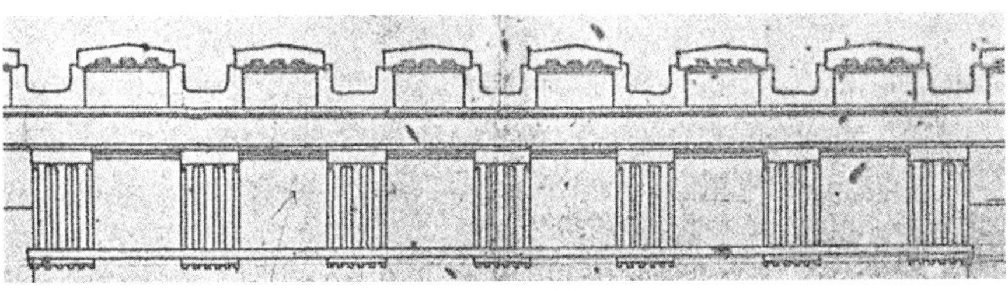

Figure 10. Temple of Apollo Epicurios at Bassai[15]. Cross section through the front porch showing U-shaped ceiling beams and coffer slabs.

Table 8. U-shaped beams - Parametric analysis - Bending stresses

Section	Hole's Dimensions B/D (m)	Beam's Weight (kN/m)	Tensile Bending Stress (MPa)	Compressive Bending Stress (MPa)
1	Full Beam	5.79	1.557	1.557
2	0.08/0.06	5.66	1.578	1.638
3	0.12/0.08	5.53	1.587	1.703
4	0.16/0.10	5.36	1.592	1.780
5	0.20/0.12	5.14	1.588	1.860
6	0.24/0.14	4.88	1.575	1.942
7	0.28/0.16	4.58	1.554	2.025
8	0.32/0.18	4.23	1.527	2.110
9	0.36/0.20	3.84	1.500	2.199
10	0.40/0.22	3.41	1.478	2.299
11	0.45/0.24	2.87	1.476	2.478
12	0.48/0.25	2.55	1.492	2.641
13	0.50/0.26	2.28	1.523	2.913
14	0.52/0.27	2.00	1.582	3.117

optimum solution. In order to clarify the optimization procedure a parametric analysis was undertaken. The investigation starting from the full section and testing 14 different holes with ascending dimensions (B: breadth, D: depth) proves that the realized hollowed beam (section No11, bold in Table 8) gives the minimum bending stress in tension, whereas the increased compressive stress remains small enough. Furthermore, it should be noted that the builders at Bassai were aware of the stiffening technique; they used it carving the limestone girders that supported the ceiling across the long flanks. For instance the bending stress of the unreinforced beam (BXD=0.64X0.31 m), which was equal to 0.53 MPa, was reduced about 60% to 0.23 MPa, by adding a rib 0.40 m thick and 0.25 m high on the top of the beam. It should also be noted that the permissible stress for the limestone blocks are much less than the marble, according to Table 3.

It seems that the ancient constructors have taken into serious consideration the cost of the transportation, since forming an U-shaped section saves no material, for the material cut away is useless. Supply of materials to different sanctuaries necessitated transport over long distances by both land and sea; it was a common practice in ancient world. The numerous temples, which were built in materials brought from elsewhere, suggest that effective means of heavy transport were known to the ancient builders. For instance, the Epidaurian circular temple (*tholos)* contained soft limestone (*poros)* from the Corinthian quarries, Pentelic marble from Attica, and black limestone from the hills of Argos. Nevertheless, taking into account the roughness and/or the inclination of the way, the expenses for the improvement of the pavement of the existing access roads, and the availability of yokes, land transport (*λιθαγωγία*) was expensive and inherently slow in antiquity. It took two and a half to three days to transport one drum, weighting 5.5 tn, from the Pentelic quarries to Eleusis (333 B.C.), and for the 40 km journey more than thirty ox-yokes were used[16]. On the other hand land transport was a risky deal; a certain Megakleidas, who undertook to bring 71 blocks of Pentelic marble to the Epidaurian sanctuary lost over 60% of his price in fines for the delay[17]. Another Athenian paid a huge amount in fines for delaying in the delivery of Pentelic marble to Piraeus for shipment to Epidauros. However, transport by sea was an easier task, despite the additional work requested for loading and unloading the ship. In Delphi, for the rebuilding of the temple of Apollo, a considerable amount of blocks was transferred from the quarries of Sicyon to the sanctuary through the harbor of Kirrha (336 B.C.). As shown by Bousquet[18], for an angular block of the cornice the cost breakdown is as follows (values in drachmas):

- Quarry delivery: 61.
- Transport from quarry to the port of Lechaion: 280.
- Transport by sea from the port of Lechaion to the port of Kirrha: 224 (=4/5x280).
- Transport from Khirra to Delphi: 420 (1.5x280). It is to be noted that Delphi stands at an altitude of 550m above sea level and the distance between the two places is approximately 15 km, which means a mean inclination of the road more than 3.50%. For the rehabilitation of the access road the financial committee spent an amount of 12600 drachmas[19]. Taking into account that the mean quarry price of a cubic meter of stone was equal to 21 drachmas[20], the renovation of the road corresponded to 600 m^3 of stone.
- Placing: 280.

From the abovementioned figures it is concluded that the quarry delivery represents only 5% of the total cost, whereas the transport more than 70%. Moreover, as shown by Rehm[21], the inscriptions from the major Ionian sanctuary dedicated to Apollo at Didyma give additional evidence (values in drachmas per cubic feet):

- Quarry delivery: 3.
- Transport from quarry to the port of Ionia Polis (<2.0 km): 1.
- Transport by sea: 5/6.
- Transport from the port of Didyma (Panormos) to sanctuary (= 3.8 km): 2.

It should be noted that the cost of the land transport in the short distance of less than 6 km is equal to cost of quarrying. Again, the cost of the ship transport is very low, less than 30% of the land transport.

Returning to the peculiar U-shaped beams of Apollo and keeping in mind the preceding presentation regarding transportation expenses one main conclusion can be drawn. In order to reduce the risk and the cost of a very long journey in such a mountainous region the architect adopted an ingenious solution. Moreover, the adapted beam is dimensioned on a basis indicating deep engineering knowledge concerning the structural behavior of simply supported beams.

REFERENCES

1. W. B. Dinsmoor, Structural iron in Greek architecture, *Am. Jour. Arch.*, ***26*/*2*,** 154-156 (1922).
2. M. Korres, in: *The Parthenon Architecture and Conservation*, edited by M. Korres, G.A. Panetsos and T. Seki (Untimely Books, Athens, 1996), pp. 58-59.
3. P. W. Lehmann, in: *Samothrace, Excavations conducted by the Institute of Fine Arts of New York University,* (Princeton University Press, Princeton, 1969), pp. 110-117.
4. A. Orlandos, *The Architecture of the Parthenon* (Library of the Archaeological Society at Athens, 86, Athens, 1978, in Greek).
5. W. B. Dinsmoor, *The Architecture of Ancient Greece* (B.T. Batsford, London, 1950).
6. A. Diamantopoulou, *The Marble in Architecture* (Melissa, Athens, 1992, in Greek).
7. K. Zambas, *Study for the Restoration of the Parthenon* (Ministry of Culture, Committee for the Preservation of the Acropolis Monuments, Athens, 1994, in Greek).
8. C. Papantonopoulos, *Temple of Apollo Epicourios. Structural Restoration Study* (Ministry of Culture, Committee for the Preservation of the Temple of Apollo Epicourios, Athens, 1995, in Greek).
9. W. B. (Jr.) Dinsmoor, The roof of the Hephaisteion, *Am. Jour. Arch.*, **80,** 221-236 (1976).
10. C. Dugas, J. Berchmans and M. Clemmensen, *Le Sanctuaire d' Alea Athena a Tegee* (Librairie Orientaliste Paul Geuthner, Paris, 1924).
11. F. Courby and C. Picard, *Recherches Archeologiques a Stratos d' Acarnanie* (Bibl. des Ecoles Francaises d' Athenes et de Rome, Boccard, Paris, 1924).
12. M. H. McAllister, The temple of Ares at Athens, *Hesperia*, **28,** 38-43 (1959).
13. K. Iliakis, The roof of the temple of Nemessis at Rhamnous, *Restoration-Conservation-Protection of Monuments*, Vol. B, (Ministry of Culture, Athens, 1987, in Greek).
14. J. J. Coulton, *Ancient Greek Architects at Work* (Cornell University Press, Ithaka, New York, 1977).
15. G. Roux, *Karl Haller von Hallerstein. Le Temple de Bassae* (Biblioheque Nationale et Universitaire, Strasbourg, 1976).
16. G. Glotz, Un transport de marbre pour le portique d' Eleusis, (333/2), *Rev. Et. Gr.* **23,** 26-45 (1923).
17. A. Burford, *The Greek Temple Builders at Epidauros*, (Liverpool, 1969).
18. J. Bousquet., *Etudes sur les Comptes de Delphes* (Bibl. des Ecoles Francaises d' Athenes et de Rome, 267, 1988).
19. M. Maass, *Das antike Delphi: Orakel, Schaetze und Monumente* (Wissenschaftliche Buchgesellschaft, Darmstadt, 1993).
20. G. Roux, Les comptes du IVe siecle et la reconstruction du temple d' Apollon a Delphes,, *Rev. Arch.*, **1966**/2, 245-296 (1966).
21. A. Rehm, *Didyma, II: Die Inschriften* (R. Harder, Berlin, 1958).

Chapter 4.6

JOINING FRAGMENTED MARBLE ARCHITRAVES USING TITANIUM BARS: A NUMERICAL ANALYSIS

Stavros K. Kourkoulis[1], Evi Ganniari-Papageorgiou[1] and Marilena Mentzini[2]
[1]School of Applied Sciences, Department of Mechanics, National Technical University of Athens, Theocharis Building, 5 Heroes of Polytechnion Avenue, 157 73 Zografou, Athens, Hellas, stakkour@central.ntua.gr; [2]Committee for the Conservation of the Acropolis Monuments, Acropolis, Athens, Hellas

Abstract: The problem of joining together fractured marble architraves is studied numerically in the present paper using the Finite Element Method. The study was motivated by the needs of the conservation project in progress of the Parthenon Temple on the Acropolis of Athens, where pioneering work is carried out during the last thirty years concerning, among others, the restoration of fragmented architraves using titanium bars. The numerical model constructed simulates the bending test of a centrally fractured prismatic marble architrave of rectangular cross section. The influence of the loading mode simulating the actual loading conditions is explored and the distribution of the stress and strain components is investigated in an effort to detect the points most prone to fail.

Key words: natural building stones; marble; Dionysos marble; restoration; monuments; architraves; bending; titanium; finite element method.

1. INTRODUCTION

The durability of interventions and the protection of the authentic material during the restoration of a monument are among the priorities of scientists working in relevant projects. The experience of previous botched restoration attempts pressed upon a thorough study of the materials used and their compatibility (both mechanical and physico-chemical) with the authentic one. In this direction the scientists working for the restoration of the monuments on the Acropolis of Athens have developed a pioneering method for joining together fragmented structural elements using titanium bars[1-3] in combination

S.K. Kourkoulis (ed.), Fracture and Failure of Natural Building Stones, 269–286.

with suitable cement mortar. The final target is to reach either the initial load-carrying capacity of the member (if it is possible) or the capacity corresponding to the maximum load expected to be exerted on the particular member after the completion of the project (taking into account all possible future interventions). The latter is the approach preferred when the reinforcing titanium bars required for reaching the initial carrying capacity can not be fitted either due to the shape of the sections of the members or due to the relatively small size of the remaining authentic parts of the member or due to the great loss of authentic material[3]. The above method is nowadays used widely for the restoration of multi-fragmented architraves (epistyles) of the monument, which are subjected principally to bending by distributed loads due to their own weight and the weight of the superimposed structural elements.

The experimental assessment of the reinforcing method requires carefully designed and executed laboratory bending experiments. Unfortunately these tests are usually carried out as either three- or four-point bending, since the simulation of a uniformly distributed loading is not easily realizable for the majority of the laboratory loading frames, unless dead-weight type loads are used. However, even the latter procedure is difficult in case large specimens are to be tested in order to take into consideration the size effect (extremely pronounced for geomaterials and especially for marble) shadowing the experimental results if relatively small scaled specimens are used.

Recently a multi-point bending arrangement (Figure 1) was designed approaching in a better manner the uniformly distributed load[4,5]. The arrangement was used in bending experiments with specimens simulating in an 1:3 scale typical fragmented architraves of the Parthenon Temple, in an effort to

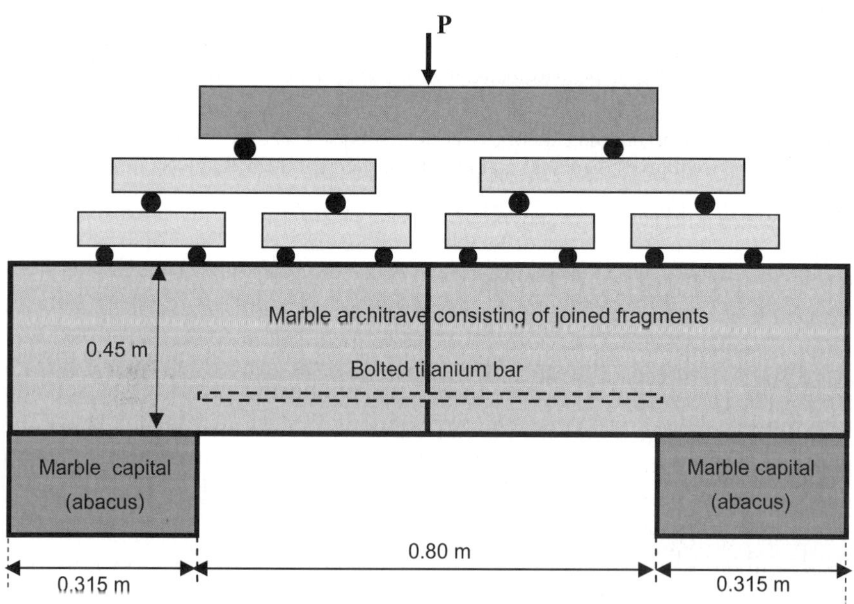

Figure 1. Schematic representation of the arrangement modeled numerically.

assess the reinforcing method of the architraves. The analysis of the results of the tests revealed some discrepancies between them and the respective ones obtained from the theoretical analysis[6]. It was thus indicated that one should consider carefully both the loading system as well as the details of the supports on which the specimen rests, in order to determine the actual span of the architrave and therefore the maximum developed bending moment.

In this sense a numerical analysis is carried out in the present paper divided into two parts: In the first part the possible ways of simulating the actual loading conditions during a typical bending test is studied and the one producing the most severe stress field is pointed out. For the shake of simplicity and CPU-time economy the analysis in this part is carried out considering an intact architrave. In the second part the stress and strain fields developed in a prismatic architrave consisting of two equal fragments joined together with one titanium bar are explored in case the specimen is loaded according to the loading way producing the most severe stress field, as it has been determined in the first part of the study.

2. BENDING UNDER HOMOGENEOUS LOAD

2.1 Bending of intact marble architraves

The influence of the exact load-application mode on the stress distribution during the bending test of an intact marble architrave was studied numerically using the Finite Element Method with the aid of the commercially available software ANSYS 9.0. The geometry of the model matched exactly that of the worst damaged architrave of the north colonnade of the Parthenon Temple, i.e the fifth external one, in a scale of 1:3 (ignoring damages and cracking).

The mechanical properties assigned to the material were these of the Dionysos marble, since it is used extensively for the construction of copies of lost members or of completions of damaged ones by the experts working for the restoration of the Temple. The material was considered as linearly elastic and isotropic with Young's modulus E_m=70 GPa, Poisson's ratio ν_m=0.3, density ρ_m=2.78 gr/cm^3 and coefficient of static friction m=0.70. The transverse isotropy of Dionysos marble and its slight non-linearity[7] were ignored.

The dimensions of the scaled model are: Length L=1.43 m, thickness w=0.18 m and height h=0.45 m. The span was set equal to s=0.80 m.

Four loading types were simulated:

- Uniformly distributed load along the total length (Figure 2a).
- Uniformly distributed load along the span (Figure 2b).
- Eight-point bending along the span (Figure 2c).
- Eight-point bending along the total length (Figure 2d).

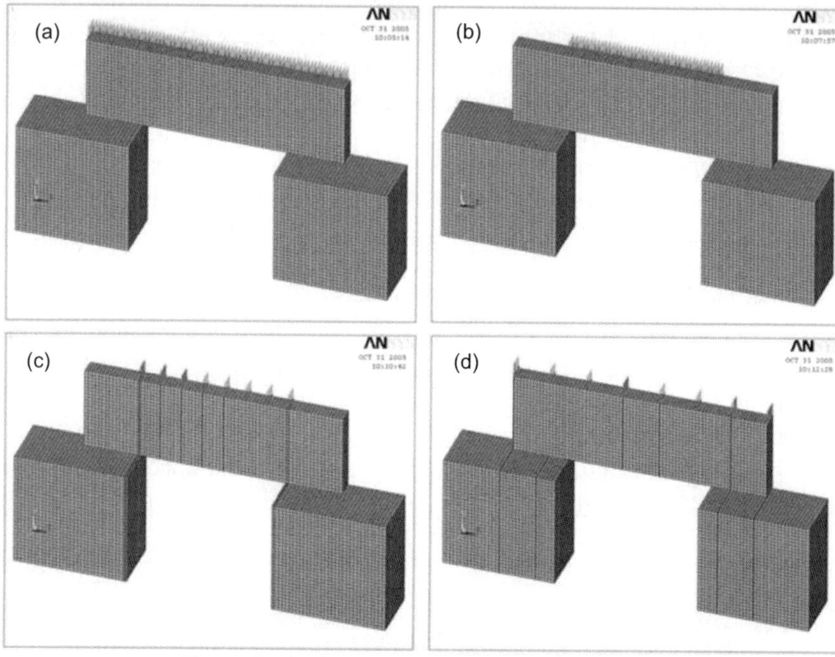

Figure 2. The four loading cases studied numerically.

Both the architrave and the abacuses (capitals) on which it rests were taken into consideration and thus the numerical model constructed consisted of three volumes of the same material, as it can be seen in Figures 1 and 2.

The mapped meshing technique was chosen. Mapped mesh consists of either all quadrilateral or all triangular elements. The volume was shaped as a brick (bounded by six areas) and the element used was the SOLID185, defined by eight nodes with three degrees of freedom at each one. Concerning the density of the mesh it is known that it plays a significant role in obtaining accurate results from a numerical analysis. Several preliminary "runs" were made with different element sizes and it was concluded that for meshes with 50000 elements and over the results converge in a satisfactory manner.

Two couples of 2-D contact elements were introduced between the architrave and the abacuses. For the needs of the present study rigid-rigid contact was assumed since the contact surfaces are made from the same material. The TARGE170 and the CONTA174 elements were used for the analysis. For the quantification of the "contact element stiffness" additional preliminary "runs" indicated that for values exceeding 20 the results are stabilized.

Concerning the boundary conditions it was assumed that the lower bases of the supporting abacuses are rigidly clamped. For reduction of the "running-time" advantage was taken of the vertical plane of symmetry including the axis of the architrave and only half of the configuration was modelled (see Figure 2). The total load exerted was equal to 65 kN, i.e. the maximum load expected for the particular architrave, after the completion of its restoration.

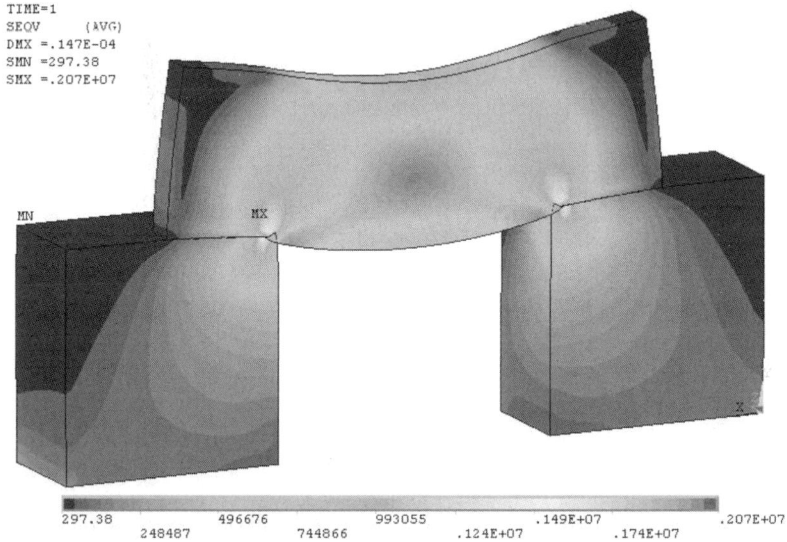

Figure 3. The distribution of the von Mises equivalent stress for the maximum load expected.

Typical results of the analysis are shown in Figure 3, in which the distribution of the von Mises equivalent stress for the whole model is plotted for the second loading type. It is observed that the maximum stress is developed at points close to the corners (fillets, "*hypotomess", υποτομές*) of the supporting abacuses rather than at the mid-span, as it was perhaps expected. The quailtative behaviour of the stress field for the other loading types is of similar nature.

The variation of the equivalent stress along the central longitudinal line of the lowest base of the architrave is plotted in Figure 4 for all four loading types. The conclusions concerning the dramatic influence of the corners of the abacuses on the stress field in the architrave are verified quantitatively: The stress at the mid-span of the architrave is in all cases almost 70% lower compared to the stress developed in the vicinity of the corners of the abacuses. It is obvious that unless the corners are rounded the failure of the architrave will start there either as local exfoliation or cracking. This conclusion was recently supported by a thorough in situ observation of the architraves of the Parthenon Temple, which have never been removed from their original place from the antiquity until the present days: More than half of the fractures and the cracks observed (excluding the ones caused by interventions) have their starting points very close to the vicinity of the corners of the abacuses[6].

From Figure 4 it is also concluded that the most intensive stress fields developed at the central cross section of the architrave are those of cases (b) and (c), i.e. when the load is restricted in the span of the member instead of its total length. However, the most striking conclusion is that in case the load is uniformly distributed the maximum stress is slightly higher compared to the respective case where the load is concentrated at eight strips. On the other hand the loading type causing the weakest stress field is the eight-point

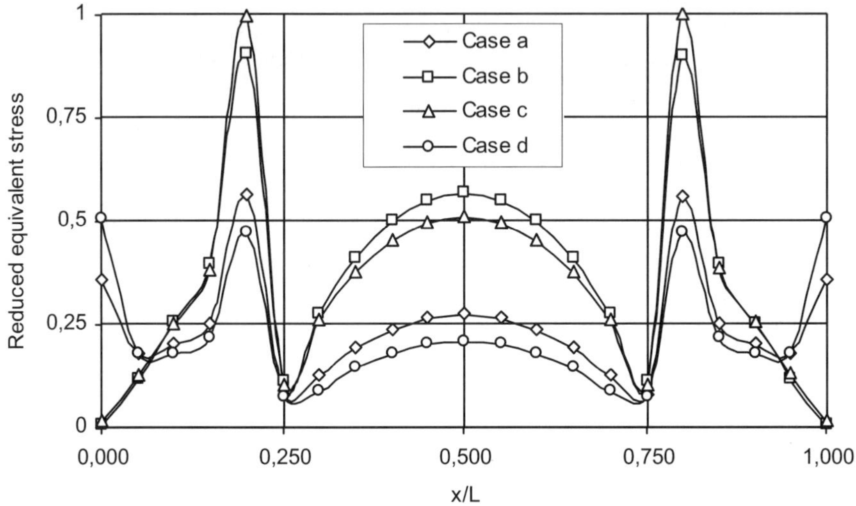

Figure 4. The equivalent von Mises stress along the bottom central longitudinal line of the architrave for the four loading cases. The distance is reduced over the length, L, of the architrave while the stress is reduced over the maximum value developed.

bending with strip loads applied along the whole beam length. It is thus indicated that the simulation of bending under uniform load with multi-point bending tests leads to underestimation of the actually developed stress field.

2.2 Bending of a fragmented and restored architrave

As a next step a symmetrically fractured and restored architrave is considered. The architrave simulates again the fifth external architrave of the north colonnade of the Parthenon Temple in a scale of 1:3. The reinforcement required for the restoration was calculated according to a recently developed method[2,3] which is described shortly in the next paragraph.

2.2.1 The calculation of the reinforcement required

The procedure followed here was introduced by Ioannidou and Paschalides[2] and Mentzini[3] for the needs of the extensive restoration project in progress of the Acropolis of Athens monuments. The architraves are simulated as composite structural members made of marble and titanium bars subjected to simple bending. The analysis is based on the following assumptions:

- The stresses developed do not exceed the linearity limit of the materials.
- Marble is assumed as transversely isotropic material, the modulus of elasticity of which is constant within the bedding planes.
- The strains developed are compatible to each other; there is no relative sliding or pull-out of the reinforcing bars from the marble body.
- The bending loads act normally to the bedding planes.

- The cross sections remain plane and normal to the longitudinal neutral axis of the beam (Bernoulli-Euler technical beam-bending theory).

The latter is perhaps the weakest assumption of the method since the ratio of the length over the height of most architraves of the monument is far from 10, the conventional limit of validity of the Bernoulli-Euler technical theory.

The approach takes into consideration all the loads that could be applied on the member after it is replaced in its initial position in the restored monument. These loads include the own weight of the member, the weights of the members that will rest on it after the restoration as well as possible dynamic loads (earthquake loading). The latter are taken into account by increasing the respective static loads using suitable safety factors.

As a simple example consider an architrave of rectangular cross section reinforced with a single titanium bar. The moment, M_{ex}, developed by the action of the externally applied loads is undertaken by an internal couple of forces, compressive in the marble and tensile in the titanium bar. Denoting by A_t the cross section area of the titanium bar, by ε_m and ε_t the strains in marble and titanium, by σ_m and σ_t the stresses in the uppermost fiber of the marble and in the titanium bars, by E_m and E_t the elasticity moduli of marble and titanium, respectively, and finally by F_t the force exerted by the titanium bar (Figure 5), Hooke's law and the strain compatibility condition give:

$$\varepsilon_m = \frac{\sigma_m}{E_m}, \quad \varepsilon_t = \frac{\sigma_t}{E_t} \Rightarrow \frac{\varepsilon_m}{\varepsilon_t} = \frac{E_t}{E_m}\frac{\sigma_m}{\sigma_t} \tag{1}$$

$$\frac{\varepsilon_m}{\varepsilon_t} = \frac{y_{na}}{h_t - y_{na}} \tag{2}$$

Combining Eqs.(1) and (2) one obtains:

$$\sigma_m = \frac{y_{na} E_m}{\left(h_t - y_{na}\right) E_t}\sigma_t \tag{3}$$

where y_{na} and h_t are the distances of the neutral axis and the titanium layer from the upper side of the beam, respectively. Denoting by the lever of the internal couple of forces and by b the width of the cross section of the architrave, the equation of equilibrium of moments yields:

$$\sigma_m \frac{y_{na} b}{2} z = F_t z = \sigma_t A_t z \tag{4}$$

Combining Eqs.(3) and (4) one obtains a simple second order algebraic equation, which if solved for y_{na} gives the position of the neutral axis as:

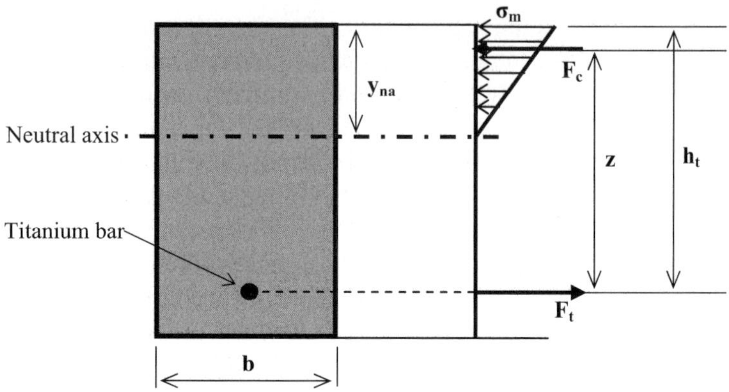

Figure 5. Reinforcing an architrave using one titanium bar.

$$y_{na} = \frac{E_t A_t}{E_m b}\left[-1 \pm \sqrt{1 + \frac{E_m 2bh_T}{E_t A_t}}\right], \quad 0 < y_{na} < h_T \tag{5}$$

Knowing y_{na} one can calculate the normal stresses developed in marble and titanium and compare them with the respective allowable ones.

2.2.2 Simulation of the bending of a restored architrave

The architrave was assumed to consist of two equal parts joined together with a single titanium bar, as shown schematically in Figure 1. The area of the cross section of the bar was determined according to the above procedure and its radius was found equal to r_t=12.7 mm. The bar is placed at a distance h_t= 0.305 m from the upper side of the architrave. The length of the bar used was L_t=0.8 m. The layer of cement used for increased adhesion was ignored.

The mechanical properties of marble are the same with the respective ones of the intact architrave while these assigned to the titanium bar were: Young's modulus E_t=105 GPa, Poisson's ratio v_t=0.32, and density ρ_t=4.51 gr/cm^3.

The dimensions of the scaled model are again: Length L=1.43 m, thickness w=0.18 m and height h=0.45 m. The span was set equal to s=0.8 m.

The specimen was discretized without employing the mapped meshing technique. Instead attention was paid for suitable specification of divisions and spacing ratios on unmeshed lines. The mesh was denser in the vicinity of the titanium reinforcement as well as in the contact section of the two parts of the architrave. The SOLID187 element was used for the mesh. It is a tetrahedral structural solid defined by 10 nodes having three translational degrees of freedom at each node along the nodal x, y and z directions. It has a quadratic displacement behaviour and it is well suited for irregular meshes. The final model consisted of 79596 such elements. An overall view of the model (architrave, supporting abacuses and titanium bar) is shown in Figure 6.

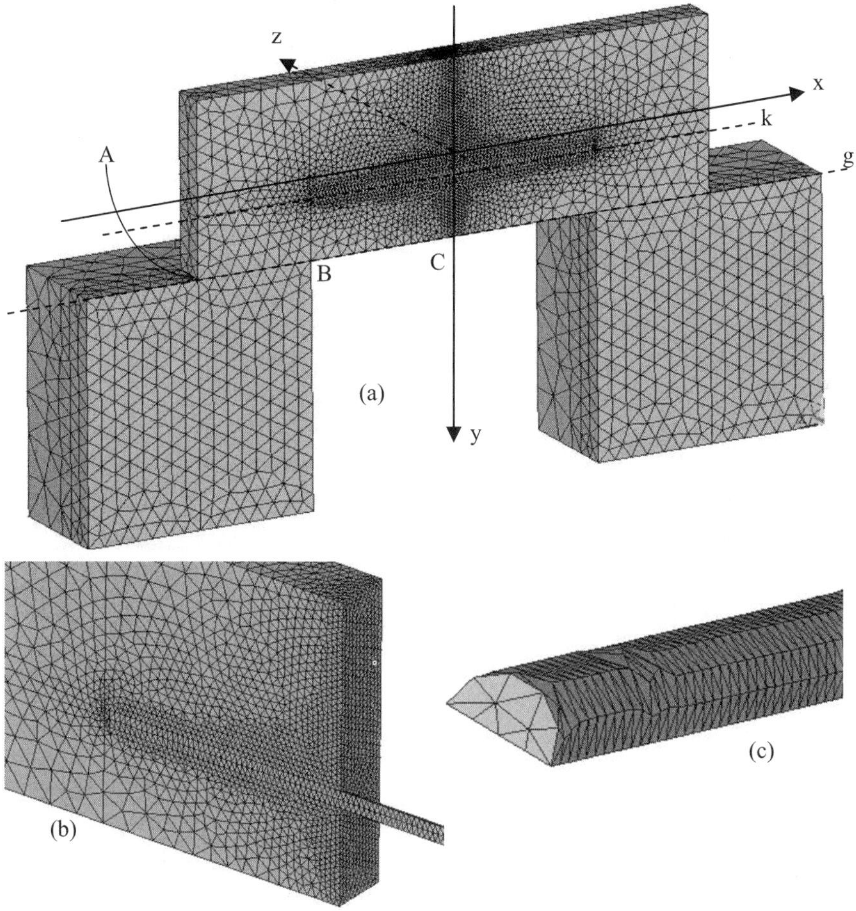

Figure 6. (a) An overall view of the model in the case of a fractured and restored architrave. (b) Detailed view of the mesh in the region of the titanium bar. (c) The mesh of the bar.

As it is seen from Figure 6 advantage was taken again of the vertical plane of symmetry containing the longitudinal axis of the architrave.

The load was simulated as uniformly distributed along the span of the architrave (case b), since according to the analysis of the previous paragraph, it corresponds to the worst case concerning the severity of the stress field developed. The total magnitude of the load exerted was again equal to 65 kN.

An overall view of the distribution of the von Mises equivalent stress in the case of the restored architrave is shown in Figure 7. As expected the situation is completely different compared to that of the intact member. The major part of the architrave is relieved and only at the region around the titanium bar the stresses approach the fracture stress of marble (~6-8 MPa), remaining however far from the respective limit of titanium (~300 MPa). Therefore the failure of the element in its restored form is expected around the section of the titanium in the cracked surface bar rather than at the abacuses' corners.

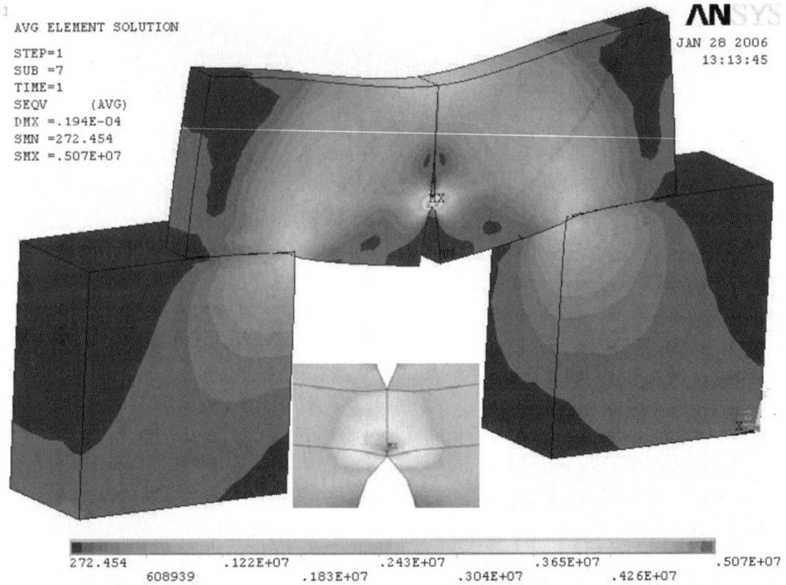

Figure 7. The distribution of the equivalent stress in the restored architrave. The embedded figure shows a detail of the central section around the titanium bar.

Some interesting conclusions can be drawn from Figure 8 in which the variation of the normal axial strain ε_{xx} is plotted along the central vertical line of the cross section (axis y) of the architrave. Axis y is reduced over the height of the beam, h, it is directed downwards and its origin corresponds to the geometrical center of the cross section (Figure 6). Concerning the intact member it is seen that the strain variation is not linear, as it is predicted by the classical Bernoulli-Euler technical bending theory, but rather it is of sigmoid nature. In addition the neutral axis does not pass from the centroid of the section: It is displaced downwards by more than 20% of the height. The

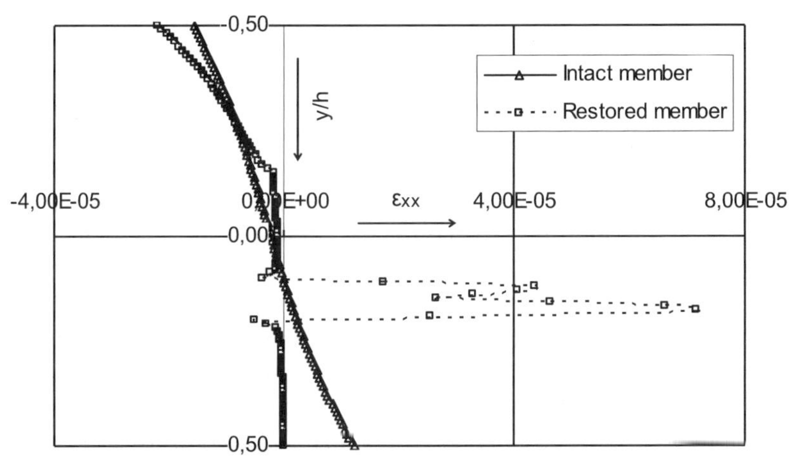

Figure 8. The variation of the axial strain ε_{xx} along the height of the architrave.

above observations are in full agreement with the respective ones by Vardoulakis et al.[8] and Kourkoulis et al.[9,10], who have pointed out experimentally both the lack of linearity of the strain variation as well as the displacement of the neutral axis again for Dionysos marble. Obviously these phenomena are attributed to the geometry of the architrave: The length over the height ratio is 3.2, far from the validity limit of the technical bending theory.

Concerning the variation of ε_{xx} along the height of the restored architrave it is seen from the same figure that for the upper one third of the section the values are of the same sign and order of magnitude with those of the intact member. However, as one approaches the titanium bars the situation changes dramatically and the strain is almost zeroed since the contact of the two parts of the architrave is lost as it was seen in Figure 7. Then in the immediate vicinity of the reinforcing bar the strain attains tensile values almost ten times higher in comparison to the maximum one developed in the intact member. From this point on the strain becomes zero since below the reinforcement the contact of the two constituent parts is lost again. In any case, since the maximum fracture strain of Dionysos marble is about 200 μstrain[7] it is concluded that even in the region of the titanium bar the member is safe.

In Figure 9 the variation of the components of the displacement vector are plotted along the height of the architrave. As it is seen from Figure 9a the vertical displacement, u_y, in the restored case is almost constant (non-zero) all along the height of the member, although it appears to increase slightly in the portion below the reinforcing bar. On the contrary, for the intact member, the vertical displacement decreases almost linearly from a maximum value at the upper side of the architrave to a zero one at the bottom side, as it is predicted by the classical approach, since this side is load free. On the other hand the horizontal displacement, u_x, for the intact member is zero all along the height of the section as it was expected for obvious symmetry reasons.

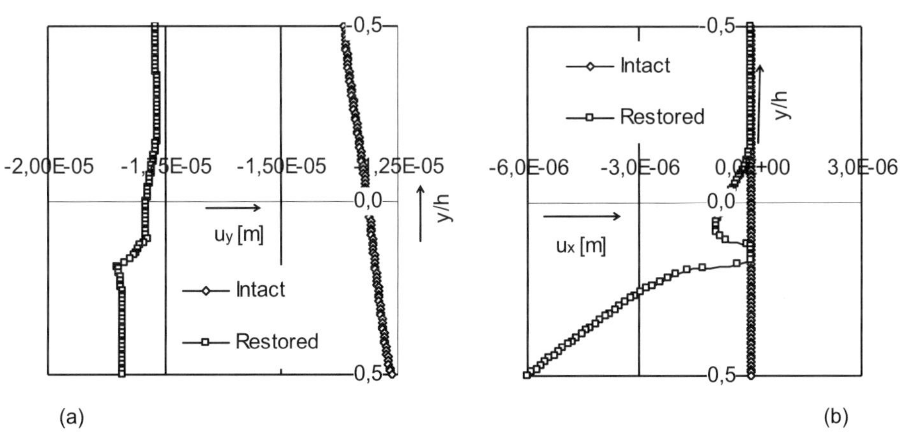

Figure 9. The variation of the vertical (a) and the horizontal (b) displacements along the height of the architrave at its central cross section.

This is not the case for the restored member for which u_x attains high values in the portions of the section where the contact of the two constituent parts is lost (both below and above the reinforcing titanium bar).

The variation of the normal components of the stress field along a horizontal line at the height of the axis of the reinforcing bar (line k in Figure 6) is plotted in Figure 10. For the intact member, the longitudinal stress, σ_{xx}, is almost zero, since by coincidence it is located, very closely to the displaced neutral axis of the architrave, as it was indicated from the analysis of Figure 8. For the restored member, σ_{xx} starts increasing abruptly as one approaches the central section and reaches a value of about 6 MPa, which is almost two orders of magnitude higher compared to the respective stress of the intact member. The behaviour of the transverse normal stress, σ_{yy}, is strongly influenced by the type and the size of the supporting abacuses: Indeed, for the portion of the architrave resting on them, relatively high compressive stresses appear, which for the intact beam tend to be eliminated towards the central section. At the same section of the restored member σ_{yy} becomes tensile with values equal to about 3 MPa, well comparable to the respective axial ones.

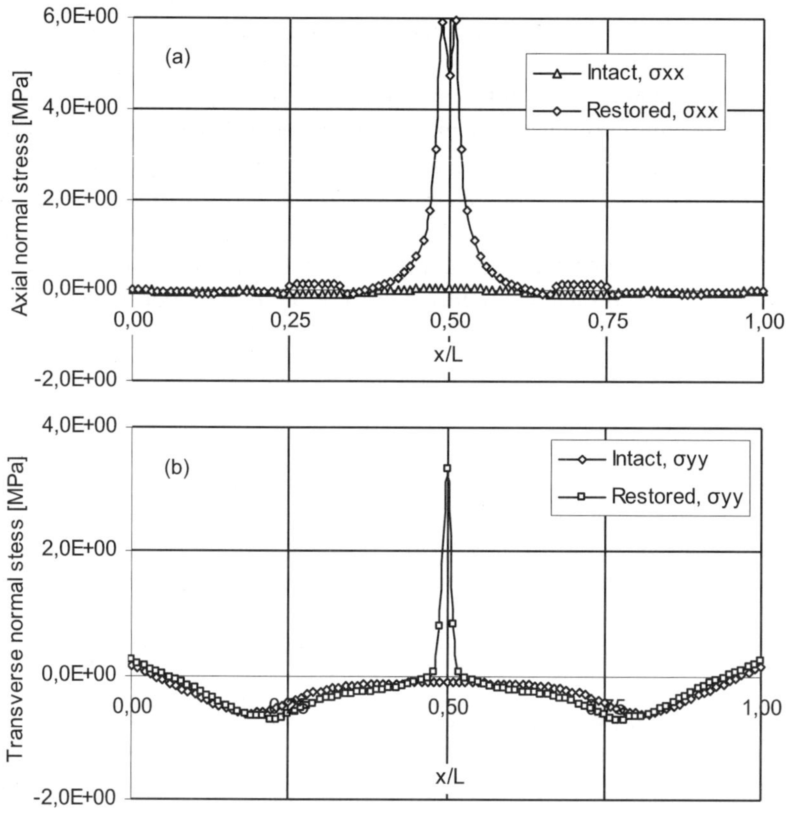

Figure 10. The variation of the normal axial, σ_{xx}, (a) and the normal transverse, σ_{yy} (b) stresses along an axial line (line k in Figure 6) passing from the center of the reinforcing bar.

As a last step of the investigation of the stress field the equivalent stress, σ_{eq}, is plotted in Figure 11 along the bottom central axial line (line g in Figure 6) of the member. As it was expected the equivalent stress along the span of the restored architrave is considerably lower in comparison to that in the intact member and (as it is expected) it becomes zero at the central section. At the points close to the corners of the abacuses the situation is reversed and the values of σ_{eq} are almost double with respect to those of the intact architrave. However, even at these points the stress field is weaker from that developed in the immediate vicinity of the titanium bar (Figure 10) indicating that in the case of a member restored with a single titanium bar failure is expected in the vicinity of the reinforcement while for the respective intact member failure is expected at the area of the corners of the abacuses supporting the epistyle. In order to avoid such failure phenomena ancient Greek engineers used to sculpture the so-called fillet (“*hypotomi*”, “*υποτομή*”) at the corners of the abacuses (capitals) supporting the epistyles (architraves), reducing the stress singularity and therefore the intensity of the stress field.

The variation of the axial strain ε_{xx} along the same line g is shown in Figure 12. It is interesting to observe that in both cases (intact and restored) the axial strain obtains negative (compressive) values as one moves from point (x,y,z)=(-L/2, h/2, 0) (point A in Figure 6) towards the corners of the abacus (point B in Figure 6). The extreme values are about -20 μstrain for the intact and -35 μstrain for the restored member. This behaviour could be explained by taking into account the fact that part of the beam, initially resting on the abacuses, tends to lift up losing contact if the load is applied only along the span of the member[4]. From point B on the strain starts increasing in an almost parabolic form until it either reaches the maximum value at the mid-span of the intact beam (~15 μstrain) or it becomes zero (restored member).

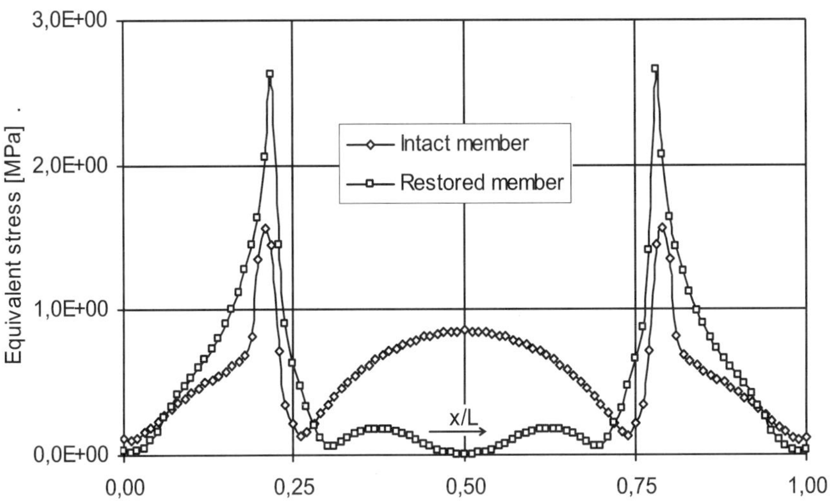

Figure 11. The equivalent stress along the bottom central axial line (line g in Figure 6).

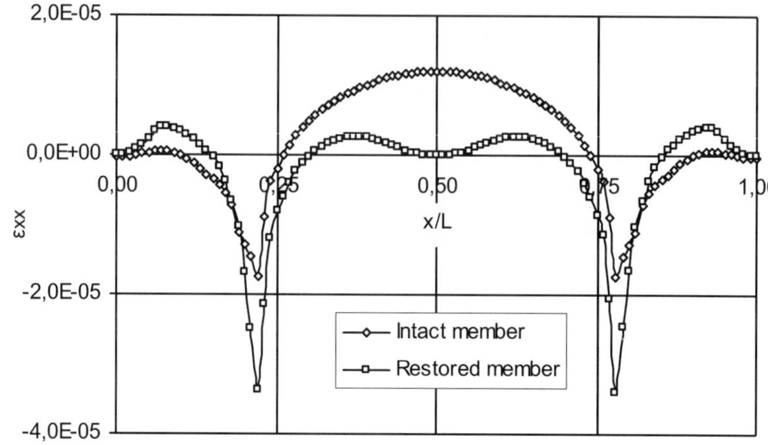

Figure 12. The axial strain ε_{xx} along the bottom central axial line (line g in Figure 6).

3. MULTI-POINT BENDING OF A RESTORED ARCHITRAVE

As it was mentioned in the introductory paragraph the realization of a bending test under homogeneous load is a difficult experimental task. Instead multi-point bending tests are carried out, as it was shown schematically in Figure 1. It is clear, however, that the stress and strain fields in the two cases are not equivalent, and as it has been proved[4] it is possible that multi-point bending results to weaker fields. Something like that, however, may be catastrophic if the design of the restoration is based on the results of the tests. In this direction the multi-point bending test of Figure 1 is studied here numerically for the same as previously restored architrave, in an effort to quantify a "loading-type" correction factor that could enable calibration of the experimental results, in order for them to be compatible with the theoretical ones.

The boundary conditions are the same as the ones described in paragraph 2.1 while the mesh is denser than that in paragraph 2.2.2. The load was again equal to 65 kN and it was divided in eight linear strip loads along the thickness of the beam. The strips were considered to be at equal distances from each other; the first one was exerted at a distance 0.09 m from the upper left edge. The width of each strip was set equal to 2 mm following the results of a recent analysis concerning the influence of the strip width on the overall strain field. The arrangement simulated the experimental set-up that will be used in future steps for the validation of the numerical model and the assessment of the method used for the calculation of the reinforcement required.

The results of the numerical analysis concerning the variation of the axial strain along the height of the architrave at the central section are plotted in Figure 13. In this figure the variation of ε_{xx} is plotted for both the intact and

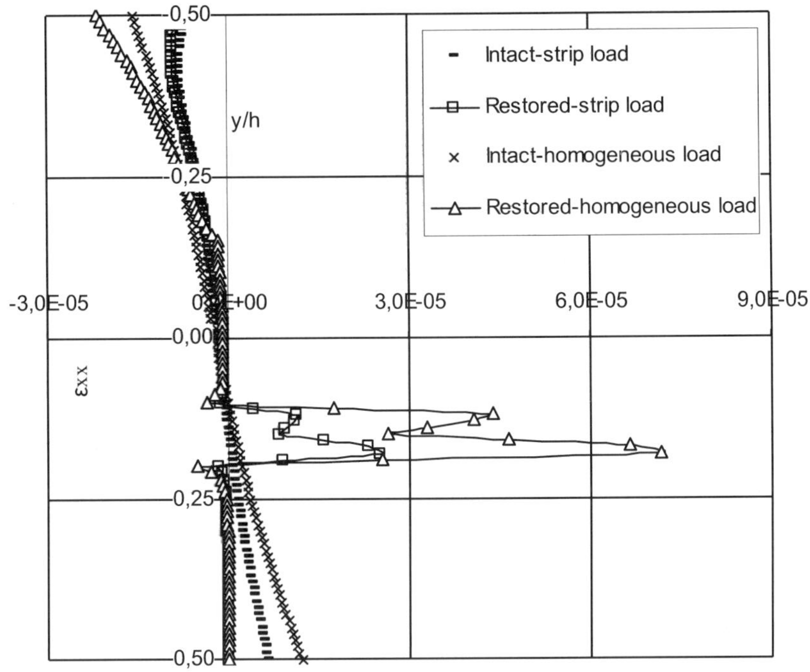

Figure 13. The variation of the axial strain, ε_{xx}, along the height of the central cross section for both the intact and the restored architrave for uniform load and eight-point bending.

the restored architraves under either homogeneous load or eight-point bending, for comparison reasons. As it can be seen, the variation of ε_{xx} for the intact member is again non-linear. It is clearly of sigmoid nature, obviously for the same reasons analysed in paragraph 2.1. Also, the neutral axis appears to be displaced downwards with respect to the longitudinal centroidal axis of the architrave. In addition, it is interesting to observe that the absolute values of the maximum strains attained at both the upper and the lower side of the beam are significantly higher in the case of the homogeneous load (compared to the respective values for multi-point bending). The same conclusions are valid for the restored architrave: The homogeneous load induces higher compressive strains (by about 30%) at the upper side of the member and, also, higher tensile strains (almost by 300%) in the vicinity of the reinforcing bar.

The variation of the axial strain along the bottom central line of the architrave is plotted in Figure 14 for both the homogeneous load and the eight-point bending. From a qualitative point of view, the graphs are of the same nature, but the absolute values of the axial strains in the vicinity of the corners of the abacuses for the case of eight-point bending are almost half the respective ones for bending under uniform load. The conclusions drawn for the variation of the equivalent stress, plotted in Figure 15, are almost identical: The stresses developed in the member subjected to bending under uniform load are almost two times higher compared to the eight point bending.

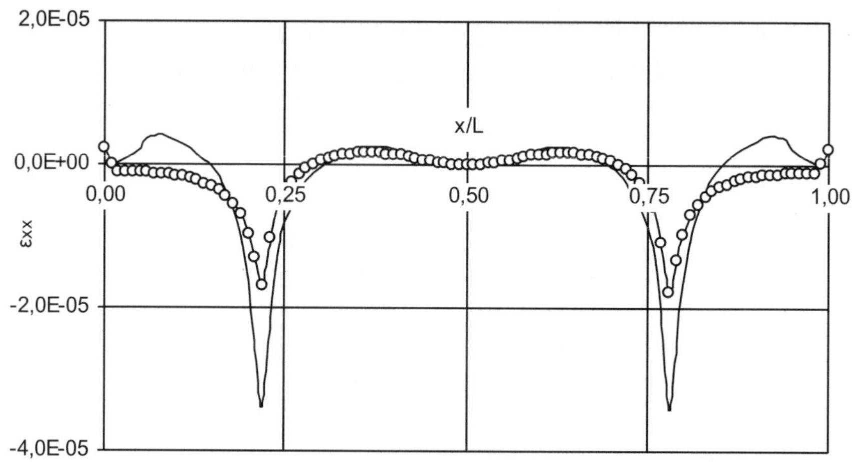

Figure 14. The variation of the axial strain along the bottom central line of the architrave. The continuous line corresponds to the homogeneous load while the one with the cyclic symbols to the eight-point bending.

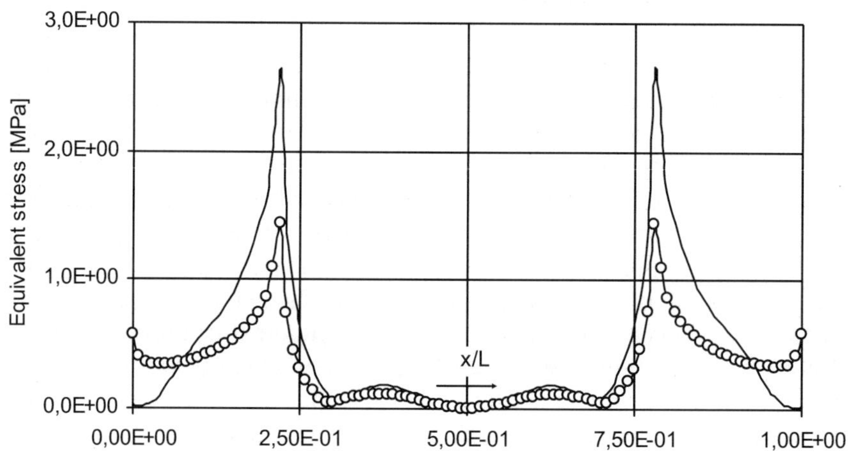

Figure 15. The variation of the equivalent stress along the bottom central line of the architrave. The continuous line corresponds to the homogeneous load while the one with the cyclic symbols to the eight-point bending.

4. CONCLUSIONS

The restoration of a symmetrically fragmented architrave was explored numerically in the present study. The reinforcement required for joining together the parts of the member was calculated according to the method followed by the scientists working for the restoration of the Parthenon Temple on the Acropolis of Athens. The analysis revealed some critical points that should be considered carefully before the design and the realization of interventions, at least on structural members carrying bending loads:

The calculation of the reinforcement is usually carried out adopting the technical bending theory (Bernoulli-Euler), which is approximately valid for relatively long beams. Most architraves, however, do not fulfill this requirement. As a result the axial strains are not distributed linearly along the height of the element but rather they follow a sigmoid distribution. In addition, the neutral axis is translated towards the bottom side of the architrave, rendering the calculations carried out considering the centroidal longitudinal axis as the neutral one rather rough approximations of the real conditions.

The points most prone to fail are the ones closest to the corners of the supporting abacuses rather than those at the mid-span of the beam, in case the architrave is intact. Such conclusion dictates that these corners should be rounded in order to prevent local exfoliations and cracking. On the contrary, in case the architrave is restored the points more prone to fail are the ones in the immediate vicinity of the reinforcing titanium bar.

The reproduction of actual bending conditions in the laboratory does not give reliable results unless the load was applied in exactly the same manner as it is applied in the real structure. More specifically the simulation of bending under uniform load by a laboratory multi-point bending test does not lead to accurate results unless the loading points are too many, something unrealizable for practical reasons. Thus the engineer who designs based on an experimentally evaluated and calibrated model should increase the safety factors of the study accordingly in order to take into account the fact that the stress field of the test is weaker compared to the actually developed one.

In the particular case of eight-point bending the extreme values of the stresses and strains developed are (at some points of the architrave) almost half the respective values developed if the load is applied uniformly. It is implied therefore that an index should be introduced (a load-correction factor) that would enable the direct comparison between the actual loading states of the architraves (the uniform load corresponding to the own weight of the architrave and structural elements resting on it) and the results obtained from laboratory bending tests.

ACKNOWLEDGEMENTS

The authors are indebted to Mr George Ferentinos (K. Liontos and Associates, ANSYS Channel Partners of Greece) for his valuable comments and his continuous support during the preparation of the numerical models. Also, the assistance of Mr Panagiotis Chatzistergos, PhD student in the Department of Mechanics of the National Technical University of Athens, is gratefully acknowledged. The present study is part of the Master Thesis, in progress, of the second author (E. Ganniari-Papageorgiou).

REFERENCES

1. C. Zambas, M. Ioannidou, and A. Papanikolaou, in: *Proc. of the IIC Congress on Case Studies in the Conservation of Stone and Wall Paintings* (The International Institute for Conservation of Historic and Artistic Works, Bologna, 1986), pp. 138-143.
2. M. Ioannidou, and V. Paschalides, in: *Proc. of the 5th Int. Symp. for the Restoration of the Acropolis Monuments*, edited by F. Mallouchou-Tufano (Committee for the Preservation of the Acropolis Monuments, Athens, 2002), pp. 291-300.
3. M. Mentzini, in: *Proc. of the 5th Int. Symp. for the Restoration of the Acropolis Monuments*, edited by F. Mallouchou-Tufano (Committee for the Preservation of the Acropolis Monuments, Athens, 2002), pp. 233- 242.
4. S. K. Kourkoulis, E. Ganniari-Papageorgiou and M. Mentzini, in: *Proc. of the Int. Conf. on Heritage, Weathering and Conservation-HWC*) (Balkema, Rotterdam, 2006) (to appear).
5. V. Paschalides, M. Mentzini, S. K. Kourkoulis and I. Vardoulakis, in: *Proc. of the 1st Nat. Conf. of the Greek Association of Mechanical Engineers* (Athens 2004), p. 72. (in Greek).
6. M. Mentzini, A study on the cracks of the epistyles of the peristasis of the Parthenon of Athens, *The Acropolis Restoration News - Anthemion* (*Committee for the Preservation of the Acropolis Monuments, Hellenic Ministry of Culture*), **6**, (2006), (to appear).
7. I. Vardoulakis, S. K. Kourkoulis, G. E. Exadaktylos and A. Rosakis, in: *Proc. of the Interdisciplinary Workshop "The building stone in monuments*, edited by M. Varti-Mataranga and Y. Katsikis (IGME Publishing, Athens, 2002), pp. 187-210.
8. I. Vardoulakis, G. E. Exadaktylos and S. K. Kourkoulis, Bending of Marble With Intrinsic Length Scales: A Gradient Theory With Surface Energy and Size Effects, *Journal de Physique IV,* **8**, 399-406 (1998).
9. S. K. Kourkoulis, G. E. Exadaktylos and I. Vardoulakis, U-notched Dionysos Pentelicon marble in three point bending: The effect of nonlinearity, anisotropy and microstructure, *International Journal of Fracture*, **98**, 369-392, (1999).
10. S. K. Kourkoulis, M. C. Stavropoulou, I. Vardoulakis and G. E. Exadaktylos, in: *Proc. of the 9th Int. Congress on Rock Mechanics*, edited by G. Vouille and P. Berest (Balkema, Rotterdam, 1999), pp. 623-626.

PART II:

Physico-Chemical and Environmental Aspects

Chapter 5: Weathering

Chapter 5.1

SUSCEPTIBILITY OF BUILDING STONES TO ENVIRONMENTAL LOADS: EVALUATION, PERFORMANCE, REPAIR STRATEGIES

Antonia Moropoulou, Kyriakos Labropoulos, Agoritsa Konstanti, Konstantinos Roumpopoulos, Asterios Bakolas, and Prodromos Michailidis
National Technical Univ. of Athens, School of Chemical Engineering, Section of Materials Science and Engineering, 9 Iroon Polytechniou, 15780, Zografou Campus, Athens, Greece

Abstract: The decay of materials is a function of intrinsic and extrinsic factors. The intrinsic factors are related to the material itself, and include the type of the material, its properties and its microstructure. The extrinsic factors refer to the effect that the environment has on the material and can be generally divided into factors related to the atmosphere and factors related to the usage of the material respectively. Salt decay is arguably the most significant deteriorating mechanism for monuments and buildings in marine environment. Salts migrate in the form of solution through the complex capillary system of porous stones towards the surface. Crystallization of soluble salts may appear in the form of efflorescences or subefflorescences. The assessment of the susceptibility of stone to salt decay can be accomplished either through the calculation of crystallization pressure or through the determination of the most probable scenario as determined by an energetic evaluation. Non-destructive techniques can also be used in situ to assess the susceptibility of the building materials and to assess the effectiveness of interventions.

Key words: decay of building stones; salt crystallization; susceptibility to environmental loads.

1. DECAY OF MATERIALS

The decay of materials can be defined as the degradation, over time, of the material's properties (physical, chemical, mechanical, etc.) and characteristics (texture, mineralogical composition, etc.), leading to the failure of the material as a building component. Decay phenomena develop at interfaces; interface between the material and the environment, or interface be-

S.K. Kourkoulis (ed.), Fracture and Failure of Natural Building Stones, 291–297.

tween materials. Such interfaces are present at various scales. In the case of monuments and buildings, the decay of materials can be studied on two scales: The macroscale (looking at the pathology at the scale of the building), and the microscale (studying the kinetics and the thermodynamics of the phenomenon).

1.1 Intrinsic factors

The decay of materials is a function of intrinsic and extrinsic factors. The intrinsic factors are related to the material itself. They include the type of the material (e.g. stones, mortars, ceramics, metals, wood, and advanced materials such as concrete, fibre reinforced composites and others), its properties (mineralogical, physical, physicochemical, chemical and mechanical), its mass distribution (macrostructure, microstructure, nanostructure), and its source and processing technology (e.g. location of quarry, mining technology, thermal history of the materials, stress history of the material, etc.). In addition, the history of the material (from the initial construction phase throughout all the applied conservation interventions/reconstructions of the structure) and its compatibility with other materials can also be classified as intrinsic factors.

1.2 Extrinsic factors

The extrinsic factors refer to the effect that the environment has on the material and can be generally divided into factors relating to the atmosphere and factors relating to the usage of the material respectively. In particular, the main extrinsic factor relating to the atmospheric effects on the material is the "type" of the atmosphere. The same material will generally behave differently in terms of resistance to decay when subjected to an urban/polluted, suburban, rural, or marine atmosphere. This is often reflected in the general longevity of structures subjected to rural environment as opposed to those subjected to heavily polluted urban environments. Another example is the severe decay of the materials subjected to marine atmospheres.

Other important extrinsic factors related to the environment include the general climate characteristics such as the distribution, orientation and amplitude of the main environmental parameters (temperature, relative humidity, pH, wind, gas pollutants' and aerosols' composition, concentration, and reactivity, and atmospheric precipitations), and the microclimate (orientation / location on the building in respect to the prevailing macroclimatic factors, scale, surface morphology and mode of attack by atmospheric precipitations). The microclimatic conditions are a key factor behind the diverse decay status portrayed by the same material placed on different locations on a building.

Another central extrinsic factor is water. Water comes in contact with building materials through the atmosphere (rain, snow, humidity condensation, aerosols) or through the ground (rising damp) and plays a significant role in salt crystallization, a major decay phenomenon for porous building materials. In addition, many reactions between the building materials and the atmospheric pollutants take place in an aqueous environment, leading, for example, to the creation of crusts or the dissolution of stones by acid rain.

The mechanical loadings the material is subjected to are extrinsic factors related to the usage of the material. These factors refer to static (compressive, tensile and shear stresses, torsion) and dynamic (earthquake, cyclic) loads, stresses due to thermal expansion/contraction, differential expansion of adjacent materials, stresses imposed by salts during crystallization, stresses imposed by the freeze-thaw procedure, and abrasion. Biological factors such as the presence of fauna and flora can also be classified as extrinsic factors related to usage.

1.3 Surface crusts

The decay mechanisms evolve in various modes. One of the most important is the occurrence of surface crusts due to atmospheric suspensions and total depositions, which reflect the development of the decay phenomenon in depth (Moropoulou et al., 1998). Acid dissolution is of major concern for calcitic materials. When there is direct contact, rainwater that has become acidic due to the CO_2 of the atmosphere reacts with the calcitic material producing soluble $Ca(HCO_3)_2$, which after water evaporation, transforms into recrystallized calcite. The recrystallized calcite layer is more porous and more susceptible to detachment than the original calcite and allows the introduction of water and other pollutants in depth.

Gypsum formation on the surface of calcite is a significant decay phenomenon, and it occurs when the surface is not in direct contact with water. It relies on the reaction of the stone with SO_2 from the atmosphere, but because the gypsum layer is more soluble than the calcite, it detaches - dissolves with rain, leading to the loss of surface details. Other surface crusts include yellow-colored oxidation crusts, cementitious crusts, loose black depositions and black-gray crusts (Moropoulou et al., 1998).

2. SALT CRYSTALLIZATION

Salt decay is arguably the most significant deteriorating mechanism for monuments and buildings in marine environment. Salts migrate in the form of solution through the complex capillary system of porous stones towards the surface. Crystallization of soluble salts may result in:

- Salt efflorescences, which deposit on the external surface of the stone. The phenomenon is observed when the rate of salt solution migration to the surface is higher than the drying rate.
- Salt subefflorescences, which appear below the stone surface. The phenomenon is observed when the rate of salt solution migration is lower than the drying rate. This case is considered the most critical for deterioration of stone, as it may lead to stress development into the pore walls (Lopez – Acevedo et al., 1997).

Susceptibility of stone to soluble salts decay is highly dependent on its microstructural and mechanical characteristics (Leith et al., 1996; Theoulakis and Moropoulou, 1997). The assessment of the susceptibility of building stones to salt decay can be accomplished with two approaches. The first one involves the evaluation of the crystallization pressure as a function of the stone's microstructure. The second approach involves the prediction, in thermodynamic terms, of the most probable development of the deterioration phenomenon.

2.1 Direct evaluation – crystallization pressure

In the first approach the crystallization pressure is estimated by a microstructural study, and is then compared to the mechanical properties of the material.

Critical microstructural parameters, such as the ratio of active to total porosity, the average pore radius, the specific surface area, and the bulk density are measured through mercury intrusion porosimetry. These microstructural parameters allow for a direct evaluation of stone susceptibility to decay due to crystallization (Moropoulou et al., 2003).

Furthermore, another characteristic value of the crystal can be calculated, the crystallization pressure, which is the excess pressure developed due to salt crystal growth. More specifically, crystallization pressure can be calculated as a function of the microstructural characteristics of the porous stone as shown in Eq. (1) (Everett, 1961):

$$P = 2\sigma\left(\frac{1}{r_i} - \frac{1}{R}\right) \tag{1}$$

where P is the crystallization pressure, σ the interfacial crystal salt-solution tension, r_i the sum of small pores radii of the porous stone, and R the radius of coarse pore of the porous stone. As far as the critical radius is concerned, that is the distinguishing radius among small and coarse pores, the recommended value is 5 μm. This means that pores with radii greater than 5 μm can be classified to the category of coarse pores (Moropoulou et al., 2003).

Compression and bending tests are performed for the evaluation of the mechanical characteristics of the stones. The values of the crystallization pressure are then correlated with the values of the compressive and tensile strength of the stone. When the crystallization pressure value exceeds the strength of the stone, mechanical failure is observed. Susceptibility of stone to salty decay is a function of:

- The compression modulus and the compression strength of the stone in conjunction with the microstructural characteristics or
- The modulus of elasticity and the tensile strength in conjunction with the microstructural characteristics. It should be noted that tensile rather than compressive strength is considered to be the most representing factor for the comparison with the crystallization pressure values. This is due to the fact that tensile strength represents the cohesive strength of the material, in other words, the maximum load per surface area, which can be supported without material failure.

An energetic evaluation is performed, combining the crystallization pressure with the compressive and tensile strength values of the stone and estimating the probability of disruption of the stone pore walls or filling of the smaller pores.

2.2 Energetic evaluation – thermodynamic scenario

The most probable scenario is determined by thermodynamic factors by the following susceptibility index (Theoulakis and Moropoulou, 1997; Moropoulou et al., 2003):

$$\begin{aligned}
&\Delta G_1 - \Delta G_2 = (\Delta G_{bulk} + \Delta W) - (\Delta G_{bulk} - \Delta G_{surf}) = \Delta W - \Delta G_{surf} \\
&\Delta W > \Delta G_{surf} \Rightarrow \text{Propagation of the salt crystals along the capillaries} \\
&\Delta W < \Delta G_{surf} \Rightarrow \text{Growth of bulk crystal and mechanical failure}
\end{aligned} \tag{2}$$

where ΔG_1 is the free energy change, which corresponds to the progressive crystal growth originating in coarse pores, leading to an increase of the crystallization pressure and eventually leading to pore walls rupture, ΔG_2 is the free energy change which corresponds to propagation of the salt crystals along the capillaries after filling the coarser pores, ΔW is the work needed to rupture the pore walls (estimated to equal to $\frac{1}{2}\,\sigma_c\,\Delta V$ for compression), ΔG_{surf} is the surface change contribution to the free energy of the salt crystal formation.

3. VALIDATION OF ESTIMATED SUSCEPTIBILITY

Following the estimation of the susceptibility of building stones to environmental loads such as salt crystallization, a simulation of the decay phenomena under accelerated ageing is performed. This is based on an analysis of the factors affecting the decay phenomena, and appropriate selection of the environmental conditions to be tested. Accelerated ageing tests include the salt spray chamber, Kosternich chamber, and hydration-drying cycles. Moreover, artificial ageing tests, where the appropriate marine atmosphere is simulated, allow for the evaluation of the durability, under aggressive conditions, of the porous building stones after various conservation and protection treatment.

4. REPAIR STRATEGIES

4.1 Intervention approaches

After the performance of a diagnostic study to assess the extent and pathology of the decay problem and after the evaluation of the susceptibility of the building stones to the prevailing environmental loads, a number of conservation approaches are employed that include the following:

- Cleaning (mechanical, physical or chemical removal of the surface depositions).
- Pre-consolidation (only in cases of severe decay).
- Consolidation (rehabilitation of the cohesion of the corroded stone)
- Surface protection (isolation and protection of the stone from the corrosive action of environmental factors).
- Application of restoration materials, compatible with the original building materials.
- Treatment of rising damp.
- Continuous monitoring and control for the evaluation of the effectiveness of the restoration interventions.

A large variety of conservation materials have been used in the past. Inorganic materials have been widely used, due to their chemical relevance to the authentic building material. Their use is limited nowadays, because they usually tend to form undesirable crusts and by-products after their application to the stones. Organic materials tend to increase the stone's mechanical strength; however, they also tend to form crusts and usually degrade when exposed to oxygen and ultraviolet radiation. In general, alkoxysilanes are the preferred materials for the treatment of siliceous sandstones, due to the formation of chemically relevant materials to the binder during polymerization

4.2 Assessment of effectiveness

In addition to the mercury intrusion porosimetry and the mechanical tests mentioned above, the assessment of the susceptibility of porous stones to environmental loads can be accomplished by a series of non-destructive methods, such as ultrasonic velocity measurements (to evaluate non-destructively the stone's elastic modulus), the fiber optics microscopy (to reveal the surface microstructure and evaluate the depth of penetration of the consolidants and protective coatings) and infrared thermography (to study the water-transport phenomena). Such non-destructive techniques are used in-situ to assess the susceptibility of the building materials but they are also used at the conservation phase to assess the effectiveness of any treatment and applied conservation materials. With the continuous development of the technology, there is a clear trend towards the use of such non-destructive techniques.

REFERENCES

Everet, D.H., 1961, The thermodynamics of frost damage to porous solids, *Trans. Faraday Soc.*, **57**:1541-1551

Leith, S.D., Reddy, M.M., Ramirez, W.F. and Heymans, M.J., 1996, Limestone characterization to model damage from acidic precipitation: effect of pore structure on mass transfer, *Environ. Sci. Technol.,* **30**:2202-2210

Lopez - Acevedo, V., Viedma, C., Gonzalez, A., La Iglesia, A., 1997, Salt crystallization in porous materials II. Mass transport and crystallization processes, *Journal of Crystal Growth,* **182:**103-110

Moropoulou, A., Bisbikou, K., Torfs, K., Van Grieken, R., Zezza, F., and Macri, F., 1998, Origin and growth of weathering crusts on ancient marbles in industrial atmosphere, *Atmospheric Environment* **32**(6):967-982

Moropoulou, A., Kouloumbi, N., Haralambopoulos, G., Konstanti, A., and Michailidis, P., 2003, Criteria and methodology for the evaluation of conservation interventions on treated porous stone susceptible to salt decay, *Progress in Organic Coatings,* **48**:259-270

Theoulakis, P., and Moropoulou, A., 1997, Microstructural and mechanical parameters determining the susceptibility of porous building stones to salt decay, *Construction and Building Materials,* **11**:65-71

Chapter 5.2

CONTROLLING STRESS FROM SALT CRYSTALLIZATION

Jason Houck and George W. Scherer
Princeton University, Civil & Env. Eng./PRISM, Eng. Quad. E-319, Princeton, NJ 08544, USA

Abstract: Salt crystals are able to exert stress on the pore walls in stone because there is a repulsive force between the salt and mineral surfaces, so that a film of supersaturated solution is in contact with the growing crystal. Damage from salt could be prevented if the repulsion were eliminated, so we have screened a variety of organic systems to find species that adsorb on carbonates and on salts. Several candidates were identified and tested for their influence on nucleation and growth of sodium sulfate; promising systems were applied to Indiana limestone, which was then subjected to cycles of soaking in sodium sulfate solution followed by drying. The treatment that offered the most protection was polyacrylic acid with very low molecular weight.

Key words: crystallization pressure; nucleation; polymer; interfacial energy.

1. INTRODUCTION

Stresses created by crystallization of salts play an important role in the shaping of the natural environment[1,2], as well as being a cause of deterioration of buildings, old and new. The goal of our research is to find a method for preventing damage to monuments from salt crystallization by attacking the cause: disjoining pressure. Salts normally repel minerals, so that growing crystals push the pore wall away and generate stress. If we could modify the surface of the stone so that the salt did not repel it, the salt crystal would grow into contact with the pore wall and stop, so there would be no stress or damage[3]. In this paper we present promising results for a water-based treatment that substantially reduces susceptibility of limestone to damage from sodium sulfate.

The origin of crystallization stress has been studied for more than a century, and the principles have recently been thoroughly reviewed[4,5]. It was

S.K. Kourkoulis (ed.), Fracture and Failure of Natural Building Stones, 299–312.

recognized by Becker and Day[6] and Taber[7] that the ability of a salt crystal to exert pressure on a confining surface indicates that a film of liquid is present between the two solids. Correns[8] argued that the two solids (viz., the growing salt and the confining mineral surface) resist contact if the energy of the new solid/solid interface exceeds the sum of the two solid/liquid interfaces. More generally, a crystal will repel the wall until the driving force for growth is so large that it can overcome the disjoining forces, which may result from van der Waals forces (for ice[4] and some organics[9]), electrostatic forces, or ordering of the solvent[10]. Only the latter two factors contribute to disjoining pressure for salts in stone, since the van der Waals forces between salts and minerals are attractive.

The pressure, p, that would have to be exerted on a salt crystal to prevent its growth is related to the supersaturation by[5,8,11]

$$p = \frac{R_g T}{V_C} \ln\left(\frac{Q}{K}\right) \tag{1}$$

where Q is the solubility product, K is the equilibrium solubility for a macroscopic crystal, R_g is the gas constant, T is the absolute temperature, and V_C is the molar volume of the crystal. For a small crystal, the solubility product must exceed K by an amount that depends on its curvature, κ_{CL}, according to the Freundlich equation[12]

$$\gamma_{CL}\,\kappa_{CL} = \frac{R_g T}{V_C} \ln\left(\frac{Q}{K}\right) \tag{2}$$

where γ_{CL} is the crystal/liquid interfacial energy. For a polyhedral crystal with different surface energies, γ_i, for each face of type i, the Wulff condition[13,14] requires that the quantity γ_i/r_i be a constant, where r_i is the distance from the crystal face to the centroid. Therefore, for a polyhedral or spherical crystal, Eqs. (1) and (2) indicate that

$$p = \frac{2\gamma_{CL}}{r} \tag{3}$$

which implies that high crystallization pressures are only expected in small pores. Estimating the crystal/liquid interfacial energy of 0.1 J/m^2 for sodium sulfate[15], Eq. (3) indicates that stresses exceeding 3 MPa would only be required to suppress growth of crystals smaller than ~70 nm. Since pores that small are relatively rare in stone, why is it so common to see damage caused by salt?

In fact, the pressure given by Eq. (1) is not the pressure exerted on the pore wall. If an irregular crystal (i.e., one having curvature that varies from place to place) is surrounded by a solution, then equilibrium can exist only at points satisfying Eq.(2). In the case illustrated in Figure 1, the curvature of the portion of the crystal labeled E, κ_{CL}^{E}, is in equilibrium with the solution that surrounds the whole crystal, which has solubility product Q^E. At point C, where the curvature is less positive than at E, the equilibrium value for the solubility product would be $Q^C<Q^E$; consequently, the crystal at P experiences a supersaturation of Q^C/Q^E, so it tends to grow and apply pressure on the pore wall equal to[5]

$$p_w = \kappa_{CL}^{E} - \kappa_{CL}^{P} = \frac{R_g T}{V_C} \ln\left(\frac{Q^E}{Q^P}\right) \tag{4}$$

If the interior of the pore is much larger than the pore entry, then $\kappa_{CL}^{C} \to 0$ and $Q^C \to K$, so Eq.(4) reduces to Eq.(1), and the pressure is given by Eq.(3) with r equal to the radius of the pore entry.

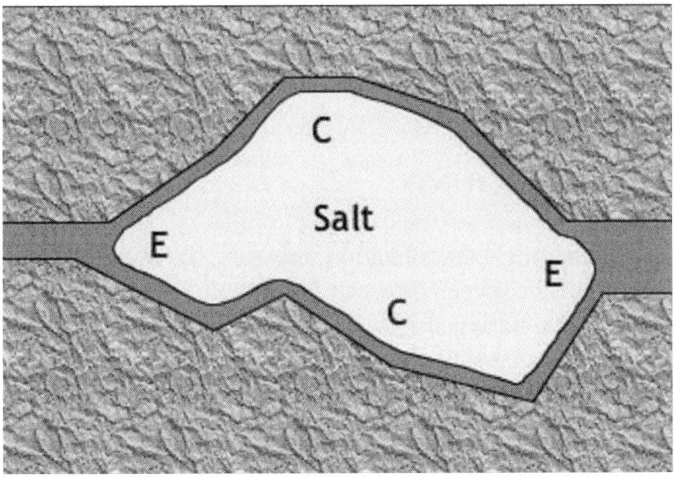

Figure 1. Crystal of salt in a large pore with small entries. The part of the salt crystal adjacent to the entry (at locations labeled "E") has curvature in equilibrium with the solution that surrounds the crystal. At point C, the curvature is positive, but smaller than at E, so Eq.(2) is not satisfied and the crystal tends to grow, exerting pressure against the pore wall.

The preceding analysis indicates that high crystallization pressure only arises in small pores, but this conclusion is based on the assumption that the crystal is in equilibrium with a solution that bathes its entire surface. However, high pressures can develop in large pores, if the solution is only present as a film between the crystal and pore wall[5,16], as in Figure 2. In that case, the supersaturation rises indefinitely as the liquid evaporates, and the crystal applies increasing pressure on the pore wall. The excess solute cannot be

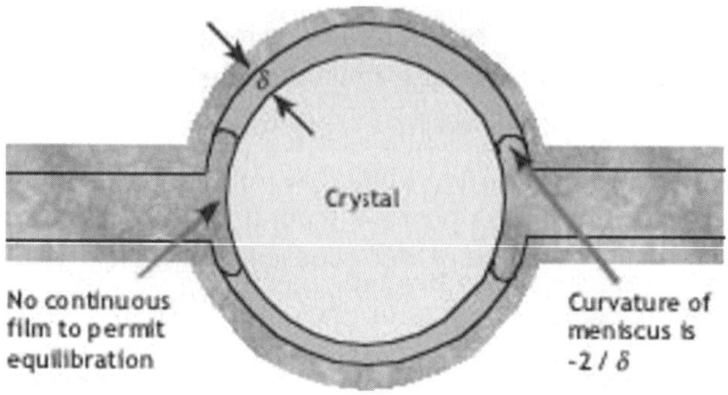

Figure 2. As water evaporates from a porous body, isolated pockets of liquid can be trapped in clusters of pores with relatively small entries. As the liquid within those pockets evaporates, the solution may retreat into disconnected films lying between salt crystals and pore walls. In this sketch, a film of solution with thickness δ is trapped between the crystal and the wall; as its concentration rises, the crystal applies increasing pressure on the wall.

consumed by growth in other directions, because the liquid does not extend around the surface. The upper bound on the pressure exerted on the pore wall is the disjoining pressure, which can amount to tens of megapascals[16]. The supersaturation is not related to the curvature of the crystal, so high stresses can be exerted by large crystals.

2. BACKGROUND

To avoid damaging crystallization pressure, it would be sufficient to eliminate the repulsive forces between the growing salt crystal and the pore wall[3]. This could be achieved by soaking stone with a solution that would alter the surface chemistry of the pores, so that it was attractive to salt. For example, certain polyelectrolytes are known to adsorb on carbonates and thereby inhibit their growth[17]. If the ligands that adsorb on the salt were attached to the pore wall, the wall might attract, rather than repel, the salt. For optimal interaction of the charged ligands with the surface of the salt, the spacing of the charges should be complementary. Therefore, coating the stone with a dense monolayer of ligands might be unfavorable, because they would not be likely to be properly spaced. However, if short polymers were attached to the wall, then they might have the necessary flexibility to adjust their spacings so as to interact with any ionic crystal that approached the wall. The optimal polymer would have a group capable of anchoring to the wall of the pore (e.g., carboxylic groups to attach to calcite) and a short chain with a ligand able to adsorb strongly onto various kinds of salt. If a salt crystal were to grow in a pore whose walls were covered with such a coating, the crystal would be attracted rather than repelled by the wall; it would grow into contact with the wall, but growth would then stop without any stress developing.

Following this idea, an early study[18] tested the effectiveness of polyacrylic acid (PAA) for protection against salt. An aqueous solution of the polymer was soaked into limestone, which was then exposed to cycles of soaking in sodium sulfate solution, then drying. The damage actually worsened compared to untreated stone, but the problem was the osmotic swelling of the polymer, which had too high a molecular weight, so that it filled the pores and generated more stress than the salt crystals. In the present study, we revisited this system using polymers of PAA with very low molecular weight, and obtained very positive results, as detailed below. Another polymer used in earlier tests[19,20] was a terpolymer (TP) containing ethylene, methacrylic acid, and isobutylacrylate groups. Unfortunately, TP is not soluble in water, and could only be dissolved in warm tetrahydrofuran. Limestone treated with this solution showed much improved resistance to sodium sulfate testing, but it was suspected that the difference resulted from reduced penetration of the salt solution into the stone, owing to the hydrophobic coating. In the present study, samples treated with TP were vacuum-saturated with the salt solution, so that there would be no ambiguity about the amount of salt present. On the basis of a survey of the literature regarding adsorption of organics on salts, a variety of other polymers were identified for testing.

Thomas et al.[21] found that fatty acids adsorbed to calcite nearly irreversibly, while carboxylated polymers and carboxylic acids bound less tightly and could be washed off, if the polymer was hydrophilic. These results suggest that fatty acids and hydrophobic (or mildly hydrophilic) carboxylic acids or carboxylated polymers are attractive candidates for surface modification of limestone. Polymers known to inhibit salt crystal and mineral growth include[17,22-24] phosphoric acid ($–OPO(OH)_2$), phosphonic acid ($–CPO(OH)_2$), and phenolic hydroxyl groups, as well as carboxylic acid (–COOH). Inhibition is most efficient when polymers have a distribution of negatively charged functional groups spaced similarly to the distribution of cations of a growing crystal surface. Various factors affect the efficiency of crystallization inhibittion: the relationship between the geometry of the polymer functional groups and the geometry of the mineral cations[23-25]; the quantity and rigidity of inhibitor functional groups as well as the molecular weight of the polymers[17]; the presence of non-inhibitor functional groups in the polymer (e.g. $–NH_2$ and –OH) that strengthen polymer adsorption by forming hydrogen bonding with crystal surface cations and water molecules in crystal hydrates[22]; and the pH of the mineral and polymer solutions, which can influence the amount of deprotonated functional groups along the polymer chain[22]. Sarig et al. conclude that polymers that are geometrically compatible with salts are the most effective crystal inhibitors of those salts, but Cody[22] notes that "very close crystal lattice fit cannot always be important, since the same organic substance is usually effective in modifying crystallization processes of a variety of minerals with different interatomic spacings."

3. EXPERIMENTAL PROCEDURE

3.1 Screening tests

On the basis of these previous studies, we selected the materials listed in Table 1 for further testing. Each of them is known to adsorb on calcite or to inhibit growth of calcite and/or gypsum. The organics were applied to Indiana Limestone (from Pasvalco, Closter, NJ) or single crystals of Iceland spar (Wards' Natural Science Establishment). The only salt used in the present study was sodium sulfate.

Table 1. Materials selected for study

Polymer	Manufacturer	CAS#	Functional Group	Density[c] (g/cm^3)
Polyvinyl alcohol (99% hydrolyzed, ave MW 16000	Acros organics	9002-89-5	Carboxylic acid	1.305
Maleic acid	Dajac Laboratories	110-16-7	Carboxylic acid	1.411
Acrylic acid (very low MW)	Dajac Laboratories	9003-01-04	Carboxylic acid	1.282
Acrylic acid (medium low MW)	Dajac Laboratories	9003-01-04	Carboxylic acid	
Xanthan gum	ICN Biomedical	11138-66-2	Carboxylic acid	
5-Phenylvaleric acid	Avocado Research Chemicals, Ltd.	2270-20-4	Carboxylic acid	1.218
Humic acid	Fluka Chemicals	1415-93-6	Carboxylic acid	1.581
Fumaric acid (99%)	Aldrich Chemical Co.	110-17-8	Carboxylic acid	1.624
TP[a]	DuPont		Carboxylic acid	0.947
Optima 100[b]	Chryso, Inc		Diphosphate and ethylene oxide	1.059
1-Hexadecane sulfonic acid (Na salt, monohydrate, 99%)	Acros Organics	15015-81-3	Sulfonic acid	1.116

[a] Terpolymer with approximate composition ethylene (70 wt%), methacrylic acid (20 wt%), and isobutylacrylate (10 wt%); experimental polymer provided by Dr. John W. Paul of DuPont Packaging and Industrial Polymers. Based on the melt index (80 g/10 min), the molecular weight is estimated to be ~10^5 g/mole.

[b] Optima 100 is a concrete superplasticizer that contains a diphosphate and ethylene oxide group.

[c] Measured by helium pycnometry (Accupyc 1330, Micromeritics).

To test the ability of the polymers to nucleate crystallization of the salt, samples of calcite were soaked in a solution containing 1.1 vol% of each polymer, then dried under cover (to prevent dust from settling on the sample). All of the solutions were prepared in water, except for TP, which had to be dissolved in tetrahydrofuran at 60 °C. The coated stone was then immersed in a solution of sodium sulfate saturated at room temperature, and the ensemble was placed in a cold stage (Physitemp TS-4) mounted under a Nikon SMZ-U zoom microscope, and cooled until nucleation was observed. Similar tests were made with a cup containing only the solution, and with a piece of calcite with no polymer coating. The tests were repeated three times for each case, but none showed any evidence of favoring nucleation compared to the solution alone (in which nucleation occurred at ~3 °C.

To evaluate the interaction between the organic coating and a growing salt crystal, a drop of solution was allowed to dry on the surface of a single crystal of calcite, then a drop of sodium sulfate solution was placed on the polymer and allowed to dry at room temperature. The contact angle of the solution was high on the samples treated with polyvinyl alcohol, hexadecane sulfonic acid and TP; the angle was intermediate on phenylvaleric acid, maleic acid, and medium molecular weight polyacrylic acid; the solution spread on Optima 100^{TM}, humic acid, and very low molecular weight polyacrylic acid (PAA-VLMW). In most cases, the crystals grew at the boundary of the droplet of solution; only on samples coated with humic acid or PAA-VLMW did the crystals spread across the surface of the polymer coating. The most striking and reproducible result was for PAA-VLMW, for which crystals with a unique morphology spread across the surface (see Figure 3). These crystals remain clear, so they are apparently thenardite (Na_2SO_4), whereas mirabilite ($Na_2SO_4 \cdot 10H_2O$) crystals turn white as they dehydrate.

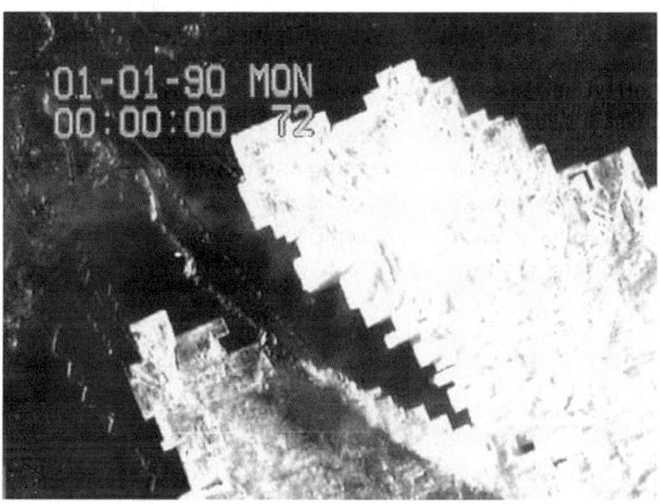

Figure 3. Thenardite (Na_2SO_4) crystals grown on the coating of PAA-VLMW on calcite.

3.2 Sulfate test

To test the influence of the organic coatings on the resistance of Indiana limestone to sulfate crystallization, two series of tests were performed. In the first two, five cubes (25 mm on each side) were soaked in each of the solutions, containing 1.1 vol% of the organic; xanthan gum and medium molecular weight polyacrylic acid were excluded, because the solutions were too viscous to penetrate the stone. The concentration was chosen so as to deposit about 1 nm of coating on the interior surface of the stone (surface area=0.7 m^2/g, by BET analysis using nitrogen, ASAP 2010, Micromeritics). In each case, the stone was dried at 60 °C, then the solution (heated to the same temperature) was poured into a jar containing the stone until the sample was half submerged. Within an hour, the sample was saturated by capillary rise, so the excess solution was discarded and the stone was left in the oven to dry for at least 2 days. The elevated temperature was necessary for coating with TP, owing to its poor solubility; the same temperature was used for all solutions for the sake of consistency. During the application of the polymer, it was obvious that some of the organics were not suitable: 1-hexadecane sulfonic acid and fumaric acid reacted visibly with the limestone, and phenylvaleric acid produced a weight loss, while humic acid turned the stones dark brown. To insure that the stones were fully saturated with the salt solution, regardless of the effect of the organic treatment on the contact angle, vacuum saturation was employed. The samples were put individually into a desiccator, which was evacuated and then back-filled with a solution of sodium sulfate saturated at room temperature. Once the sample was covered with liquid, the atmosphere was admitted, so that the solution was driven into the sample by atmospheric pressure, in addition to capillary suction. The sample was allowed to soak for an hour before being placed in an open container, then transferred to an oven at 60 °C for drying.

In the first series of tests, the first two cycles of soaking and drying were performed at 60 °C, following the work of Tsui et al.[26], which indicated that the growth of thenardite under these conditions would not cause damage. The advantage of the higher temperature is that more concentrated solutions can be used, so the pores of the stone are filled more quickly with salt. Subsequent cycles of soaking were done at room temperature with a solution saturated at that temperature (~22°C). Humic acid, hexadecane sulfonic acid, and polyvinyl alcohol all led to rapid deterioration relative to the untreated stone in the first tests, so they were not included in the next series. In the second series, all of the soaking cycles were performed at room temperature, and the samples were dried in a convection oven at 60°C for one day. The weight of a sample was found to stabilize after about 30 hours in the oven, as shown in Figure 4. Since some of the samples were removed after only 24 h, they might have retained as much as 5% of the original solution in liquid form. When samples were split in half after drying, slight efflorescence formed

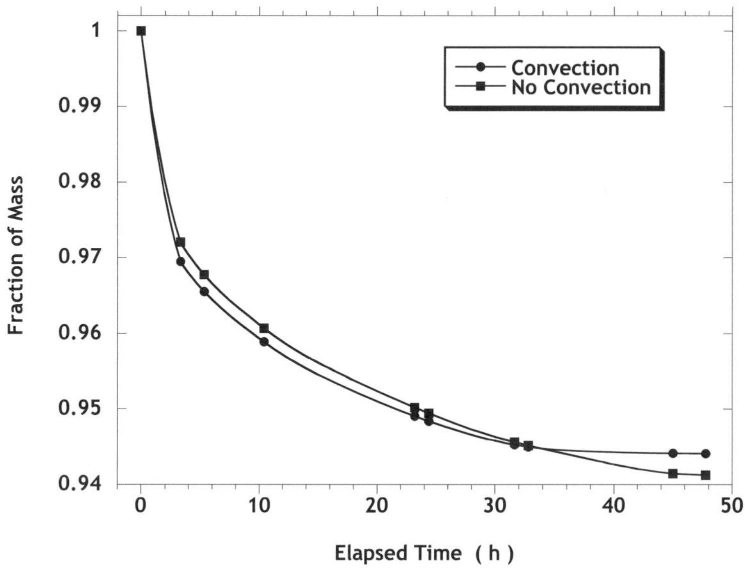

Figure 4. Weight versus time at 60 °C for stone sample in convection oven.

on the fracture surface, confirming the presence of some liquid in the pores. Incomplete drying compromises the results[27], so future studies will employ more lengthy drying treatment.

Damage to the stone was evaluated by weighing the debris lost from each sample, collected from the individual containers in which they were dried. At the end of the series of cycles, representative cubes were split to reveal the distribution of salt inside.

4. RESULTS

The initial soaking at 60 °C succeeded in slightly raising the amount of salt in the samples of Series 1, as indicated in Figure 5. The additional salt may be responsible for the earlier loss of weight (at cycle 16) in Series 1.

Measurements of acoustic velocity (PUNDIT, 54 MHz) showed no significant change after application of the organic coatings, so no consolidation effect was produced. The only organic treatments that improved the resistance to the sulfate test were very low molecular weight polyacrylic acid (PAA-VLMW) or the terpolymer (TP). In the first series, TP was clearly superior, as shown in Figure 6, showing slow deterioration only after the untreated samples had collapsed. In this series, the PAA-VLMW samples suddenly failed after cycle 11, which coincided with resumption of the cycles after a pause of two weeks (during which the samples remained in the oven at 60 °C). It is possible that the sample had not dried completely in the previous cycles, and that the damage resulted from additional crystallization during the extended drying period. This seems unlikely, however, in view of the excellent performance of this polymer in the second series, shown in Figure 7. Treatment with maleic acid also provided some benefit in this series.

The physical appearance of the samples after cycle 16 in Series 2 is shown in Figure 8. The untreated stones are rounded owing to substantial damage on the faces and edges, while samples treated with PAA-VLMW show only minor damage on the edges.

After the 16th cycle in Series 2, several stones were split open to reveal the distribution of the salt. As shown in Figure 9, the salt was in a band just below the surface in the untreated stone and in the one treated with PAA-VLMW, as is typical for drying in a hydrophilic solid: evaporation removes

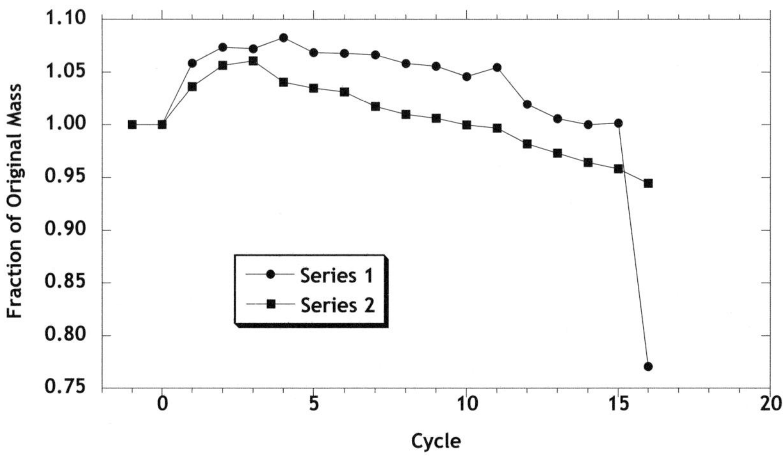

Figure 5. Average mass change for untreated stones in Series 1 and 2.

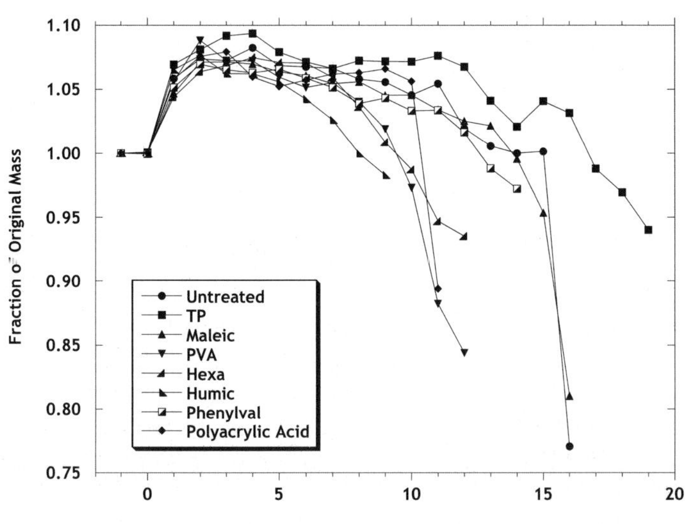

Figure 6. Average mass change of samples in Series 1.

liquid first from the larger pores, but capillary suction retains the solution in the smaller pores, preserving a contiguous network that allows the solution to flow toward the outer surface where it evaporates[28]. The salt accumulates in the location where evaporation occurs, so it appears near the surface in the stone treated with a hydrophilic polymer, such as PAA. The fact that the salt is concentrated at the surface, but causes less damage in the treated stone, is strong evidence that the polymer coating reduces the crystallization pressure. In contrast, the sample treated with the relatively hydrophobic TP had salt only in the center. This is to be expected when the contact angle approaches 90°, in which case there is not significant capillary pressure to draw the liquid toward the surface. Instead, it retreats into the interior and the salt accumulates there. This may mean that the protection provided by TP is illusory: the crystallization pressure is exerted only in the interior of the stone, where cracks are prevented from growing by confinement within the surrounding undamaged stone.

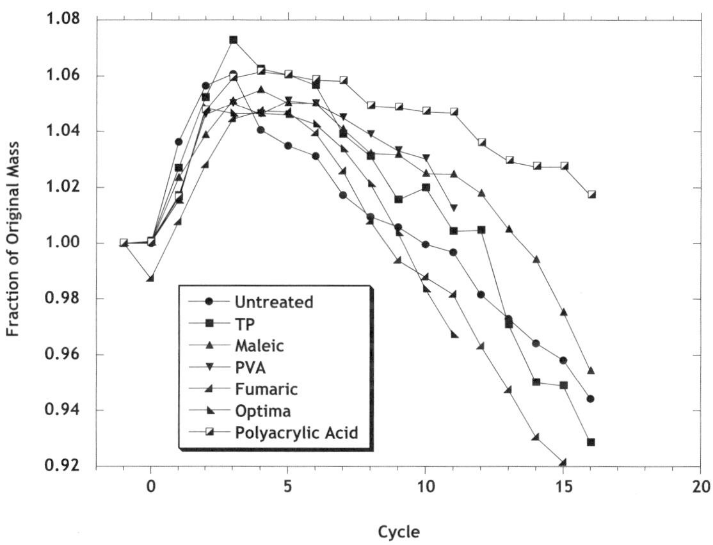

Figure 7. Average mass change of samples in Series 2.

Figure 8. Trial 2 Cycle 16: Untreated stones (left two) and stones treated with PAA-VLMW (right two). Untreated stones begin to round off while the treated stones show only slight deterioration at corners and edges.

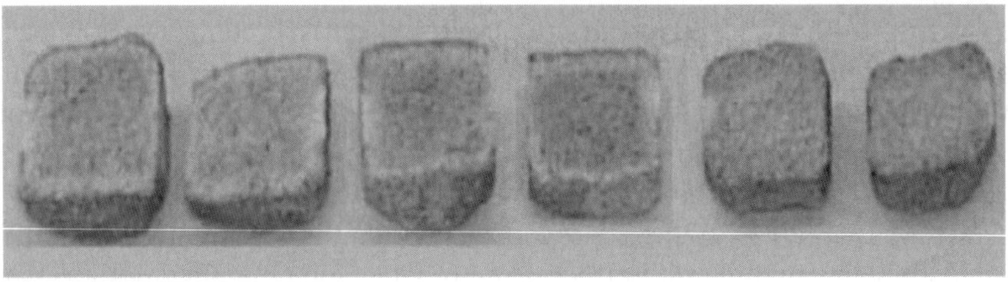

Untreated **PAA-VLMW** **TP**

Figure 9. Salt distribution in samples split open after cycle 16 in Series 2. Efflorescence appeared after 10-15 minutes, indicating location of salt. Untreated stone and sample treated with PAA-VLMW has ring of salt within ~1 mm of surface, whereas all salt has retreated to the center of the sample treated with the terpolymer.

5. CONCLUSIONS

Crystallization pressure can only be exerted on the pore walls when disjoining forces create a film of liquid between the crystal and the pore walls, which allows ions to attach to the growing surface. If the disjoining forces could be eliminated, then the crystal would grow into contact with the pore wall and growth would stop, so no pressure would be exerted. In this work, we identified several organic coatings with the potential to absorb on both salt and carbonate stone, and thereby reduce or eliminate disjoining forces between them. When a film of each coating was applied to a calcite crystal and exposed to a supersaturated solution of sodium sulfate, none of them showed enhanced ability to nucleate crystallization. This means that the distribution of charged sites in the organic layer was not initially similar to that in a crystal of sodium sulfate so that it could act as a substrate for heterogeneous nucleation. However, this does *not* mean that the layer of polymer could not adjust its structure to accommodate an existing crystal that approached the surface. That is, the polymers may not have the appropriate structure when no salt is present, but they have enough flexibility so that they could shift their charged sites into contact with those in an adjacent salt crystal when it approaches the coating.

When crystals were forced to grow on the polymers (by evaporation of a drop of solution), only polyacrylic acid clearly interacted with the crystals, inducing them to grow with a unique morphology along the coated surface. This polymer also provided substantial protection to the stone in Series 2 of the sodium sulfate tests, although it suddenly failed in Series 1 for reasons that are not clear. Another polymer that has shown promise in earlier tests, a relatively hydrophobic terpolymer, performed well in Series 1 of the sodium sulfate tests, but was less helpful in Series 2. The effectiveness of this polymer seems to be related to its effect on contact angle, rather than crystallization pressure, since it was found that the salt retreated to the interior of stones coated with this polymer. In contrast, the salt accumulated near the outer surface of stones treated with PAA-VLMW, as it did in untreated stone. This implies that the PAA coating indeed reduced the crystallization pressure by reducing the disjoining forces between limestone and sodium sulfate. More detailed tests are underway to explore the potential of this treatment to protect limestone. Other anchoring groups and polymer architectures will also be explored, and wetting/drying cycles will be imposed to ascertain the durability of the polymer coatings.

For silicate stones, anchoring groups other than carboxylates will be necessary. The use of amine groups will be explored, since propylamine ligands seem to be effective at coupling silicate consolidants to stone[29,30].

If one succeeds in making the surface of the pores attractive to salt, the possibility exists that the pore will fill with a dense volume of salt during repeated wetting/drying cycles. This could lead to significant stresses from thermal expansion mismatch between the stone and salt. The risk from this phenomenon will be evaluated through cycling tests (to monitor pore filling) and dilatometry (to evaluate thermal stress development).

REFERENCES

1. I.S. Evans, Salt crystallization and rock weathering, *Rev. Géomorphologie dynamique* **XIX** (4) 153-177 (1969-70)
2. Goudie and H. Viles, Salt Weathering Hazards (Wiley, Chichester, 1997)
3. G.W. Scherer, R. Flatt, and G. Wheeler, Materials Science Research for Conservation of Sculpture and Monuments, *MRS Bulletin*, Jan. 2001, pp. 44-50
4. G.W. Scherer, Crystallization in pores, *Cement Concr. Res.* **29**(8) 1347-1358 (1999); Reply to discussion of Crystallization in pores, *Cement and Concr. Res.* **30**(4), 673-675 (2000)
5. G.W. Scherer, Stress from crystallization of salt, *Cement Concr. Res.* **34**, 1613-1624 (2004)
6. G.F. Becker and A.L. Day, Note on the linear force of growing crystals, *J. Geology* **XXIV**(4) 313-333 (1916)
7. S. Taber, The growth of crystals under external pressure, *Am. J. Sci.* **41**, 532-556 (1916)
8. C.W. Correns, Growth and dissolution of crystals under linear pressure, *Disc. Faraday Soc.* **5**, 267-271 (1949)
9. A.W. Neumann, S.N. Omenyi, and C.J. van Oss, Attraction and repulsion of solid particles by solidification fronts. 3. van der Waals interactions, *J. Phys. Chem.*, **86**, 1267-1270 (1982)
10. Israelachvili, Intermolecular & Surface Forces, 2nd ed. (Academic, London, 1992)
11. R.J. Flatt, Salt damage in porous materials: how high supersaturations are generated, *J. Cryst. Growth* **242**, 435-454 (2002)
12. J. Freundlich, Colloid & Capillary Chemistry (Methuen, London, 1926) pp. 154-157
13. G. Wulff, Zur Frage der Geschwindigkeit des Wachsthums und der Auflösung der Krystallflächen (On the question of the growth and dissolution rates of crystalline faces), *Z. Krist.* **34**, 449-530 (1901)
14. A.W. Adamson, Physical Chemistry of Surfaces, 5th ed., (Wiley, New York, 1990)
15. L.A. Rijniers, Salt crystallization In porous materials: an NMR study, Ph.D. thesis, Technische Universiteit Eindhoven, Nov. 2004

16. G.W. Scherer, Stress from crystallization of salt in pores, pp. 187-194 in *Proc. 9th Int. Congress on Deterioration and Conservation of Stone*, Vol. 1, ed. V. Fassina (Elsevier, Amsterdam, 2000)
17. S. Sexmith and E.Q. Petrey, The use of polyelectrolytes for scale control in sea water evaporators, *Desalination* **13**, 87-90 (1973)
18. L. Gustafson, Polymer applications in stone: An attempt to decrease deterioration due to salt crystallization, Senior thesis, Dept. Chemical Eng., Princeton University, 1998
19. N. Alleyne, Art conservation: Testing interfacial energies, Senior thesis, Dept. Chemical Eng., Princeton University, 2001
20. L. Tangeman, The effects of polymer coatings in decreasing the interfacial energy between salt and stone, Senior thesis, Dept. Chemical Eng., Princeton University, 2002
21. M.M. Thomas, J.A. Couse, and J. Longo, Adsorption of organic compounds on carbonate minerals, *Chemical Geology*, **109**, 201-213 (1993)
22. R.D. Cody, Organo-crystalline interactions in evaporite systems: The effects of crystallization inhibition, *J. Sedimentary Petrology*, **65** (5) 704-715 (1990)
23. S. Sarig, A. Glasner, and J. Epstein, Crystal habit modifiers, *J. Crystal Growth*, **28**, 295-299 (1975)
24. S. Sarig and F. Kahana, On the association between sparingly soluble carbonates and polyelectrolytes, *J. Crystal Growth*, **35**, 145-152 (1976)
25. S. Sarig, F. Kahana, and R. Leshem, Selection of threshold agents for calcium sulfate scale control on the basis of chemical structure, *Desalination*, **17**, 215-229 (1975)
26. N. Tsui, R.J. Flatt, G.W. Scherer, Crystallization damage by sodium sulfate, *J. Cultural Heritage*, **4**, 109-115 (2003)
27. C.A. Price, The use of the sodium sulphate crystallisation test for determining the weathering resistance of untreated stone, in Proc. UNESCO-RILEM Symposium on the Deterioration and Protection of Stone Monuments, Vol. I. Sect. 3-6 (UNESCO/RILEM, Paris, 1978) pp. 1-24
28. G.W. Scherer, "Fundamentals of drying and shrinkage", in Science of Whitewares, eds. V.E. Henkes, G.Y. Onoda, and W.M. Carty (Am. Ceram. Soc., Westerville, OH, 1996) pp. 199-211
29. E. Wendler, D.D. Klemm, and R. Snethlage, Consolidation and hydrohobic treatment of natural stone, in *Proc. 5th Int. Conference on Durability of Building Materials and Components*, eds. J.M. Baker, P.J. Nixon, A.J. Majumdar, and H. Davies (Chapman & Hall, London, 1991) pp. 203-212
30. K.L. Nagy, R.T. Cygan, C.S. Scotto, C.J. Brinker, and C.S. Ashley, Use of coupled passivants and consolidants on calcite mineral surfaces, in Materials Issues in Art and Archaeology V (Mater. Res. Soc., Pittsburgh, PA, 1997) pp. 301-306

Chapter 5.3

WEATHERING OF BUILDING STONE: APPROACHES TO ASSESSMENT, PREDICTION AND MODELLING

P.A. Warke, J. McKinley and B.J. Smith

School of Geography, Archaeology and Palaeoecology, Queen's University Belfast, Belfast BT71NN, Northern Ireland, United Kingdom, p.warke@qub.ac.uk

Abstract: In the natural environment, weathering and breakdown of stone is an accepted part of long-term landscape development but the same acceptance of change and deterioration is not extended to stone used in construction especially when such deterioration affects historically and/or culturally important structures. The value of an integrative approach to improve understanding of weathering and failure of building stone is examined through review of three investigative approaches. First, condition assessment of the structure is an essential component of any remedial programme as this facilitates identification of the nature, extent and severity of deterioration and provides a measure of the degree of intervention required. Summary data are reported from a Staging System that seeks to impose a more formal structure on condition assessment providing a commonality of methodology, terminology and meaning whilst also providing a procedure for forecasting extent of treatment required and likely outcome in terms of 'life expectancy' of the structure. The second approach to investigation is assessment of the appropriateness of replacement stone. This involves many analytical procedures but gas permeametry in particular is becoming an increasingly useful portable, non-destructive analytical technique for predicting potential durability. Summary data are presented from analysis of Dumfries sandstone and Leinster granite illustrating spatial variability in stone surface permeability and the implications of this for post-emplacement weathering response and long-term durability. The final approach involves modelling potential stone decay pathways. The ability to more accurately model stone behaviour has significant implications for the design of conservation strategies and the avoidance of inappropriate treatments that may, inadvertently, trigger the sequence of decay and failure.

Key words: sandstone; building stone; weathering; condition assessment; staging system; permeametry; modelling.

S.K. Kourkoulis (ed.), Fracture and Failure of Natural Building Stones, 313–327.

1. INTRODUCTION

In terms of mineralogy and structure, stone is an extremely complex material – a complexity that is reflected in its weathering response to both natural and urban environmental conditions. An awareness of the potential complexity of stone weathering response and an understanding of the factors controlling this is essential for successful and long-lasting conservation work. Stone is one of the oldest building materials used by man, selected because of its perceived durability especially when compared to the human lifespan. This perception of durability has given rise to a somewhat flawed assumption on the part of those unfamiliar with its performance in the natural environment that, as a construction material, stone should remain fundamentally unchanged during the lifetime of a building. However, as with any other natural material, change is an inherent characteristic as component minerals degrade through exposure to complex subaerial conditions comprising a combination of frequent low-magnitude weathering episodes (e.g. wetting and drying, heating and cooling) interspersed with the effects of rarer high-magnitude events such as a severe frost or extreme heating as a result of fire in the building. Overtime the combined effects of these processes act to weaken the fabric of the stone with associated physical disruption and loss of material.

Consequently, deterioration in mechanical properties such as stone strength/durability should be expected during the lifetime of a building, especially in historically and archaeologically important structures that have stood for hundreds or possibly thousands of years. The nature and rate of this deterioration is controlled by complex interactions between a variety of intrinsic and extrinsic factors (Table 1) with the significance of each of these varying from building to building and even across different façades on the one building. Further complexity arises from the spatial and temporal variability of the rate, nature and extent of stone deterioration. Rates of deterioration can change overtime in response to changes in conditions of exposure through, for example, an alteration in groundwater conditions or through the use of inappropriate repair techniques such as replacing lime mortar with harder cement-based material during repointing. Spatial variability in the nature and extent of stone deterioration is a common feature affecting outcrops of stone in the natural environment and is primarily a response to differences in structural characteristics and micro-environmental conditions. On buildings similar spatial differences in the condition of stone are common and can make decisions regarding selection of conservation strategies difficult especially with regard to non-selective procedures such as cleaning and application of surface treatments.

Approaches to building stone conservation should be underpinned by an understanding of the potential complexity of stone deterioration and recognition of the controlling variables, the interactions between these and how these interactions change over space and time. In the following sections the value of adopting an integrative approach to improve understanding of weathering and failure of building stone is examined through review of three investigative approaches. The first involves the initial condition assessment of a building. This is an extremely important exercise that forms an essential component of any remedial programme through identification of the nature,

Table 1. Examples of intrinsic and extrinsic factors that may influence the nature, rate and extent of building sandstone deterioration.

Intrinsic Factors	**Examples**
Mineralogy	Calcareous sandstones may be susceptible to long-term chemical dissolution resulting in changes in pore dimensions and generalised disintegration through removal of inter-granular calcite cement. The presence of clay minerals may increase deterioration as clays can adsorb moisture and act as loci for salts.
Structural characteristics	Structures such as bedding planes and soft seams provide access points for moisture increasing weathering. Porosity characteristics have a major control on ingress and subsequent mobility of moisture and pollutants in stone.
Pollution legacy	Particulate pollutants absorbed into the stone prior to introduction of 'Clean Air' legislation can continue to influence stone deterioration despite improvements in air quality.
Legacy of previous conservation	Accelerated degrading of susceptible minerals due to inappropriate chemical cleaning and changes in surface porosity through over-aggressive mechanical cleaning such as grit blasting with possible removal of protective case-hardened outer layers of stone.
Extrinsic Factors	
Micro-environmental conditions	Aspect related differences in micro-environmental conditions at the rock/air interface influence the nature and effectiveness of weathering processes. Stone in shaded areas can remain damp for longer periods after wetting thus facilitating chemical weathering and increasing mobility of agents such as salt within the fabric of the stone.
Exposure history	Change in exposure conditions since original construction can alter receipt of atmospheric pollutants and change micro-environmental conditions. A building originally standing as an isolated structure but subsequently surrounded by urban development will have a comparatively complex exposure history. Also relevant is exposure to extreme events such as fire, earthquake and bomb blasts.
Availability of weathering agents	Proximity of building to the coast and the associated availability of marine aerosols. Also the proximity of buildings to industrial complexes, busy roads and junctions may increase concentrations of gaseous and particulate pollutants within the local environment.
Groundwater characteristics	A rise in groundwater can result in accelerated deterioration of stone in the lower few metres of a building with an increased risk of salt penetration especially in locations close to the coast where marine water incursion may be a problem.
Quality of original workmanship	Inappropriate mortar replacement, inaccurate positioning of sub-surface cramps and face-bedding of blocks.

degree and type of intervention required. The second involves assessment of the appropriateness of replacement stone with particular emphasis on the benefits of gas permeametry, a portable, non-destructive analytical technique. Summary data are presented from analysis of Dumfries sandstone and Leinster granite illustrating spatial variability in surface permeability and the implications of this for post-emplacement weathering response and long-term durability. The final investigative approach involves creating models of stone behaviour which build upon a better understanding of the factors that trigger deterioration to provide an overview of potential decay pathways. The ability to more accurately model stone behaviour has significant implications for the design of conservation strategies and the avoidance of inappropriate treatments that may, inadvertently, trigger the sequence of decay and failure.

The methodologies outlined primarily focus on the weathering response of sandstones but the issues raised are equally applicable to other types of dimension stone.

2. CONDITION ASSESSMENT

Condition assessment of building stone is an essential component of any conservation programme. A number of descriptive schemes have been developed to provide a framework for determining the nature and extent of stone deterioration through reference to a pre-defined list of decay and alteration features. Such schemes emphasize a block-by-block classification that requires considerable operator time and expertise[1,2]. Although such schemes have been successfully applied to internationally important archaeological and historical sites where each building block is a valued part of the whole[3], they are not generally appropriate for the less historically significant but more common structures that comprise the urban fabric of many towns and cities. In such instances, there are usually financial and manpower constraints that necessitate a more rapid and less detailed form of condition assessment usually undertaken by individual project managers who, in the absence of a formal methodology, are guided by their own experience. This lack of a common condition assessment scheme has contributed to an often, confusing mix of terminology used by different "experts" and the absence of meaningful long-term records against which to compare the condition of other buildings and the success or failure of different conservation strategies[4].

System assessment and classification are essential tools in many disciplines such as medicine where reliable condition assessment schemes have been developed as a necessary means of conveying clinical information in an unambiguous way. One of the most widely used of these schemes is the TNM Staging System developed for assessment and treatment planning for patients with cancer[5]. This is a deceptively simple system that "...provides an indication of prognosis and the extent of medical intervention required with the latter seen in the context of 'life expectancy' set against treatment cost, the probability of cure or disease recurrence and the presence and potential influence of predisposing factors"[4].

The apparent simplicity of this scheme belies the many decades of research that have informed the selection of classification categories and weighting of notation values. The principles that underpin the TNM Staging System are similar in many respects to those that govern the weathering response of stone. For example, in both cases, system response and outcome is determined by complex interactions between system components with many different factors influencing the dynamics of deterioration. In addition, the 'prognosis' and extent of remedial intervention required is normally closely associated with the nature and extent of 'malignant' change. In the case of the cancer patient, the larger the tumour and the greater the evidence of spread, the poorer the long-term prognosis whilst on a building, the greater the extent of damage to individual blocks and the greater the area affected on a façade, the more invasive and costly the conservation treatment required to prolong the 'life' of the building.

There are limits to the parallels that can be drawn between these two systems but because of the many common elements the TNM Staging System was adopted as the basis for the development of a similarly robust scheme for condition assessment of stonework – a scheme that goes beyond purely descriptive classification to include a forecast of likely long-term outcome with classification criteria and notation weightings based on current understanding of the decay dynamics of sandstone.

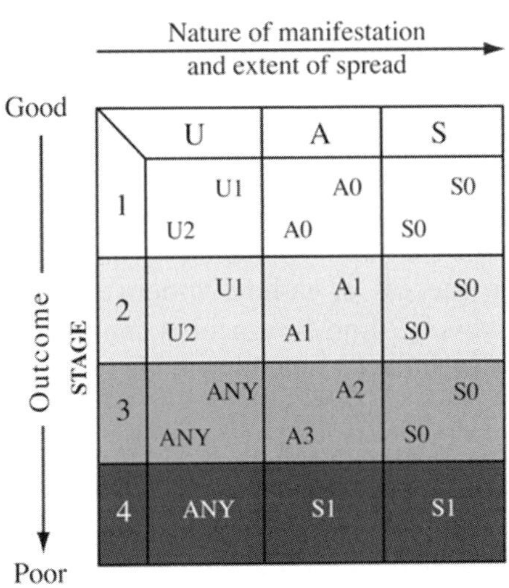

UNIT

U0 – No deterioration detectable.
U1 – Some surface alteration with minimal breakdown affecting only parts of individual blocks.
U2 – Well-developed surface alteration and/or obvious breakdown involving whole blocks.
U3 – Well-established surface breakdown with loss of original surfaces affecting approximately 10% of the façade.

AREA

A0 – No detectable involvement of surrounding blocks.
A1 – Positive involvement of adjoining blocks affecting less than 10% of the façade.
A2 – Positive involvement of adjoining blocks affecting 10–20% of the façade.
A3 – Extensive localised involvement of adjoining blocks affecting more than 20% of the façade.

SPREAD

S0 – Deterioration restricted to to specific sections of the façade.
S1 – Deterioration affecting distant unconnected portions of the façade involving more than 50% of the total surface area.

Figure 1. UAS Staging System developed for condition assessment of stone buildings from the TNM Staging model[4].

A necessary part of adapting the TNM Staging System was a change of terminology and revision of the definition of each of the Stage categories. For example, the original TNM notation, which is shorthand for 'Tumour' (T), 'Node' (N) and 'Metasteses' (M), was changed to UAS representing 'Unit' (U), 'Area' (A) and 'Spread' (S). 'Unit' refers to individual blocks of stone, 'Area' refers to sections of stonework where blocks are adjoining and 'Spread' refers to the extent of façade affected by deteriorating stonework. Once an assessment of which value is most appropriate for each of the U, A and S categories has been made one of 4 Stage classifications can be assigned (Table 2). For example, a façade showing evidence of well-developed surface alteration involving whole blocks of stone (U2) with connecting blocks affected over more than 20% of the façade (A2) but with deterioration restricted to specific parts of the façade (S0) would be classified as being Stage 3. Although Stage 3 indicates the necessity for significant remedial intervention (Table 2) it also notes that such intervention should prolong the serviceable life of the structure. Stage classification can be done for the building as a whole or for individual facades.

The benefits of developing a formal condition assessment scheme that is quick and easy to use like the UAS Staging System are fourfold:

- The criteria for condition assessment used in the Staging System method are deliberately limited in number and extremely generalised placing emphasis on identifying the presence or absence of obvious surface deterioration/alteration features coupled with an estimate of the percentage area of façade affected thereby enabling a more rapid evaluation of the condition of the building as a whole.
- Because the UAS Staging System is less detailed it should provide a relatively uncomplicated assessment tool for everyday use that is quick and easy to use with consequently less demands placed on operator time. Consequently it should be a valuable tool for condition assessment of the less exceptional but much more common stone buildings.
- Use of such a system would also provide the basis for a common condition assessment terminology providing an easily understood record of original condition against which the success or failure of conservation treatments could be assessed. Alternatively, if no treatment is considered necessary, the Stage classification of each façade would provide baseline data against which to compare subsequent assessments in a long-term monitoring program.
- The Staging System approach goes beyond the basic descriptive form of classification to provide an indication of likely outcome and the degree of intervention required based on current understanding of the decay dynamics of sandstone.

Detailed discussion and preliminary data from initial field trials of the UAS Staging System are reported in a paper by Warke et al.[4] but the system

Table 2. Summary guidelines for the extent of remedial treatment indicated by each of the four condition stages identified[4]

Stage	Summary of required intervention
Stage 1	A façade in this condition would require only localised remedial treatment concentrating on individual stone blocks. A staging classification of 1 may also indicate that no active intervention is required with only periodic reassessment of the façade advised.
Stage 2	Section specific remedial action would be required in this case but the extent of intervention should be relatively limited because of the lack of distant involvement within the façade boundaries.
Stage 3	Significant intervention will be required with up to 40% of the total façade surface showing evidence of deterioration. Although the extent of deterioration is severe, appropriate conservation treatment should prolong the life expectancy of the structure.
Stage 4	Serious deterioration affecting more than 50% of the total façade surface with stone decay detected on unconnected, distant portions of the façade. On a stage 4 category of façade, considerable intervention will be required to restore the stonework. If the structure is of limited historic and/or architectural merit then consideration should be given to the provision of palliative rather than restorative treatment.

still requires considerable refinement and field-testing especially with regard to the 'certainty' status of initial stage classifications and to the development of classification criteria for other stone types such as limestone and granite.

The Staging System model is an attempt to provide a rapid, relatively simple to use, assessment scheme that provides a formalised recording method for everyday use and is not intended to replace or devalue other existing but more detailed descriptive classification schemes. Both approaches have their place, the former for the stone structures that make up much of the historic urban fabric of contemporary towns and cities and the latter, for assessment and monitoring of internationally important stone built structures.

3. RESPONSE PREDICTION

Deterioration and failure of building stone can result from the action of many different weathering processes operating both sequentially and cumulatively. For example, salt weathering is primarily a mechanical process that disrupts the physical integrity of stone whereas chemical weathering processes degrade specific chemical elements within the crystal lattice of constituent minerals. Although disparate, nearly all weathering processes rely on the presence of moisture to transport salts or other contaminants into stone and to enable chemical reactions to occur with subsequent removal or relocation of solutes.

Whilst the presence of moisture is important, the ease with which it can penetrate the fabric of stone is a critical factor in determining the susceptibility of stone to the action of weathering processes. The movement of moisture into and within stone reflects its porosity and permeability characteristics. It is important to note that porosity and permeability are fundamentally different properties with the former describing the fractional space between solid particles in a given volume while the latter is a measure of how easily a fluid flows through stone under a pressure gradient[6, 7].

Many studies have identified a link between porosity characteristics and susceptibility of stone to weathering related damage with particular emphasis on 'effective' porosity whereby the abundance of interconnected pore spaces increases moisture penetration and movement within substrate material[8–11]. Although porosity characteristics have been used as a means of predicting resistance of stone to weathering[9,12] there are limitations associated with this property because of the problems of identifying 'effective' as opposed to 'total' porosity and the different rates of moisture loss associated with different sizes of pores. McGreevy[8] notes that there are just too many anomalies associated with porosity that limit the extent to which generalisations can be made between porosity characteristics and predicted weathering response.

Small-scale spatial variations in effective porosity can have a significant influence on permeability leading over time to differential surface weathering and loss of material[11,13]. Recognition of the importance of permeability and the implications of this for long-term weathering response and the uptake of stone surface treatments such as consolidants and biocides has only come to the fore in recent years because of the transfer of technology developed for the oil industry which allows detailed measurement of permeability under both field and laboratory conditions[7].

The ability to quantify permeability may prove to be a more reliable means of predicting long-term weathering response of stone because it can provide a more accurate representation of the potential for moisture movement in stone. However, as with any stone property it is important to recognise that permeability characteristics can change over time in response to the blocking of some pores through salt accumulation or by the creation of a secondary porosity through mechanical breakdown and microfracture development[14]. To illustrate the role of permeability as a potential indicator of the likely weathering response of stone, summary data are presented from a laboratory study in which the durability of five stone types (Table 3) was tested using a complex weathering simulation experiment comprising salt weathering and freeze/thaw cycles. A total of sixty-six 75mm^3 blocks of each stone type were used in this study and a full explanation of the experimental procedure and detailed discussion of Stanton Moor sandstone data is given in Warke et al.[11].

At the end of the simulated weathering experiment weight-loss data indicated that Dumfries sandstone sample blocks had undergone the greatest amount of breakdown and Leinster granite, the least with the other three stone types falling between these two end points (Table 4). Dumfries sandstone samples lost on average over one third of their initial weight but given the combination of a comparatively high porosity and permeability and the abundance of clay minerals (smectites) that form interstitial clay laminae, this was not unexpected. More significant was the existence of a clear association between the range of permeability values for each of the five stone types and the extent to which samples failed under experimental conditions. Data indicate that the smaller the range of permeability values the less susceptible this stone was to weathering under the experimental conditions used in this study (Table 4). It is also evident that the lower the mean permeability the more durable the stone type although the reliability of this relationship breaks down when comparisons are made between the response of Stanton Moor A, Stanton Moor B and Portland limestone (Table 4).

Table 3. General description of stone types used in the experimental study.

Stone Type	General Description
Leinster granite	Tertiary feldspar-rich granite with interlocking crystalline structure. Feldspars comprise 42.5% (plagioclase 31.5%), quartz 26.7%, mica 19.7% and clays 11%.
Portland limestone	Jurassic oolitic limestone with bimodal pore size distribution (30–8μm and 0.3–0.08μm). Calcite comprises 57.8%; quartz 2.3% and clays 13.5%. Porosity can be extremely variable especially with regard to larger pores.
Dumfries sandstone	Permian quartz and iron-rich red sandstone with well defined bedding and a unimodal pore size distribution of approximately 100μm diameter. Quartz comprises 52%, feldspars 10.5%, clays 18% (smectites) and mica 1%.
Stanton Moor A (fine to medium-grained)	Carboniferous fine–medium grained (<300μm) quartz-rich sandstone (Millstone Grit series) with well-developed interlocking quartz and feldspar overgrowths. Quartz comprises 62%, feldspars 14.25%, mica 0.75% and clays (chlorite) 11.25%.
Stanton Moor B (medium to coarse-grained)	Carboniferous medium–coarse grained (>300μm) quartz-rich sandstone (Millstone Grit series) with well-developed interlocking quartz and feldspar overgrowths. Quartz comprises 62.25%, feldspars 21.5%, mica 2% and clays (chlorite) 7.5%.

Table 4. Durability ranking results based on mean weight change data from weathering simulation experimentation with the most durable stone type being Leinster granite and the least durable Dumfries sandstone.

Stone type	Initial Porosity (%)	Mean weight change (%)	Permeability range (mD)	Mean permeability (mD)
Leinster granite	0–3	-0.93	3.7 (0.4–4.1)	1.75
Stanton Moor sandstone B	13.5	-10.26	109.3 (4.7–114)	58
Portland limestone	13–26	-22.63	149 (1–150)	15
Stanton Moor sandstone A	17	-27.02	198.3 (7.7–206)	61
Dumfries sandstone	18–25	-33.46	800 (200–1000)	600

While general permeability data were collected from each of the five different sets of 66 blocks used in the simulation study, a more detailed investigation of the spatial variability of permeability was undertaken for each of the five stone types used in this study. Space restrictions preclude the discussion of all five stone types but an overview of data from Dumfries sandstone and Leinster granite is given.

Detailed analysis of permeability was made using an unsteady-state portable gas permeameter following a regular grid scheme with 10 mm sample spacing applied to one face of a block of Dumfries sandstone and one face of a block of Leinster granite. The dimensions of each of the two blocks were 75x75x75mm. A total of 49 measurements were obtained from each block face. Sequential Gaussian Simulations (SGS) were performed using the grids of permeability data to produce representations of permeability variation for each of the two block faces (Figure 2). The SGS method is a form of conditional simulation in which simulated values are conditional on the original data and previously simulated values. The simulated realisations of permeability were made to 1mm grid-point spacing. SGS was conducted using algorithms supplied as a part of the Geostatistical Software Library[15] and further detail regarding the geostatistical procedure can be found in[7,15-17].

The simulations clearly illustrate marked differences in the permeability characteristics of both stone types with the Dumfries sandstone showing a much greater range of values in comparison to the Leinster granite (Table 5). Whilst permeability of Dumfries sandstone is an order of magnitude greater than that of the Leinster granite, the patterns of spatial variability across each block also exhibit significant differences.

Permeability data from Dumfries sandstone show several clearly identifiable areas of higher values (Figure 2a) that may be associated with the surface expression of clay laminations that are characteristic features of this sandstone. Data from Leinster granite show that areas of higher permeability tend to be less well defined and are possibly related to the presence of mica and feldspar minerals that comprise significant proportions of this stone (*c*.19% and *c*.42%, respectively) (Figure 2b). It is important to note that because of the specialised software used to produce the permeability grids, the shading scales shown in Figure 2 are block-specific and therefore the same degree of shading will not indicate the same permeability value.

Table 5. Permeability statistics for single unweathered block faces of Dumfries sandstone and Leinster granite used in durability testing (values are expressed as MilliDarcys (mD)).

Permeability Statistics	Dumfries sandstone	Leinster granite
Maximum	298.0	34.8
Median	103.0	22.0
Minimum	49.7	7.2
Mean	127.2	22.1
Range of Values	248.3	27.6
Standard Deviation	61.0	7.9

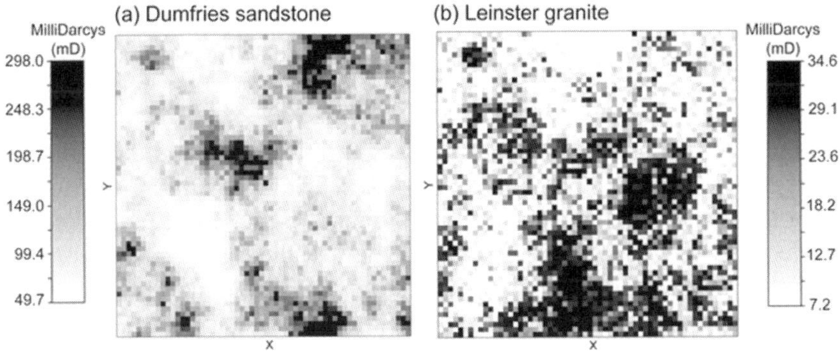

Figure 2. Sequential Gaussian Simulations (SGS) of single block faces of (a) Dumfries sandstone and, (b) Leinster granite, highlighting the spatial variability of surface permeability. Note the differences in shading scales for each block.

These data demonstrate that under laboratory conditions stone types with a large range of permeability values, where greater spatial uncertainty occurs in (simulated) permeabilities, tend to be less durable. These data also suggest that the spatial distribution and variability of permeability is more important in predicting the overall strength and weathering properties of stone than mean permeability and absolute minimum and maximum values. However, much more research is required to better understand the factors that control permeability and how these may change overtime as stone 'ages' in response to long-term weathering.

4. MODELLING DECAY PATHWAYS

The final investigative strand in an integrative approach to improved understanding of the weathering and failure of building sandstones is the modelling of decay dynamics. The importance of this aspect cannot be over-emphasized as it is only through development of such models that research is made effective. Price[18] notes that although the number of papers relating to aspects of building stone deterioration has increased markedly in the last few decades, many authors fail to follow-up on the significance of their data or to set their research in context. Consequently, despite an ever-increasing knowledge base, its actual effectiveness in dealing with 'real-world' stone conservation issues is somewhat limited. It is only through using these data to inform development of conceptual models that the relevance of research is assured and meaningful advances in understanding made.

Research has shown that sandstone deterioration in particular, is characterised by threshold decay phenomena such as granular disintegration, scaling and flaking – features that are triggered by the crossing of intrinsic and/or extrinsic stress/strength thresholds[19,20]. In order to identify the most appropriate and cost effective conservation treatment a sound understanding of the factors that can trigger the development of such features, and the conditions that promote continuation of deterioration and failure through establishment of positive feedback conditions, is essential so that every effort can be made to avoid them[21].

Our understanding of the factors that control sandstone deterioration has rapidly advanced in the last two decades primarily through the combination of laboratory-based experimental work linked with field site studies with considerable progress made and insights gained through the adoption of an interdisciplinary approach to investigation. For example, there is now a much improved understanding of the complexity of factors that contribute directly to the weathering and surface retreat of sandstone. In particular, research regarding the formation, mobilisation and accumulation of salts within stone demonstrates their spatial variability on the surface and at depth and the implications of this for differential surface retreat and the long-term success of conservation treatments[19,22]. For example, the concentration of salts at depth in 'hot-spots' has been shown to have significant implications for the continuation of deterioration once failure has been initiated because these 'hot-spots' act as reservoirs of salt that continue to 'fuel' the decay process[21,22].

Advances in technology have also enabled more detailed short and long-term characterisation of micro-environmental conditions at the rock/air interface. These data are important because they illustrate the complexity of conditions and enable identification of the weathering processes that operate and how these in turn can change overtime in response to differential surface weathering[23]. Micro-environmental data (e.g. air and stone temperature and atmospheric humidity) have also been used to create laboratory-based weathering experiments that more accurately simulate 'real-world' conditions thus producing results that have a direct relevance for predicting the performance of stone on buildings.

These advances in understanding, together with progress in many other areas have enabled the development of conceptual models of stone breakdown that identify potential decay pathways (Figure 3). The model shown in Figure 3 illustrates the episodic nature of failure in sandstone where, for example, failure through the initiation of surface scaling of a block with an apparently stable crusted surface but with a weakened substrate may be triggered by an extreme event such as a severe frost acting on previously wetted stone. Following this loss of surface material the sandstone can take one of two decay pathways. First, the newly exposed surface may stabilise sufficiently to allow development of a secondary crust with associated substrate weakening until some point in the future when the cycle of failure and stabilization is repeated. Alternatively, the newly exposed but weakened substrate may continue to degrade through development of multiple flaking and granular disintegration with accelerated surface retreat. These two different responses can be seen on many sandstone buildings and demonstrate the complexity of interactions between factors such as moisture and salt avail-

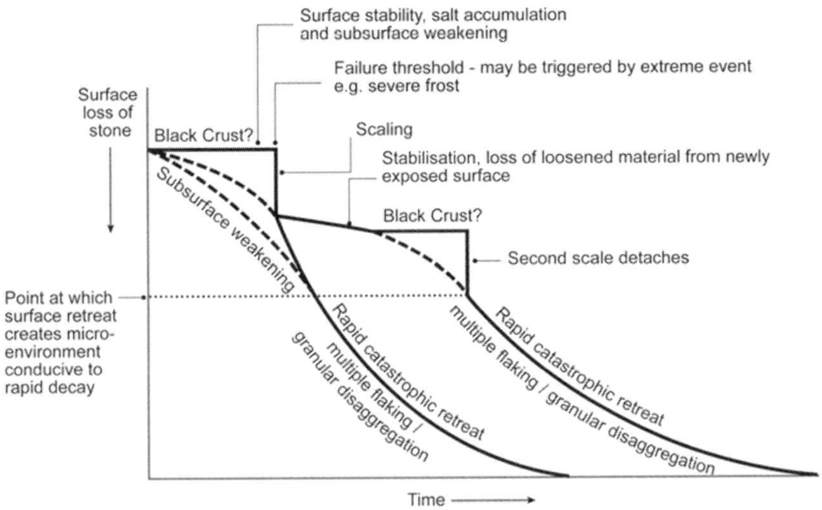

Figure 3. Schematic diagram modelling the possible decay pathways associated with sandstone building blocks exposed to the effects of salt weathering[21]. (Reproduced with permission from the Geological Society, London).

ability, permeability, porosity and mineralogy. These interactions can vary from block to block and will determine whether conditions of positive or negative feedback become established.

It is this improved understanding of the factors that control the decay dynamics of sandstone that enables informed decisions to be made regarding the selection of conservation treatments and will hopefully help prevent the use of inappropriate techniques that have resulted in so much damage in the past.

5. CONCLUSION

This brief overview of three investigative approaches to the problem of the weathering and failure of building sandstones highlights the value of an integrative strategy and the importance of linking laboratory-based study with 'real-world' conditions. However, there are still many gaps in our understanding and much work still to be done especially with regard to testing and refining predictive models and condition assessment methodologies.

ACKNOWLEDGEMENTS

The authors are deeply indebted to Gill Alexander in the School of Geography, Archaeology and Palaeoecology Cartographic Unit who prepared the diagrams and to financial support provided by EPSRC grants GR/L99500/ 01, GR/L57739/01 and GR/R54491/01.

Figure 1 and Table 2 are reprinted from: *Building and Environment*, **38**, Warke, P.A., Curran, J.M., Turkington, A.V. & Smith, B.J. Condition assessment for building stone conservation: a staging system approach: 1113-1123, 2003, with permission from Elsevier.

Figure 3 is reprinted from: *Geological Society Special Publication, 205*, Smith, B.J., Turkington, A.V., Warke, P.A., Basheer, P.A.M., McAlister, J.J., Meneely, J. & Curran, J.M. Modeling the rapid retreat of building sandstones: a case study from a polluted maritime environment. In: S. Siegesmund, T. Weiss & A. Volbrecht (eds) *Natural Stone, Weathering Phenomena, Conservation Strategies and Case Studies*, 347–362, 2002, with permission from The Geological Society.

REFERENCES

1. Fitzner, B., Heinrichs, K. & Kownatzki, R. Classification and mapping of weathering forms, in: *Proceedings of the 7th International Congress on Deterioration and Conservation of Stone.* 15–18 June, Lisbon. 1992; Vol.2: 957–968.
2. Fitzner, B. & Heinrichs, K. Damage diagnosis at stone monuments – weathering forms, damage categories and damage indices, in: Prikryl, R. & Viles, H. (eds) *Understanding and managing stone decay, Proceedings of the International Conference Stone Weathering and Atmospheric Pollution Network (SWAPNET 2001).* The Karolinum Press, Prague: 11–56, 2002.
3. Heinrichs, K. & Fitzner, B. Deterioration of rock monuments in Petra, Jordan, in: *Proceedings of the 9th International Congress on the Deterioration and Conservation of Stone.* 19–24 June, Venice, 2000; Vol.2: 53–61. Elsevier, Amsterdam.
4. Warke, P.A., Curran, J.M., Turkington, A.V. & Smith, B.J. Condition assessment for building stone conservation: a staging system approach. *Building and Environment*, **38**: 1113–1123, 2003.
5. Hermanek, P. & Sobin, L.H. *TNM classification of malignant tumours*. Springer-Verlag, Berlin, 1987.
6. Allen, D., Coates, G., Ayoub, J. & Carroll, J. Probing for permeability: an introduction to measurements with a Gulf Coast case study. *The Technical Review*, **36**(1): 6–20, 1988.
7. McKinley, J.M., Warke, P.A., Lloyd, C.D., Ruffell, A.H. & Smith, B.J. Geostatistical analysis in weathering studies: case study for Stanton Moor building sandstone. *Earth Surface Processes and Landforms*, (in press).
8. McGreevy, J.P. Pore properties of limestones as controls on salt weathering susceptibility: a case study, in: B.J. Smith & P.A. Warke (eds) Processes of Urban Stone Decay. Donhead Publishing, Shaftsbury: 150–167, 1996.
9. Goudie, A.S. Experimental salt weathering of limestone in relation to rock properties. *Earth Surface Processes and Landforms*, **24**: 715–724, 1999.
10. Smith, B.J. & Kennedy, E. Moisture loss from stone influenced by salt accumulation, in: M.S. Jones & R.D. Wakefield (eds) *Aspects of stone weathering, decay and conservation*: 55–64, 1999. Imperial College Press, London.
11. Warke, P.A., McKinley, J.M. & Smith, B.J. Variable weathering response in sandstone: factors controlling decay sequences. *Earth Surface Processes and Landforms* (in press).
12. Nicholson, D.T. & Nicholson, F.H. Physical deterioration of sedimentary rocks subjected to experimental freeze–thaw weathering. *Earth Surface Processes and Landforms* **25**: 1295–1307, 2000.

13. Rodriguez-Navarro, C. & Doehne, E. Salt weathering: influence of evaporation rate, supersaturation and crystallisation pattern. *Earth Surface Processes and Landforms* **24**: 191–209, 1999.
14. Smith, B.J., Warke, P.A., McGreevy, J.P. & Kane, H.L. Salt weathering simulations under hot desert conditions: agents of enlightenment or perpetuators of preconceptions? *Geomorphology*, **67**: 211–227, 2005.
15. Deutsch C.V. & Journel, A.G. *GSLIB: Geostatistical Software Library and User's Guide* (2nd edition). 1998, Oxford University Press, New York.
16. Lloyd, C.D., McKinley, J.M. & Ruffell, A.H. Conditional simulation of sandstone permeability, in *Proceedings of IAMG 2003 Portsmouth*, UK September 7–12, 2003. International Association for Mathematical Geology.
17. McKinley, J.M., Lloyd, C.D. & Ruffell, A.H. Use of variography in permeability characterisation of visually homogeneous sandstone reservoirs with examples from outcrop studies. *Mathematical Geology*, **36**(7): 761–779, 2004.
18. Price, C.A. *Stone Conservation: an overview of current research.* The Getty Conservation Institute, Los Angeles, 1996.
19. Warke, P.A. & Smith, B.J. Salt distribution in clay-rich weathered sandstone. *Earth Surface Processes and Landforms*, **25**: 1333–1342, 2000.
20. Smith, B.J., Magee, R.W. & Whalley, W.B. Breakdown patterns of quartz sandstone in a polluted urban environment: Belfast, N. Ireland, in: D.A. Robinson & R.B.G. Williams (eds) *Rock Weathering and Landform Evolution*, 131–150. Wiley & Sons, Chichester, 1994.
21. Smith, B.J., Turkington, A.V., Warke, P.A., Basheer, P.A.M., McAlister, J.J., Meneely, J. & Curran, J.M. Modelling the rapid retreat of building sandstones: a case study from a polluted maritime environment, in: S. Siegesmund, T. Weiss & A. Volbrecht (eds) *Natural Stone, Weathering Phenomena, Conservation Strategies and Case Studies, Geological Society Special Publication 205.* The Geological Society, London: 347–362, 2002.
22. Turkington, A.V. & Smith, B.J. Observations of three-dimensional salt distribution in building sandstone. *Earth Surface Processes and Landforms*, **25**: 1317–1332, 2000.
23. Turkington, A.V., Smith, B.J. & Basheer, P.A.M. The effect of block retreat on subsurface temperature and moisture conditions in sandstone, in: R. Prikryl & H. Viles (eds) *Understanding and Managing Stone Decay, Proceedings of the International Conference Stone Weathering and Atmospheric Pollution Network (SWAPNET 2001).* The Karolinum Press, Prague: 113–126, 2002.

Chapter 5.4

NATURAL AND ACCELERATED WEATHERING OF TWO COLOURED SICILIAN BUILDING STONES

Rizzo Giovanni[1], Ercoli Laura[2], Megna Bartolomeo[1]
[1]*Dipartimento di Ingegneria Chimica dei Processi e dei Materiali, Università di Palermo;* [2]*Dipartimento di Ingegneria Strutturale e Geotecnica, Università di Palermo, Viale delle Scienze 90128 Palermo, Italy*

Abstract: Among many coloured stones employed as "marbles" in the rich baroque architecture in Sicily, the dark grey *Breccia di Billiemi* and the *Rosso di Piana* degli Albanesi are of widespread use. Nowadays they are also exported in Northern Europe and in Arabian countries. Both stones were employed outdoor and indoor as slabs and for structural elements, such as columns, but when exposed outdoor, they undergo significant alteration processes, affecting not only their chromatic aspect but also the state of aggregation. Aim of this experimental work is to investigate some alteration processes induced by weathering in order to understand the mechanism of their evolution. The effects of accelerated weathering in laboratory are compared with natural ageing in the urban environment. The study of such stones includes the description of lithofacies, the characterisation by means of petrographic and physical-chemical techniques - optical and electronic microscopy, XRD - and finally accelerated weathering experiments on both lithotypes - UV radiation plus dew, thermal treatments, washout, exposition to both oxidative and acid environment.

Key words: carbonatic stones; artificial weathering; natural weathering.

1. GEOLOGICAL SETTING AND USE OF THE LITHOTYPES

Billiemi grey stone, known since XI century, has been widely quarried starting from the XVI century to nowadays from the southwestern mountains surrounding Palermo. This lithotype was employed, mainly in the Baroque architecture, as monolithic high columns (up to 10.50 m high) and for monu-

S.K. Kourkoulis (ed.), Fracture and Failure of Natural Building Stones, 329–341.

mental basins because of its good physical and mechanical properties (apparent density 2.71 g/cm^3, compressive strength (dry) 115-158 MPa). The rock formed in Upper Trias-Early Lias (Jurassic) during the tectonic deformation of a platform margin, and could be considered as a resedimented limestone because of its multigenerational composition: in fact the breccia allochems of all shapes and sizes are in their turn made of breccia clasts of polimictic composition (intraclasts).

The quarries of coloured “marbles" of Piana degli Albanesi, located on the northern cliff and on the eastern walls of Kumeta mountain, were exploited since XVII sec., providing significant varieties of sedimentary stones with very different textural and chromatic characteristics, i.e. nodular and brecciated, red white and green coloured, the widespread used one being the Ammonitic red, apparent density 2.72 g/cm^3, compressive strength (dry) 126-132 MPa. The *Ammonitic red* of Kumeta mountain, ascribed to Toarcian - Callovian age (Late Lias-Dogger), sedimented in a seamounts environment[1], identified as a zone of the Trapanese pelagic platform[2], that was tectonically deformed during the Tortonian age.

The red stones had a significant use in baroque architecture for grand staircases and columns, often associated with *Breccia di Billiemi* in order to create bichromatic effects, e.g. in the amazing courtyard of Palazzo Ganci with a staircase of *Breccia di Billiemi* and columns of *Rosso di Piana*, and in Palazzo Comitini and Palazzo Cutò, where the staircase is made of *Rosso di Piana* and the dark grey columns of *Breccia di Billiemi*[3]. Such stones were used also in some important historic palaces elsewhere in Italy, like Palazzo Montecitorio in Rome, the Italian Chamber of Deputies, so that the most uniform, deep red, variety of *Rosso di Piana degli Albanesi* is often named *Rosso Montecitorio*.

2. EXPERIMENTAL

2.1 Mineralogical analysis

Thin sections were studied by optical microscopy in polarised transmitted light with the aim of characterizing the microtexture.

The mineralogical characterisation was carried out by X-ray diffraction (XRD) using a Philips PW 1130/1050 equipment with CuKa radiation, tube conditions 40 kV and 30 mA. The identification of the clay content was carried out by XRD pattern after separation treatment in order to obtain oriented aggregates. Samples of orientated aggregates were also analysed after a treatment with ethylene-glycol or after heating at 550 °C [4]. EDS microanalysis was carried out on fragments of different parts of the rocks, i.e. both veins and grains.

2.2 Artificial weathering

Experimental tests were performed using at least two slabs of each lithotypes, 2 cm thick. The exposed surfaces, 2.5x10 cm, were polished in order to evidence the degradation effects. The light induced fading was simulated in a QUV apparatus, produced by the Q-Panel Company. Weathering was induced by the combined effects of both UV radiation and water condensation on the sample surface. Treatment cycles were settled consisting of 8 hours dry UV irradiation at 70°C followed by 4 hours water condensation at 50°C. The maximum duration of such tests was 5.6×10^3 hours.

The effect of thermal ageing was investigated using two different procedures. Some samples were heated up to high temperature, with the purpose of inducing both degradation of organic matter and microcracking as effect of thermal stresses. Samples were heated in an oven from room temperature up to 150 °C, then up to 600 °C in steps of 50 °C; samples were kept at each temperature for 24 hours. Other samples underwent thermal cycles performed to induce stresses at temperature close to the environmental one. The thermal cycles were performed in a sealed dry box with silica gel, moved alternatively between an oven at 70 °C and a freezer at -13 °C.

The effect of acid chemical agents, present in any urban environment, was simulated by exposure of samples to moisture and HCl vapours at room temperature. Samples were kept in a sealed glass box in presence of HCl 12M for ten days.

The effect of rain washout was simulated by means of an ad-hoc apparatus consisting of two tanks (Figure 1). CO_2 was bubbled in water in order to enhance the rock dissolution. A film of water, with a flow rate of about 3.5 lt/min, flowed continuously on the surface of the rock samples for a maximum time of 10^3 hours.

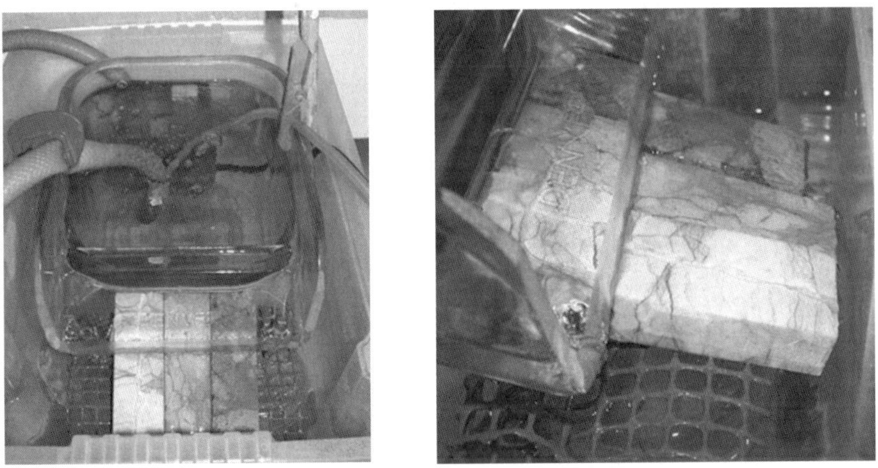

Figure 1. Details of the washout equipment.

3. PETROGRAPHICAL AND MINERALOGICAL CHARACTERISATION OF THE LITHOTYPES

3.1 Microtexture

The texture and the composition of *Breccia di Billiemi* at both macroscopic and microscopic scale were thoroughly described in previous works[5,6]. It is worth reminding that the rock shows elements of light grey coloured limestone (allochems) of all shapes and size, from a few millimetres to one meter, in an aphanitic darker matrix and cement, generally dark grey or black, somewhere ochre or red (Figure 2a). In thin section the elements of breccia, in their turn, show a clastic texture - calcarenitic to fine calciruditic - made of carbonatic intraclasts. The composition of intraclasts is both micritic (calcite) and sparry, the cement is microsparitic calcite and/or dolomite; the dolomite somewhere forms a drusy texture (sparry dolomite surrounded by rims of sparry magnesian calcite). The cement between allochems of the breccia is an argillaceous-micritic-apatitic mud, containing a dense intergrowth of dolomite crystals, somewhere cutted by sedimentary microdikes and sparry veins. The contact between allochems and the ground mass is somewhere stylolitic and exhibits a residual concentration of pyrite and organic matter (Figure 2c).

As for *Rosso di Piana*, the study concerns two varieties of a lithotype - referred as "pink" and "red" - with a bulk appearance very similar to other Jurassic "Ammonitic reds", e.g. the better known *Rosso di Verona* and *Rosso di Castellammare*, another widespread sicilian red stone. All of these Jurassic red limestones appear very similar to each other, even if some are characterized by a nodular structure and others by a flaser macrotexture.

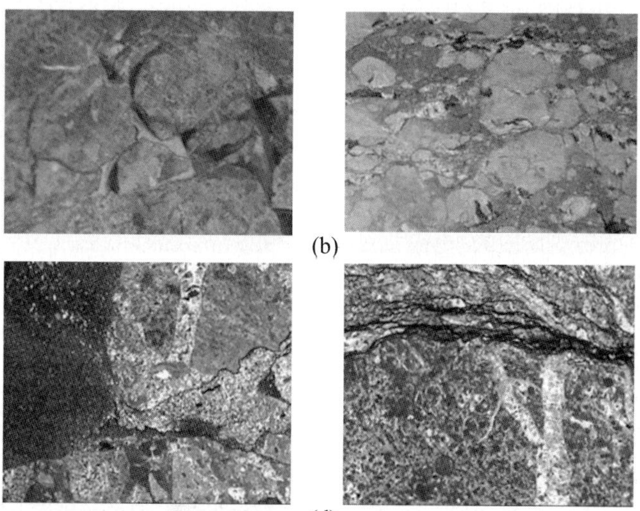

(a) (b) (c) (d)

Figure 2. Slabs and thin sections of Breccia di Billiemi (a and c) and Rosso di Piana (b and d).

The pink variety of *Rosso di Piana* at a macroscopic scale looks like an aggregate of calcareous roughly rounded lithoclasts, cream white coloured, wrapped in thin red-violet patinas; in the red variety the lithoclasts are pale brick red and the patinas are dark-ochre and dark-red, the latter darker material forming also thicker veins and zones (Figure 2b). Optical microscopy shows that both kinds of *Rosso di Piana* are made of centimetric elements of two different textural grainstones, welded by grain to grain pressure-solution contacts that originated stylolithic interfaces with residual concentration of clays and oxides. Somewhere the veins are forked, with calcite crystals in-between. In particular for both varieties of *Rosso di Piana* two different grainstones were observed: a) Grainstone with 70% bioclasts, (esoscheletal fragments of echinoderms, mollusca, brachiopods) and 30% of micritic botroids and peloids, with a grain size varying between 30 and 450 μm. b) Grainstone characterised by elements between 30 and 350 μm, made of micritic botroids and peloids, more than 80%, and about 20% of bioclasts, mainly lamellibrancs fragments, streaked by very thin sparitic veins.

The grainstone of type (b) is more abundant in the "pink", whereas the "red" is richer of sparitic zones. The thickness of patinas wrapping the nodular elements ranges from 25 to 150 μm in the "red" and from 10 to 150 μm in the "pink". The areas appearing as dark red zones in fact are made by smaller grainstone lithoclasts wrapped by patinas, which are so close each other that appear like cement zones. These textural characteristics play a major role in the evolutive mechanism of decay phenomena.

3.2 Mineralogical composition

3.2.1 Breccia di Billiemi

Because of the heterogeneity of Billiemi stone, each textural component was individually analysed: i) allochems; ii) black matrix plus cement; iii) red matrix plus cement; iv) ochre matrix plus cement. In Table 1 the crystalline components, as individuated by XRD analysis, are compared with results[7].

Table 1. Crystalline components of *Breccia di Billiemi*

Crystalline phases (°enriched samples)	Rizzo, Ercoli[5,6] Allochem	Black matrix	Red matrix	Ochre matrix	Bellanca[7] (Tout venant)
Calcite	***	***	***	**	***
Dolomite	*	**	***	***	*
Apatite-Fluorapatite °	-	*	-	-	tr
Kaolinite °	-	-	tr	*	-
Smectite °	-	*	*	*	-
Illite °	-	*	*	*	-
Quartz °	-	-	-	-	tr
Pyrite °	-	*	-	-	-

Table 2. EDS results for *Breccia di Billiemi*, atomic percentages

	O	Ca	Mg	Si	Al	P	F	Na	K
Grain	63.80	32.73		0.63		2.84			
Vein	46.21	27.64	1.01	4.44	1.87	11.88	4.55	1.25	0.58

The results of EDS analysis in Table 2, performed on both grain and black matrix plus cement scanning an area of about 150x200 μm, are in agreement with the XRD results. The absence of iron and sulphur is due to the fact that there was no pyrite crystal within the analysed spot. As reported by Rizzo and Ercoli[5], EDS analysis performed on the residual ground powder (<180 μm), attacked by HCl, confirms the presence of Si and Al as main components, followed by S and Fe, P, K and Mg; all these elements are compatible with the XRD evidence of apatite, phyllosilicates and pyrite, the latter two not resulting from Bellanca XRD analysis, but compatible with his chemical analysis[7]. As reported[8], the colour of the black matrix plus cement is due to the presence of aliphatic hydrocarbons with both linear and branched chain, whereas the available analytical apparatus was not right for evidencing heavy aromatic components.

3.2.2 Rosso di Piana

The same analytical procedure was performed on samples of *Rosso di Piana*, after separating the different macrotextural areas (light coloured "nodules" and darker red veins). The XRD results show the presence of predominant calcite in the bulk sample; a further analysis after a weak acid attack reveals the presence of traces of quartz, feldspar and clay minerals, identified as illite and swelling chlorite[4] in the XRD patterns of oriented aggregates.

Table 3. Mineralogical composition of some Jurassic Ammonitic Reds

Crystalline species	*Rosso di Piana*		*Rosso di Castellammare*	*Ammonitic red, Tuscany*	*Rosso di Verona*
	Present work	Bellanca[7]	Bellanca[7]	Fazzuoli et alii[9]	Spadea & Perusini[10]
Calcite	(1) ****	96.19	93.21	>95	95.8-84.5
Quartz	(2) *	(3)	0.82	0-5	(3)
Kaolinite		(3)	traces	traces	traces
Illite	(2) ***	(3)	1.50	2-7 (4)	(3)
Vermiculite		-	-		
Chlorite	(2) *** (swelling)	-	-	-	(3)
Apatite and Phospates		(3)	3.62	-	-
Fe minerals	(2) traces	traces	traces	-	(3)
Feldspar	(2) traces	-	-	(5)	-
Magnesium carbonate		2.05	-	-	-

(1) tout-venant, (2) enriched sample, (3) not quantified, identified by XRD, (4) plus accessories minerals; (5) lower than instrumental detectability

In Table 3, the composition of other Italian Jurassic Ammonitic Reds is compared with *Rosso di Piana*. Fazzuoli et al.[9] describe the Ammonitic Red employed in the monumental complex of S. Maria del Fiore in Florence, as made of bioclastic wackstones and micritic nodules with stylolitic joints and thin spatic veins. The *Rosso di Verona* studied by Spadea and Perusini[10] is a rosy nodular biomicrite in a brick-red matrix, formed of hematite and clay minerals bearing biomicrite. EDS analysis was performed on both nodular elements and stylolitic veins, scanning an area of about 150x200 μm. The compositions, summarized in Table 4, are in agreement with the XRD results.

The above results show that grains are made of calcite, whereas the veins contain also clay minerals, quartz, hematite and feldspar. Even if the XRD patterns do not provide evidence for the existence of apatite the presence of phosphorous, as revealed by EDS, confirms the results of Bellanca[7]. The main differences are the absence of kaolinite and the presence of swelling chlorite.

Table 4. EDS results for the two varieties of *Rosso di Piana*, atomic percentages

		O	Ca	Mg	Si	Al	Fe	P	K	Others
Red	Grain	63.05	36.95							
	Vein	54.22	6.99	2.49	21.32	8.25	2.67	0.65	3.05	0.36
Pink	Grain	59.13	40.87							
	Vein	52.14	25.03	2.00	9.91	3.68	2.19	0.37	1.51	3.18

4. EFFECTS OF ACCELERATED WEATHERING

4.1 UV radiation and water condensation on the sample surface

After weathering in the QUV apparatus, the *Breccia di Billiemi* samples show opacifying of surfaces and a noticeable chromatic inversion, i.e. the allochems preserve their grey original colour, whereas the black matrix plus cement becomes whitish, due to UV degradation of organic matter. As for *Rosso di Piana,* no detectable colour variation due to UV radiation occurs; the only remarkable effect is a slow dissolution of the boundary between lithoclasts and stylolitic joints, produced by dew in a relatively few cycles, even in the absence of any aggressive polluting agent.

4.2 Thermal ageing

Samples of each lithotype, after heated from room temperature up to 600 °C, show considerable modification in colour and texture. Starting from 250 °C, the bioclastic texture of elements (both allochems and nodules) becomes sharper, as well as the interfaces between lithoclasts and surrounding elements, probably due to thermal stresses. Furthermore, in *Breccia di Billiemi*, the opaque spots around the pyrite crystals become more circumscribed since they lose the halo, which is compatible with the degradation of organic matter, whereas *Rosso di Piana* keeps its red colour, thus indicating that the red pigmentation is not related to the presence of organic matter. At 600°C thermal decomposition of carbonates starts for *Breccia di Billiemi*, due to the presence of dolomite.

The thermal cycles were performed in isostatic conditions, i.e. without rigid bonds, between +70 °C and – 13 °C. They don't produce any significant effect even after 50 cycles, as no new cracks seem to grow or propagate.

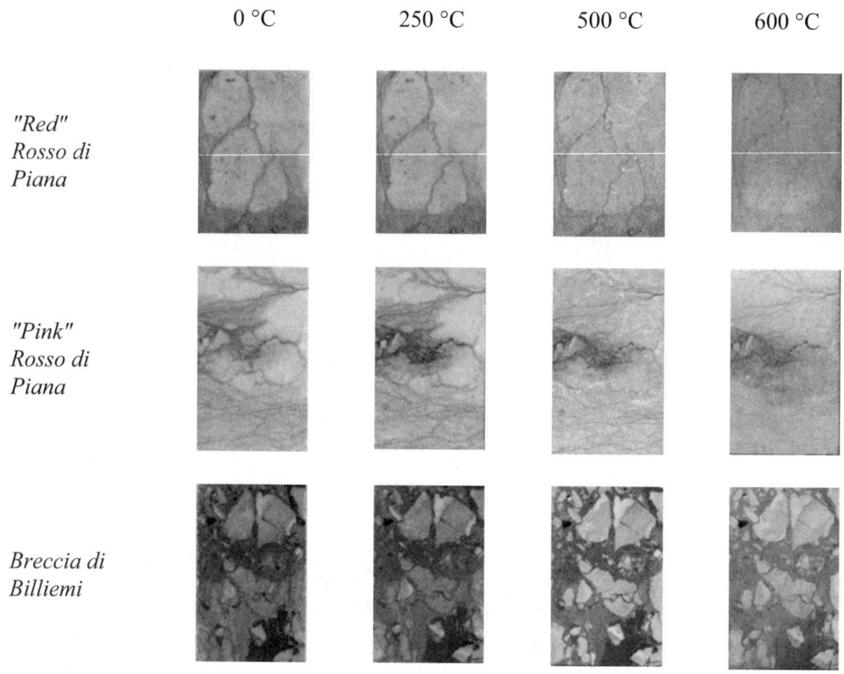

Figure 3. Effects of step by step heating on *Breccia di Billiemi* and *Rosso di Piana* slabs.

4.3 Acid chemical agents

The acid attack produces similar degradation effects on both lithotypes, *Breccia di Billiemi*[6] and *Rosso di Piana.* After ten days exposure to vapours of hydrochloric acid and moisture an highly deliquescent brine drops out from the veins on both the varieties of *Rosso di Piana*, Figure 4, and the grain surface appears to be dull due to a mild corrosion. The samples were photographed after rinsing with fresh water, showing that red veins were corroded.

Figure 4. Deliquescent brina due to the acid attack on red *Rosso di Piana* slabs.

In both kinds of *Rosso di Piana* and in *Breccia di Billiemi* a severe aggression of the cement occurs, so that veins withdraw with respect to the elements in the more decayed zones, as illustrated in Figure 5 for "pink". The red variety underwent a particular decay: the red inclusions within the whitish grain were dissolved leaving holes and cracks, whereas the surface of grains is in a quite good condition, Figure 6.

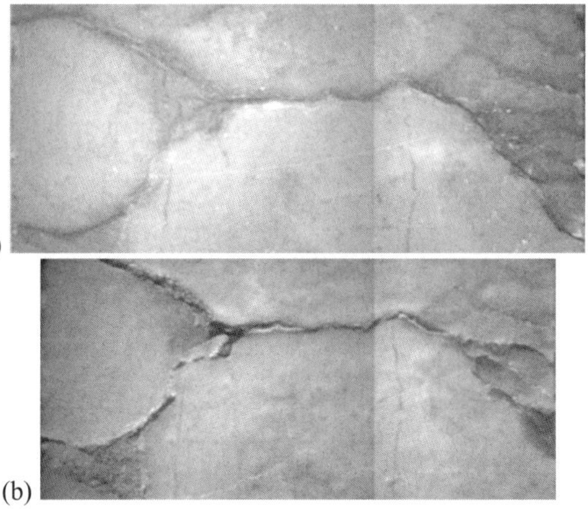

(a)

(b)

Figure 5. The pink *Rosso di Piana*: (a) before and (b) after acid attack. The dissolution of vein between clasts is evident.

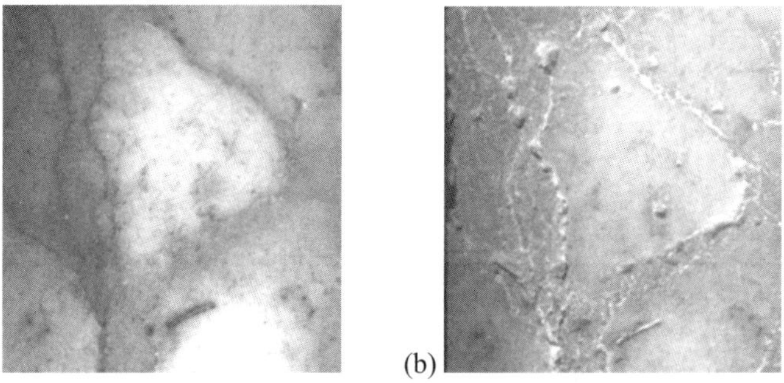

(a) (b)

Figure 6. The red *Rosso di Piana*: (a) before and (b) after acid attack. Dissolution of vein between clasts can be observed and some holes replace the red inclusions in the clast.

4.4 Accelerated washout

As a result of the washout process all of the tested materials show significant differential dissolution phenomena of lithoclasts with respect to welding contacts. This process goes on differently for each lithotype and, in some cases, growing of microcracks can be observed.

As for *Breccia di Billiemi*, the surface is strongly faded, with an evident strong inversion of the grey scale, after 51 days of treatment[5,6].

Both kinds of *Rosso di Piana* underwent a quite different decay as effect of accelerated washout. As a first step, dissolution is more severe in the grain, which withdraws with respect to the stylolitic joints, so that the red veins appear in relief, showing new microporosity due to the dissolution of calcite microcrystals within the red matter, Figure 7. As a second step the red veins collapse resulting in withdrawing of stylolitic joints with respect to the grains. This second step of decay occurs when the first step has gone ahead enough to allow mechanical removal of the clay minerals present in the veins as a result of washing water and the vein becomes a furrow. As a consequence of these steps, for the red veins it appears to be equally possible either relief or withdrawn (Figures 8 and 9).

The dissolution of the lithoclasts proceeds uniformly but at different rate for the two kinds of grainstone as shown in Figure 9.

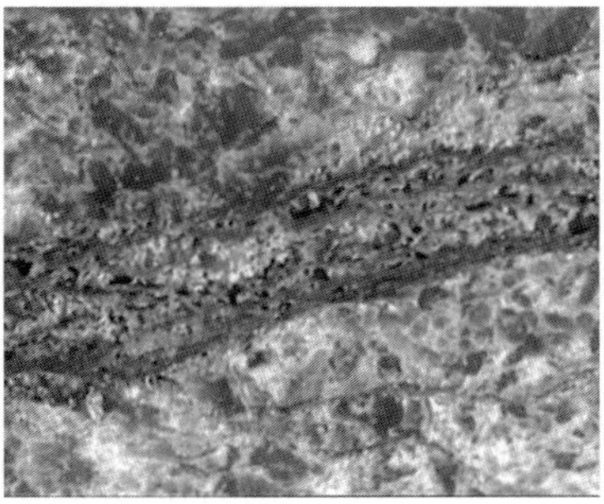

Figure 7. Dissolution of calcite microcrystals within the red veins on *Rosso di Piana* due to accelerated washout.

Figure 8. Accelerated washout of *Rosso di Piana*. First step of decay: the vein appears in relief.

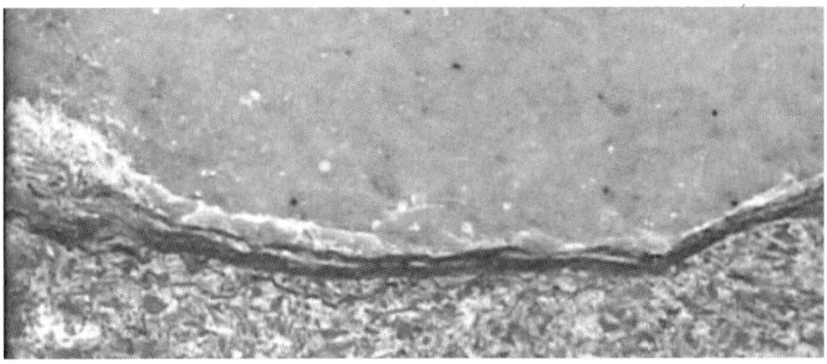

Figure 9. Accelerated washout of *Rosso di Piana.* Second step of decay: the vein withdraws with respect to the grain. Note the different behaviour of the two grainstones above and below the vein.

5. TYPOLOGIES OF NATURAL WEATHERING

As for *Breccia di Billiemi* several typologies of decay can be observed on monuments exposed outdoor and indoor[5,6]: i) the fading of exposed outdoor surfaces, due to synergy of two degradation processes that have been simulated in laboratory, i.e. UV photodegradation of organic matter and development of neater rims between the allochems and the background, probably due to microcracking; ii) the differential withdrawing of allochems and matrix plus cement due to both acid attack and rain washout, Figure 9a.

As for *Rosso di Piana* the main typology of decay is deepening of the red veins, that results in a rough surface. No evident fading or colour change is observed, the veins keep their red colour even when exposed to very aggressive environment - i.e. marine spray and heavy traffic pollutants (Figure 9b). A similar behaviour is also described[10] for *Rosso di Verona*, a stone appearing very similar to *Rosso di Piana*: the nodules are reported to become prominent with respect to the internodular areas as a result of loss of material by scaling, exfoliation and pulverisation, with swelling when covered by black crusts. The authors relate such swelling to an increase of volume due to formation of both gypsum and mixed layers of illite-smectite, the latter originated from the illite present in the unweathered rock.

(a)  (b)

Figure 9. (a) *Breccia di Billiemi* slab in S. Matteo church, Palermo; (b) ammonitic red column in Palazzo Jung, Palermo.

6. CONCLUDING REMARKS

The main features of natural weathering of *Breccia di Billiemi* and *Rosso di Piana* were documented by observation of monuments in the area of Palermo. The major environmental factors assumed as possible decay agents were thermal excursion, water washout, UV irradiation, acid attack by polluted atmosphere. Some of these aggressive agents were reproduced in the laboratory performing accelerated ageing tests on stone slabs. The expected effects on the stone were modification of surface microtexture, microroughness and/or formation of new species as effects of chemical attack, degradation of organic matter.

The accelerated washout treatment showed for *Breccia di Billiemi* a differential dissolution rate between lithoclasts and cementation contacts. The higher solubility of calcite grains with respect to dolomite, pyrite and organic matter of matrix plus cement produces withdrawing of grain. For *Rosso di Piana* a more complex dissolution process was evidenced: the red veins are less soluble than grain, but they can appear alternatively withdrawn or in relief depending on the progress of the washout. In fact the withdrawal of the "nodules" is the first step, followed by opening of stylolitic contacts, due to mechanical removal after dissolution of limy particles.

The acid attack produced in both lithotypes a preferential dissolution of veins, as observed for natural weathering. The conditions of acid attack are closer to the natural ones than for washout test. In the washout test a steady state dissolution condition is simulated, so that the major factor in producing decay is the solubility of the different parts of the rocks. In the acid attack the pollutant is a gas and the dissolution reaction can take place only when both gaseous hydrochloric acid and moisture dew on the surface of the rocks; in both lithotypes grains have lower porosity than the elements so that, even if they consist of more soluble minerals, they can resist to the aggressive environment. The soluble salts formed during the acid attack are then removed by rain washout, without any steady state conditions.

The different effects of UV radiation, fading of *Breccia di Billiemi* versus no significant change in colour of *Rosso di Piana*, suggest a different role of organic matter in the pigmentation of the two stones. Similar indication comes from thermal treatments. Such a detailed analysis of the role played by each component of the rocks in producing decay could direct further research towards both products and techniques which are the most suitable for a durable conservation process.

ACKNOWLEDGEMENTS

This work has been financially supported by the Italian Ministry of University and Scientific Research (MIUR ex 60%). Special thanks are due to A. Cancelliere for the contribution to analyses of materials within her thesis.

REFERENCES

1. H. C. Jenkyns, *The Jurassic of western Sicily in Geology and History of Sicily*, (Petroleum Exploration Society of Libya, Tripoli, pp. 45-254, 1970).
2. B. Abate, R. Catalano and P. Renda, I monti di Palermo in Guida alla Geologia della Sicilia Occidentale, *Memorie Società Geologica Italiana* **XXIV**, pp. 43-48, Palermo, (1982).
3. F. Rodolico, *Le Pietre delle Città d'Italia* (Le Monnier, Firenze, Italy, 1953).
4. G. Brown, *The X ray identification and crystal structures of clay minerals* (Mineralogical Society, London U.K., 1961).
5. G. Rizzo and L. Ercoli, in *Proceedings of the Fifth International Symposium on Conservation of Monuments in the Mediterranean Basin "Protection and Conservation of the Cultural Heritage of the Mediterranean Cities*, edited by E. Galan and F. Zezza, (Balkema Publishers, 2002) pp. 235-241.
6. L. Ercoli and G. Rizzo, in *Proceedings of the 2nd Congresso Nazionale di Archeometria*, (Pàtron Editore, Bologna, 2002) pp.123-133.
7. L. Bellanca, *Marmi di Sicilia*, (IRFIS-Sicilcamere, I.R.E.S. publishers, Palermo 1969).
8. G. Rizzo, L. Ercoli and P. Agozzino, in *Proceedings of 6th Convegno Nazionale AIMAT*, (Modena, 8-11 Settembre 2002).
9. M. Fazzuoli, R. Sartori and S. Vannucci, in *Proceedings of the 79th Congresso Nazionale della società Geologica Italiana*, edited by Giovanna lo Cicero (Offset studio, Palermo, 1998) pp. 446-448.
10. P. Spadeam and T. Perusini, in *Proceedings of the Fifth International symposium on Conservation of Monuments in the Mediterranean Basin "Protection and Conservation of the Cultural Heritage of the Mediterranean Cities*, edited by E. Galan and F. Zezza, (Balkema Publishers 2002) pp. 257-267.

Chapter 6: Freeze-Thaw

Chapter 6.1

ON-SITE AND LABORATORY STUDIES OF STRENGTH LOSS IN MARBLE ON BUILDING EXTERIORS

John M. Logan
Department of Geological Sciences, University of Oregon, Eugene, OR 97403, USA

Abstract: This study integrates on-site studies of two marble-clad buildings with laboratory tests in an attempt to elucidate the causes of deterioration of the panels on these buildings. The buildings each contain about 45,000 panels and are 83 (ACB) and 72 (TC) storeys tall respectively. The taller building was 17 years old at the time of study and the latter is presently 27 years old. There are minor differences in the panel sizes, anchoring and arrangement on the buildings but they have not been found to be significant. Similar marble types from the vicinity of Carrara, Italy were used on both buildings. The primary manifestation of panel deterioration was the progression from flat panels upon installation to bowed or more accurately dish-shaped geometry. On ACB 31% of the building developed bows over 1 cm in 17 years and 26% on TC in 20 years. All panels on both buildings were measured periodically, demonstrating a demographic distribution of the bowing. On both buildings the bowing is most extensive in those regions exposed to the most thermal heating, either from direct sunlight or from reflected sunlight, regions of shading showed negligible bowing. Therefore, the south- and west-facing panels and particularly those on the upper floors generally showed the most bowing. Strength measurements from 122 panels removed from ACB show an average strength loss of 26% when compared to Virgin panels that had been kept in temperature controlled storage for the same period. Comparable tests on 158 panels from TC show average strength losses of about 35% over the longer period of time. The strength distribution in panels containing significant bow is not uniform. At the top and lower edges of panels from TC the strength loss is only about 19%, however, where the amplitude of the bow is largest the loss averages 50% for the 158 panels. Additionally, test specimens from areas of bow curvature show the extended portion of the bow to be 25 to 30% weaker than the inside of the bow. Petrographic observations document that microfractures are concentrated in the extended portions contributing to the significant strength loss in those areas. Thermocouples to record air temperatures were placed outside and behind the panels, as well as on outside and inside panel faces and in the center of the panel. These clearly demonstrate that thermal heating, not

S.K. Kourkoulis (ed.), Fracture and Failure of Natural Building Stones, 345–362.

air temperature correlate with panel bowing. Recognizing that differential thermal expansion of calcite can result in weakening of marble led to controlled laboratory experiments to attempt to define the strength-loss curve. Portions of thirty-one panels were subjected to a cycle of twelve hours in an oven at a given temperature followed by twelve hours at room temperature. Three temperatures of 12, 66 and 107 °C above 24 °C room temperature were used. Initial data contained 200 cycles, but has now been extended to 450 cycles. Initially, panels show a sharp decrease in strength during the first 25 cycles, with the rate of decrease becoming significantly less with increasing cycles. The strength decrease is interrupted by periodic, temporary increases, which appear to be related to increases in the residual elastic strain of the marble. Upon release of this strain, the strength drops to low levels again. By about 120 cycles the panels appear to reach a basic strength beyond which further loss is minimal. Porosity measurements made on panels removed from TC show a sub-population of panels that have values significantly higher than the general panels. Average porosities for the general population are about 1.5% with a standard deviation of 0.3%. The smaller group of panels has values ranging from 2.1 to 5.8%. The porosity measurements are all made close to the upper or lower edges of the panels to minimize the influence of pore space associated with fractures created during bowing. As would be predicted, this group has average strengths 72% lower than the general population. Some of the high-porosity panels were in shaded areas on the building and had not developed significant bows, nor do they presently show any ordered distribution on the building. It is hypothesized that this group of panels had intrinsically high porosities when quarried. These panels do not have any macroscopically recognizable features to distinguish them, so with their low strength they pose a significant threat to the cladding integrity. An acoustic instrument, which measures compressional-wave travel time between two transducers, was developed and is used to detect the panels of higher porosity. A correlation between longer travel times and higher porosity offers a non-destructive method of evaluating panel strength on panels attached to the building and the potential for correlating strength changes with time.

Key words: marble; bowing; building demographics; thermal radiation; thermal cycling; residual strain; porosity; acoustic delay meter.

1. INTRODUCTION

During the 1960's as urban land became rare and very expensive in the United States and Canada, new construction often focused on high-rise buildings. At the same time marble cladding on building exteriors and interior entrances became popular. Marble panels were used in a non-structural context to provide an architecturally clean, white appearance. Panels were hung independently so as not to load or impinge adjacent ones or windows. Unfortunately, in these and many other cases adequate investigation of the physical and mechanical properties of the selected marble was not extensive, particularly with respect to the anticipated environment of the building. The result has been the replacement of marble cladding on a number of buildings world wide including the Finlandia Hall in Helsinki[1], the Grand Arche de la Defense in Paris, the Lincoln Memorial in the United States[2] and the widely publicized Amoco Corporation Building, Chicago, United States. As a result, the obvious degradation of marble under a range of environmental conditions has led to growing research into its properties[3-7].

The investigation and evaluation of the Amoco Building Chicago (ABC) and a building in Toronto, Canada (TC) provides a unique opportunity to compare the two structures (Figure 1). Both buildings were constructed between 1971 and 1978, designed by the same architect, built by the same construction firm, and the marble was supplied by the same company. ABC was roughly 17 years old at the time of the investigation, while TC is now about 27 years old and still under study. The ABC building is over 83 stories, and about 346 meters tall. The TC building is 72 stories and about 300 meters tall. Both buildings have reentrant corners.

They differ in architecture with ACB having vertical triangular columns separated by windows (Figure 1a). The marble-clad faces on the columns roughly face at right angles to each other. The panels are about 1.0 by 1.32 m and generally 32 mm thick. In the upper floors of the reentrants, however they are 38 mm thick. On TC, the panels are generally 0.76 by 1.19 m, and nominally 32 mm in thickness, although some panels as thin as 28 mm have been measured. Here they are aligned with the floors and thus distributed horizontally across the building (Figure 1b). On both buildings the panels are individually hung by anchors top and bottom and separated on all sides by about 10 mm of grout to avoid vertical stacking, but the anchoring systems

a

b

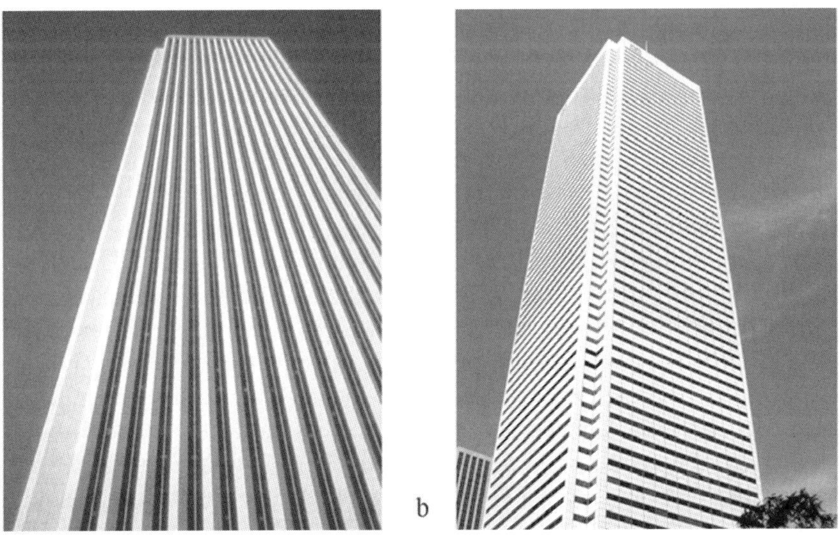

Figure 1. Photographs of buildings under study. (a) ABC showing vertical windows and columns with panels on both sides. (b) TC with each floor containing a row of windows and two rows of panels above and below.

differ on the two buildings. TC has access holes between panels to allow air circulation behind the panels.

In both cases the marble came from the Carrara area of Italy. The Chicago building was clad in Canalgrade material, which was termed “Alpha Gray”. For the building in Toronto, “Edward White” was initially used, but as the quarry material became darker, a shift was made to Canalgrade. Samples from both buildings show the marble to be 99% pure calcite. Petrographic studies indicate that the crystallographic axes are randomly orientated, the grains are roughly equant in shape, and generally between 0.1 and 0.2 mm in diameter.

2. PANEL SHAPE AND THERMAL EXPOSURE

The primary manifestation of panel deterioration on both buildings was the progression from flat panels upon installation to bowed, or more accurately, a dish-shaped geometry (Figure 2a). On ACB 31% of the building developed bows over 1 cm in 17 years and on TC, 26% in 20 years. All panels on both buildings were measured periodically and demonstrated a demographic distribution of bowing. On ACB there are 14 chevrons on each side of the building and values for bows of 1.13 cm or more are averaged for each side of them (Figure 2b). The north side of the building showed only about 7% of the panels bowed to this extent. In comparison, the east side had an average of 44% of the panels bowed that much. Within the corner reentrants the panels also reflected a demographic position, although the thicker panels on the upper floors were not bowed as extensively as the lower thinner ones. In the southeast reentrant about 61% bowed this amount, although in the north-east and north-west reentrants only about 30% of the panels met this criterion. Most of the panels are bowed convexly outward, with just 2% inward or concave bows. It was concluded that difference in the number and magnitude of bows was directly correlated with the panel exposure to sunlight or thermal gain.

On TC a similar demography was found above floor 59. Below that shading from surrounding buildings reduced the thermal exposure, but panels on that floor and higher also showed a correlation to sunlight exposure. On the west and south side 7% of the panels were bowed 1.13 cm or more, while only 4% were measured on the east side and none on the north side. Highest values were also found in the sout-heast and south-west reentrants with no panels bowed to this extent in the northeast reentrant. In the north-west reentrant only 1% of the panels on the north-facing side were bowed 1.13 cm while on the west facing side 15% were bowed this amount. On the entire building the direction of bowing is in contrast to ABC, for although 71% are bowed outward, 27% are bowed inward. This number of inward bowed panels is larger than has been found on other buildings investigated.

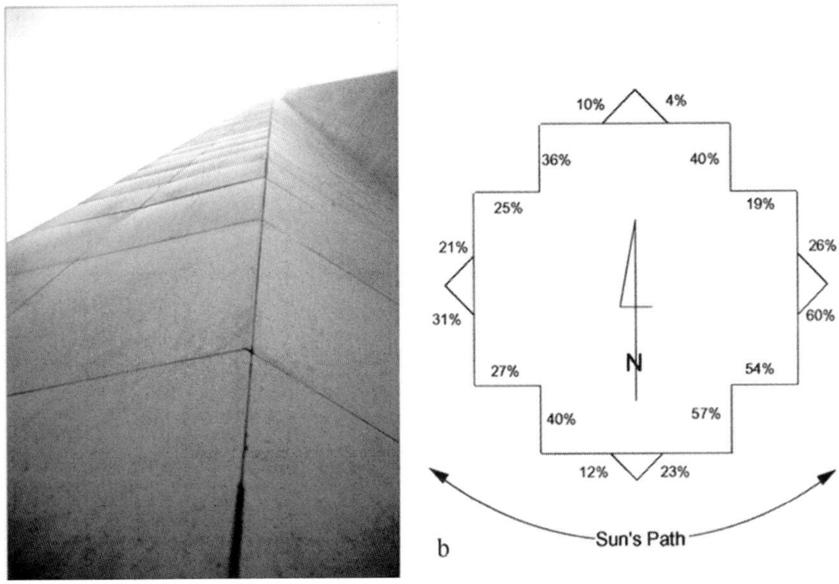

Figure 2. (a) Photograph of reentrant corner on ABC show panels with convex outward bows. (b) Schematic foot print of ABC showing percentage of panels bowed 1.13 cm or more. Values are averages for all the panels on a specific reentrant side or all panels facing a given direction on the 14 columns on each side of the building. The general daily path of the sun is shown.

Given the correlation of degree of panel bowing and exposure to sunlight, temperature measurements were made on ABC for a period of one year. Monitoring stations were installed on opposite sides of selected columns on all four sides of the building at floor 35. Thermocouples measured outside air, and temperatures of the air space behind the panels as well as outer, middle and inner faces of the panels[5]. An average difference between the maximum and minimum daily values was calculated for measurements of the air and middle panel temperatures (Figure 3).

Insights into the direct effects of solar radiation can be gained from the temperature measurements: (1) Panel temperatures are close to air temperatures at night but during the day are raised by radiant heating. (2) Temperatures in the middle of the panels are generally warmer than the outside air temperatures. This difference is greatest on sides of the building receiving the most solar radiation. This sensitivity is seen on the east side of the building where the difference between air and panel temperatures is greatest on the south faces of the chevrons compared to the north-facing panels. This reaction to solar radiation is emphasized on the north side of the building were there is little exposure to direct sunlight. The result is that there is little difference between air and panel temperatures. (3) Although there are temperature differences between the outer and inner faces of the panels, with the outside always warmer, the gradient is less than 6 °C and never over 11 °C.

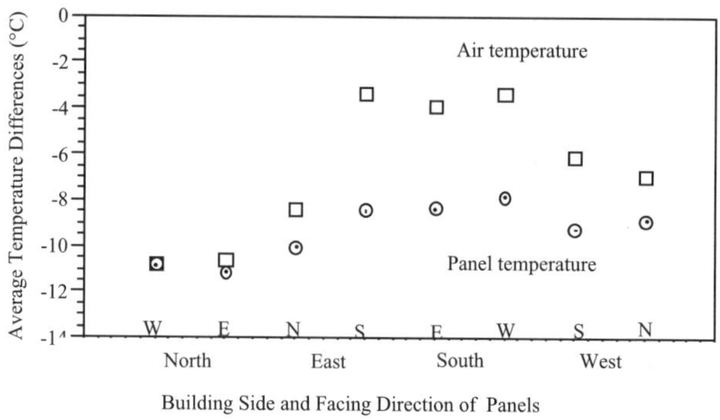

Figure 3. Average difference between measured temperatures of the air and in the middle of panels on floor 35 of ABC from March 1986 through February 1987.

(4) The temperature in the middle of a panel usually responded within an hour to changes in the outside air; that is, the panels appear to be close to thermal equilibrium most of the time.

The correlation between panel bowing and thermal radiation was not unexpected as in 1934 Lord Raleigh[8] published his seminal work on the effects of heating marble. He found that after heating, pieces lengthened, increased in porosity, and their shape changed from an initially flat geometry to a curved configuration. With increasing temperature the deflection, which was initially elastic and recovered upon cooling, now became inelastic and permanent.

3. STRENGTH LOSS WITH BOWING

The recognition of inelastic bows on both buildings suggested an associated loss of strength that leads to bowing. To assess this hypothesis tensile strength was measured by point-load tests[9] on over 122 panels from ABC using specimens 2.5 cm in diameter with supplemental uniaxial, Brazilian and C880 flexural strength tests.

The panels were divided into three groups. Twenty-four Virgin panels were sampled. These panels had been stored under controlled temperature conditions so they presumably approximated the strength of the material as originally installed. Some 48 full panel flexural tests were done on Test panels randomly selected on the building and were subsequently re-tested in the laboratory. Examination of the panels suggested that damage during the full-panel tests made the results of subsequent point-load tests questionable and they are not considered here. Fifty-five Access panels were removed adjacent to the Test panels for equipment installation and were also tested. The panel strength is an average of 900 to 1330 measurements for each panel.

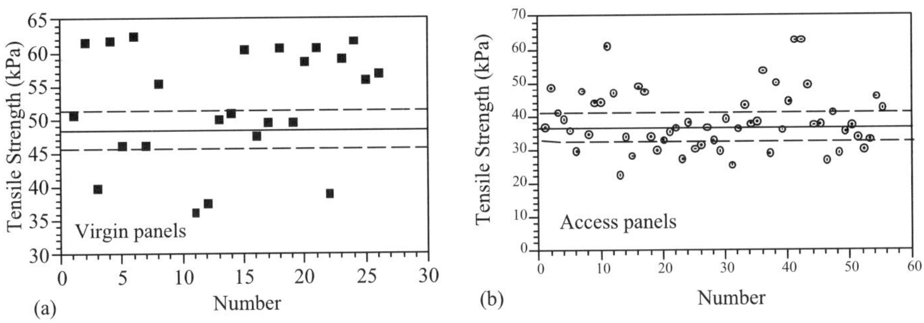

Figure 4. Tensile strength measurements from ABC. (a) Twenty-four Virgin panels showing a mean of 48.4 kPa with plus and minus one standard deviation of 2.8 kPa. (b) Fifty-five Access panels with a mean value of 35.6 kPa and a standard deviation of 3.9 kPa.

Individual values for the Virgin panels show a mean of 48.4 kPa with a standard deviation of 2.8 kPa (Figure 4a). The mean value for Access panels is 35.6 kPa with a standard deviation of 3.9kPa (Figure 4b). Thus the Access panels have lost an average of 26% strength over about 17 years when compared to the average Virgin strength.

Similar tensile strength measurements were made on 8 Virgin panels from TC and 83 panels removed from the building (Figure 5a). Most of the panels were removed as they had measured dishes of 18 mm or more so the sampling is not the same as on ABC. The strength distribution in panels containing significant dish is not uniform. At the top and lower edges of panels from TC the strength loss is only about 19%, however, where the amplitude of the bow is largest the loss averages about 53% for the 83 panels (Figure 5b).

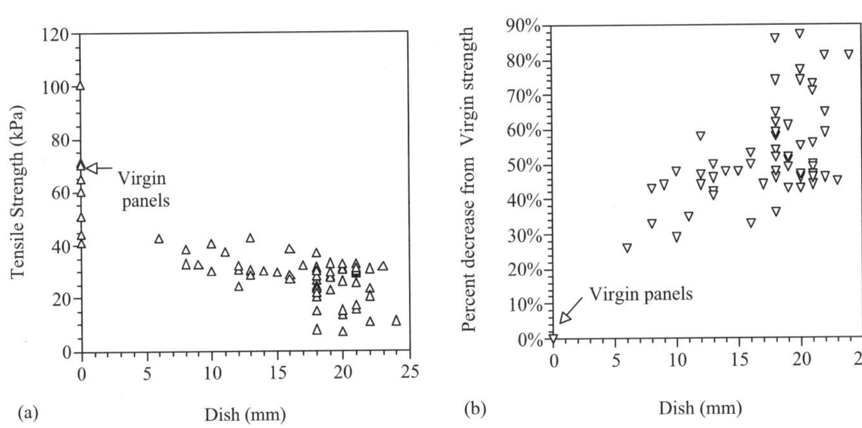

Figure 5. Tensile strength measurements from TC where each data point is an average of ten specimens measured from the center of 83 panels removed from the building and 8 Virgin panels. (a) Tensile strength versus the dish measured for each panel. (b) The percentage decrease of the measured tensile strength from an average of those of the Virgin panels.

Additionally, test specimens from areas of bow curvature show the extended or outside portion of the bow to be 25 to 30% weaker than the inside part. Petrographic observations document that microfractures are concentrated in the extended portions contributing to the significant strength loss in those areas. Thus after 25 years or more the average for measurements covering the whole panel show a loss of about 35%. The latter value is more comparable to data from ABC.

4. THERMAL CYCLING EXPERIMENTS

The strong correlation between panel bowing and radiant heating provided a base for doing controlled laboratory experiments of heating and cooling cycles of selected panels removed from ABC. Measurements on heating single crystals of calcite from 20 to 100 °C show that the crystals expand about 0.91% parallel to the c-axis but in contrast to most other minerals contract about 0.042% perpendicular to it[10]. Given the random orientation of c-axes observed in the material used on these buildings it was hypothesized that heating and cooling cycles would lead to differential expansion and contraction of adjacent grains producing cyclic fatigue of the grain boundaries which would lead to a loss of strength with time. The cumulative effect of the microscopic fatigue would result in general loss of strength of the whole panel, and development of bow or dish geometry. The laboratory cycling experiments were designed to evaluate this hypothesis and possibly provide some basis for an extrapolation to longer times.

Pieces from thirty one Virgin, Access and Test panels were placed in an oven at room temperature of 24 °C and the temperature raised at about 1°C per minute to the desired level for 12 hours. They were then removed to cool to room temperature for 12 hours, and the cycle repeated. Cycles were done at temperatures of 12, 66 and 107 °C and generally stopped after 200 cycles[4], but six Virgin and Access panels were continued to 450 cycles. Tensile strength measurements were made at the beginning and periodically during the cycles, with 24 to 180 measurements made per panel. Although the test pieces became smaller as cycling progressed, no noticeable differences in the test results were found. Temperatures were measured with a thermocouple in the middle of each panel and compared to oven thermocouples. The panels reached oven temperature after 60 minutes of heating and cooled to within 10% of each other after 4 hours, thus closely approximating natural equilibrium times.

Although there were variations within the panel groups, they generally all showed similar behavior patterns (Figure 6). All panels showed a relatively rapid loss of strength for the first 15 to 30 cycles independent of the cycling temperature. This loss generally increased with the temperature difference going from an average of 20% to 32% to 40% in Virgin material and 18% to 34% to 39% in Access material, with Virgin panels showing the largest strength loss.

This rapid loss in strength was replaced by one of gradual strength decline and appears to approach a residual value. This was interrupted periodically by temporary increases in strength, which is evident in both the mean values for the panels (Figure 6), but also for individual panels[4].

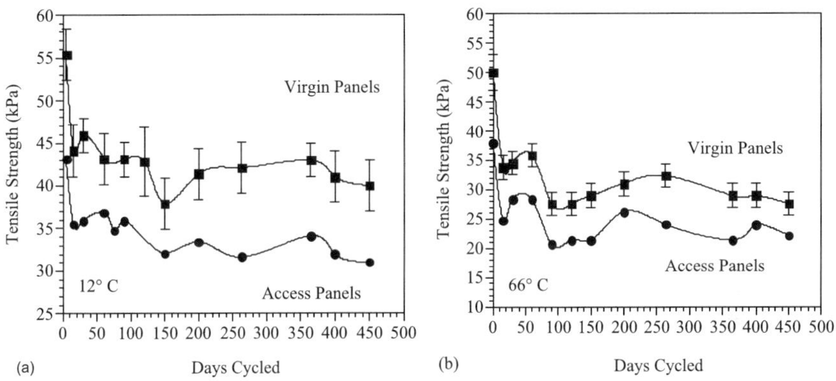

Figure 6. Tensile strength versus thermal cycles for panels from ABC. One standard deviation is shown for Virgin panels and is typical for Access panels. (a) Temperatures cycles 12 °C and (b) 66 °C above 25 °C room temperature.

In some cases the increase in strength persisted for a number of cycle, but eventually decreased to a near residual level. A few individual Virgin panels loss 50% of their initial strength and one Access panel 40%.

Petrofabric analyses of thin sections taken from panels before and after thermo cycling show significant changes in the microstructure of the marble (Figure 7). Virgin panels from ABC have very tight grain boundaries and in many cases are hard to define (Figure 7a). There is a minor amount of intracrystalline twinning and microfractures as remnants of the stress sustained during the geologic history that was not annealed out at higher temperatures. There is a little recognizable porosity reflected in average laboratory measurements of 0.81%. In contrast with thermocycling the microfabric changes so that the grain boundaries become well defined as a result of the differential thermal expansion and contraction (Figure 7b). Intracrystalline twins and microfractures become abundant as a result grain-boundary stresses produced by the differential expansion and contraction parallel to the crystallographic axes. The microfractures generally follow grain boundaries leading to an increase in porosity and permeability.

These results support the contention that the loss of strength in the marble appears to be a result of the crystallographic response to the thermo cycles. These cycles may be laboratory controlled or naturally occurring. In both cases they have regularity with natural cycles developing on a daily, seasonal and annual basis. They vary in intensity and duration, but all contribute to pro-

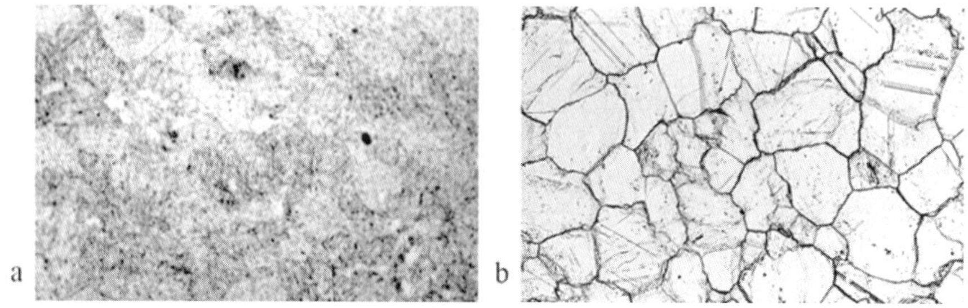

Figure 7. Photomicrographs of thin sections illustrating the effects of thermal cycles. Photos are about 1 mm across. (a) Virgin panel where the grain boundaries hardly visible and show little evidence of damage. (b) Panel after 130 days of thermal cycling at 66°C, where the grain boundary separation is sharply defined and sufficiently open in some cases for epoxy cement to penetrate.

ducing fatigue of the marble grain boundaries. With each cycle an increment of damage is created at every grain boundary. Because of differing grain orientations and mismatches of thermal expansion, some grain boundaries are damaged more than others in a given cycle. The amount of damage accrued is directly proportional to the magnitude of the temperature difference and duration. This process is similar to mechanical cycling well documented in damage mechanics studies involving metals and characterized by exponential losses of strength[11]. Potential macroscopic failure of the marble may occur when the sum of the damage, or grain-boundary breakdown reaches a critical value lowering the strength so that panels may no longer be able to sustain the external wind loads. This value would be constant regardless of the temperature differences or number of cycles. Other studies have shown that the extent to which calcite crystallographic axes are parallel, the grain size and shape, and the orientation of grain boundaries, may strongly influence the magnitude and extent of strength loss[12].

The question that arises from the laboratory thermal cycling experiments is whether the strength decay reaches some asymptotic value for natural thermal cycles? To obtain some estimation of this exponential decay, curves were fitted to thermo cycle data obtained from the laboratory experiments at cycles of 12°C above room temperature (Figure 8). Obviously, other statistical curves might be applied, but the fit in Figure 8 suggests that some general minimal asymptotic strength may be sustained over long intervals.

5. POSSIBLE FLUID ABSORPTION

At the time of the study of ABC some investigators were placing considerable importance on the role of water absorption and potential freeze-thaw cycles and the degradation of marble strength[13]. Laboratory studies were conducted to assess such a possibility. Porosity measurements made on Virgin

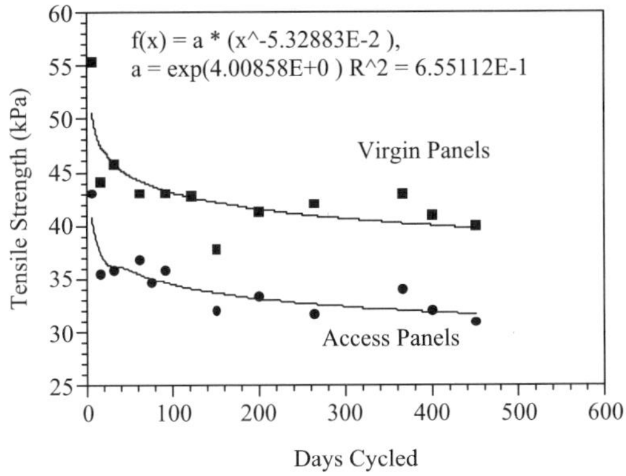

Figure 8. Exponential curve fit to thermal cycling of Virgin and Access panels presented in Figure 6 for cycles at 12 °C above room temperature.

and Access panels have average values of 0.81% and 1.11% respectively, with permeability values of 0.19 md and 0.36 md respectively.

These reflect the damage accrued by the Access panels due to radiant heating for 17 years. Tests were done on 15 Access panels, as they potentially would show more effects. Water was used containing a colored dye so the effects would be visible. Initially a reservoir of fluid was formed by a tube on the upper surface of a horizontal panel and the reservoir replenished as necessary. The reservoir was on the convex side of the bow. Within 24 hours the dye had moved laterally about 8 cm in all directions and vertically almost completely through the panel (Figure 9a). Interestingly no evidence of dye penetration was found on the outer 3 mm of concave faces of the panels of the bow where grains are more tightly packed due to the bowing.

The preceding conditions are seldom met on the sides of a building, however, as the panels are vertical and the only reservoir is water running down the outside of the panels. A simulation of this more realistic condition was made with marble pieces placed in a vertical position and the dye-saturated water allowed to flow continuously down the outside of the panel. After 24 hours there is little evidence of absorption (Figure 9b), and even after 14 days of continuous "rain" the dye had generally only penetrated about 9 mm into the panel. These results combined with the demography of thermal radiation suggest that although water may contribute to some deterioration of the panel surface, the general effect on panel strength is secondary.

6. RESIDUAL ELASTIC STRAIN

The exposure to thermal radiation cycles may significantly decrease the strength of the marble framework in panels on these two buildings but another factor may also contribute to the resulting bows and dishes. This is the release of residual strain. Many of the quarries in the Carrara area are situ-

a

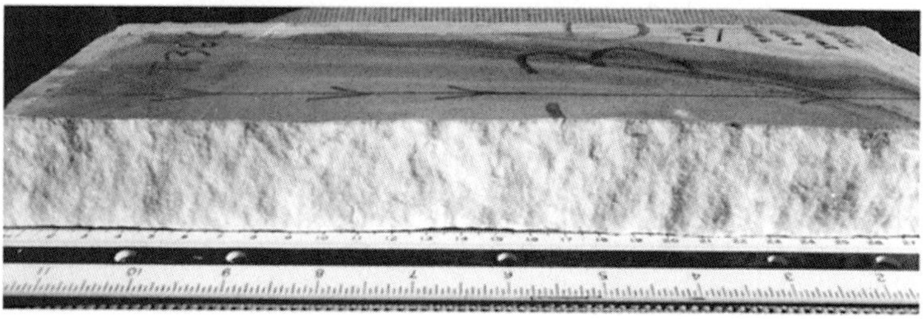

b

Figure 9. Photographs of marble following absorption tests. (a) Twenty-four hours after dye-water mix flooded the convex surface of a panel. The dye penetrated almost all of the panel thickness except about 3mm on the bottom, concave surface. (b) A panel after standing vertically for 24 hours during which water ran down the one face. Note the lack of appreciable penetration of dye under these more realistic conditions.

ated in locations of extensive geologic deformation. Frictional heating from extensive movement along faults has been postulated to produce the green-schist metamorphism observed in the region[14]. That the regional heating did not completely anneal the deformation is evidenced by the extensive folding, distortion of inclusions, brecciation and mineralized fractures present in many of the quarries. On-site observations of blocks recently removed from quarries show corners fractured, and in some cases, lying adjacent to the blocks suggesting the release of stored elastic strain. Anecdotal discussions with quarry operators confirm this hypothesis, with many quarries identified as to the presence or absence of stored strain in the marble taken from them. There is no present evidence that in-situ measurements have been made in the area of the quarries.

To evaluate this factor, residual strain measurements were made on all panels removed from ACB. A strain gage rosette was mounted on the surface of the panel and overcored with a 5-cm drill, and water conditioned so that it's temperature and that of the air were the same. The release of residual strain was monitored for at least 24 hours or until no further changes were found. An initial elastic response was measured followed by an inelastic strain recovery. The average differential strain, or the algebraic maximum minus minimum, with elongation taken as positive, was found in all panels. These values increased from 20 microstrains in the Virgin to 50 in Access panels and to 85 in Test panels that were further damaged during testing on the building. Maximum differential strains of over 900 microstrains were found in a few panels. That the average strain measured increased with panel exposure or potential damage, suggest that microfracturing or grain-boundary separation resulting from thermal radiation cycles or mechanical stress from wind loads facilitats the development of the residual strain.

This hypothesis led to making periodic residual strain measurements on 43 panels as they were thermally cycled up to about 90 cycles. Most panels were found to release strain throughout the cycling process. Virgin and Access panels generally released the same amount of differential strain throughout the cycling. There were, however, some Access panels that showed cyclic behavior where periods of larger strain release alternated with intervals of less strain release (Figure 10a). The larger strain release was usually accompanied by temporary strength decline.

The most significant relationship found was that of the tensile fracture orientation produced during the associated strength testing with the maximum strain, or applied stress axis (Figure 10b). This anisotropy decreased as thermocycling progressed, though was still found at 300 cycles in a few instances. Beyond this the fracture orientation appeared to become random. The residual strain could have multiple origins. As mentioned, the geological environment and history are probably the most significant factor, however, a second contributor could be the thermal cycling of panels on the building as suggested by the laboratory experiments. Mechanical flexing of the panels under wind load on the building may also have contributed. Possible elastic straining of the grains during natural thermal cycles may also have been released upon overcoring[15].

Notable is the change in some panels that were stored outside after delivery from the quarry. They were placed vertically on A-frames, but not secured, and left unattended. After some 15 years, some of the panels were

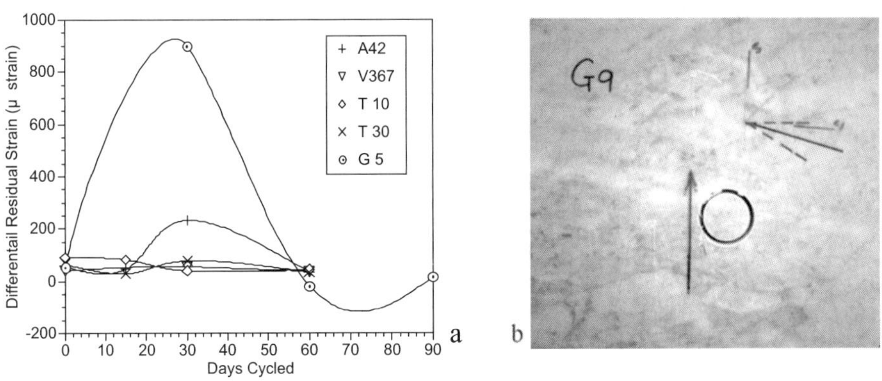

Figure 10. (a) Differential residual strain measured for thermo cycles for the five panels shown. Computer curves generated to pass through the data points do not necessarily represent values between the measured ones. Panel T10 and 30 were cycled at 12 °C above room temperature, while the other panels were cycled at 66 °C above it. (b) Photograph of a panel from ABC, which has been overcored, and residual strain measured. The maximum and minimum principal strain axes are shown with spread of induced tensile fractures shown by dashed lines on either side of the maximum strain axis. Twenty fractures were measured from separate tensile tests.

found to have developed over 2 cm of bow. They also released a significant amount stored strain when over-cored. It is concluded that the combination of cycling thermal radiation and the release of inherent residual strain in the panels are synergistic in producing bowing in the panels.

7. POROSITY AND ACOUSTIC ASSESSMENT

Analysis of data collected from panels on TC revealed panels of higher than average porosity that had correspondingly lower strengths (Figure 11). The porosity measurements were made on specimens taken from the panel edges where the effects from exposure on the building would be minimal, suggesting that these panels had about the same porosities when installed on the building some 25 years previously. Porosity measurements on 59 specimens from 37 panels showed most panels had average porosities of about 1.5% with a standard deviation among the panels of 0.3%. Some 13 panels, however, had porosities ranging from 2.8 to 5.8%, with an average of 3.3%. The pore spaces are clearly recognizable in petrographic thin sections taken from the panels (Figure 12). These high porosity-panels composed a major portion of the weakest panels removed, although their measured dish was not significantly larger than the other panels. It is suggested that these panels not only had higher porosities than others upon installation, but that they also lost strength faster than other panels. This is presently under laboratory investigation.

There are no distinguishing features on the panel surfaces to indicate their higher porosities and potentially lower strengths even to the most experienced observers. In an effort to recognize these panels a prototype acoustic delay meter was employed (Figure 13). A transducer sends a compressional

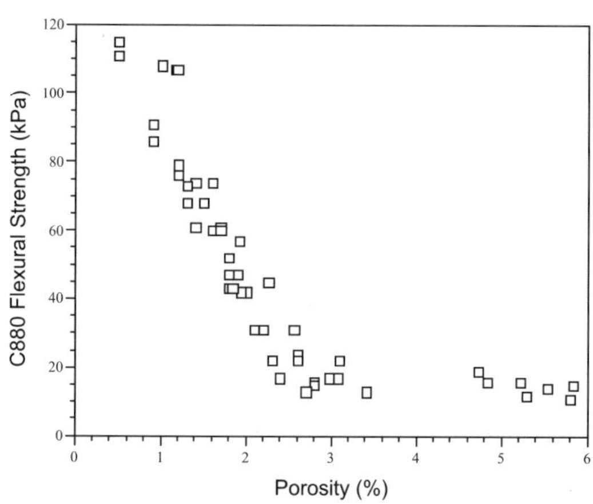

Figure 11. C880 flexural strength vs. porosity for 59 specimens from 37 panels removed from TC. In some cases there are multiple test specimens from different places in the same panel.

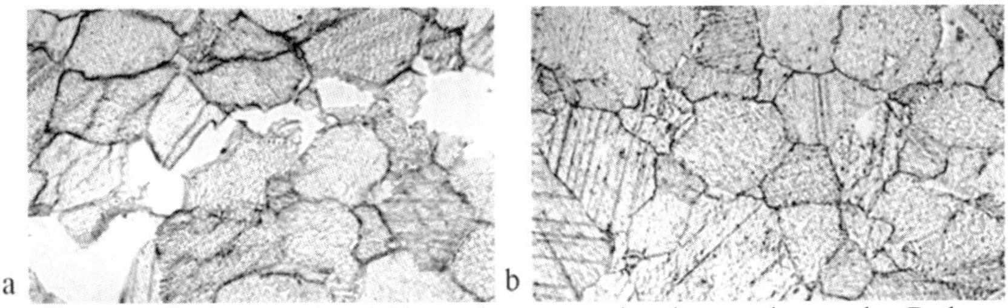

Figure 12. Photomicrographs of thin sections comparing the porosity panels. Both panels had been on same building side and floor on TC for about 17 years. (a) Photomicrograph of a panel with porosity of 3.4% where the large open pore spaces are clearly visible. (b) A thin section of a panel with a porosity of 1.2%, which shows grain-boundary damage but no large pore space.

wave at 250 kHz through the marble that is received by a second transducer. The travel time is recorded in microseconds. The travel time is primarily a function of the porosity of the marble although it is also influenced by the orientation of the calcite grains, their size, and orientation of grain boundaries. The longer the travel time, the higher the porosity and lower the corresponding strength of the marble. A suction cup is activated to bring the transducers into contact with the marble surface and facilitate supporting the delay meter on the vertical panel face.

Acoustic travel times were measured for the same 37 panels documented in Figure 11 upon which porosity values were obtained and they show an increase of travel time with increasing porosity (Figure 14a). As this technique is non destructive and does not require removal of panels already installed on a building nor disturbance of the surrounding caulk, it provides an important method to supplement measurement of bow or dish magnitude in evaluating the panel strength on a building.

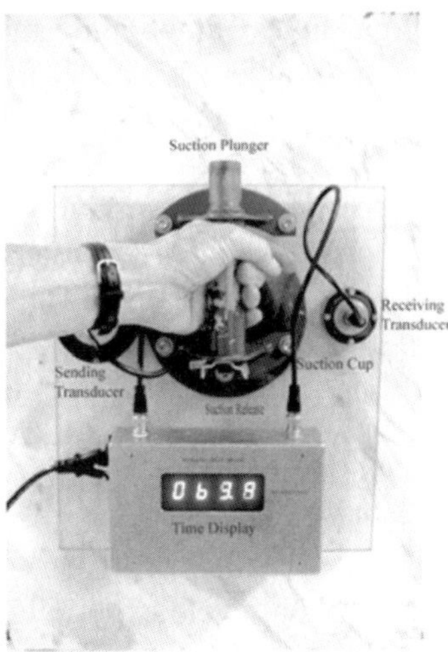

Figure 13. Photo of acoustic delay meter with travel time in the marble shown on the display in microseconds.

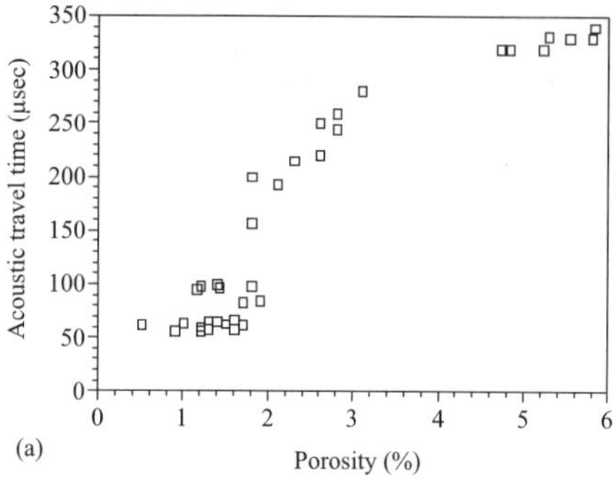

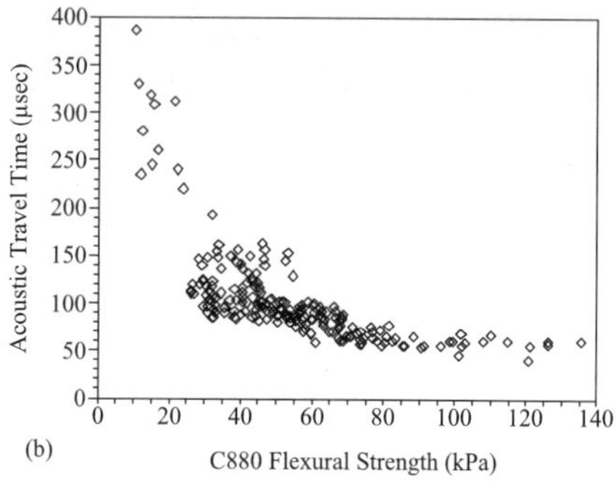

Figure 14. (a) Acoustic travel time versus porosity for 37 panels from TC. (b) Acoustic travel time versus C880 flexural strength from 234 measurements on 158 panels.

To enhance its capability 234 measurements of acoustic travel times were made on 158 panels removed from TC (Figure 14b). Here the panels of very low strength are separated from the general population, and these are also panels of abnormally high porosity.

8. SUMMARY

On site and laboratory investigations of the two buildings have provided some insight into the processes leading to degradation of the marble exterior on these structures. Although different in architectural design, panel size and anchoring systems, the exterior panels have apparently responded to the same processes. The on-site relationship between cyclic thermal radiation and magnitude of panel dish or bow indicates a correlation of loss of panel strength with thermal radiation. Daily, seasonal and annual cycles of thermal exposure all apparently contribute. Temperature measurements made on the buildings demonstrate that air temperature is not a factor, rather it is the exposure to direct sunlight, with panels coming to equilibrium temperature in about one hour. The anisotropic thermal expansion and simultaneous contraction of calcite grains is hypothesized to produce mechanical fatigue to the granular framework. This fatigue eventually breaks down the grain-to-grain cohesion and general marble strength. This provides the basis for development of bows or dishes in the panel. Measurements of panel bow and associated strength show very strong correlation between these two parameters. The resulting magnitude of the bows clearly reflects the extent of the thermal activation, with shaded places of both buildings showing little effect. Other possible contributing factors such as freeze-thaw cycles appear to be unimportant.

The loss of marble panel strength with thermal cycles is further correlated with the laboratory thermal cycle experiments lasting for almost two years. The use of three cycles of temperatures above room conditions document an initial rapid loss of strength over the first 25 to 30 cycles followed by a sharp decrease in the strength reduction. The comparison of the number of cycles and associated strength loss when compared to the 17 to 25 year age of the two buildings suggests that one laboratory thermal cycle at 12 °C may approximate one natural annual cycle. This second and extended phase is often punctuated by temporary strength recoveries but they generally recede to an approximately asymptotic strength level. The laboratory data are best fitted with an exponential decay curve suggesting minimal strength loss over longer periods of time.

A second parameter of stored residual strain is also suggested to contribute to the panel bowing. Measurements on both Virgin panels which were not installed and stored in controlled temperature and humidity conditions, with panels removed from the building show the presence of stored residual strain. It is suggested that this primarily reflects the geologic origin of the material, but contributions may have come from thermal and mechanical cycling on the buildings. Their release upon panel strength loss appears to be a significant contribution of energy leading to panel dishing and bowing. Their presence also produces a mechanical anisotropy within the panels that acts to influence the orientation of induced fractures.

A group of panels showing anomalously low strength was found on one building which appears to be the result of higher than average porosity. It is suggested that these panels had initially higher porosity as a result of their geologic history. These panels do not show any diagnostic surface features indicating their porosity so that they are not recognizable with ordinary monitoring techniques. An acoustic delay meter was developed to measure the compressional wave travel time, which shows a strong correlation between longer travel time, higher porosity, and low panel strength. In that this technique is non-destructive, and the instrument is applied to the exposed flat surface of a panel it has become a significant method to supplement bow measurements and determine changes in panel strength and possible replacement.

ACKNOWLEDGEMENTS

The study of the Amoco Building was supported by Amoco Corporation with contributions to the data acquisition by M. Hasdedt, D. Lehnert, and M. Denton. Study of the building in Toronto, Canada was supported by O&Y Enterprise, Toronto and benefited from discussions with M. Wertheim and J. Shelemay and Michael Lewis of Façade Forensics. The invitation to publish this work by S. Kourkoulis is appreciated.

REFERENCES

1. H. Ritter, "Die Marmorplatten sind falsch dimensioniert", Stein, Vol. 1, 1992, pp.18-19.

2. E. S. McGee, "Colorado Yule marble-building stone of the Lincoln Memorial", U. S. Geological Survey Bulletin 2162, 1999.
3. G. Simmons and D. Richter, "Cracks in building stone", International Journal Rock Mechanics Mineral Science Geomechanical Abstracts, Vol. 30, 1993, pp.1553-1557.
4. J. M. Logan, M. Hastedt, D. Lehnert, M. Denton, "A case study of the properties of marble as building veneer", International Journal Rock Mechanics Mineral Science Geomechanical Abstracts, Vol. 30, 1993, pp.1531-1537.
5. J. M. Logan, "Laboratory and case studies of thermal cycling and stored strain on the stability of selected marbles", Environmental Geology, Vol. 46, 2004, pp. 456- 467.
6. R. Dunakin, Mechanical aspects of weathering in marble crack growth, Master of Science Thesis, Department of Material Science and Engineering, Royal Institute of Technology, Stockholm, Sedem, 1999.
7. A. Zeisig, S. Siegesmund, T. Weiss, "Thermal expansion and its control on the durability of marbles", Geological Society of London Special Publication, Vol. 205, 2000, pp. 57-72.
8. L. Rayleigh, "The bending of marble", Proceedings of the Royal Society A, Vol. 144, 1934, pp. 266-279.
9. ISRM, "Suggested method for determining point load strength", International Journal Rock Mechanics Mineral Science Geomechanical Abstracts, Vol. 22, 1985, pp 53-60.
10. S. P. Clark Jr, Ed., Handbook of Physical Constants, Geological Society of America, New York, 1966.
11. J. A. Collins, Failure of Materials in Mechanical Design, John Wiley &Sons, New York, 1981.
12. R. Dal Pino, P. Narducci, G. Royer-Carfagni, " A SEM investigation on fatigue damage of marble", Journal of Material Science Letters, Vol. 18, 1999, pp. 1619-1622.
13. S. Bortz., J. Stecich, B. Wonneberger, I. Chin, "Accelerated weatherin in building stone", International Journal Rock Mechanics Mineral Science Geomechanical Abstracts, Vol. 30, 1993, pp.1559-1566.
14. L. Carmignani, G. Giglia, R. Klingfield, "Structural evolution of the Apuane Alps: an example of continental margin deformation in the Northern Apennines, Italy, Journal of Geology, Vol. 86, 1978, pp. 487-504.
15. S. Siegesmund, K. Erlenmeyer, T. Weiss, E. K. Tschegg, "Physical weathering of marbles caused by anisotropic thermal expansion", International Journal of Earth Science Vol. 89, 2000, pp. 170-182.

Chapter 6.2

THE INTEGRITY LOSS OF PHYSICO-MECHANICAL PROPERTIES OF BUILDING STONES WHEN SUBJECTED TO RECURRENT CYCLES OF FREEZE–THAW (F-T) PROCESS

Raşit Altindag[1], Nazmi Şengün[1], Avni Güney[2], Mahmut Mutlutürk[3], Remzi Karagüzel[3], Turgay Onargan[4]

[1]SDU Engineering and Architecture Faculty, Department of Mining Engineering, 32260 Isparta, Turkey; [2]Mugla University Engineering Faculty, Department of Mining Engineering, 48000 Muğla, Turkey; [3]SDU Engineering and Architecture Faculty, Department of Geological Engineering, 32260 Isparta, Turkey; [4]DEU Engineering Faculty, Department of Mining Engineering, Bornova-İzmir, Turkey

Abstract: In this study, a total of nearly 900 samples were obtained from 7 different rock types, also recognised as marbles of sedimentary and metamorphic origin and commercially used in the building of various monuments, art works as well as the interior/exterior covering of the structures. In order to determine the physical and mechanical properties of these rocks at the end of freeze-thaw (F-T) cycles, experiments were conducted on a set of nearly 130 samples prepared in different sizes for each rock type. The samples from each rock type were subjected to a routine of recurrent F-T cycles with an increment of 10 cycles for each routine, amounting to a total of 60 cycles (6 routines). Prior to the F-T cycles and at the end of each routine of 10 cycles up to 60 cycles, the physical rock properties (unit volume weight, apparent porosity, *P*-wave velocity and Shore hardness) and the mechanical rock properties (Brazilian tensile strength, point load index, block punch index and bending strength) were determined up to 50 cycles and then the values of test results were evaluated graphically. Moreover, the results of this study were associated with the Decay Function Model (DFM), introduced by Mutlutürk et al.[1], in order to determine the integrity loss and half-life of the 7 different rock types tested. Besides, it was investigated whether the DFM can be validly used to determine the half-life of the rocks based on their physical properties.

Key words: natural stones; freeze-thaw (F-T); integrity loss; half-life of rocks; mechanical properties; physical properties.

S.K. Kourkoulis (ed.), Fracture and Failure of Natural Building Stones, 363–372.

1. INTRODUCTION

Natural stones, which have long been used in ancient history are even today utilised widely in the building of various structures, monuments, art works and interior/exterior covering of the structures. Ancient structures built of natural stones are progressively subject to long term effects of atmospheric phenomena such as wind, rain, humidity, freeze-thaw, changes in temperature etc. Their building materials, in the long run, tend to lose integrity and consequently change in natural colour and more importantly weaken in mechanical strength. However, the rates of progressive disintegration may differ depending on the rock type, or even within the same rock type.

Several researchers have studied the cyclic effect of F-T process on various building stones, ancient monuments and art works. Broms and Yao[2] have investigated the effect of F-T cycles on shear strength of soils. Binal[3] and Binal et al.[4,5] have studied the F-T cycles to determine the changes in physical and mechanical properties of volcano-sedimentary rocks cropped out near Yazlikaya-Eskisehir, Turkey. Simonsen and Isacsson[6] have examined the thaw weakening of pavement structures in cold regions. Binal and Kasapoglu[7] have conducted 30 F-T cycles on Selime ignimbrite from Ihlara Valley in Aksaray, Turkey and concluded that an increase in the number of F-T cycles resulted in an increase of the values of some physical properties and a decrease of the values of some mechanical properties.

Topal and Sözmen[8] investigated the effect of F-T cycles on the changes in the original state of Midas tuffs based on detailed regional climatic data of monthly average temperature, rainfall and the number of days snow stayed on the ground and reported a decrease in the *P*-wave velocities and uniaxial compressive strength and an increase in the porosity of the rock. The F-T cycles induced the rocks to lose integrity, namely a considerable decrease in their mechanical strength (half-life). Thus, Mutlutürk et al.[1] introduced the DFM to determine the half-life of the rocks subjected to F-T cycles.

Altindag et al.[9] observed a decrease in the mechanical strength of Isparta andesite in the end of 50 F-T cycles and determined its mechanical half-life applying the DFM. Also, 55 F-T cycles have been conducted on Dereboğazı ignimbrite (Isparta) to determine its mechanical half-life using the DFM[10].

2. EXPERIMENTAL STUDY

2.1 Rock Samples and Preparation

Seven different rock types (Table 1) from various locations in Turkey were selected and the samples were prepared in compliance with the standards of ISRM[11], Ulusay et.al.[13] and standards suggested by Altindag and Güney[12].

2.2 Physical Properties

The suggested procedures of ISRM[11] and Altindag and Güney[12] for the sample preparation and the tests were used for the following:

a) Dry unit weights were determined by saturation and calliper method.
b) Pore volume values were obtained by the water saturation procedure.
c) *P-wave* velocity values were determined by applying ultrasonic compression wave pulses to the rock samples.
d) Shore hardness values were determined by C-2 type Shore Sclerescope.

The results of the tests are given in Table 2.

Table 1. Name and location of the rocks selected for this study

Rock no	Rock type	Rock class	Location
1	Limestone	Sedimetary	Burdur
2	Travertine	Sedimetary	Burdur-Bucak
3	Travertine	Sedimetary	Kütahya
4	Limestone	Sedimetary	Antalya-Finike
5	Limestone	Sedimetary	Kastamonu
6	Limestone	Sedimetary	Antalya-Finike
7	Limestone	Sedimetary	Kastamonu

Table 2. Pre-test physical properties of the rocks used in this study

Rock no	Dry unit weight (g/cm^3)	Apparent porosity (%)	P-wave velocity (m/s)	Shore hardness
1	2.704	0.093	6603.3	58.8
2	2.453	2.658	4144.7	38.3
3	2.444	5.301	5402.8	42.6
4	2.382	7.544	4905.1	36.5
5	2.665	1.178	5908.1	54.7
6	2.430	3.203	5114.8	33.1
7	2.609	2.603	5637.9	54.6

2.3 Mechanical Properties

Some of the mechanical properties of the rocks (Brazilian tensile strength, point load index, block punch strength index, bending strength) were initially determined prior to the F-T process as shown in Table 3. In order to observe the changes in mechanical properties of the rocks as a result of F-T process, the same properties were re-determined starting from the 10th period and following each routine of 10 periods until the end of 50th period, e.g. initial, 10th, 20th, 30th, 40th, 50th periods.

2.3.1 Brazilian tensile strength test

Brazilian tensile strength tests were conducted on core samples of diameter 54.7 mm and height-to-diameter ratio of 0.5. The load was applied continuously at a constant stress rate, so that failure would occur within 5 min of loading. The tests were performed on 4–5 specimens at the end of each 10 cycles.

2.3.2 Point load test

The diametrical point load tests were carried out on the block samples. The results were corrected to a sample diameter of 50 mm. The diametrical point load tests were performed on 4 samples at the end of each routine for each rock type.

2.3.3 Block punch strength index test

The block punch strength index was determined on thin disc samples having a diameter of 54.7 mm and a disc thickness of 10 mm. The test apparatus used in this study was fabricated in-house. The tests were carried out in compliance with the draft ISRM suggested method[13]. The load was applied in such a rate that failure would occur within 10–60 seconds, as suggested by ISRM[11]. Then, the BPI values were determined by using Eq.(1). The tests were performed on 4–5 specimens at the end of each 10 cycles.

$$\mathrm{BPI} = 3499\mathrm{D}^{-1.3926} t^{-1.1265} F_{t,D} \tag{1}$$

where, D is the diameter of specimen, t is the thickness, $F_{t,D}$ is the failure load. D and t are in mm and $F_{t,D}$ is in kN.

2.3.4 Bending strength

The tests were performed according to the classical three-point bending procedure. The beams, 50 x 100 x 200 mm in size, were tested over a simply supported span of 180 mm. The failure load was recorded and the bending strength value was determined using Eq.(2).

$$BS = \frac{3P}{2bh^2} \tag{2}$$

where, P is the failure load in kg, b is the thickness in cm and h is the height in cm.

Table 3. Pre-test mechanical properties of the rocks tested in this study

Rock no	TS (MPa)	BS (MPa)	BPI (MPa)	$Is_{(50)}$ (MPa)
1	7.65	12.452	12.387	4.47
2	3.25	8.755	7.718	3.15
3	5.63	13.569	8.175	3.816
4	4.96	10.086	9.98	3.274
5	9.35	15.534	12.94	5.933
6	6.98	12.301	12.65	3.714
7	7.79	15.992	14.2	5.263

2.3.5 Freeze-thaw tests

Freeze-thaw tests were applied in compliance with TSE (TS 699) standard[14]. Saturated specimens were placed into a freezer and conditioned at –20°C for two hours. Then, they were taken out of the freezer and placed into a water bath at 20 °C, where they were allowed to thaw for 2 h. Since the Newton's Law of Cooling applies to this F–T cycle as well, the temperature change will follow almost the same route at each cycle (Figure 1). For each test, this routine was repeated 60 times. The mechanical properties of ignimbrite were measured at the 10th and every succeeding 10 cycles up to the 60th cycle of the F–T process.

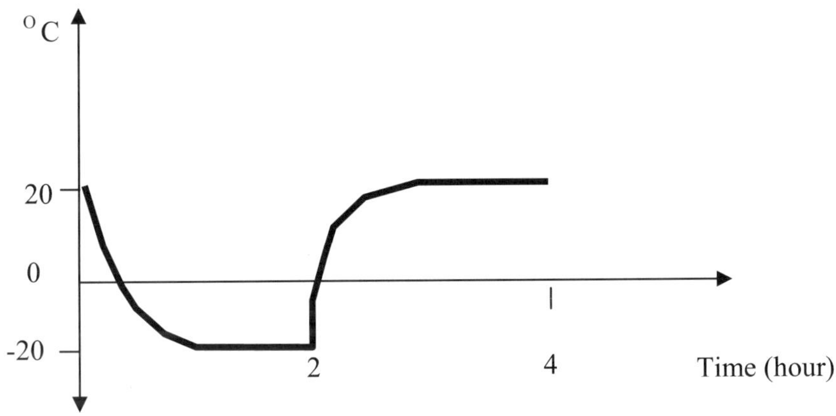

Figure 1. Generalised temperature curve for 40 °C F-T cycle[1].

2.3.6 Decay Function Model (DFM)

In the DFM proposed by Mutlutürk et al.[1], the process of integrity loss due to F–T was modelled as a first-order process. That is, the rate of integrity loss (or disintegration rate) is proportional to the rock integrity present at the beginning of each cycle.

This proposition can be mathematically expressed as follows:

$$-\{dI/dN\}=\lambda I. \quad (3)$$

where {dI/dN} is the disintegration rate. The minus sign in Eq. (3) indicates a decrease in the integrity. λ is the decay constant and *N* is the number of cycles.

By integrating this equation between I_0 (the original integrity of the rock) and I_N (the integrity after *N* cycles) the following equation is obtained in logarithmic form:

$$\ln(I_0 = I_N) = \lambda N \quad (4)$$

This relation can be expressed in an exponential form as follows:

$$I_N = I_o e^{-\lambda N} \quad (5)$$

The term $e^{-\lambda N}$ is the *decay factor*, indicating the proportion of the remaining integrity after N^{th} cycle, i.e. (I_N/I_0). Also, the *decay constant,* λ, indicates the mean relative integrity loss by the action of any single cycle. In addition, as a measure of rock durability, the *half-life*, $N_{1/2}$, of the rock is defined as the number of cycles required to reduce the integrity to half its initial value. This durability parameter is inversely related to the decay factor. In fact, by replacing $(I_O/2)$ with I_N in Eq. (4) the following equation can be obtained:

$$N_{1/2} = \ln 2/\lambda \cong 0.693/\lambda \quad (6)$$

The half-life values ($N_{1/2}$) of the physical and mechanical properties of the tested rocks subjected to F–T process were calculated based on this model (see Table 14)

3. RESULTS AND CONCLUSION

The changes in the physical properties of rocks when subjected to F-T cycles were determined and they are shown in Tables 4-7. Similarly, the changes in the mechanical properties were also determined and they are illustrated in Tables 8-11.

Table 4. The changes in unit weight depending on the F-T cycles

Rock no	Initial	10th cycle	20th cycle	30th cycle	40th cycle	50th cycle	60th cycle
1	2.704	2.714	2.704	2.705	2.703	2.703	2.705
2	2.453	2.429	2.396	2.391	2.398	2.395	2.378
3	2.444	2.438	2.430	2.404	2.392	2.390	2.393
4	2.382	2.381	2.377	2.376	2.371	2.369	2.369
5	2.665	2.660	2.662	2.662	2.658	2.658	2.655
6	2.430	2.426	2.410	2.408	2.400	2.396	2.397
7	2.609	2.606	2.606	2.605	2.597	2.597	2.591

At the end of each routine, statistical relationships were established between the number of cycles and the mechanical and physical properties of the rocks and the best fit curve was determined to be exponential. The equations and the correlation coefficients are displayed in Tables 12 and 13. These relationships were then evaluated using the $I_N = I_o e^{-\lambda N}$ form of DFM and the half life of the rock properties was determined by Eq. (6). Table 14 illustrates the number of half-life time cycles of physical and mechanical properties of the rocks studied.

Table 5. The changes in apparent porosity depending on the F-T cycles

Rock no	Initial	10th cycle	20th cycle	30th cycle	40th cycle	50th cycle	60th cycle
1	0.093	0.111	0.107	0.100	0.108	0.107	0.121
2	2.658	3.018	2.894	3.064	3.196	3.226	3.601
3	5.301	5.551	5.638	5.656	5.717	5.775	5.804
4	7.544	7.700	7.546	7.665	7.759	7.642	7.827
5	1.178	1.356	1.278	1.288	1.379	1.376	1.531
6	3.203	3.204	3.217	3.231	3.231	3.230	3.224
7	2.602	3.465	3.283	3.211	3.659	3.568	3.984

Table 6. The changes in P-wave velocity depending on the F-T cycles

Rock no	Initial	10th cycle	20th cycle	30th cycle	40th cycle	50th cycle	60th cycle
1	6603.3	6172.1	6158.7	6057.9	6042.8	6043.49	6015.3
2	4144.7	3951.0	3981.3	4330.1	4438.4	4667.6	4838.5
3	5402.8	4954.1	4953.9	4897.5	4794.9	4791.6	4745.1
4	4905.1	4430.2	4341.3	4300.4	4295.8	4259.1	4233.4
5	5908.1	5170.0	5089.8	4963.9	4905.9	4929.8	4825.9
6	5114.8	4818.1	4755.0	4729.1	4759.9	4755.4	4722.7
7	5637.9	5329.6	5082.9	5010.6	4916.9	4797.8	4698.1

Table 7. The changes in Shore Hardness depending on the F-T cycles

Rock no	Initial	10th cycle	20th cycle	30th cycle	40th cycle	50th cycle	60th cycle
1	58.8	57.3	56.7	56.7	56.1	54.4	54.1
2	38.3	37.8	37.8	37.1	37	36.4	35.9
3	42.6	41.7	41.9	41.6	41.5	40.6	40.2
4	36.5	35.72	36.4	35.5	35.2	34.7	33.5
5	54.7	52.5	51.1	51.5	51.6	51.2	50.7
6	33.1	32.7	32.3	32.4	32.2	31.9	31.1
7	54.6	52.9	51.8	51.1	50.5	48.4	47.6

Table 8. The changes in TS depending on the F-T cycles

Rock no	Initial	10th cycle	20th cycle	30th cycle	40th cycle	50th cycle
1	7.65	7.03	6.77	6.37	4.87	4.77
2	3.25	3.22	2.96	3.08	2.12	2.17
3	5.63	5.07	4.87	4.56	3.96	3.75
4	4.96	4.80	4.34	4.81	3.22	2.69
5	9.35	8.61	7.08	7.73	6.40	6.67
6	6.98	5.51	5.86	5.80	4.20	3.66
7	7.79	6.94	6.46	6.6	5.26	4.11

Table 9. The changes in $I_{s(50)}$ depending on the F-T cycles

Rock no	Initial	10th cycle	20th cycle	30th cycle	40th cycle	50th cycle
1	4.47	4.352	4.419	4.544	3.565	4.087
2	3.15	2.632	2.293	2.32	2.365	2.215
3	3.816	3.254	3.159	3.137	3.815	2.744
4	3.274	3.158	3.08	3.114	3.159	3.179
5	5.933	5.818	5.783	5.799	5.81	5.7
6	3.714	3.638	3.672	3.658	3.643	3.524
7	5.263	3.14	2.812	2.68	2.394	2.05

Table 10. The changes in BPI depending on the F-T cycles

Rock no	Initial	10th cycle	20th cycle	30th cycle	40th cycle	50th cycle
1	12.387	10.02	8.89	8.61	7.57	7.03
2	7.718	6.66	5.90	5.50	5.79	5.63
3	8.175	6.98	6.87	6.08	5.81	5.70
4	9.98	5.29	5.03	4.93	4.94	5.23
5	12.94	11.10	8.39	8.20	8.10	7.59
6	12.65	10.54	9.75	8.09	7.84	6.15
7	14.20	10.87	8.72	6.77	6.09	6.99

Table 11. The changes in BS depend on the F-T cycles

Rock no	Initial	10th cycle	20th cycle	30th cycle	40th cycle	50th cycle
1	12.452	11.770	9.971	8.491	8.116	8.105
2	8.755	8.104	7.000	7.012	6.838	4.763
3	13.569	13.337	14.547	9.547	9.042	10.84
4	10.086	8.104	10.753	9.239	8.220	8.691
5	15.534	15.590	14.567	12.332	12.878	11.800
6	12.301	11.962	11.396	10.744	9.815	8.987
7	15.992	13.559	16.243	12.543	13.651	11.298

Table 12. Relationship equations for physical properties depending on F-T cycles

Equation no	Property	Equation	Correlation coefficient (r)
1	Unit weight	----	---
2		$Y=2.4369e^{-0.0004x}$	0.867
3		$Y=2.4438e^{-0.0004x}$	0.940
4		$Y=2.3824e^{-0.0001x}$	0.977
5		$Y=2.6641e^{-0.000005x}$	0.884
6		$Y=2.4277e^{-0.0003x}$	0.953
7		$Y=2.6103e^{-0.0001x}$	0.950
1	Apperent porosity	$Y=0.0984e^{0.0026x}$	0.681
2		$Y=2.7266e^{0.0041x}$	0.930
3		$Y=5.4163e^{0.0013x}$	0.917
4		$Y=7.5678e^{0.0004x}$	0.698
5		$Y=1.2152e^{0.0032x}$	0.845
6		$Y=3.2062e^{0.0001x}$	0.808
7		$Y=2.8877e^{0.0052x}$	0.828
1	P-wave	$Y=6382.3e^{-0.0012x}$	0.805
2		----	---
3		$Y=5195.2e^{-0.0017x}$	0.860
4		$Y=4647.1e^{-0.0019x}$	0.802
5		$Y=5523.1e^{-0.0026x}$	0.831
6		$Y=4944.3e^{-0.0009x}$	0.724
7		$Y=5506.2e^{-0.0028x}$	0.971
1	Shore Hardness	$Y=58.521e^{-0.0013x}$	0.967
2		$Y=38.355e^{-0.001x}$	0.983
3		$Y=42.502e^{-0.0008x}$	0.938
4		$Y=36.693e^{-0.0012x}$	0.909
5		$Y=53.398e^{-0.001x}$	0.806
6		$Y=33.076e^{-0.0009x}$	0.938
7		$Y=54.404e^{-0.0022x}$	0.988

It can be seen in this study that the DFM can be used to determine the number of half-life cycles of the physical properties of rocks besides determining the number of half-life cycles of the mechanical properties of rocks.

The results of this study reveals a fact that determination of integrity loss of physical and mechanical properties of rocks when subject to freeze-thaw conditions is very important if natural stones are to be used in the restoration of ancient structures such as castle walls, monuments and statues etc.

Table 13. Relationship equations for the mechanical properties depending on F-T cycles

Equation no	Property	Equation	Correlation coefficient (r)
1	Tensile strength	$Y=7.9047e^{-0.0101x}$	0.953
2		$Y=3.4731e^{-0.0092x}$	0.876
3		$Y=5.628e^{-0.0081x}$	0.989
4		$Y=5.4266e^{-0.0119x}$	0.872
5		$Y=9.0432e^{-0.0071x}$	0.896
6		$Y=6.9646e^{-0.0116x}$	0.907
7		$Y=8.0765e^{-0.0114x}$	0.933
1	Point load strength	$Y=4.5443e^{-0.0029x}$	0.599
2		$Y=2.8718e^{-0.0059x}$	0.842
3		$Y=3.5883e^{-0.0034x}$	0.495
4		----	---
5		$Y=5.891e^{-0.0006x}$	0.837
6		$Y=3.7099e^{-0.0007x}$	0.795
7		$Y=4.3385e^{-0.0159x}$	0.919
1	BPI	$Y=11.63e^{-0.0106x}$	0.975
2		$Y=7.1345e^{-0.0059x}$	0.855
3		$Y=7.8164e^{-0.0071x}$	0.962
4		$Y=7.2793e^{-0.0099x}$	0.667
5		$Y=11.932e^{-0.0104x}$	0.913
6		$Y=12.475e^{-0.0134x}$	0.985
7		$Y=12.681e^{-0.0158x}$	0.913
1	Bending strength	$Y=12.345e^{-0.0098x}$	0.958
2		$Y=8.9666e^{-0.0101x}$	0.905
3		$Y=14.103e^{-0.0077x}$	0.724
4		$Y=9.7064e^{-0.0024x}$	0.402
5		$Y=15.933e^{-0.006x}$	0.933
6		$Y=12.661e^{-0.0063x}$	0.982
7		$Y=15.855e^{-0.0056x}$	0.755

Table 14. The half-life values of physical and mechanical properties of the rocks tested.

Rock no	Unit weight	Effective porosity	Shore hardness	P-wave velocity	TS	$I_{s(50)}$	BPI	BS
1	---	267	533	578	69	239	65	71
2	1733	169	693	---	75	118	117	69
3	1733	533	866	408	86	204	98	90
4	6930	1733	578	365	58	---	70	289
5	13860	217	693	267	98	1155	67	116
6	2310	6930	770	770	60	990	52	110
7	6930	133	315	248	61	44	44	124

REFERENCES

1. M. Mutlutürk. R. Altindag. G. Türk, A decay function model for the integrity loss of rock when subjected to recurrent cycles of freezing-thawing and heating-cooling, *Int. J. Rock Mech.& Min. Sci.* **41** (2), pp.237-244 (2004).
2. B. Broms. L.Y.C. Yao, Shear strength of a soil after freezing and thawing, *J. Soil mechanics Foundation Division*. ASCE, pp.1-25 (1964).
3. A. Binal, Investigation of the instability mechanisms abserved in volcanosedimentary rocks at Aksaray-Ihlara valley, (in Turkish). *H.U. Institute of Science. Ms. Thesis,* 1996, p.95, Ankara. Turkey.
4. A. Binal. K.E. Kasapoğlu. C. Gökçeoğlu, The Surficial Physical Deterioration Behaviour of Neogene Volcanosedimentary Rocks of Eskişehir – Yazılıkaya, NW Turkey. *Engineering Geology and the Environment.* (Marinos. P.G., Tsiambaos. G.C., Stournaras. G.C., Eds), Greece, pp. 3065-3069 (1997).
5. A. Binal. K:E: Kasapoğlu. C. Gökçeoğlu, Variation of some physical and mechanical parameters of the volcanosedimentary rocks around Eskişehir-Yazılıkaya under frezing-thawing effects (in Turkish), *Bulletin of earth sciences application and research centre of Hacettepe University,* **20**, pp.41-54, (1998).
6. E. Simonsen. U. Isacsson, Thaw weakening of pavement structure in cold regions, *Cold regions science and technology,* **29**, pp.135-151 (1999).
7. A. Binal. K.E. Kasapoğlu, Effects of freezing and thawing process on physical and mechanical properties of selime ignimbrite outcops in Aksaray-Ihlara Valley, (In Turkish), in *Proceedings of VIth Regional Rock Mechanics Symposium.* (Şensöğüt. C., Özkan. İ. Eds.). Konya, Turkey, pp.189-196 (2002).
8. T. Topal. B. Sözmen, Deterioration Mechanisms of Tuffs in Midas Monument. *Engineering Geology,* **68**, pp.201–223 (2003).
9. R. Altındağ. M. Mutlutürk. R. Karagüzel, The effects of freezing-thawing cycles on the useability of Isparta andesite as a building stone, in *Proceeding of Int. Symp. of industrial minerals and building stone,* İstanbul, Turkey, pp. 289-292 (2003).
10. R. Altindag. İ.S. Alyıldız. T. Onargan, Mechanical property degradation of ignimbrite subjected to recurrent freeze-thaw cycles, *Int. J. Rock Mech.& Min. Sci.,* V.**41**, pp.1023-1028 (2004).
11. International Society for Rock Mechanics. Rock Characterisation. Testing and Monitoring, In: Brown ET (Ed.). ISRM suggested methods, (Pergamon Press, Oxford, 1981); p. 211.
12. R. Altindag. A. Güney, ISRM suggested Method for determining the Shore Hardness value for rock. *Int. J. Rock Mech.Min.Sci.* **43**:19-22 (2006).
13. R. Ulusay. C. Gökçeoğlu C. S. Sülükçü, Draft ISRM suggested method for determining block punch index (BPI), Int J. Rock Mech. Min. Sci., **38**:1113–9 (2001).
14. TSE. Methods of testing for natural buildings stones (TS 699), Institute of Turkish Standards, p.82(1987).

Chapter 6.3

MONITORING OF THERMAL CONDITIONS IN BUILDING STONE WITH PARTICULAR REFERENCE TO FREEZE-THAW EVENTS

Kevin Hall

Department of Geography, Geoinformatics and Meteorology, University of Pretoria, Pretoria, 0002, South Africa, Present address: Geography Program, University of Northern British Columbia, 3333 University Way, Prince George, BC, Canada V2N 4Z9, hall@unbc.ca

Abstract: Although attempting to consider the impact of freeze-thaw, the monitoring of associated thermal conditions must, effectively, be context free. This is important for several reasons. First, although freeze-thaw is being evaluated, data must be of a nature that also allows determination of the spatial and temporal role of other processes. Second, by being 'holistic' rather than (assumed) 'specific' in character the data do not pre-determine the outcomes. Third, by being able to be used for evaluation of multiple processes the data are, paradoxically, of a nature that facilitates a more detailed understanding of freeze-thaw activity itself. In considering freeze-thaw it is important to recognize that the process is not a singularity but rather comprises a range of mechanisms, each determined by an interaction between thermal and moisture conditions with the properties of any given building material. In the absence of (the required) moisture data, the thermal data need to be adequate to validate, or invalidate, specific mechanisms, as well as to offer indirect proxy information indicating whether or not some form of freeze-thaw weathering indeed took place. Significant in this regard are not just freeze-thaw amplitudes and durations but also the associated rate of change of temperature ($\Delta T/\Delta t$). In addition, the record rate must be fast enough to monitor (should they occur) 'exotherms' – the latent heat released as water turns to ice; this being a proxy identification that water was present and did indeed freeze (the finding of 'zero curtains' can also be used as a proxy for the existence of water that froze within the material). In reality, the thermal data acquisition requirements for monitoring of both $\Delta T/\Delta t$ rates and exotherms are effectively the same; namely high-frequency thermal monitoring at an interval of at least one-minute. Thermal monitoring at one-minute intervals may have produced logistical problems in the past but modern data loggers with multiple channels, long-life battery power and large storage capacities can handle such requirements with ease. The resulting data are of a temporal nature that allows for the evaluation of thermal stresses, especially those associated with $\Delta T/\Delta t$ events ≥ 2 °C·min^{-1}. Equally, such data are able to resolve the short-interval heat transfer associated with latent heat release at phase transfer. Significantly, such data not only identify the occurrence of exo-

S.K. Kourkoulis (ed.), Fracture and Failure of Natural Building Stones, 373–394.

therms but also show the temperature at which freezing took place. Further, given sub-zero rock temperatures, the absence of exotherms shows when thermal conditions may have been suitable but no water was available to freeze or despite water being present it did not freeze. Thus, this approach provides objective data allowing for the true counting of actual events rather than the subjective counting based on the assumptions that (a) water was indeed present and (b) that it froze within a certain thermal range. This latter approach (assumed counting) has now been shown to suffer from potentially massive error, especially within a spatial context. In respect of thermal monitoring, the key prime requirements are: large data capacity loggers with multiple channels, high resolution loggers, high-frequency logging capacity, high resolution transducers, fast response time transducers, large spatial distribution of transducers (including, where possible, with depth within the material being monitored). In terms of transducers, experiments have suggested that 40 gauge thermocouples satisfy resolution (0.1°C) and response time (0.04 sec) while at the same time, due to the almost invisible nature of the wire, not impacting on the aesthetics of a site. If drilling of the building material (for the emplacement of transducers) is possible, holes are less than 0.2 mm in diameter, or, if a predrilled block is situated at the site, the visual impact is still very small. Where any form of attachment or drilling is prohibited, infra-red (IR) sensors are now of a resolution (0.1 °C), response time (0.002 sec) and monitored area (1 mm^2) that can provide excellent data; but at an aesthetic cost during monitoring. The use of infrared sensors is ideal for monitoring of surface pigments (e.g. in cave art) or fragile components unsuitable for direct contact sensors. Once collected, data have shown that many pre-conceived notions, especially in respect of freeze-thaw, are in error. Despite cold temperatures (the common "indicator" for the assumed occurrence of freeze-thaw) data have shown that either water was not present in the rock to freeze or it simply did not freeze at the available temperature; equally the temperature at which freezing occurs has been found to often be colder than the assumed value. Sometimes the freezing of water was found to be progressive with depth while at other times it was instantaneous over the outer several centimetres of the rock. Spatial and temporal variability of freeze-thaw events were both extremely large. Thermal stress events often, in magnitude, frequency and spatial distribution, exceed freeze-thaw in terms of number of occurrences. Further, moisture and thermal conditions show that, in cold environments, chemical weathering can occur for long periods – perhaps all winter. Finally, as much as these data help us to go forward in our understanding of weathering, they still need to be *directly* linked to actual breakdown – we cannot simply assume that because freeze-thaw may occur it is (in the absence of *proof*) the cause of the damage we observe. This is the next step – the connectivity of material failure with specific process.

Key words: freeze-thaw weathering; thermal conditions; processes; scale; monitoring; building stone.

1. INTRODUCTION

With respect to freeze-thaw weathering of building stone, the 'freeze-thaw' attribute is usually discussed as if it were a singularity. That is to say, the term is used in a manner that suggests the weathering ('freeze-thaw') comprises but a single mechanism as indicated by the descriptor used. Further, in many studies the action of 'freeze-thaw' appears to be assumed rather than proven – the considered role being based on a climate that experiences "cold" coupled with observed physical weathering of the stone. Equally, in many studies where the building material resides in a zone of cold winter conditions so the monitoring protocols, assuming the activity of freeze-thaw, are set up in such a manner as to monitor just this and, in so doing, are inadequate for determination of other processes. Thus, the "proof" of freeze-thaw becomes one of a self-fulfilling prophecy rather than scientific evaluation. In truth, the proof (as opposed to assumption) of actual freeze-thaw events occurring *within* the stone are almost non-existent and such proof connected to actual weathering outcomes within a field situation even less so (Hall et al., 2002). While some laboratory undertakings have linked weathering with the freeze-thaw events, their veracity is questioned based on the unrealistic (to the real world situation) simulation conditions (Thorn, 1992). Paradoxically, the ability to show freeze-thaw events *did* occur while, at the same time, being able to evaluate other weathering mechanisms comes from attempts to monitor processes *other* than freeze-thaw – and in so doing also indicates the data-bound weakness of most freeze-thaw directed studies.

Perusal of many building stone weathering studies suggests unqualified assumption of both the freeze-thaw mechanism and the nature of the ensuing damage. The examples in which process, especially freeze-thaw is assumed, are too numerous to cite, and it would really be inappropriate as the goal is not criticism of studies but rather to consider the way forward based upon sound principles. To give an example, *without prejudice*, one could note the discussion of Lawrence (2001, p. 26) regarding the deterioration of Canada's Federal Parliament Building, where it is stated that among the many causes of breakdown some "Damage was attributed to .....penetration of water into open joints and subsequent freeze-thaw activity…". The key word in this otherwise excellent review by Lawrence (2001), and this pertains to subsequent discussion, is "attributed"; this attribution, I would argue, is largely a function of the cold climate and thus assumed role of freeze-thaw. Equally, Mitchell, et al. (2000) calculate the number of freeze-thaw cycles affecting Lichfield Cathedral (England) based on recorded temperatures and humidity, but without actual evidence that the water did indeed freeze and, in so doing, effect damage. Apropos later discussions, the study of Mitchell, et al. (2000) recorded temperatures at 15-minute intervals and these were used for evaluation of thermal stresses; it will later be shown that such a record interval is inadequate for any such evaluation except in the broadest terms.

Indeed, from the perspective of the "bigger picture" a major stumbling block to a realistic dealing with practical problems of building stone weathering, is that of our application of discipline-specific knowledge and/or approaches. Sad to say but, despite our accessibility to literature, there is little cross-fertilization between disciplines: engineers continue with their standard techniques (whether or not those have any relationship whatsoever to actual processes), geomorphologists work in the landscape but frequently without recourse to non-landscape studies, building conservation and asphalt road studies stay within their topic-specific literature, and specialists such as rock fracture mechanics scientists stay within their, often theoretical, sphere. Reading of the literature within any discipline is often as if none of the others even exist; sometimes with amazing conceptual outcomes or decisions. Obviously there are exceptions across the board to this: the paper by Whalley and McGreevy (1984) berating geomorphologists for not using the wealth of information from engineering studies being a classic example (however, nearly 20 years on the paper might not have existed!). Pragmatically, this is a serious issue and one that hinders studies and, if I were allowed a personal note, it would be a cry for researchers to look outside of their discipline for techniques, theory, equipment, approaches etc. Failing that, the answer would be to create multi-disciplinary teams to deal with the issues. Here, from my own perspective, I hope to suggest some issues to consider within the context of freeze-thaw affecting building stone. Obviously the discussion cannot be all inclusive, but the ideas and approaches for the findings and data presented were, in the first instance, derived from a wide range of literature. Here, rather than a plethora of data and graphs, many of which may suffer from the very ills noted above, the aim is to consider the underpinning issues, some of the practical problems, and the complexities in dealing with freeze-thaw within building materials – and only to use data to exemplify those issues.

2. FREEZE-THAW PROCESSES

As noted above, it is not a single process that comprises 'freeze-thaw' but rather there are a range of possible mechanisms (Table 1). McGreevy (1981) provides an excellent review of the multiplicity of mechanisms available within the catch-all term of "freeze-thaw" (or any of its many synonyms) and, despite its early date, subsequent to that paper the only significant addition to the mechanical aspect of rock breakdown due to the freezing of water is that of the (theoretical) model by Hallet (1983). Combined, the mechanisms require a range of different moisture conditions, freezing rates, freezing temperatures and rock properties (Table 1). Further, dependent upon mechanism, the spatial location within the material where the process may operate can differ. Thus, it is indeed critical, to both obtain adequate data on the controlling attributes and to determine the actual mechanism(s) operating; noting that there can be temporal and spatial variability in mechanism for any given site (Figure 1). The data shown in Figure 1 are simply surface temperatures recorded on each of the cardinal aspects (on a vertical brick) for one winter's day (March, 2004). Consideration of this one day shows the large spatial variability in respect of temperature and hence possible weathering processes, including freeze-thaw. South experiences very high surface temperatures coupled with large short-term fluctuations while North barely gets above freezing and has many crossings of 0°C during the day (probably *not* effective in terms of freeze-thaw but may influence water availability and/or thermal stresses). This spatial complexity is further exemplified by Figure 2 which shows the surface temperatures of two South-facing bricks but with one vertical and one at 45°. The vertical brick, being almost normal to the low angle winter sun, experiences temperatures up to +20 °C while the 45° brick, although facing the same direction, never gets much warmer than -10 °C (Figure 2); the low air temperatures (-16 °C) during that day clearly influencing the angled brick more. Thus, not only can the vertical brick, given available water, experience chemical weathering during that day but also it would be the only one able to experience freeze-thaw action; any water within the 45° brick remained frozen. While these two figures clearly illustrate the spatial variability of temperatures, there is also a temporal influence insofar as, with the changing angle of the sun through the seasons, the impact of slope angle changes – and hence the weathering potential.

Consideration of Table 1 shows that the two key elements in freeze-thaw are the moisture and temperature conditions (and their spatial and temporal variability); material properties, especially with building stone, may be well known or even able to be controlled. Regarding the two attributes of temperature and rock moisture, thermal conditions have dominated data in most studies while that for moisture are highly limited. However, with regard to the thermal conditions, data of the critical ΔT/Δt (rate of change of temperature) have been inadequate as a result of low-frequency data collection (Hall, 2003). This low-frequency data acquisition, partly as a result of process presumption, has meant that other possible processes could not be evaluated. Available data are actually inadequate to "prove" freeze-thaw, but with the assumption of freeze-thaw activity the data become a self-fulfilling prophecy as to its existence and role (see Hall, 2004 for a discussion).

Table 1. Freeze-thaw mechanisms and their controls

Person(s)	**Date**	**"Process"**	**Moisture**	**Temperature**	**ΔT/Δt**
Bridgeman	1912	Water freezing	>90% sat		none
Powers	1945	Hydrofracture	Low % sat		none
Taber	1950	Capillary action	>90% sat		Slow
Battle	1960	Volumetric	>90% sat	-5 to -10 °C	Fast
Dunn & Hudec	1966	Ordered Water	Any[1]	>-40°C	none
Mellor	1970	Freezing rate	>50% sat		V. fast
Connell & Tombs	1971	Crystal Growth	Low % sat		V. slow
Hallet	1981	Ice lenses	>90% sat	-10 to -15 °C	V. slow
Grawe[2]	1936	Questions	How much?	How cold?	

[1] Rock properties play a larger role as the available water must be in pores <5 microns in size.
[2] Grawe (1936) is included here, but put last, as he questioned the components of the freeze-thaw system from the perspectives of physics and reality rather than introduced a new mechanism (see text).

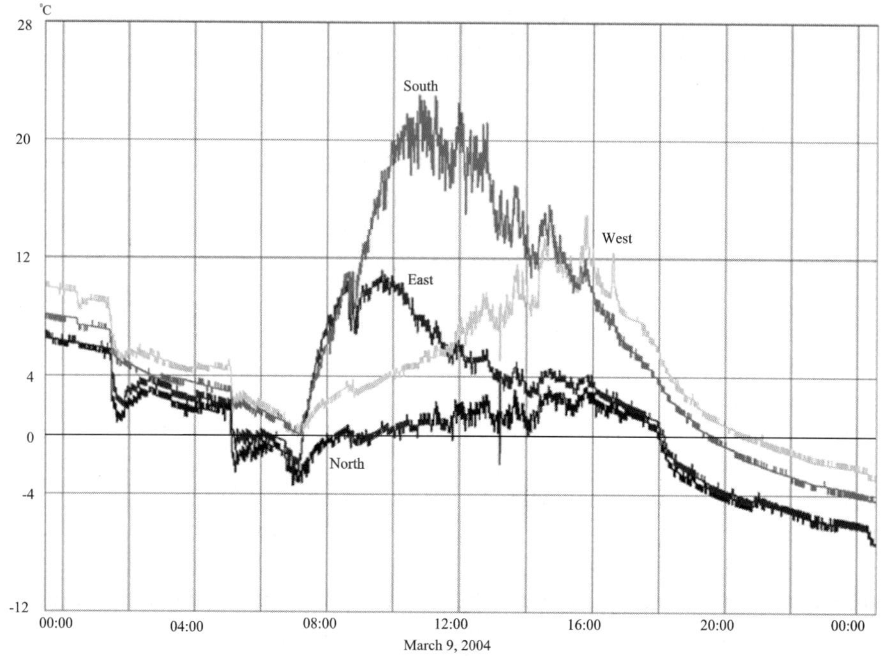

Figure 1. Temperature data for the surface of bricks facing the four cardinal directions to show the large spatial and temporal variability in temperatures, especially for the day-time.

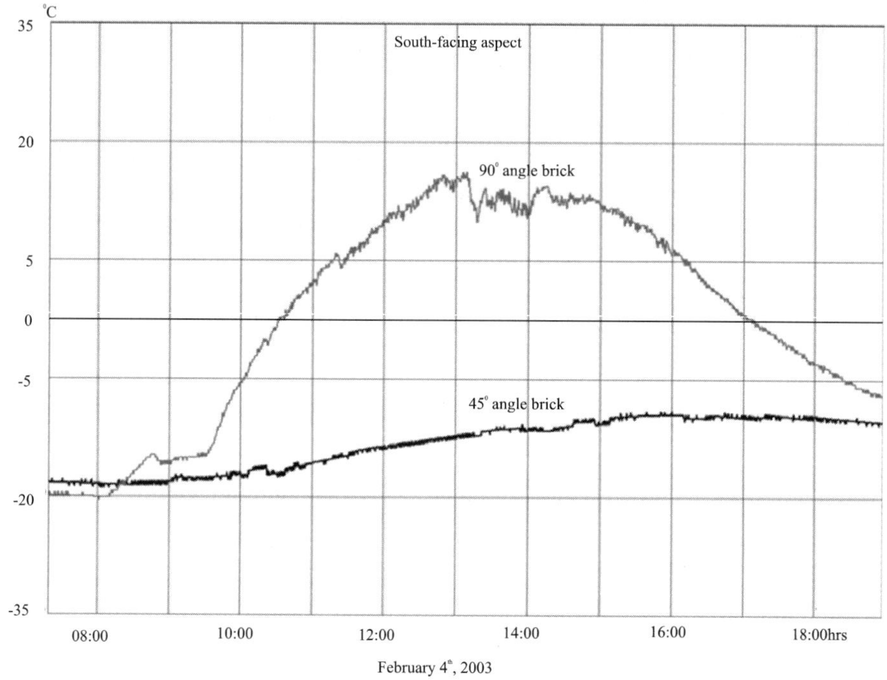

Figure 2. To show the impact of slope angle on building brick temperature in northern Canada.

The paper by Grawe (1936) is cited within Table 1 insofar as it questioned the many held beliefs with regard to freeze-thaw weathering, notably the possible stresses that can be exerted by freezing water, and despite being published over 80 years ago the questions raised are no less valid today. Indeed, the veracity of freeze-thaw might be less strong had greater recognition been given to this paper in many studies. Grawe (1936) raised the questions surrounding the greater-than-any-rock-can-withstand pressures that freezing water can generate, and thus, in principle, the overall possible effectiveness of the process. To generate the high potential pressures, the rock body needs to be totally saturated, have a temperature of -22 °C, and to comprise a completely closed system; of course, to attain the maximum possible pressure the body would also have to be able to withstand that pressure. Thus, as effective as freezing water can be in generating destructive pressures, there are significant constraints and limitations. The pressures are temperature dependent and thus values less than -22°C have commensurately lower possible maximum pressures, and this is lowered more still by less-than-saturated conditions, and, if extrusion can occur because it is not a closed system, so the pressures may then be negligible. Thus, even though the water may indeed freeze, there need not be any weathering taking place. Hence Table 1 shows variations on a theme to try and overcome some of these issues to exploit the destructive potential of freezing water.

Thus we see a range of possible freeze-thaw mechanisms requiring varying degrees of moisture, different rates of change of temperature, and each finding its own way to overcome the necessity for a closed system. Hidden within this are also the attributes of the material itself and, in respect of freeze temperatures, the period to which the material is subject to sub-zero temperatures. In respect to the latter, Battle (1960) showed that, for effective freezes, temperatures had to be between -5 °C and -10 °C, figures not greatly dissimilar to those suggested by Hallet (1983). Along the same lines, studies from France (Lautridou, 1971) have shown that to be effective, freezing temperatures had to be maintained for periods of at least 10 hours. Later studies (Lautridou and Ozouf, 1978) also showed that the rate of change of temperature was important, with rapid freezes negating water loss from the rock; conversely, Hallet (1983) suggested that, for his model, very slow rates of freeze were required to allow continued water migration to the freezing location. At the same time, the properties of the material are also important in influencing thermal changes; for example, permeability as well as the pore sizes and shapes can influence freezing (note Dunn and Hudec, 1966 re pore size and shape). Clearly, rocks with very low porosities and permeability are going to respond to available water, and hence the effects of freezing, in quite a different manner to the more porous and permeable rocks; note Trenhaile and Mercan (1984) and Hall (1986) have shown that even under conditions seemingly conducive to high moisture contents some rocks remain with very low degrees of saturation.

Thus, despite the almost casual application of the freeze-thaw concept, it is actually a very complex mechanism dependent upon an association of critical levels of moisture with specific temperature conditions, all mediated (in part) by the rock properties themselves. Thus, for any evaluation, particularly of damage actually ensuing from a freeze-thaw mechanism knowingly taking place, it is essential to obtain data in respect of thermal and moisture conditions in the material under study. At the same time, to assess the actual role, if any, of freeze-thaw it is imperative to monitor any *other* processes occurring at the same site through the annual cycle. Fortunately, many of the other processes, including chemical weathering, are (at least in part) able to be evaluated by the same combinations of temperature, moisture and rock property data as are required for an effective study of freeze-thaw. Thus, in monitoring adequately for freeze-thaw it is possible to gather a better picture of the weathering synergies and the place of freeze-thaw within this bigger picture; a more scientific approach than the ill-founded attempt to monitor just freeze-thaw.

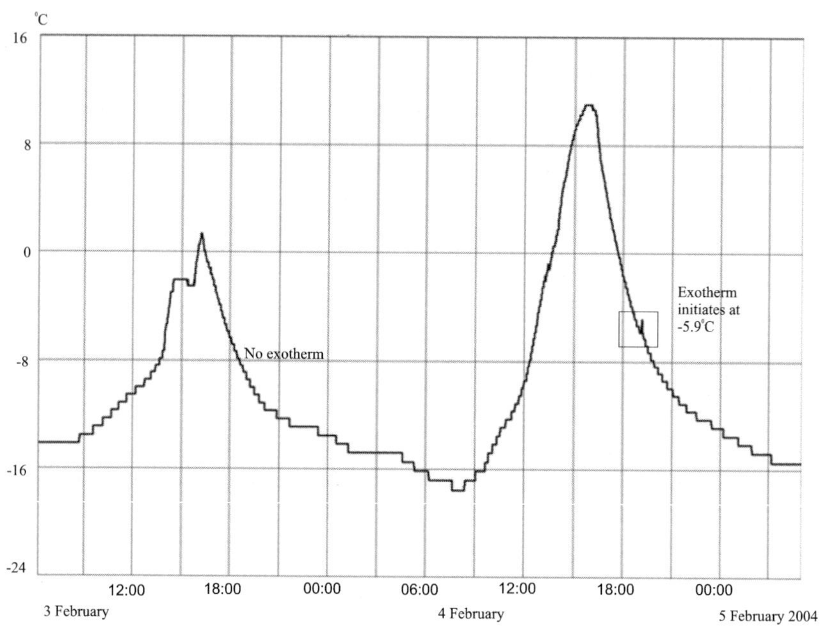

Figure 3. Example of two consecutive days with comparable freeze temperatures but only one shows evidence of water freezing (an exotherm).

3. MONITORING

In most field situations where freeze-thaw has been assumed to be active the weathering protocols were of a nature that, ironically, were unable to validate this process except by recourse to gross assumptions. Paradoxically, it has been the attempt to prove processes *other* than freeze-thaw that have generated data sufficient to validate the freezing and thawing of water within building stone (Hall, 2003, 2004). Freeze-thaw has usually been assumed operative based on geographic location – a site where cold conditions occur

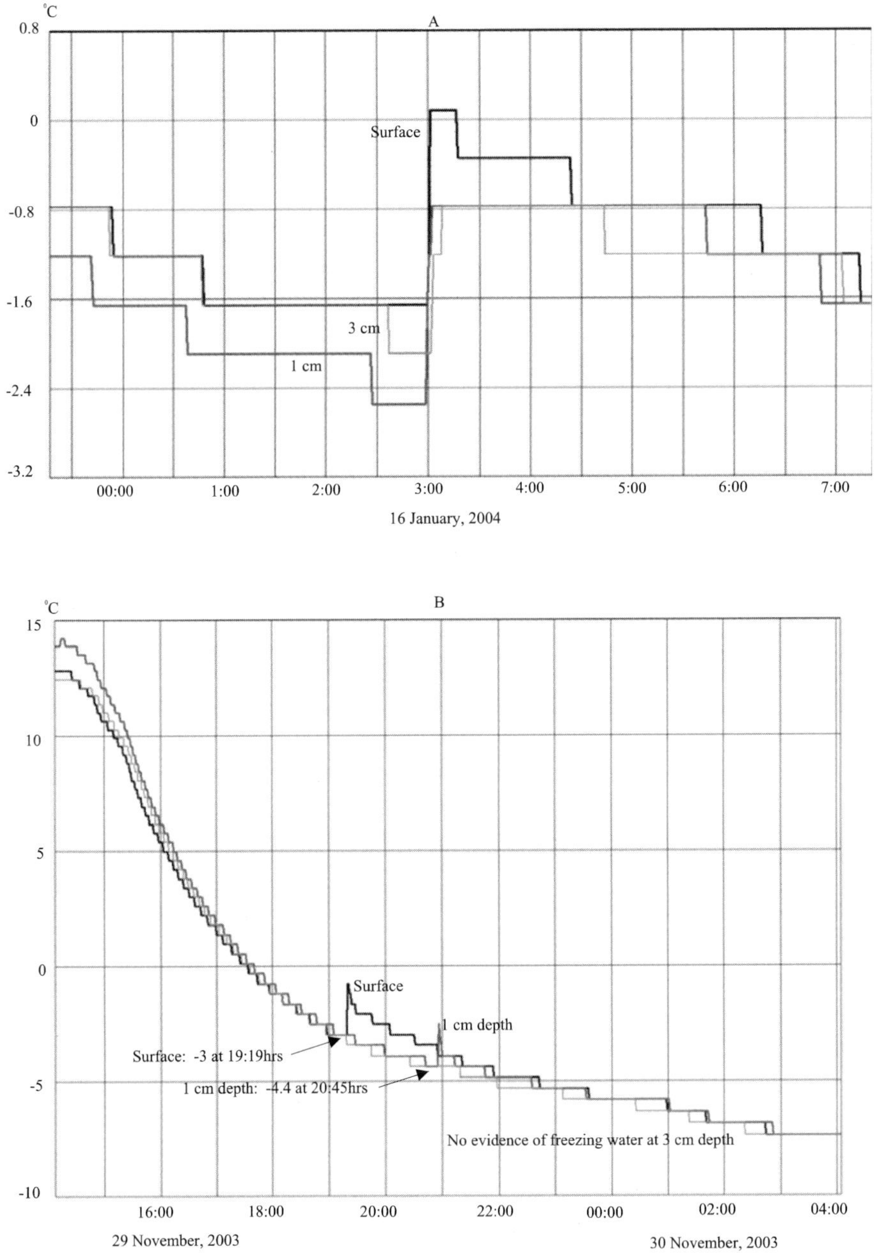

Figure 4. A. Example of all levels (surface, 1cm depth and 3cm depth) freezing simultaneously. B. Example of a progressive freeze down to 1 cm depth in the brick.

during the winter. In such studies, the general approach has been to assume the presence of water within the building material and thus when sub-zero temperature conditions (usually below a threshold of such as -3 °C: Matsuoka, 2001) could be monitored so each such event was thought to generate a freeze cycle. However, as will be discussed below (see Figure 3), where data are available, it has been shown that despite the sub-zero thermal conditions no transformation of water to ice took place; presumably (simply) because no water was present to freeze. Thus, it becomes critical that, in the case of freeze-thaw, monitoring procedures are such that it can be clearly shown that water did indeed freeze.

Thus, in 'monitoring' for freeze-thaw there are a number of attributes that need to be considered. With respect to solely thermal conditions there needs to be recognition of the large spatial and temporal variability (Figures 1-2). For the evaluation of freeze-thaw weathering impact on buildings, so the spatial variability in freeze-thaw operation, including as a function of slope, must be taken into account. To further help evaluate the spatial effectiveness of freeze-thaw so there needs to be monitoring of temperature at different depths within the material (see Figure 4 below). The rate of change of temperature is critical for evaluating the freeze-thaw mechanism itself as well as for studies of thermal stress. In the absence of moisture data, so it is important to have thermal data at a sufficient frequency that exotherms can be resolved (Figures 4-5), thereby providing proxy information of both the presence of water and that it actually froze. As building material albedo can influence thermal conditions (see below and Figure 7) so monitoring should also take this into account where there is variability in the brick and/or stone used. Lastly, as with any study, recognition must be given to processes operating at a range of scales and, where possible, an attempt made to monitor accordingly.

3.1 Monitoring suggestions - The 'holistic' approach

To monitor freeze-thaw weathering is to monitor weathering – not, in reality, to monitor just 'freeze-thaw'. There is no singular attribute that needs to be monitored that is not a requirement of some other process – even the freezing of water indicates when, for example, chemical weathering is *not* taking place. It is quite a different thing to monitor the pressure exerted by the freezing water, but no less so than the pressure exerted, at the same location but during the summer, by hydration of salt. The pressure exerted by the freezing water is an attribute *within* the freeze-thaw mechanism and is not a factor *controlling* whether freeze-thaw takes place or not; it may only control the effectiveness of freeze-thaw as a weathering mechanism, not its occurrence. Thus, to set up an effective monitoring system, the attributes of rock properties, rock moisture and rock temperatures need to be ascertained - and how these change in both space and time. The monitoring protocols need to be able to evaluate all of the weathering mechanisms and thus data acquisition times need to take this into account. In other words, if the data acquisition for monitoring thermal conditions in respect of thermal stress are adequate to evaluate freeze-thaw then they can be used; the corollary (hypothetically) that the basic data for freeze-thaw are inadequate for evaluating thermal stress thus prohibits their use. This becomes quite a complex issue, but then weathering *is* complex!

Problems, practical, aesthetic and technical, can arise in determining quite what to monitor, where and at what rate. This becomes a much bigger issue when scale attributes are taken into account. However, outside of data management, consideration of most attributes at the micro-scale will provide much data also suitable for macro-scale determinations of weathering. Pragmatically, many building stone studies will not be able to monitor at all scales, or with regard to all attributes, or with spatial or temporal variability in an adequate fashion. This is just a reality. However, it becomes less of an issue *if* the inadequacy is recognised and thus weathering evaluations take this into account. That said, most building construction, considering the thickness of the brick or stone, may well be impacted mainly by micro- or meso-scale weathering. Indeed, that very recognition may suggest specific types of measurement that can be requested for future studies or might be able to be simulated in a laboratory situation. The key is to monitor as extensively as possible the moisture and thermal attributes affecting the various building materials and, where possible, to have as much data as possible on those building materials – both the initial nature of the material and its present condition. Then, with an understanding of process and hence what information is needed to truly determine specific process operation (or not), it is possible to evaluate the available data in an objective manner.

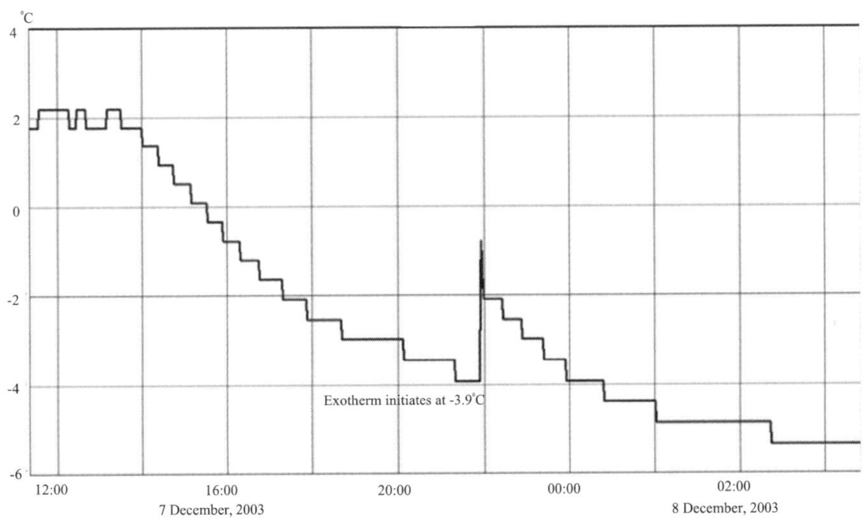

Figure 5. To show an example of an exotherm from the south-facing, vertical brick.

3.2 Monitoring techniques and technology

Recognizing the importance of both scale and aesthetics, a number of recent studies (Hall, 2003, 2004) have used ultra-small, ultra-responsive 40 gauge Type-T thermocouples with a precision of 0.1 °C. With a diameter of less than 0.15 mm the units are almost invisible to the naked eye and so can be used in aesthetically sensitive situations. The small size also allows for implanting of the sensors within materials without any significant damage; this facilitates the opportunity to monitor thermal conditions at different depths. The fast response time (>0.02 secs) also allows for high-frequency data logging sufficient for monitoring of both exotherms and rates of change of temperature that will facilitate meaningful evaluation of thermal stress. High-frequency data from such sensors (Figure 6) are ideal for evaluating processes at the micro-scale (granular disintegration) while the long-term data can still provide information suitable for macro-scale determinations (e.g. seasonal thermal variability). There is no reason why freeze-thaw cannot be happening at the micro-scale ("microgelifraction") so such data may be imperative for an understanding of rates and timings at this level; indeed, much building damage due to frost action may be at this surficial scale.

Where it is not possible to affix transducers to the building material it is still possible to obtain the same high-quality data by use of infrared thermometers. Measuring over an area of ≤0.5 mm with a resolution of 0.1 °C and a response time of 0.05 seconds the resulting data are almost comparable to those from the thermocouples. The distinct advantage is that the infrared systems can be used to monitor such as frescoes or components of a mosaic without any contact with the materials being studied. Such monitoring may be critical for evaluation of thermal differences between components with different albedos. Technological advances are now beginning to bring ever better degrees of resolution to thermal imaging systems that will soon be adequate for thermal monitoring of extensive areas (e.g. a whole wall with multi-point display of chosen areas within that) through time. The trade-off is that infrared monitoring is not so aesthetically acceptable and is, currently, more expensive.

In essence, logger and transducer technology are now such that long-term, high-frequency monitoring with adequate resolution (0.1 to 0.4 °C) and large storage capacity is now readily available at relatively low cost. Some units, such at the ACR TrendReader from Canada, can monitor eight channels at intervals down to 0.01 seconds with 12-bit precision, an internal battery providing 10 years of continuous operation and storage for one million data points – and the unit will fit in a shirt pocket! There are also practical attributes in that the units have a small aesthetic impact and, in many situations, the small size can make for easier security. Thus, outside of the overall costs ramifications, there are few technological reasons to inhibit thermal data collection of a quality commensurate with that required for process evaluation.

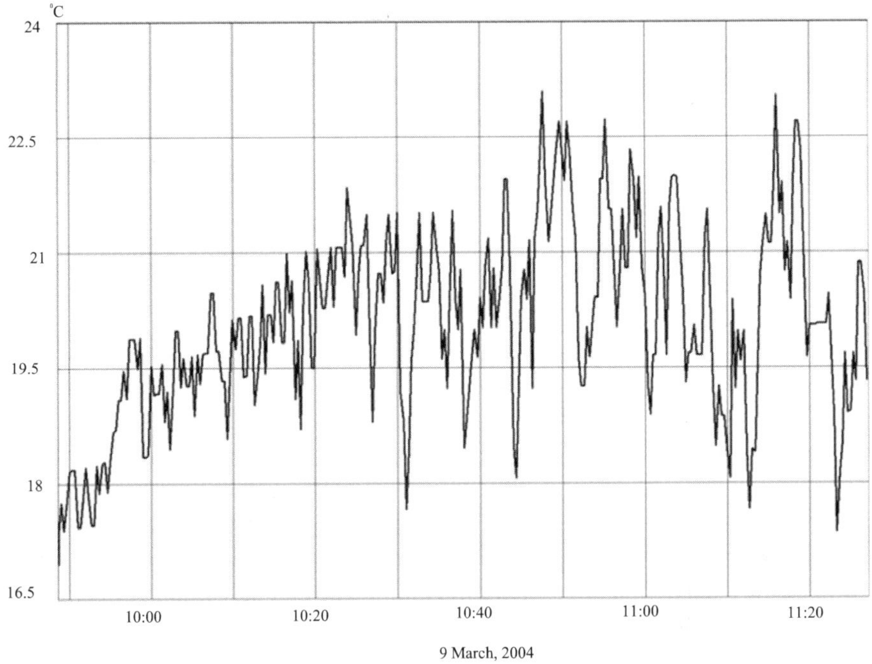

Figure 6. Surface temperatures collected at 20 second intervals on a south facing, vertical brick to show the short term thermal variations.

Data collection regarding rock moisture is substantially behind that of the thermal conditions and almost negligible with respect to such as rock moisture distribution and chemistry. The reality is that monitoring of moisture has been technologically far harder than thermal conditions. That said, this too has benefited from technological changes, and the work of such as Sass (2004, 2005) shows what can be done, particularly within the context of freeze-thaw evaluations. Moisture really is a critical factor and one so often, and in a variety of ways, 'taken for granted' within freeze-thaw studies. Not only is it necessary to understand the temporal and spatial variability of moisture, but it is also required that some understanding of the chemistry be known. The chemistry not only affects the freeze temperature (Hall et al., 1986) but can also be important in respect of other weathering processes occurring when the water is not frozen. Data on rock water chemistry are, however, all but non-existent (Hall et al., 1986; Thorn, 1988). In terms of moisture distribution within the building stone, this is critical for a number of reasons. In respect of freeze-thaw, if the mechanism is controlled (at least in part) by the degree of saturation then this is key to understanding process. While, for instance, a brick may have only 20% saturation when viewed as a whole, it is possible that the outer zone of that brick is 100% saturated while the inner part is dry. Equally, in such a scenario, this would imply the location of weathering by wetting and drying is at the inner boundary of the outer wetted zone – inside the brick. As wetting and drying can change pore properties (Hall and Hall, 1996) so this may be the precursor to freeze-thaw or it may be working synergistically with freeze-thaw.

Thus, for any meaningful evaluation of freeze-thaw it is imperative that data regarding moisture be available. Failing that, then the proxy information provided by such as exotherms (Hall, 2004) becomes all the more critical for without it there can be no way of discerning whether there was any water available to freeze and, if water were present, whether it did indeed freeze (Figure 3). This, in turn, leads to the last attribute – to monitor whether any weathering (i.e. the actual breakdown, *in situ*, of the material) took place as a result of the process occurring. Again, it is often assumed that because freezing of water did occur then weathering resulted. This need not be so. Perhaps the ice simply extruded or there was space available for the volumetric expansion to be accommodated and so no stress was exerted on the rock; even if there was a stress exerted, it may have been far less than the strength of the material. Thus there needs to be connectivity between the weathering process and the effect – an attribute missing in almost every study. Even where blocks are shown to be moved by ice wedges during the freeze event (e.g. Matsuoka, 2001) this need not, as noted above, imply weathering *per se* but rather may simply be movement of the already freed block by the ice (i.e. transport). That said, quite how to monitor, especially at an aesthetically sensitive site, is another question. Strain gauges may help and can be kept small, and non-destructive ultrasonic monitoring may be viable in some situations. As difficult as it may be to undertake monitoring of the linkage between weathering process and their effect, it is crucial for building studies if a realistic appraisal of process spatial and temporal variability is to be obtained.

3.3 Use and Interpretation of the Data

In broad terms, if the protocols outlined above have been implemented, the resulting data are essentially unconstrained by process, but can be used to *deduce* process. The data are holistic and can be used to evaluate thermal stress as well as to discern exotherms; they are also adequate for evaluation of chemical weathering processes. In other words, the data should be viewed from the perspective of "what processes are occurring" rather than "this proves what I thought" – although the latter can clearly be an outcome of the former. Essentially the call is for obtaining data that are not simply a self-fulfilling prophecy because they are inadequate to do otherwise, but rather the data show freeze-thaw occurred because they are able to prove other processes could not have occurred. The reality is that, the most likely outcome is a variety of weathering processes that, in both series and parallel, change through time and as a function of aspect. Thus, when evaluating the data, give due consideration to *all* processes and anticipate some degree of spatial and temporal variability in these. In other words, while freeze-thaw may certainly occur, the questions might be as to its timing, extent, location, and its relationship to the other processes taking place through the year. Where freeze-thaw is clearly an operative process, so the data might be able to suggest the mechanism(s) and this, in turn, may lead to effective remedial or preventative measures.

A brief example of the importance of considering data regarding freeze-thaw in respect of other processes illustrates the points above. In essence, freeze-thaw is a function of thermal cycling that, because the thermal changes cross a threshold that causes available water to freeze, may produce stresses in excess of the strength of the material. Thus, in terms of recognizing the action of freeze-thaw and determining the specific mechanism it is critical to obtain high-frequency thermal data. Thermal stress and thermal shock (Yatsu, 1988) are also the product of thermal cycling, but here the threshold is a function of the rate of change of temperature ($\Delta T/\Delta t$); the rate needing to be $\geq 2°\ min^{-1}$. Thus, both mechanisms require thermal cycling, it is primarily the 'thresholds' that differ. However, and extremely important in the context of building materials in cold environments, it is a rapid *fall* in temperature that is more stressful than a rapid rise in temperature (Mavorelli et al., 1966). Thus, in cold climates where solar radiation warms a wall, possibly transforming water into an unfrozen state and readying it for another freeze-thaw cycle (and providing a short-term opportunity for chemical weathering), so as the heat source is removed the resulting values of $\Delta T/\Delta t$ can be very high due to the cold air; thermal stress/shock can result and be particularly effective as it is a falling temperature $\Delta T/\Delta t$ event. The thermal stress precedes the freezing of water and any subsequent freeze-thaw effect; the scale (depth) of effectiveness depends on the specific thermal conditions and the nature of the building material(s). In essence, the two weathering mechanisms are operating in series and, depending on the situation, the overall weathering impact may be misinterpreted as solely freeze-thaw; equally the thermal stresses may help enhance the freeze-thaw impact and so exacerbate its effectiveness. Thus, the data need to be considered from several perspectives, including in a temporal context, to fully comprehend the weathering environment and process relationships.

4. SCALE

Scale is a significant issue and one often ignored – both in terms of monitoring protocols and with respect to the actual process (Hall, Subm). The considerations with respect to monitoring have been dealt with above – sensors and monitoring frequencies appropriate to the scale under investigation. In respect of the process(es) themselves, the level at which they operate and the temporal variability within this, less empirical recognition has been given; the best has been qualitative recognition that scale differences may exist. This is a very important consideration as, by the time damage (ostensibly due to freeze-thaw) has been identified there may already have been a process - scale change. Equally, if identification of freeze-thaw can be made at the micro- or nano- scale, then damage may be able to be inhibited before if becomes an aesthetic or remedial issue. As recently discussed by Inkpen (2005 p. 129-130), scale need not be an absolute quantity but rather relativistic – varying as a function of the entities being studied. Here, the scale attribute would be that of processes causing breakdown of the building material at the nano- and micro- scales, and the relative and temporal role of freeze-thaw within that.

In terms of mechanical weathering processes, it has been argued (Hall and Andre, 2003; Hall, Subm.) that little or no recognition has been given to either the processes or the measuring of attributes influencing the weathering processes, at the micro-scale, let alone the nano-scale. Although the qualitative judgement regarding the occurrence of "granular disintegration" or "flaking/scaling" may be made, and sometimes attributed to frost action (Matsuoka, 2001; Ballantyne, 2002; Curry and Morris, 2004), data on this are rare (Hall and André, 2003). Thus with the increasing recognition of chemical processes at the nano-level (Butenuth, 2001; Hochella, 2002), coupled with developments in technology that facilitate measurement at smaller and smaller scales (spatial and temporal), so it is really incumbent upon us to start considering the role of mechanical weathering (including freeze-thaw) at the micro-scale; as Van der Giessen and Needleman (2002, p. 141) state, "…fracture spans several length scales from the atomistic to the macroscopic scale". This is important at two levels other than that of simply process recognition. First, because of the ever-progressive damage to buildings that can occur from flaking and granular breakdown itself, and second, because the activity of micro-scale weathering may change material properties in a manner that allows macro-scale weathering processes to occur. The first of these two attributes (flaking/granular disintegration) is a practical (and aesthetic) issue and one that can impact some sites/lithologies in a detrimental manner such that knowledge of the process is imperative in order to effect inhibiting or remedial action. The second may be more significant. By identifying the macro-scale breakdown as due, for instance, to freeze-thaw activity, this may misinterpret the initial causative processes and possibly suggest remedial or inhibiting procedures that could actually exacerbate the true initial weathering process(es). In some low porosity/low permeability materials, initial weathering processes causing changes to those properties may be other than freeze-thaw; the observable breakdown from freeze-thaw is a later stage in the weathering continuum and was only able to occur because those changes had taken place.

In respect of changes to material properties through time, and the resulting process transitions, this may be a key factor not adequately considered in building stone weathering. For example, studies (Hall and Hall, 1996) have shown that wetting and drying can cause changes to the pore size distribution within sandstone as well as granular disintegration itself. In many environments, the number of wet-dry cycles may significantly exceed effective freeze-thaw events. Further, under some conditions the wet-dry weathering effects may be sub-surface in location (Hall and Hall, 1996) such that later freeze-thaw events may be misinterpreted as the actual causative mechanism. These are important considerations. If a better understanding of the processes initiating weathering can be obtained then it may be possible to inhibit the latter, perhaps more detrimental, processes. Equally, because a block of building material, of whatever size, is seen to be "moved" (released) by freeze-thaw this does in no way imply that the freeze-thaw caused any actual weathering *sensu stricto.* The block may already have been weathered 'free' (i.e. all boundaries had been disassociated) by other processes and the block was only *moved* by the effect of ice growth within the building material – thus it was transport not weathering. Further, while the block may have been viewed as the product of macro-scale action by freeze-thaw (macro-gelivation), the actual weathering and disassociation along the boundaries may have been the product of micro-scale processes (micro-gelivation) - the product of which was a block (Hall, Subm). Again, misinterpretation may lead to the wrong remedial methods.

5. A FEW LAST THOUGHTS

In recognizing the bigger picture and the need to consider freeze-thaw within a spectrum of processes that operate within a spatial and temporal framework, a framework that is itself to be considered as a function of scale, so a number of practical problems may arise to confound. Based on recent studies it is germane to indicate a few things for consideration. First, studies (Hall, 2004) have shown that even in 'cold' climates, the rock surfaces on the equator-facing aspect may achieve high (+30 °C) temperatures during the day, despite air temperatures of -20 °C. A significant implication of this is that, given the presence of water, so, within the heated zone, chemical weathering can continue throughout the winter in high latitudes that still receive winter sun. Obviously there is a spatial aspect as well as a scale one (the depth of thaw), but these too will vary through the seasons and as a function of latitude. Nevertheless, the significance is that, processes *other* than mechanical may have to be considered.

In terms of freeze-thaw, studies have shown that the temperature at which freezing begins (Hall, Submitted) can vary significantly (between -1.8 °C and -9°C for the same point in this study) such that any approach that uses counting of thermal events across a given threshold may be in error. Further, by identifying freeze events within building stone by means of exotherms

(Figure 5) it was possible to show the number of times when thermal conditions were suitable but *no* freeze occurred as there simply was no water present to freeze. Thus, the situation is clearly not 'simple'. The available data (Hall, 2004) have shown that sometimes there is a progressive freeze from the rock surface downwards and sometimes freeze occurs simultaneously from the surface to a specific depth (Figure 6); in some cases freeze begins sub-surface first. All of this has significant spatial and temporal variability, not the least as also a function of slope (Hall, 2004). Again, this sort of information may have ramifications for protective/remedial approaches, particularly to large buildings that have variation in building materials and extensive exposures, at varying slopes, on each aspect (e.g. Lawrence, 2001).

Dependent upon the scale under consideration, there can be a significant difference in process perception. For example, at the micro-scale the thermal regime is significantly different to that at the macro-scale and, despite the macro-level operation of freeze-thaw, there may be micro-level granular disintegration resulting from thermal stresses. Data (Hall, 2003; Hall and Andre, 2003) have shown that that the grain scale there can be thermal fluctuations >10 °C in intervals down to 20-seconds. These create thermal stresses at the grain boundaries and/or within the outer shell of the rock to cause fatigue leading to flaking or grain release. Even here, this can be further complicated by albedo effects (Hall et al., 2005) with dark-coloured stones being hotter than the light-coloured; however, under some circumstances the lighter-coloured minerals or building stones that can be hotter than the dark (Figure 7). Thus, again, the simple assumptions we all often apply, without testing, may be false and so generate erroneous data. Finally, while preoccu-

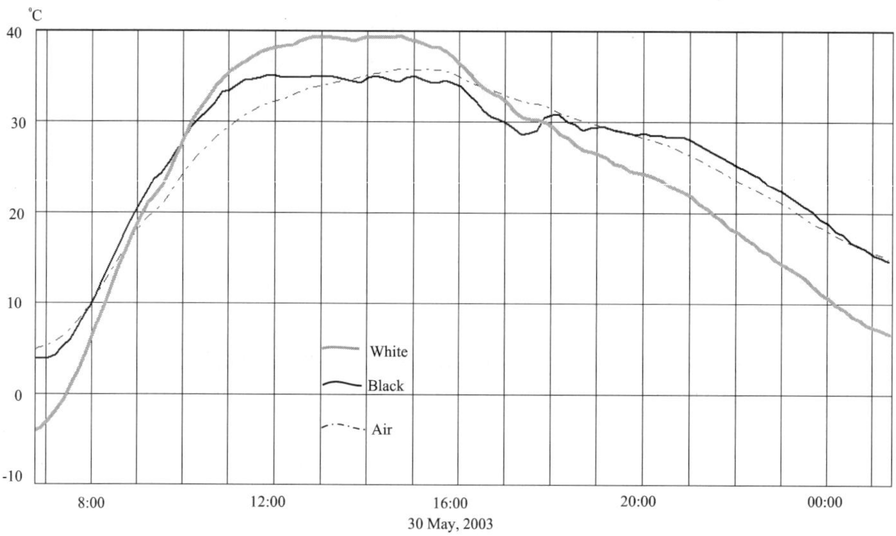

Figure 7. Example of the white brick being hotter than the black brick on a day when solar radiation heated the bricks to temperatures higher than the air.

pation has been with the thermal conditions, the reality is that much more data are required regarding moisture conditions – amount, spatial and temporal distribution, and chemistry. Very few studies have attempted to consider any of these attributes but the recent work of Sass (2004, 2005) suggests that techniques are available and that we should be applying these with the same spatial and temporal variability, and considering data across the same range of scales, as have been applied to the thermal conditions.

6. CONCLUSIONS

It seems, in some ways, that the study of the role of freeze-thaw on building stone is at a 'crossroads', or perhaps a 'roundabout' is a better metaphor. We have theory upon which the mechanisms of freeze-thaw are founded but that theory is, in part, the outcome (or, at least, affirmed by) the available data. At the same time, data acquisition technology is improving and becoming cheaper such that there should now be data better able to test the theory. However, the notion of 'freeze-thaw' is so entrenched that rarely is it felt necessary to obtain truly adequate data. As a corollary, the idea of thermal shock has been so discounted (at least in geomorphological literature) that data to test whether it does actually occur or not are rarely seen as necessary. Thus the ideas regarding what weathering 'is taking place' remain entrenched, affirmed by data that can do no other, and rarely is the weathering of the building stone ever truly researched without prejudice – i.e. in an objective manner. And so the roundabout continues to go round, stopping only to let on those who have bought in to the concept, and thus it never actually gets anywhere (scientifically). To quote the recent discussion by Inkpen (2005, p. 25): "*An important point, however, is how do you select which theories are the most coherent? What criteria do you use and who has the final say? The scientific community seems to develop and agree theories, but this is not necessarily a guide to how true those theories are*". This then extends in to the applied aspects associated with the preservation of building stone where he states (Inkpen, 2005, p. 26): "*The success of a theory is judged on how successful it is in its practical implementation. Success...lies in its successful implementation. Functionalist pragmatism is not concerned with the truth of reality, but with how truths about reality are thought to be identified, with the processes of scientific endorsement. Success is defined by how well those practices provide answers to our particular goals*". Thus, if it is assumed freeze-thaw is the causative mechanism for breakdown and the only available data support this, then it must be so. If, at least in the short-term, remedial or restorative practices work, then this only but affirms that the theory must have been correct. Sadly, however, this means neither do we ever get a proper appraisal of the weathering processes and their effects at a given site – we just think we have. As Inkpen concludes (2005, p. 26): "*...goals and pur-*

poses are important and can be matters of taste, evaluation is not. Evaluation requires a rationale, it requires rules and procedures that permit comparison of 'truths' between individual". We need to 'get off of the roundabout', having enjoyed the ride, with the goal of looking at things anew, of asking ourselves quite what are we measuring and what does it really tell us?

Following the metaphor of the 'crossroads' – we are now in a position to change direction, to truly evaluate processes, their connections and interactions, and their individual and combined effects. Whether we see fit to do this remains the question. Nevertheless, the technology is now available. True, to carry out a comprehensive study is daunting and requires a substantial undertaking of time and effort, but why should we do less? Pragmatically we are not always able to do all that which we may wish as resources are often an issue. However, recognition of the limitations regarding the data we are able to collect are imperative if our deductions are to be meaningful; consider what the data can really tell us, not necessarily what we would like it to say. It may also be beneficial to a better understanding for us all to look outside of our narrow disciplines and consider comparable theory and studies in other fields. And so, with thermal evaluation of weathering and freeze-thaw in particular, as McCarroll (1997, p.1) states: "*We do not improve our theories or models by admiring them, or by proclaiming how well they seem to fit our observations. The only way to improve them, and therefore to make progress, is actively to seek conflict between our models and the real world.*"

ACKNOWLEDGEMENTS

While there are many who have helped influence and develop my interest in weathering, especially freeze-thaw, I would like to recognize the following: Jim McGreevy, Brian Whalley, Colin Thorn, Jean-Pierre Lautridou, David Walton, Marie-Françoise André, Ian Meiklejohn and Jan Boelhouwers. That is not to say that any of them would necessarily endorse my thoughts outlined above!

REFERENCES

Ballantyne, C.K. (2002). Paraglacial geomorphology. *Quaternary Science Reviews*, 21, 1935-2017.

Battle, W.R.B. 1960. Temperature observations in bergschrunds and their relationship to frost shattering, in: *Norwegian Cirque Glaciers*, Lewis, W.V. ed; Royal geographical Society Research Series 4, 83-95.

Bridgeman, P.W. 1912. Water in the liquid and five solid forms, under pressure. *American Academy of Arts and Sciences*, 47, 439-558.

Butenuth, C. 2001. *Strength and Weathering of Rock as Boundary Layer Problems.* Imperial College Press, London, 270pp.

Connell, D. and Tombs, J. 1971. The crystallization pressure of ice - a simple experiment. *Journal of Glaciology*, 10, 312-315.

Curry, A.M. and Morris, C.J. (2004). Lateglacial and Holocene talus slope development and rockwall retreat on Mynydd Du, UK. *Geomorphology*, 58, 85-106.

Dunn, J.R. and Hudec, P.P. 1966. Clay, water and rock soundness. *Ohio Journal of Science*, 66, 153-168.

Grawe, O. R. 1936. Ice as an agent in rock weathering: a discussion. *Journal of Geology*, 44, 173-182.

Hall, K.J. 1986. Rock moisture content in the field and the laboratory and its relationship to mechanical weathering studies. *Earth Surface Processes and Landforms*, 11, 131-142.

Hall, K. (2003). Micro-transducers and high-frequency rock temperature data: changing our perspectives on rock weathering in cold regions, in: *Permafrost*, M. Phillips, S.M. Springer and L.U Arenson eds; Balkema, Lisse, Vol. 1, 349-354.

Hall, K. (2004). Evidence for freeze-thaw events and their implications for rock weathering in northern Canada. *Earth Surface Processes and Landforms*, 29, 43-57.

Hall, K. Subm. Perceptions of rock weathering: a discussion on attributes of scale. *Geomorphology*.

Hall, K. and André, M-F. (2003). Rock thermal data at the grain scale: Applicability to granular disintegration in cold environments. *Earth Surface Processes and Landforms,* 28, 823-836.

Hall, K. and Hall, A. 1996. Weathering by wetting and drying: Some experimental results. *Earth Surface Processes and Landforms*, 21, 365-376.

Hall,K., Verbeek, A., & Meiklejohn, K. 1986. The extraction and analysis of solutes from rock samples and their implication for weathering studies: an example from the maritime Antarctic. *British Antarctic Survey Bulletin*, 70, 79-84.

Hall, K., Thorn, C., Matsuoka, N. and Prick, A. (2002). Weathering in cold regions: Some thoughts and perspectives. *Progress in Physical Geography*, 4, 576-602.

Hall, K., Lindgren, B.Staffan, and Jackson, P. 2005. Rock albedo and monitoring of thermal conditions in respect of weathering: Some expected and some unexpected results. *Earth Surface Processes and Landforms,* 30, 801-811.

Hallet, B. 1983. The breakdown of rock due to freezing: a theoretical model. *Proceedings of the 4th International Conference on Permafrost*. National Academy Press, Washington D.C., 433-438.

Hochella, M.F. 2002. There's plenty of room at the bottom: Nanoscience in geochemistry. *Geochimica et Cosmochimica Acta*, 66, 735-743.

Inkpen, R. 2005. *Science, Philosophy and Physical Geography*. Routledge, Oxford, 164pp.

Lautridou, J-P. 1071. Conclusions generales des experiences de gelifraction experimentale. Recherches de gelifraction experimentale du Centre de Geomorphologie, V. *CNRS, Centre de Geomorphologie de Caen Bulletin*, 10, 84pp.

Lautridou, J-P. and Ozouf, J-C. 1982. Experimental frost shattering: 15 years of research at the Centre de Geomorphologie du CNRS. *Progress in Physical Geography*, 6, 215-232.

Lawrence, D.E. 2001. Building stones of Canada's Federal Parliament Buildings. *Geoscience Canada*, 28, 13-30.

Marovelli, R.L., Chen, T.S., and Veith, K.F. (1966). Thermal fragmentation of rock. *American Institute of Mining, Metallurgical and Petroleum Engineers*, 235, 1-15.

Matsuoka, N. (2001). Microgelivation versus macrogelivation: Towards bridging the gap between laboratory and field frost weathering. *Permafrost and Periglacial Processes*, 12, 299-313.

McCarroll, D. (1997). 'Really critical' geomorphology. *Earth Surface Processes and Landforms*, 22, 1-2.

McGreevy, J.P. 1981. Some perspectives on frost shattering. *Progress in Physical Geography*, 5, 56-75.

McGreevy, J.P. and Whalley, W.B. 1984. Weathering. *Progress in Physical Geography,* 8, 543-569.

Mellor, M. 1970. Phase composition of pore water in cold rocks. *US Army Core of Engineers, Cold Regions Research and Engineering Laboratory, Research Report,* 292, 61pp.

Mitchell, D.J., Halsey, D.P., Macnaughton, K. and Searle, D.E. 2000. The influence of building orientation on climate weathering cycles in Staffordshire, UK, in: *9th International Congress on Deterioration and Conservation of Stone, Proceedings Volume 1,* Fassina, V. ed; Elsevier, Amsterdam, 357-365.

Powers, T. C. 1945. A working hypothesis for further studies of frost resistance of concrete. *Journal of the American Concrete Institute*, 16, 245-272.

Sass, O. (2004). Rock moisture fluctuations during freeze-thaw cycles: preliminary results from electrical resistivity measurements. *Polar Geography*, 28, 13-31.

Sass, O. (2005). Rock moisture measurements: techniques, results, and implications for weathering. *Earth Surface Processes and Landforms*, 30, 359-374.

Taber, S. 1950. Intensive frost action along lake shores. *American Journal of Science*, 248, 784-793.

Thorn, C.E. (1988). Nivation: a geomorphic chimera, in: *Advances in Periglacial Geomorphology*, M. J. Clark ed; Wiley, Chichester, 3-31.

Thorn, C.E. (1992). Periglacial geomorphology: What, Where, When? in: *Periglacial Geomorphology,* J. C. Dixon and A. D. Abrahams eds;Wiley, Chichester, 1-30.

Trenhaile, A. S. and Mercan, D. W. 1984. Frost weathering and saturation of coastal rocks. *Earth Surface Processes and Landforms*, 9, 321-331.

Van der Giessen, E. and Needleman, A. 2002. Micromechanics simulations of fracture. *Annual Review of Material Research*, 32, 141-162.

Yatsu, E. (1988). *The Nature of Weathering: An Introduction.* Tokyo, Sozosha, 624pp.

Chapter 7: Thermal Stresses

Chapter 7.1

THE EFFECT OF THERMAL STRESSES ON THE MECHANICAL BEHAVIOUR OF NATURAL BUILDING STONES

Bellopede R.[1], Ferrero A.M.[2], Manfredotti L.[1], Marini P.[1], Migliazza M.[2]
[1] *Politecnico di Torino, DITAG, Corso Duca degli Abruzzi 24, Torino, Italy;* [2] *University of Parma DICATeA, Parco area delle scienze 181/A Parma, Italy*

Abstract: The physical and mechanical properties of natural stones used in buildings and monuments are affected by degradation processes due to weathering agents such as thermal actions. The effect of this degradation on rock results in increasing microcrack density and, consequently, in different strength and deformability features of the rock materials. In order to investigate the influence of different crack densities in calcitic stones several experimental and theoretical studies have been carried out. Both destructive and non destructive test procedures have been adopted to evaluate the capability of different experimental techniques to estimate the rock material integrity. On the basis of experimental data and subsequently of their mathematical simulations through the analytical models this paper proposes an experimental methodology which, by means of non destructive measurements, would give a theoretical prediction of the rock material deformability in relation to the crack density.

Key words: rock degradation and weathering; analytical models; destructive and non destructive laboratory tests.

1. INTRODUCTION

Stone is universally known as one of the hardest material, because it bears the highest strength during its life, but in reality there are many factors that can cause a change of the initial properties of a stone work and a loss of its mechanical characteristics. The aim of the research presented in this work is to study the decohesion of the grains, that is the more frequent consequence of the stone decay, by means of laboratory test and theoretical analysis. In fact, the final condition of a stone subjected to natural or induced ageing - for instance frost action, thermal shocks, loading stress, etc. - is the loss of

S.K. Kourkoulis (ed.), Fracture and Failure of Natural Building Stones, 397–425.

the contacts between the grains with a consequent increase in porosity/water absorption. This fact is more evident in calcareous stones, namely limestones and marbles, because they are more sensitive to weathering than silicatic stones and, moreover, they are widely used all over the world as ornamental stones thanks to their best workability and their finest aesthetical properties.

In order to reproduce natural ageing conditions of the rock, specific laboratory treatments and tests can be performed. In this research, thermal action has been reproduced in laboratory by submitting stone specimens to thermal cycles or to high temperature in a control system.

In this way, specimens with different weathering degrees have been obtained and the variations of the physical and the mechanical properties of the rock materials, due to the increasing weathering, have been compared. This comparison has been done by performing tests in order to determine both the physical (e.g. water absorption) and the mechanical properties (e.g. uniaxial compressive strength) through both destructive and non-destructive methods, like Ultrasound Pulse Velocity measurements on rock specimens with different crack density.

The degree and nature of fracturing have been quantified by both direct observations at the microscope and micro-hardness tests for the evaluation of punctual rock hardness and, in an indirect way, by the determination of water absorption on rock specimens at different degrees of weathering.

Experimental works[1,2] have confirmed that a larger damaged area in the rock material (generated in both cases by heating) determines variations of rock properties, such as the permeability and the ultrasonic wave velocity, strictly connected to the degree of fracturing of the rock material, that is to say that characteristics such as compression and tensile strength and rock deformability are influenced by the presence of cracks[3].

The behaviour of rock material with different crack densities can also be evaluated by means of theoretical studies[3-5]. The results of numerical simulations confirm that the rock behaviour is mainly ruled by the number of cracks within a specific rock volume (crack density).

Analytical solutions proposed by S. Nemat-Nasser and M. Hori in 1999[6] have been applied in this paper in order to better understand the deformability variation of the rock material at different crack densities.

2. EXPERIMENTAL WORKS

2.1 Tested materials

The four rocks (Figure 1) used for the tests have been chosen among those rock materials commonly used as ornamental stones in Italian buildings and monuments. Their petrographic features are:

- White-pink limestone (Figure 1a) exploited in Sicily. Its mineralogical composition is mainly calcite with a small percentage of dolomite. It is a fossiliferous limestone made of micrite (60%), fossil fragments (30%), recrystallised grains with average dimension of 0.8 mm (5%) and pellets with 0.5 mm diameter (5%). In the following sections it will be quoted as white limestone.
- Black marble (Figure 1b), is a fine grained marble coming from South-Piedmont. The analysis of the thin section shows that the rock is made by calcite (95%) and quartz (5%). It shows a preferred orientation of the long axis of grains and a layering of fine grained levels (average grain size of 0.4 mm) and medium grained levels (0.8 mm). The crystal boundary is irregular and most of them show oriented twinning. All thin sections show a diffused carbonaceous matter. Quartz is in 0.1 mm diameter rounded grains. In the following sections it will be quoted as Black marble.
- Venato marble (Figure 1c), exploited in Carrara Apuan Alps, is made of calcite (99%) and dolomite in traces, with homogeneous grain size (0.06 -0.4 mm), subeuhedral habit of crystals, often with straight boundaries, cleavages and twinning. The grey veins in the white-grey background are due to the presence of opaque minerals and minute inclusions concentrated in levels. In the following sections it will be quoted as Venato marble.
- Ordinary white marble (Figure 1d) exploited in the Monte Altissimo basin of Apuan Alps is characterized by a quite pure white colour. It is made of

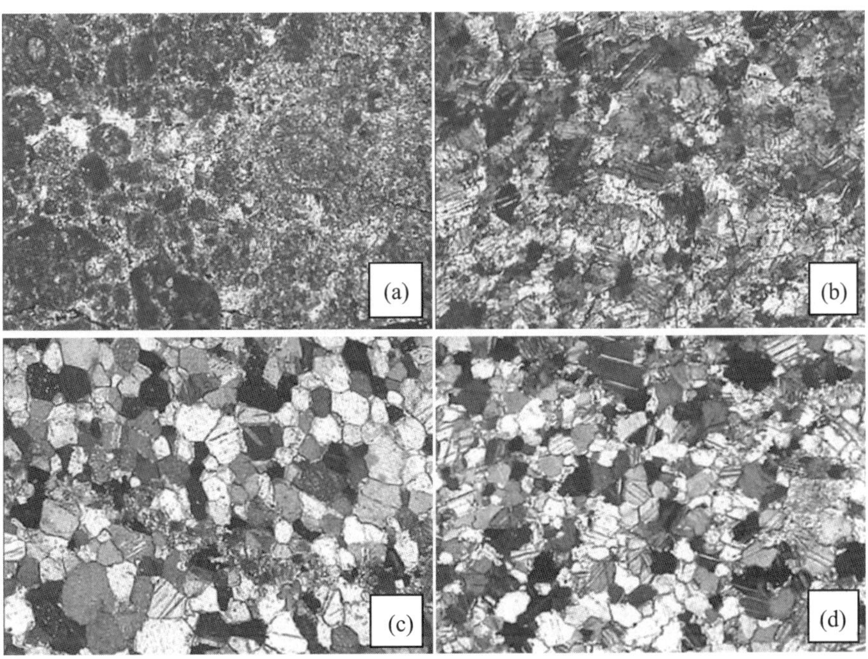

Figure 1. Thin sections of: (a) White limestone, (b) Black marble, (c) Venato marble, (d) Ordinary marble. The real dimensions of the microphotographs are 1.8 x 2.4 mm.

calcite 95% and dolomite 5%. The crystals show anhedral habitus and a slight crystal shape orientation. The grain size is quite homogenous (0.01 - 0.2 mm). In the following sections it will be quoted as Ordinary marble.

2.2 Thermal treatments

The thermal cracking of calcareous rocks, as marble and limestone are, is caused by the internal stresses arising from the anisotropy of the thermal expansion coefficient of the calcite crystals[7]. Two kinds of thermal treatments have been carried out, connected with two different dedicated researches.

White limestone and Black marble - 60 cubic specimens, of 70 mm edge each, have been heated in an oven, at a rate of 170 °C/h to reach the temperature of 400, 470, 600 °C for White limestone and 230, 400, 500 °C for Black marble. Temperatures reached by Black marble are lower, because this stone is more sensitive to heating (Figure 2 and Table 1). The maximum temperature was monitored for one hour and the subsequent cooling followed different controlled rates. Sets formed by at least five specimens that have undergone each treatment in order to obtain better statistical evaluation. Moreover, a set of 5 specimens for each stone has been kept in natural conditions.

Venato marble and Ordinary marble - 55 specimens 50x50x100 mm (30 of Venato marble and 25 of Ordinary marble) have undergone freeze-thaw cycles following the methodology prescribed for McDUR Project[8]. The saturated specimens were frozen for 4 hours at T=(-10±2) °C and then thawed in water at T=(20±5) °C for 4 hours. 120 freeze-thaw cycles divided in six steps have been performed. At the beginning of the treatment and after each step, 5 specimens have been carried out and tested.

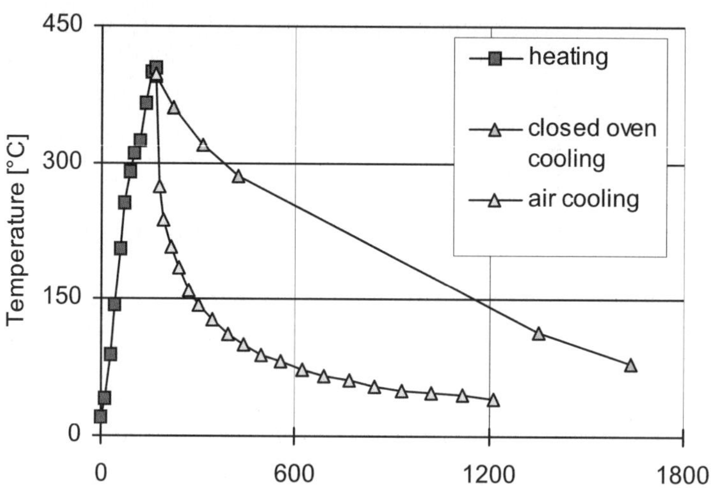

Figure 2. Example of heating – cooling curves referred to the thermal treatment performed on White limestone and Black marble.

Table 1. Thermal treatments on White limestone and Black marble.

	WHITE LIMESTONE		BLACK MARBLE	
Set	Temperature heating [°C]	Cooling procedure	Temperature heating [°C]	Cooling procedure
1	Not treated			
2	400	Closed oven	230	Closed oven
3	470	Closed oven	400	Closed oven
4	470	Half-close oven	500	Closed oven
5	470	air	--	--
6	630	air	--	--

2.3 Laboratory tests

The effect of the weathering processes and their influence on stone properties have been evaluated by measuring physical, mechanical and fracturing characteristics of the specimens at different accelerated ageing steps. After each heating - cooling phase or freeze-thaw step, the specimens have undergone laboratory destructive or non-destructive tests, in order to evaluate the variations of their physical, mechanical and fracturing features.

Physical (water absorption, apparent and real density) and mechanical (uniaxial compressive strength, Young's modulus, tensile strength) characteristics of the rocks have been determined according to the European Standards for natural stones (CEN TC 246)[9].

The degree of fracturation has been evaluated measuring the variations of water absorption (WA) values and ultrasound pulse velocities (UPV), and by means of microscopical observations on polished sections.

2.3.1 Water absorption

Water absorption at atmospheric pressure represents the quantity of water (in mass) that can seep into the open pores of the stone surface, as percentage of the dried specimen mass. This measurement can be considered a good index of the weathering degree of the stone, but it is important to know the WA value of the sound stone. This test method is non destructive and can be performed on the same specimens, before and after the artificial ageing. According to EN 13755[10] the mass of the water that saturates the pores of the specimens (see Eq. 1) is calculated as the difference between the mass of the specimen saturated to constant mass (m_s) and the mass of the specimen dried in oven at a temperature of 70°C until constant mass (m_d):

$$WA = \frac{m_s - m_d}{m_d} \times 100 \qquad (1)$$

Water absorption coefficient refers to open porosity (p_0) in the sample: in fact, the first one indicates the water mass percentage needed to saturate the open pores, while the second one indicates the volume percentage of open pores. In other words, p_0 can be considered to be the product of WA multiplied by the apparent density (ρ_b) of the specimen:

$$p_o = WA\rho_b \tag{2}$$

The average values and the standard deviation of the water absorption measurements referred to Black marble and White limestone specimens submitted to different heating-cooling phase are reported in Table 2, while the variations of water absorption values measured after each freeze-thaw step (referred to Venato marble and Ordinary marble) are reported in Table 3.

The linear regression reported in Figure 3, representing WA vs. the ageing step, shows a good agreement; the correlation coefficient (R) is close to one.

Table 2. WA measured after each heating-cooling phase on White limestone and Black marble specimens.

Litotype	Temp (°C)	WA (%)	
		Average	Stand. dev.
White limestone	Natural conditions	0.04	0.017
	400	0.53	0.118
	470	0.66	0.133
	630	0.93	0.023
Black marble	Natural conditions	0.05	0.01
	230	0.27	0.04
	400	0.36	0.07
	500	0.83	0.21

Table 3. Venato and Ordinary marble WA variations in percentage after each freeze-thaw step of 20 cycles in comparison with the fresh material.

Freeze-thaw steps	Ordinary marble	Venato marble
T_{20}	+8%	+20%
T_{40}	+15%	+30%
T_{60}	+23%	+50%
T_{80}	+31%	+70%
T_{100}	+46%	+80%
T_{120}	+46%	+100%

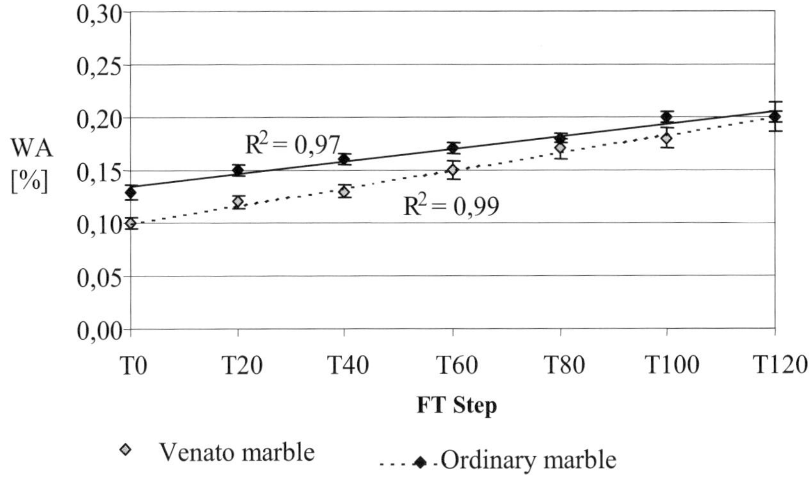

Figure 3. Trend of the linear regression of the WA values with increasing the ageing steps.

2.3.2 Ultrasonic pulse velocity

UPV tests have been performed on rock specimens artificially weathered by different thermal treatments on both dry and saturated rock materials by means of a portable apparatus (PUNDIT) connected with an oscilloscope laptop. Depending on the different arrangements of transducers on the samples surfaces, it is possible to distinguish three different test methods: direct, indirect and semi-direct method (Figure 4). Flat transducers with the frequency of 54 kHz for direct method and exponential transducers of 33 kHz for indirect method have been used in these researches.

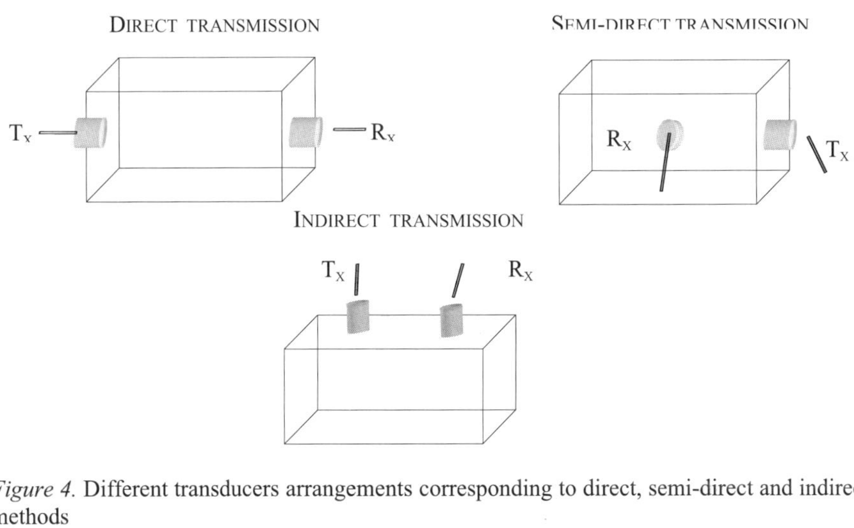

Figure 4. Different transducers arrangements corresponding to direct, semi-direct and indirect methods

The ultrasonic tests have been executed according to EN 14579:2004[11].

The UPV of the rock sample measured by means of the direct method is calculated as the ratio between the transducers distance (d) and the travelling time of the wave (t) from the transmitting to the receiving transducer (Eq.3).

$$v = \frac{d}{t} \qquad (3)$$

where v is the velocity of the ultrasonic wave inside the rock in m/s, d is the distance between the transducers measured in mm and t is the first arrival time measured in μs.

The accuracy of UPV measurements by the direct method has been studied and it was found that it depends by the dimensions and the resistance to ageing of the specimen tested[11].

In the measurements performed with the indirect method, the UPV is calculated by a linear interpolation of the times of first arrivals and the progressive distances between the transmitting transducer and the receiving one. In Figure 5 an example of linear regression used to calculate the ultrasonic pulse velocity of a marble is shown.

Table 4 shows the ultrasonic velocity percentage variation of saturated specimens in comparison with the velocities of dried specimens (ΔUPV), Eq.(4). The velocities were measured by means of direct UPV method on both White limestone and Black marble samples. The specimens have been saturated in water for 48 hours.

$$\Delta UPV = \frac{UPV_s - UPV_d}{UPV_d} \cdot 100 \qquad (4)$$

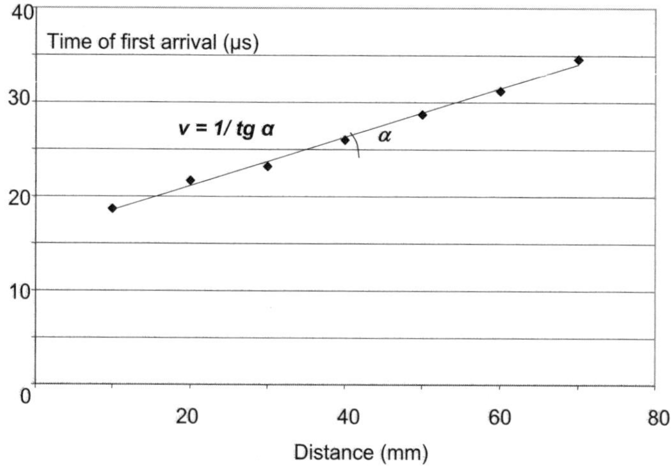

Figure 5. UPV by indirect method: graph time/distances and linear regression.

The symbols used in Eq.(4) denote:

- *ΔUPV:* UPV percentage variation of saturated specimens in comparison with dried ones;
- UPV_S: ultrasonic pulse velocities measured in saturated samples, in m/s;
- UPVD: ultrasonic pulse velocities measured in dried samples, in m/s.

The plot of UPV versus WA (Figure 6) shows that ultrasonic wave velocity measured on dry specimens is very sensitive to micro-cracking.

The average UPV variations, as a percentage of sound stone value, are shown in Table 5. Data refer to measurements carried out by the indirect method on Ordinary and Venato marble specimens having different degree of fracturation caused by a progressive number of freeze-thaw cycles (steps of 20 cycles). A comparison between UPV values obtained with direct and indirect method on Ordinary and Venato marbles is shown in Figure 7.

Data collected on Venato marble and other stones tested during McDUR European Project allowed identifying a correlation (Eq.5) between the direct and the indirect UPV test method:

$$v_{dir}=1.7(\pm 0.2)v_{ind} \qquad (5)$$

Table 4. Percentage increase of UPV for saturated specimens in comparison with the dried ones, measured using direct method on White limestone and Black marble.

White limestone		Black marble	
Temp [°C]	ΔUPV [%]	Temp [°C]	ΔUPV [%]
Nat. conditions	~ 0	Nat. conditions	0.3
400	4.0	230	11.1
470	20.1	400	21.5
		500	75.4

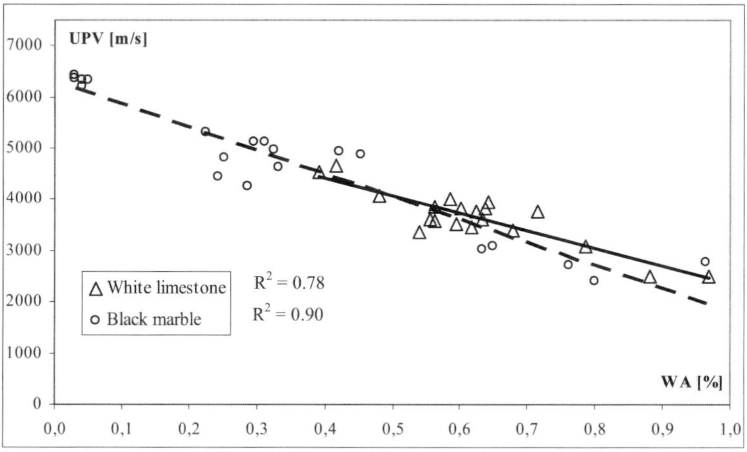

Figure 6. UPV vs. WA for White limestone and Black marble.

Table 5. Average UPV percentage variations after each freeze-thaw step (each step is made of 20 cycles) by indirect method for Venato and Ordinary marbles.

Freeze-thaw steps	Venato marble	Ordinary marble
T_{20}	-28%	-5%
T_{40}	-29%	-11%
T_{60}	-30%	-13%
T_{80}	-26%	-21%
T_{100}	-37%	-42%
T_{120}	-51%	-39%

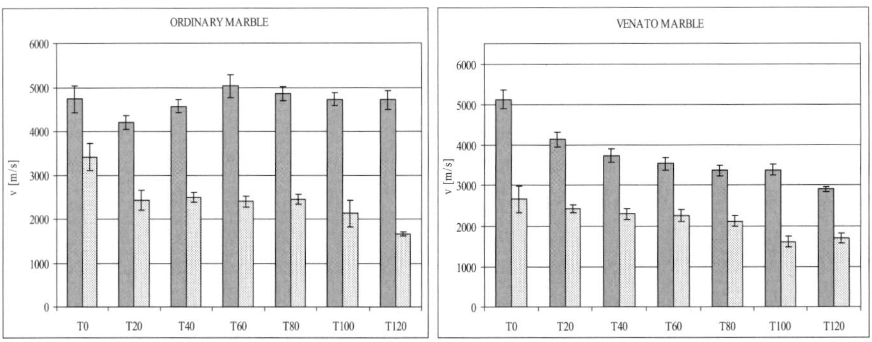

direct method indirect method

Figure 7. Ultrasonic velocities measured both with direct method and indirect method at the increasing of the accelerated ageing for the Ordinary and Venato marbles.

According to the results obtained, Eq.(5) can be used for stone materials with mesoporosity percentage on total open porosity lower than 85%[13].

In the case of the Ordinary marble only the indirect method seems to recognise the ageing. The trend of the results has been explained[12] as a consequence of the too high ultrasonic pulse velocity of the sound stone compared with the too little dimension of the specimens. Moreover, this variety of Carrara marble shows a very good resistance to weathering and, as a conesquence, the weathered layer produced is too thin to cause a decrease of ultrasonic pulse velocity measured with direct method.

2.3.3 Microscopically observations and linear expansion

Polished sections of White limestone and Black marble have been thermally treated together with the cubic specimens used for the mechanical tests and have undergone the microscopically analysis by means of a polarising optical microscope with a maximum magnification of 500X, to quantify the thermal expansion and to evaluate the degree of fracturing.

Micro-hardness has been determined by Knoop microdurimeter, under a 1.96 N load, on polished sections, according to the EN 14205[14].

The behaviour of micro hardness values on the sections treated at different temperatures shows an evident decrease of hardness due to thermal treatment (Figure 8). The Knoop marks were impressed on the polished surface of Black marble specimen through vertical marks before heating and horizontal marks after heating at 600 °C. The different dimensions of the marks, indicating the change in the surface cohesion of the stone due to the thermal treatment are shown in Figure 9.

Permanent linear expansion has been determined on the probes by directly measuring the sides, to within 0.01 mm; at a micro-scale permanent expansion has been determined on polished sections, on which a 1000 μm spacing grid was previously carved (on 7 mm x 7 mm square area), by reading the spacing changes at the microscope micrometer, to within 0.001 mm (Figure 9,10)[15].

The measurement of the permanent expansion on the polished and engraved specimens of Black marble (Table 6) shows that, at millimetric and submillimetric scale, the permanent expansion is not uniform. Local preferred expansion directions are observed (the average effect on the 70 mm side cubes

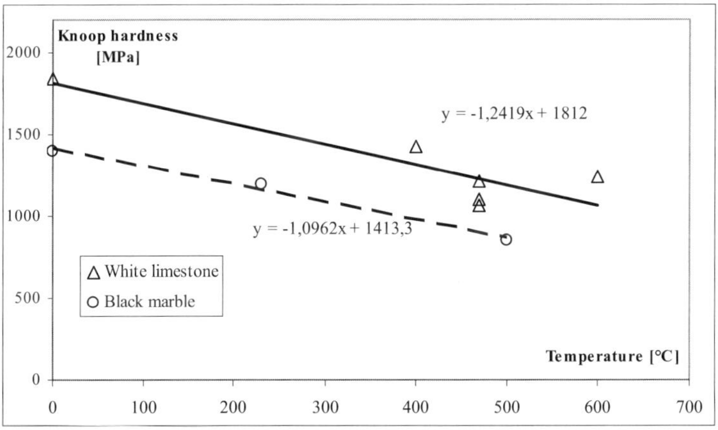

Figure 8. White limestone and Black marble micro hardness variation to temperature variation.

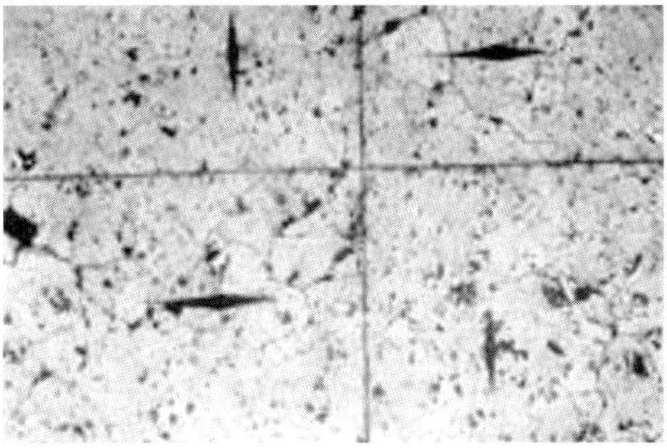

Figure 9. Knoop marks on the polished surface of Black marble; vertical marks were impressed before heating, horizontal marks after heating at 600 °C.

does not reveal such local features). Measured values at 230 °C are roughly twice the measured value for the larger cube, which could be simply a consequence of the local inhomogeneity of the stone. But, when the same specimen undergoes further heating and cooling cycles (at 400 °C) a summation effect of the permanent expansion becomes evident (the cumulated effect of a 230 °C cycle and of a 400 °C cycle produces an expansion of 1.4÷1.6%). The permanent expansion at a very small scale seems to be more uniform for White limestone than for Black marble and not so different from the average, large scale value measured on the 70 mm cubes, which indicates a more homogeneous material - at a small scale.

On the engraved grid the number of fracture along one line has been computed. The fracture spacing distribution is reported in Figure 11. It is a negative exponential for both materials at different crack densities.

Table 6. Permanent expansion at the micro-scale and at the specimen's scale of Black marble and White limestone at different heating temperatures.

	Black marble		White limestone
Micro-scale permanent expansion (%)	**230 °C**	**400 °C**	**600 °C**
direction X mean	0.18	0.23	0.05
direction X range	0.2-0.4	0.2-0.4	0.68-0.77
direction Y mean	0.368	0.532	0.042
direction Y range	0.5-0.6	0.7-0.8	0.57-0.61
Permanent linear expansion (%)	0.15	0.25	1.09

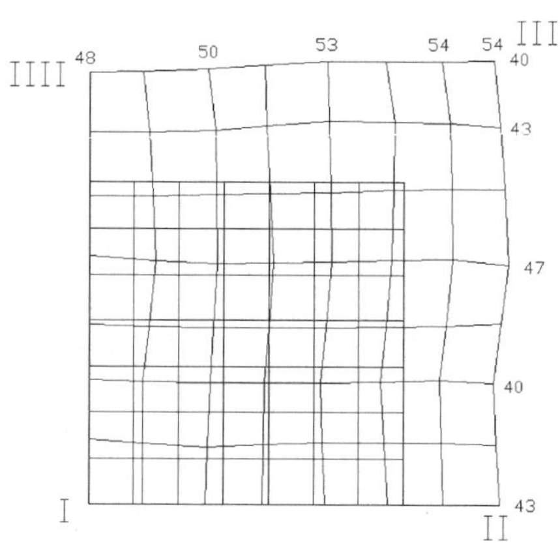

Figure 10. Interrupted line: 1000 micro meter spacing grid engraved on a polished section of the white limestone (on 7 x 7 mm square area). Continuous line: deformed grid upon heating at 600°C. Expansions are exaggerated and sides I-IIII and I-II supposed to remain straight. Numbers refer to the overall expansion, in μm, of the base lines (initial length 7000 μm).

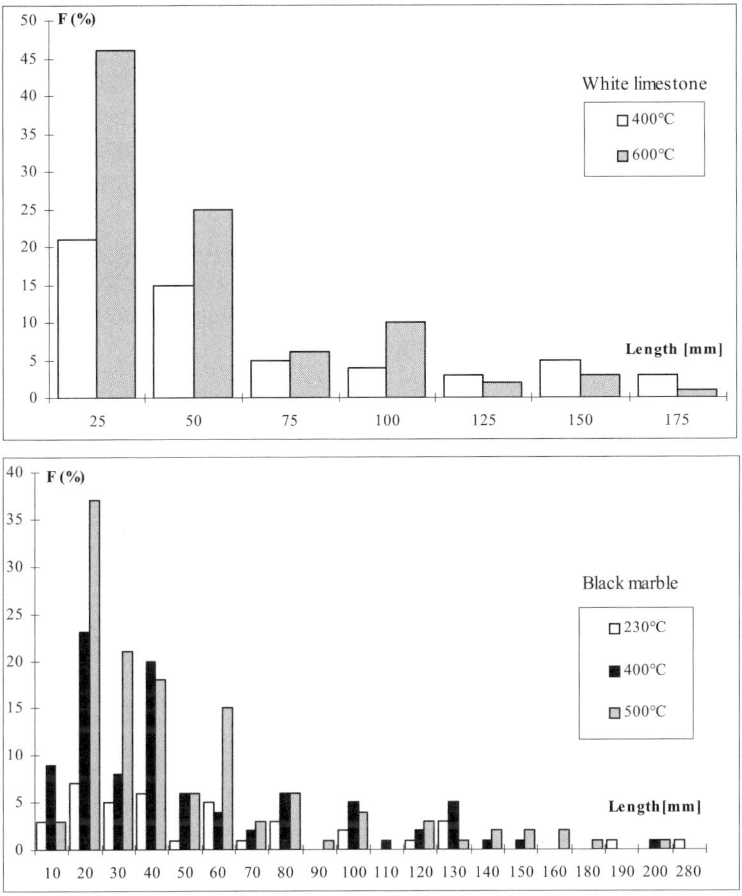

Figure 11. Crack length distributions measured on polished sections of White limestone and Black marble treated at various temperatures.

2.3.4 Strength and deformability of the cracked rock

Uniaxial compressive tests have been performed on cubes (70 mm edge), while flexural tests have been carried out with prismatic specimens of the following size: 120 mm length and 30x20 mm sections. Table 7 shows the main properties of the sound and heated stones while Figure 12 shows stress-strain curves obtained for specimens with different crack densities.

The uniaxial compressive strength versus the degree of fracturation -represented by water absorption - of all the tested specimens is plotted in Figure 13 for both stones. The effect of micro-cracking does not seem to be well marked on the ultimate rock strength, while stress-strain curves are very sensitive to micro-cracking: they show a well marked decrease in stiffness with increasing crack densities as reported in Figure 14 where Young's modulus (E) versus WA is plotted. A decrease up to 50% of the modulus is evident for the White limestone and even higher for Black marble, where a decrease up to the 35% of the original value is reached.

The values of the tangential elastic modulus (E_{stat}) obtained from the uniaxial compression tests carried out on White limestone and Black marble samples were compared to those obtained through the measurement of the UPV compression ultrasonic waves on the same dried samples.

If the rock sample is assumed to be an ideal elastic, isotropic solid having small diameter compared to the length, the elastic quantities could be computed in relation to the shear and compressional wave velocities, Vs and Vp respectively[16]. In particular, the value of the dynamic elastic modulus (E_{dyn}) could be obtained from the following equation:

Table 7. Average values and Standard Deviation of uniaxial compression strength (Co), tangential (Et) and secant (Es) Young modulus and tensile strength (To) measured on White limestone and Black marble.

Litotype	Temp [°C]	Co [MPa]		E_t [MPa]		E_s [MPa]		To [MPa]	
		Av.	S.D.	Av.	S.D.	Av.	S.D.	Av.	S.D.
White limestone	Sound stone	91	15.6	29.4	6.27	20.8	6.61	18.7	4.6
	400	115	14.8	22.2	5.89	17.8	6.94	11.8	2.3
	470	116.5	19.4	19.40	5.00	12.53	3.25	---	---
	630	---	---	---	---	---	---	6.8	2.0
Black marble	Sound stone	87	28.1	29.33	1.53	20.00	1.73	28.1	7.9
	230	106.2	10.1	23.40	1.82	15.80	0.84	---	---
	400	99	6.0	19.40	3.85	6.24	6.71	16.3	2.1
	500	92	9.5	12.20	1.48	9.80	2.17	11.6	1.4

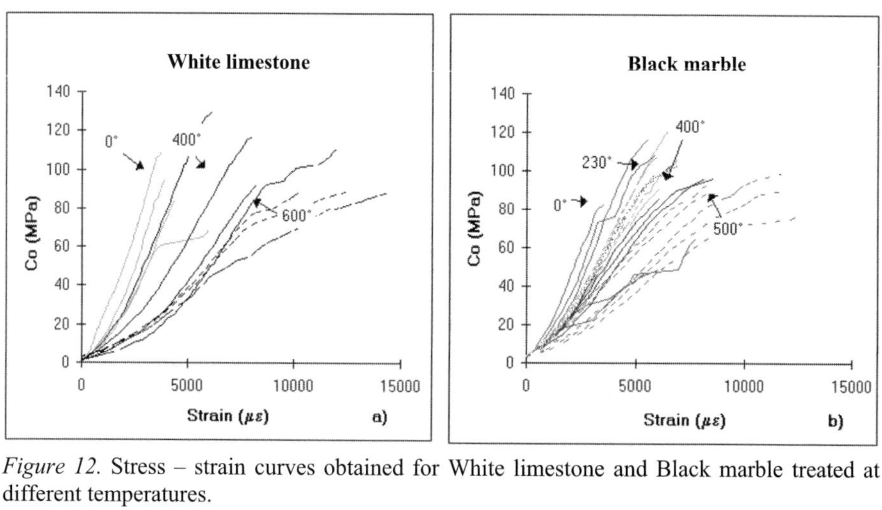

Figure 12. Stress – strain curves obtained for White limestone and Black marble treated at different temperatures.

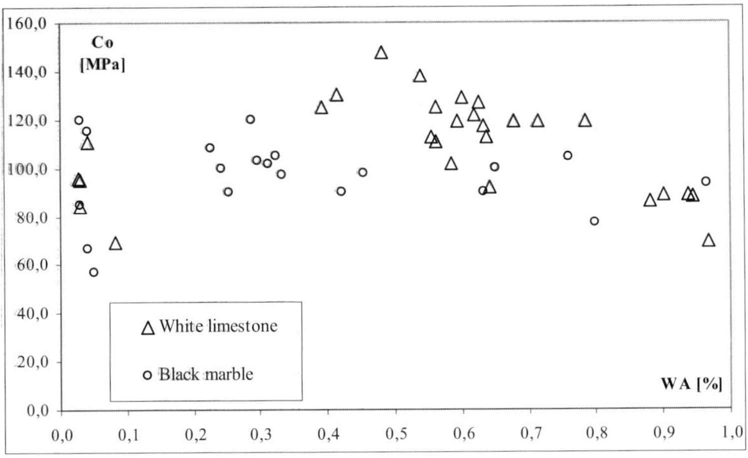

Figure 13. Uniaxial compression Strength vs. water absorption measured in White limestone and Black marble specimens.

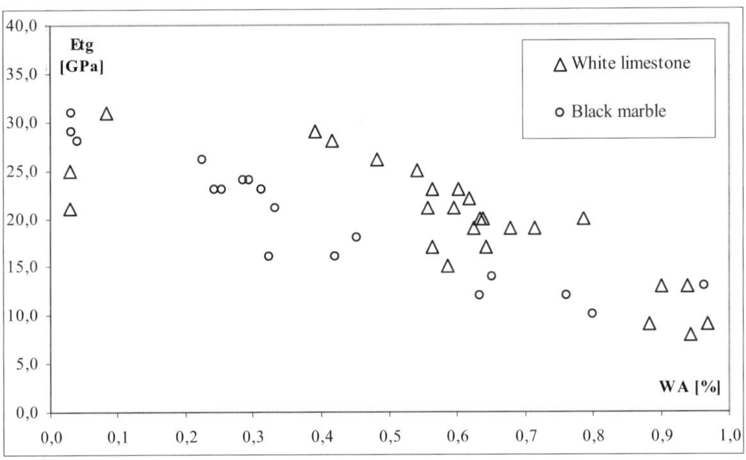

Figure 14. Tangential elastic modulus vs. water absorption measured in White limestone and Black marble samples.

$$E_{dyn} = \frac{(1-2\nu)(1+\nu)}{(1-\nu)} \rho \cdot V_p^2 \tag{6}$$

For the dynamic elastic properties of rock to be used it is necessary to convert the dynamic properties to the static properties. In literature several authors have tried to correlate dynamic to static elastic properties of rocks[17,18]. There are several correlations but no one seems to have a general validity since it depends on the nature of rocks. Consequently, in this work, experimental data have been treated employing different equations in order to find the best fit for this set of data.

The comparison between the E_{stat} values, which were measured, and the E_{dyn} ones, which were computed, as well as the logarithmic and linear correlations that better describe this ratio are reported in Figure 15. In this case it seems that the logarithmic equation is the best correlation choice.

$$E_{stat} = 0.86 \cdot Ln(E_{dyn}) + 0.94, \quad R=0.59 \tag{7}$$

$$E_{stat} = 0.20 \cdot E_{dyn} + 1.23, \quad R=0.50 \tag{8}$$

$$E_{stat} = 0.445 \cdot E_{dyn}, \quad R=0.47 \tag{9}$$

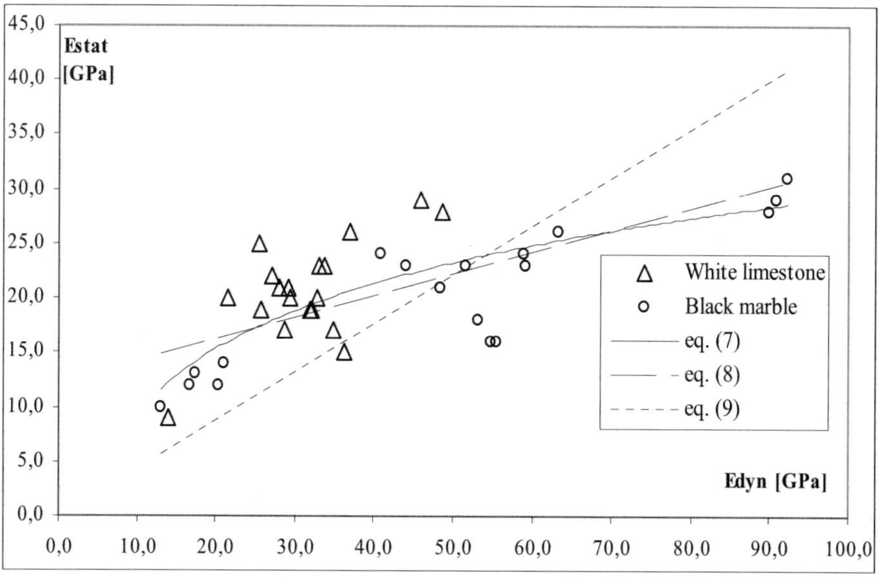

Figure 15. E_{sta} vs. E_{dyn} measured in White limestone and Black marble samples.

The trend of the ratio between E_{dyn} and E_{stat} is reported in Figure 16 together with WA degree showing how this ratio increases as the degree of the induced fracturation increases.

The linear correlation, expressed through Eq.(10) has been extended and used to reckon - once the dynamic elastic modulus is known - the static elastic modulus value of the Ordinary and Venato marbles fractured samples. The values then determined are reported in Table 8.

$$\frac{E_{dyn}}{E_{stat}} = -2.41 \cdot WA + 3.20 \tag{10}$$

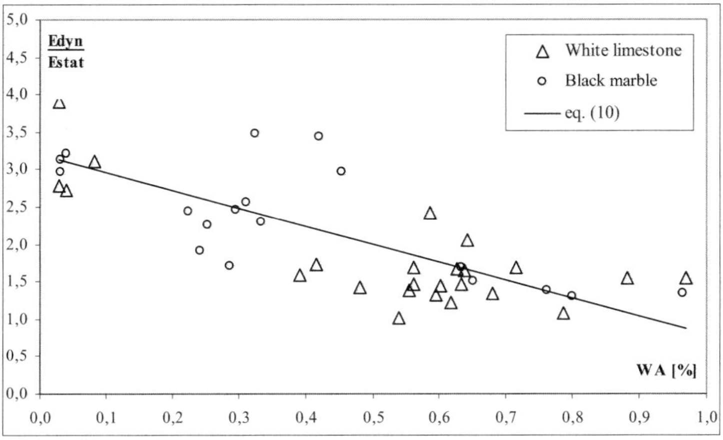

Figure 16. $E_{dyn} - E_{stat}$ ratio vs. water absorption degree.

Table 8. Average values of water absorption degree, E_{dyn}-E_{sta} ratio, E_{dyn} measured and E_{stat} computed for each step of freeze-thaw treatment of Ordinary and Venato marbles.

Marble	Freeze-Thaw Cycle	WA [%]	E_{dyn}/E_{stat}	E_{dyn} [GPa]	E_{stat} [GPa]
Ordinary marble	T0	0.13	2.88	50.66	17.6
	T20	0.15	2.83	39.88	14.1
	T40	0.15	2.83	47.07	16.6
	T60	0.17	2.80	57.11	20.4
	T80	0.18	2.77	53.15	19.2
	T100	0.20	2.71	50.46	18.6
	T120	0.20	2.73	56.88	20.9
Venato marble	T0	0.10	2.96	59.43	20.1
	T20	0.12	2.91	38.80	13.3
	T40	0.13	2.88	31.66	11.0
	T60	0.15	2.84	28.28	10.0
	T80	0.17	2.79	25.54	9.2
	T100	0.18	2.77	25.75	9.3
	T120	0.20	2.71	19.04	7.0

3. ANALYTICAL MODELS

Experimental observations reported above have shown rock material inhomogeneity due to the presence of microcracks. Analytical models commonly applied for the theoretical forecasting of the mechanical properties of heterogeneous materials are basically derived from the continuous mechanics approach. At the microscopic scale rock is a discontinuous material but it can be studied by elasticity models if applied on rock volumes that are representative, from a statistical point of view, of rock including defects (in particular microcraks).

The idea of REV (Representative Element Volume) is now introduced: it is defined as the minimum volume of the material including a number of micro elements being sufficiently high to statistically represent the local properties of the continuous corresponding material. One of the hypotheses adopted in this work is that the rock volumes utilised in the experimental campaign are always larger than the REV and, consequently, the micro mechanical models, below described, can be applied to interpret the results.

The appendix shows how REV mechanical properties are expressed by the tensor D, reported in Eq. (A15), in presence of microcracks under a constant macroscopic state stress. Tensor H, in the same equation, corresponds to the variation of the average value of the deformation field of the REV due to microcracks presence, as shown in Eq.(A13). Eq.(A15) is rewritten as:

$$\bar{\varepsilon}_{ij}^{mc} = H_{ijkl}\sigma_{kl}^{o} \qquad (11)$$

H components explicit expression depends upon the physical model adopted to describe the REV microcracks presence. More specifically, the overall elasticity tensors of the linear elastic matrix containing microcracks are estimated using two single models for the averaging procedure: the dilute distribution model, which assumes that the inhomogeneities are small and far apart, so that their interaction can be neglected and the self-consistent model, which takes into account the corresponding interaction in a certain, overall, approximate manner. Both models consider a random distribution of cracks.

For the 2D and 3D random slits, an energetic formulation of the Self-Consistent model reported by S. Nemat-Nasser and M. Hori[6] has been adopted. For simplicity, we considered, in all adopted models, the additional assumption that all cracks are open and remain so for the considered class of loading.

3.1 Dilute Distribution (2D random slits)

This model is based on the hypothesis that the distribution of the microcracks is such that their interactions can be ignored in the estimation of the overall elastic moduli. The H components can be obtained by the following expression[6]:

$$H_{ijkl} = h_1\delta_{ij}\delta_{kl} + h_2\frac{1}{2}(\delta_{ik}\delta_{jl} + \delta_{il}\delta_{jk}) \qquad (12)$$

where $h_1=0$, $h_2=f\pi/E'$ and $E'=E_0$ or $E'=E_0/(1-\nu_0^2)$ stand respectively for plane stress or plane strain. If a more usual matrix formulation is adopted (where subscripts *b* and *c* are respectively the subscripts *ij* and *kl*), tensor H can be written in the following way:

$$\left[H_{bc}\right]=f\frac{\pi}{E'}\begin{bmatrix}1&0&0\\0&1&0\\0&0&2\end{bmatrix}\tag{13}$$

3.2 Self-Consistent (2D random slits)

This model supposes that the distribution of microcracks is random and it is desirable to include their interaction to a certain extent. In this case global properties of the microcracked REV are determined, considering each microcrack contained in a linear elastic material. This material has the unknown properties of the REV itself. Components of the tensor H are dependent on these properties, and, in particular the following is true[6],

$$H_{ijkl}=\frac{f\pi(k+1)}{16G}(\delta_{ik}\delta_{jl}+\delta_{il}\delta_{jk})\tag{14}$$

where

$k=(3-\nu)/(1+\nu)$ or $k=3-8G(\nu/E+\nu_0^2/E_0)$

stand for plane stress or plane strain, respectively.

In matrix formulation, H becomes

$$\left[H_{bc}\right]=\frac{f\pi(k+1)}{8G}\begin{bmatrix}1&0&0\\0&1&0\\0&0&2\end{bmatrix}\tag{15}$$

In each model the material has been considered isotropic and, consequently, the tensor *D* in a matrix formulation is expressed by:

$$\left[D\right]=\frac{1}{E_o}\begin{bmatrix}1&-\nu_o&0\\-\nu_o&1&0\\0&0&2(1+\nu_o)\end{bmatrix}\tag{16}$$

in case of plain stress and by:

$$[D]=\frac{1+\nu_o}{E_o}\begin{bmatrix} 1-\nu_o & -\nu_o & 0 \\ -\nu_o & 1-\nu_o & 0 \\ 0 & 0 & 2 \end{bmatrix} \quad (17)$$

in case of plain strain.

For the Dilute Distribution model, it has been verified that the Young's modulus of the microcracked REV (to be used in the above reported expressions of the tensors D and H) is given as:

$$E/E_0=(1+f\pi)^{-1} \quad (18)$$

in case of plain stress and

$$E/E_0=[1+f\pi(1-\nu_0^2)]^{-1} \quad (19)$$

in case of plain strain.

Concerning the Self-Consistent model, it has been verified that the same as above ratio can be determined as:

$$E/E_0=(1-f\pi) \quad (20)$$

in case of plain stress and as:

$$E/E_0=(1-f\pi)(1-f\pi\nu_0^2)^{-1} \quad (21)$$

in case of plain strain.

Eqs.(18-21) have been employed for comparison with experimental values as shown in Figure 18 where f is computed by the following definition:

$$f=(\sum_\alpha N_\alpha l_\alpha^2)/4L^2 \quad (22)$$

where l_α is the fracture length, N_α is the number of fractures with a certain length l_α considering a cubic specimen of size L. Parameters l_α and N_α have been obtained by direct measurements through an optical microscope for the Black marble and through an indirect correlation with the imbibition degree for the other tested marbles.

3.3 Self-Consistent (energetic formulation, 2D and 3D random slits)

The Self-Consistent model has been developed according to the classical failure mechanics hypothesis. A fundamental parameter in this approach is the global energy realized by microcracks. This energy can be quantified as the difference between the energy released by an integer solid and the energy released by a cracked solid. Cracked solid energy is expressed in terms of tensor D. An essential step in this method is the evaluation of the energy released by a single microcrack placed in a solid of infinite dimension with mechanical properties being equal to the cracked solid properties. From the single crack energy it is then possible to define, from a statistical point of view, the global energy released by the crack distribution, considering different possible crack orientations. The tensor D is given by the sum of two terms, see Eq.(A15), but the contribution of the cracks has a different structure if compared with the formulation reported in Eq.(14).

Crack dimensions are explicitly taken into account in this particular version of the Self-Consistent model, in which the fracture density is defined as:

$$e = \frac{2N}{\pi}\langle A^2 / P\rangle = \frac{4N}{\pi}\langle a^2 l\rangle \tag{23}$$

where N is the number of cracks in unit volume, A, P, a, and l are the area, the perimeter, the semi-width, and the length of the cracks respectively, and symbol $\langle\ \rangle$ is the average value of the quantity reported in brackets.

For Young's modulus of the cracked material the following expressions are given[6]:

$$E / E_o = (4 - \pi^2 e)(4 - \pi^2 e \nu_o^2)^{-1} \tag{24}$$

for 2D random slits, and

$$E / E_o = 1 - \frac{\pi^2 e}{30}(1 + \nu)(5 - 4\nu) \tag{25}$$

for 3D random slits.

In the present work a simplified version of Eq.(25) is applied considering ν as ν_0. This assumption is considered to be perfectly consistent for most of the applications[6]. Eqs.(24) and (25), in their simplified formulation have also been applied for comparison with the experimental values as shown in Figures 19 and 20.

4. APPLICATION OF THE MODELS TO THE EXPERIMENTAL WORK

Young's modulus of rock specimens containing microcracks is determined by applying the above reported techniques. Cracks are supposed to open throughout all of the compression tests. The procedures adopted during this work to experimentally determine the parameters required for the model application are reported in this chapter. For this aim some hypotheses have to be adopted. Young's modulus of the intact rock is known and this value is constant since the intact rock sample is considered to be a linear elastic material. For this purpose the Young's modulus measured on natural material (without thermal treatment) is taken into account since natural material showed a very low degree of fracturing both at microscopic observations and through indirect macroscopic evaluation. Microscopic counting of the cracks on polished sections has shown the random crack distribution both by the centre crack position and by their orientation and length.

Elastic modulus variation can be computed through the above described DD or SC methods for the 2D random slits case for a given fracture density. Fracture density can be computed for a given distribution of fracture. Comparison between experimental and computed values is shown in Figure 17.

In particular, fracturation density f is computed in relation with an area identified on a polished section. This section is considered to be representative of the whole specimen or, in other words, it is assumed that the same degree of fracturing is present on the specimens.

For the application of the method described by Nemat-Nasser and Hori[6] crack aperture $2a$ must be estimated. Crack aperture has been determined on the hypothesis that the aperture of all cracks during heating concurs to the specimen dilatation in equal way. Aperture $2a$ can then be determined as an average value by knowing the number of cracks per unit volume and the rock thermal dilation. The influence of different aperture values $2a$ (16 μm, 20 μm) has been investigated for White limestone while for Black marble the value of 15 μm has been assumed. The value of Poisson's ratio, ν_0 of the rock material has been assumed to be equal to 0.25.

Table 9 reports crack density values computed with the aid of the various methods described by analysing the experimental data collected by microscopic observation of two White limestone and three Black marble polished sections, previously treated at different temperatures.

The comparisons between crack density measures experimentally determined and the corresponding water absorption values are reported in Figure 17. The linear correlations, which have been determined and expressed by the equations reported in the same diagrams, have been considered to be valid for both Ordinary and Venato marbles, so that their crack density has been indirectly reckoned according to the values of water absorption degree.

Finally, Table 10 reports the crack density values which have been measured and/or computed in relation to the thermal treatment the marble samples have undergone.

Table 9 Crack density f and e for different treatment temperatures.

Litotype	Temp [°C]	Crack density		
		f	e	
			a =0. 16 μm	a =0. 20 μm
White limestone	400	0.189	0.121	0.188
	600	0.542	0.149	0.233
Black marble	230	0.165	0.069	
	400	0.229	0.132	
	500	0.315	0.179	

Table 10. Crack density values measured or computed.

Litotype	Temp [°C]	WA [%]	f	e	
				a = 0.16 μm	a = 0.20 μm
White limestone	400	0.529	0.189	0.1206	0.1884
	470	0.655	0.330	0.1378	0.1766
	600	0.928	0.542	0.1490	0.2327
Black marble	230	0.268	0.165	0.0688	0.0688
	400	0.362	0.229	0.1324	0.1324
	500	0.762	0.315	0.1793	0.1793
Ordinary marble	T0	0.135	0.068	0.0284	0.0363
	T20	0.152	0.076	0.0319	0.0409
	T40	0.152	0.076	0.0319	0.0409
	T60	0.167	0.084	0.0351	0.0450
	T80	0.180	0.091	0.0379	0.0485
	T100	0.204	0.103	0.0429	0.0550
	T120	0.196	0.099	0.0412	0.0528
Venato marble	T0	0.100	0.050	0.0210	0.0270
	T20	0.119	0.060	0.0251	0.0322
	T40	0.134	0.068	0.0283	0.0362
	T60	0.150	0.075	0.0315	0.0403
	T80	0.171	0.086	0.0359	0.0460
	T100	0.178	0.090	0.0375	0.0480
	T120	0.202	0.102	0.0425	0.0544

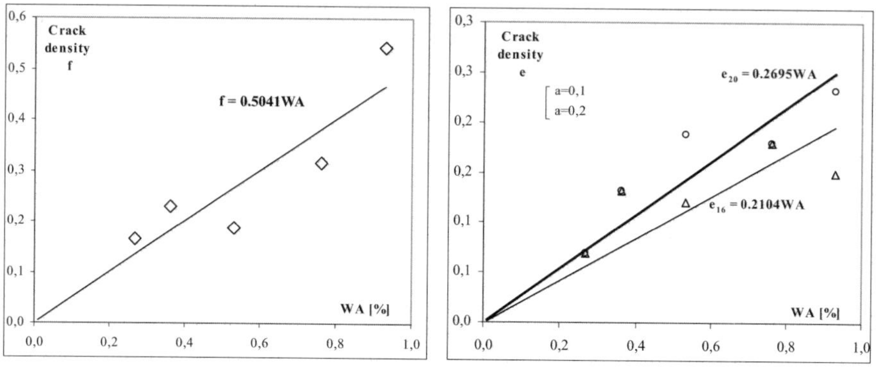

Figure 17. Crack density computed by experimental measures vs. water absorption degree and linear correlations.

The above mentioned theoretical models have been compared with data obtained from the experimental tests by following two different methods. The first comparison has been carried out by considering the experimental data in terms of static Young's modulus and crack density. These values have been directly obtained – as described above – for White limestone and Black marble samples. The previously explained empirical correlations related to water absorption have instead been used for Ordinary and Venato marble samples. The results of this comparison are reported in Figures 18-20.

The second approach aimed only at verifying the prospective use of those tests that are not destructive for the characterization of the material. In this regard the theoretical-experimental comparison has been carried out by taking into account the dynamic elastic modulus and by giving the same value to both the crack density and the water absorption degree. Figures 21 and 22 report this comparison.

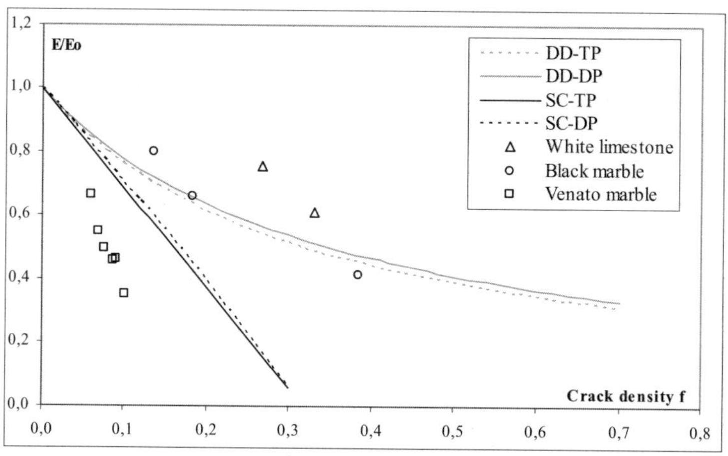

Figure 18. Comparison between experimental and theoretical results (models DD, SC 2D random slits).

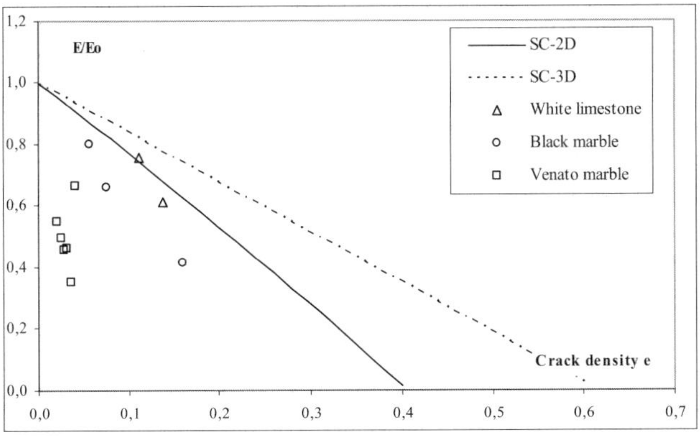

Figure 19. Comparison between experimental and theoretical results (energetic model SC - 2D, 3D random slits) obtained by considering an aperture value equal to 0.16 μm for White limestone marble and to 0.15 μm for Black marble.

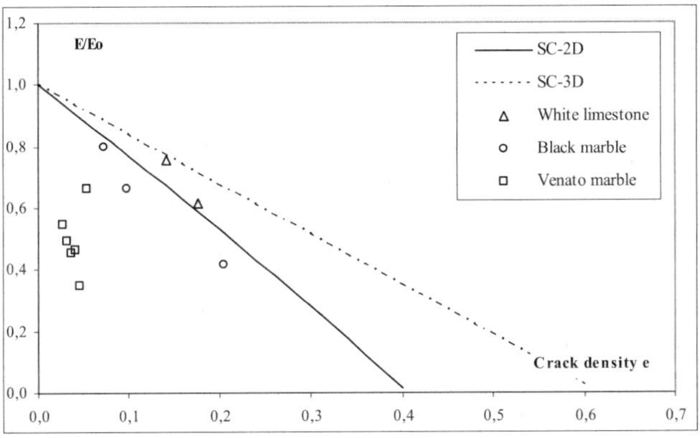

Figure 20. Comparison between experimental and theoretical results (energetic model SC - 2D, 3D random slits) obtained by considering an aperture value equal to 0.20 μm for White limestone marble and to 0.15 μm for Black marble.

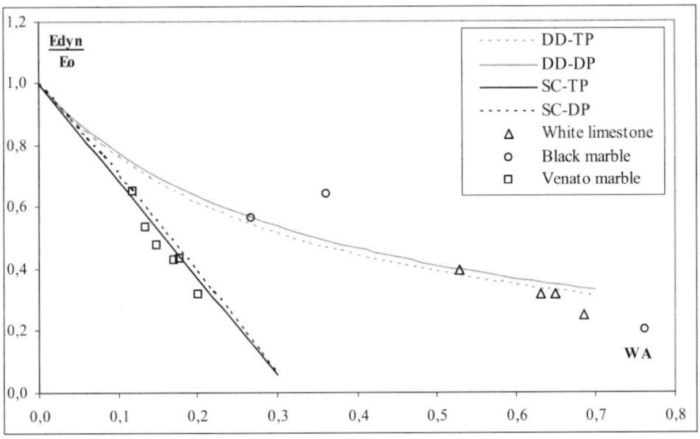

Figure 21. Comparison between not destructive experimental theoretical and results (models DD, SC – 2D random slits).

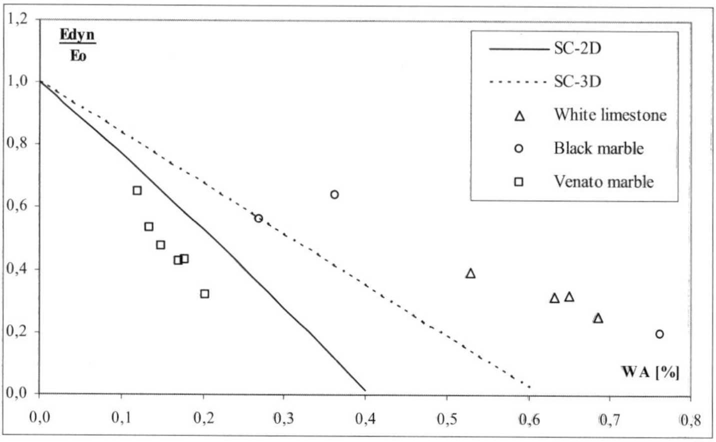

Figure 22. Comparison between non destructive experimental and theoretical results (energetic model SC - 2D, 3D random slits).

5. CONCLUDING REMARKS

Experimental results have shown that the Young's modulus, the velocity of propagation of elastic wave and the water absorption degree are strongly influenced by the degree of fracturing.

Analytical models based on continuous mechanics have been applied to determine cracked material elastic modulus. The results of the models have shown a reasonable agreement between computed and measured values. The results of the SC and DD models diverge with increasing fracturing degree *f*. Models do not show strong differences for different boundary conditions (plane strain or plane stress).

Experimental data are closer to the DD forecasting at low *f* values whilst they diverge at increasing *f* values for both materials. This fact is possibly due to the basic DD assumption (non interacting cracks) that becomes less realistic at increasing *f* values.

Concerning the SC model, it is strongly influenced by the crack aperture (*2a*), which is a parameter difficult to be determined experimentally.

Finally, the crack shape (2 or 3 dimensional) is important for both materials which, in all cases show a better agreement with the 3D assumption.

The comparison between the different results reported in Figures 19-22 shows how the best fit between experimental and theoretical values can be obtained when WA is considered to assess the degree of fracturing. This fact is possibly due to the experimental procedures adopted within this work since thermal treatments have been carried out in order to induce different crack densities. Thermal treatments affect rock specimens in different ways with increasing depth within the rocks. The superficial part of the specimens are characterised by a higher degree of fracturing. This inhomogeneity cannot be taken into account by the theoretical definitions of degree of fracturing (either f or *e*) that consider the same statistical crack distribution within the rock. WA instead is strictly related to the connected porosity hence it is more likely to be present on the sample surface than in sample depth. For this reasons WA results are considered to be a better indication of the degree of fracturing when this kind of phenomena (e.g. frozen-thaw cycles) occur.

ACKNOWLEDGEMENTS

This research has been in part developed within the European Project McDUR-Acoutherm ("Effects of the Weathering on Stone Materials: Assessment of their Mechanical Durability" - Contract no. G6RD-CT-2000-00266 – GRD3-CT2001-6001).

REFERENCES

1. T. Rotonda, Mechanical behaviour of an artificially microcracked marble, *Atti 7° Congresso ISRM Aachen*, **1**, 16-20 September 1991, pp. 345-350.
2. M. Franzini, Stones in monuments: natural and anthropogenic deterioration of marble artefacts, *Eur. J. Mineral.* **7**, 735-743 (1995)
3. H. P Yin and A. Ehrlacher, Size and Density Influence on Overall Moduli of Finite Media with Cracks, *Mechanics of Materials* **23**, pp. 287-294 (1991).
4. N. Laws and J. R. Brockenbrough, The effect of micro-crack system on the loss of stiffness of brittle solids, *Int. J. Solids Structures* **23**(9), pp. 1247-1268 (1987).
5. H. Robina, C. Wong, K. T. Chau and P. Wang, Microcracking and grain size effect in Yuen long marbles, *Int. J. Rock Mech. Sci. & Geomech. Abstr.* **33**(5), pp. 479-485, (1996).
6. S. Nemat-Nasser, M. Hori, *Micromechanics: overall properties of heterogeneous materials*, North-Holland Elsevier, 2nd Rev. Ed., 1999.
7. S. Siegesmund, J. Ruedrich and T. Weiss, in *Proceedings of Dimension Stone* 2004, edited by R. Prikryl, 2004.
8. European Project McDUR (Effects of the Weathering on Stone Materials: Assessment of their Mechanical Durability - Contract no. G6RD-CT-2000-00266).
9. CEN TC 246 "Natural Stone" Status of work - 2005
10. EN 13755 – Natural stone test methods – Determination of water absorption at atmospheric pressure - 2001.
11. EN 14579: Natural stone test methods – Determination of sound speed propagation - 2004.
12. R. Bellopede, L. Manfredotti and P. Marini, About accuracy on ultrasonic measurements on stone, in *Proceedings 10th Int. Congr. on deterioration and conservation of stone*, Stockholm, 2004, pp.659-666, ISBN 91-631-1458-5.
13. P. Marini, R. Bellopede, C. De Regibus and L. Manfredotti, Stone weathering evaluation with UPV measurements: a comparison between direct and indirect method, *ART '05,* Lecce May 2005, Cd rom ISBN 88-89758-00-7.
14. EN 14205 Natural stone test methods – Determination of Knoop hardness – 2003.
15. A. M. Ferrero and P. Marini, Experimental studies on the mechanical behaviour of two thermal tracked marbles, *Rock Mechanics & Rock Engineering*, **34**(1), pp. 157-166 (2001).

16. R.E. Goodman, *Introduction to Rock Mechanics*, John Wiley & So, 1989.
17. Z. Wang, Dynamic versus Static Elastic Properties of Reservoir Rocks, in, *Seismic and acoustic velocities in reservoir rocks*, edited by Z. Wang, and A. Nur., **3**: Recent development, published by Society of Exploration Geophysicists, 2000.
18. L. Sambuelli, A. Vernier and E. Vernier, Evaluation of the ultrasonic dynamic characteristics of some lithotypes from Sardinian quarries, *8° Congress IAEG*, Vancouver, Canada, 1998, pp.2911-2917.

APPENDIX

A volume V of a linear elastic material (defined REV) with and external boundary equal to ∂V and containing microcracks is considered. The volume is not under tensile stresses. Ω is the global volume of the cracks. The REV region obtained by subtracting volume Ω is called matrix and its volume is V-Ω. If the REV is under a uniform macro state of stress σ^o a corresponding uniform stress state is acting on the boundary ∂V:

$$t^o=\lambda.\sigma^o \quad \text{(A1)}$$

For a homogeneous REV, without cracks, the filed deformation will be still uniform and given by:

$$\varepsilon^o=D:\sigma^o \quad \text{(A2)}$$

The presence of cracks induces a variation on the uniform distribution of both stresses and deformations determining variable fields of matrixes $\sigma=\sigma(x)$ and $\varepsilon=\varepsilon(x)$, with $\sigma=0$ in Ω. In these cases, the evaluation of the REV mechanical properties has to be done in terms of average values of σ and ε:

$$\bar{\sigma}=\frac{1}{V}\int_V \sigma dV, \qquad \bar{\varepsilon}=\frac{1}{V}\int_V \varepsilon dV \quad \text{(A3)}$$

The computation of σ and ε leads to the following important results[7]:

$$\bar{\sigma}=\sigma^o \quad \text{(A4)}$$

$$\bar{\varepsilon}=\varepsilon^o+\bar{\varepsilon}^{mc} \quad \text{(A5)}$$

The meaning of Eq.(A4) is that the presence of the microcracks does not modify the average fields of stress in comparison with the stress field acting on an homogeneous REV. Eq.(A4) is obtained in the following way. Tensor σ can be written in the following form:

$$\sigma=1^{(2)}\cdot\sigma=(\nabla\otimes x)^T\cdot\sigma=[\nabla\cdot(\sigma\otimes x)]^T \quad \text{(A6)}$$

The second part of Eq.(A6) is verified by developing it in components. $\nabla=\partial_i e_i$, $x=x_j e_j$ can be written by taking into account the implicit sum convention and thus obtaining:

$$\nabla\otimes x=\partial_i x_j e_i\otimes e_j=\delta_{ij}e_i\otimes e_j=e_i\otimes e_i=1^{(2)} \quad \text{(A7)}$$

The last equivalence of (A6) is due to the fact that σ is a tensor with nil divergence. The divergence of a third order tensor $\sigma\otimes x$ (considering $\sigma=\sigma_{kl}e_k\otimes e_l$, with $\sigma_{il}=\sigma_{li}$, $\partial_i x_j=\delta_{ij}$) becomes:

$$\nabla\cdot(\sigma\otimes x)=\partial_k(\sigma_{kl}x_j)e_l\otimes e_j=[(\partial_k\sigma_{kl})x_j+\sigma_{jl}]e_l\otimes e_j=\sigma \quad \text{(A8)}$$

This came as a direct consequence of $\partial_k\sigma_{kl}=\nabla\cdot\sigma=0$. If Eq.(A6) is substituted in the first one of Eqs.(A3) and the Gauss theorem is applied, the following is obtained:

$$\bar{\sigma} = \frac{1}{V} \int_{\partial V} [\lambda \cdot (\sigma \otimes x)]^T dS = \frac{1}{V} \int_{\partial V} x \otimes t^o dS \tag{A9}$$

A similar computation in component terms shows that :

$$\lambda \cdot (\sigma \otimes x) = t^o \otimes x .$$

It is straightforward to verify that: $(t^o \otimes x)^T = x \otimes t^o$. Using Eq.(A1), Eq. (A9) can be written in the following form:

$$\bar{\sigma} = \frac{1}{V} \int_{\partial V} x \otimes t^o dS = \frac{1}{V} \int_{\partial V} (x \otimes \lambda dS) \cdot \sigma^o \tag{A10}$$

In order to verify Eq.(A10) the same kind of computation in component terms can be carried out. If: $\sigma^o = \sigma_{jk}{}^o e_j \otimes e_k$, $\lambda = \lambda_i e_i$, $x = x_l e_l$ is considered, and Eq.(A1) is rewritten in component terms as $t_k{}^o = \lambda_j \sigma_{jk}$, the following is obtained:

$$(x \otimes \lambda) \cdot \sigma^o = x_l \lambda_i \sigma^o_{jk} (e_l \otimes e_i) \cdot (e_j \otimes e_k) = x_l \lambda_i \sigma^o_{jk} \delta_{ij} e_l \otimes e_k = x \otimes t^o \tag{A11}$$

Since $\lambda \otimes x = \lambda \cdot (1^{(2)} \otimes x)$ the Gauss theorem can be applied:

$$\frac{1}{V} \int_{\partial V} \lambda \cdot (1^{(2)} \otimes x) dS = \frac{1}{V} \int_V \nabla \cdot (1^{(2)} \otimes x) dV = \frac{1}{V} \int_V 1^{(2)} dV = 1^{(2)} \tag{A12}$$

The validity of Eq.(A12) is based on the equation $\nabla \cdot (1^{(2)} \otimes x) = 1^{(2)}$. This equation also determines the following divergence of the third order tensor $1^{(2)} \otimes x$. Finally, since $x \otimes \lambda = (\lambda \otimes x)^T$, the above reported results from Eq.(A10) can be obtained.

As far as deformation is concerned, Eq.(A5) shows that, in general terms, the average value of the deformation fields of cracked REV is not equal to ε^o.

Additional deformation $\overset{-mc}{\varepsilon}$ can be determined by applying Betti theorem on the cracked REV and by using the Green function to express $\overset{-mc}{\varepsilon}$ as a function of σ^o. This development required a long series of computations that have not been reported in this appendix.

The important result obtained for this kind of application is[6]:

$$\overset{-mc}{\varepsilon} = H : \sigma^o \tag{A13}$$

For a REV containing microcracks, the relation between $\bar{\varepsilon}$ and $\bar{\sigma}$ can be defined in the following way:

$$\bar{\varepsilon} = \bar{D} : \bar{\sigma} \tag{A14}$$

If Eqs. (A2, A4, A5, A13) are considered, tensor $\bar{D}$ is the sum of the two tensors D and H, which, from the physical point of view, corresponds to homogeneous REV and cracks influence contribution respectively:

$$\bar{D} = D + H \tag{A15}$$

Chapter 7.2

DIGITAL IMAGE ANALYSIS CONTRIBUTION TO THE EVALUATION OF THE MECHANICAL DECAY OF GRANITIC STONES AFFECTED BY FIRES

Miguel Gómez-Heras[1], Carlos Figueiredo[2], María José Varas[1], Antonio Maurício[2], Mónica Álvarez de Buergo[1], Luis Aires-Barros[2], Rafael Fort[1]
[1]*Instituto de Geología Económica (CSIC-UCM), C/ José Antonio Novais 2. Facultad de CC. Geológicas, Universidad Complutense. 28040 Madrid, Spain;* [2]*Centro de Petrología e Geoquímica do Instituto Superior Técnico. Av. Rovisco Pais 1049-001 Lisboa, Portugal*

Abstract: The study of the decay promoted in building stones by fire is important in the context of the conservation and restoration of historic buildings. As opposed to granular stones, which are more sensitive to chemical changes, tough stones present a noticeable mechanical decay after being affected by fires. The direct observation of decay features at a micro-scale could be the key for understanding the decay processes at greater scales, especially in the cases where extensive sampling is not recommendable or even possible. Microscopic techniques allow quantifying fracture system and establishing differences among different types of fractures. Results obtained so far have demonstrated image analysis and processing techniques as a useful tool to help establishing, in a qualitative and quantitative way, the fracture system variations resulting from mechanical stresses induced by fire.

Key words: granitic stones; stone decay; thermal behaviour; petrophysics; fracture systems; image analyses.

1. INTRODUCTION

Fire is a significant decay agent of stone-made historic buildings, as the temperature gradients generated during and after fire can be enough to promote a short-term irreversible decay in building stones. Moreover, it is thought that fire causes the damage of about one historic building in the European Union each day (COST C17, 2001).

S.K. Kourkoulis (ed.), Fracture and Failure of Natural Building Stones, 427–437.

Fire may generate both chemical and physical changes in building stones. The type and intensity of these damages will depend mainly on the temperatures attained and, specially, of the type of burnt material (Gómez-Heras, 2006). Granular stones are more sensitive to chemical changes while tough and dense stones present a noticeable mechanical decay after being affected by fires. Mechanical decay consists mainly of the generation or growth of fissures due to the thermal stresses. Studies on burnt stones had focused the attention to the bulk mechanical changes (Allison and Goudie, 1994; Chakrabarti et al., 1996), whose testing is relatively sample-costly. This is not in consonance with the sampling of heritage buildings that must try to deal with samples as inconspicuous as possible. In the last years, this trend has shifted to the observation of variations at a more detailed scale, that is to asses mineralogical and micro-morphological changes (Ehling and Kohler, 2000; Hajpál, 2002; Gómez-Heras et al., 2004; Hajpál and Török, 2004; Gómez-Heras, 2006).

The direct observation of decay features at a micro-scale shall be a key for understanding the processes at greater scales both in terms of space (i.e. fracture mechanisms and patterns within the stone) and of time (i.e. future physical behaviour of the stone). Microscopic techniques allow quantifying the fracture system and establishing differences among different types of fractures according to Griffith fracturing criteria. Microscopic techniques have also the advantage of being compatible with small amounts of destructive sampling. These techniques at the micro-scale are complementary to direct and indirect bulk dynamic testing, that enable to compare the 'quality' of bigger samples of stone before and after fires.

Image analysis is both a low-cost and easy to handle non-destructive methodology that is suitable for the assessment of building stone decay (Mauricio and Figueiredo, 2000). Several papers concerning the use of digital image processing and analysis techniques in field and laboratory studies in order to contribute to an accurate assessment of environment-induced damage could, for instance, be referred: Aires-Barros et al. (1991); Ortiz et al. (2000); Zezza (1989).

In this research, some numerical parameters and indexes are proposed to evaluate the mechanical state of the rock at the micro-scale. These numerical indexes are based on different parameters such as fracture length/perimeter area, density or orientation, obtained after refining the images by applying some linear filters and morphological processing techniques for fractures and grain joints detection.

2. MATERIALS AND METHODS

Samples of granite that had been burnt by limited bonfires in two different historic buildings were studied. These buildings are placed in rural areas within the region of Madrid: The *Coracera* Castle (14th–15th century,

San Martín de Valdeiglesias, Madrid) and the *Infante Don Luis* Palace (18th century, Boadilla del Monte, Madrid). Fire temperature was estimated in around 500 ºC due to the mineralogical changes found (Gómez-Heras, 2006).

The granitic stones from the Guadarrama's mountain range (which is located in the north part of the region of Madrid, Spain) are the most important building stone types in the central area of the Iberian Peninsula. These types of granitic stones have been historically included within the term *Piedra Berroqueña*, which include compositions from leucogranites to granodiorites and monzogranites.

Two samples of about 30 cm^3 were taken for the microscopic evaluation. Samples were impregnated with an epoxy resin containing fluorescent dye and cut perpendicular to the burnt surface. Two thin sections were made from each sample, both comprising the external burnt area and the internal area unaffected by the fire. The unaffected portion of the rock was considered as the standard to compare the microcrack pattern before and after the fire.

Sample images in colour RGB (for Red, Green and Blue colour component) format of fracture patterns of the granite samples were acquired using Oxford Inca 4.09 software coupled to a JEOL JSM 6400 Scanning Electron Microscopy (SEM) in Secondary and Backscatter Electrons modes and with a 3.34 million pixel resolution. DP12-BSW *Olympus* Camera connected to an *Olympus* Polarizing Microscope coupled with an *U-RFL-T* Ultraviolet Fluorescence Microscopy unit. *Olympus DP-Soft* v. 3.2 was used for image acquisition. Several of these colour digital images, with different size ranging from 994x744 to 1024x768 pixels, were then processed for quantitative analysis of the fracture system patterns. Four types of images were used for the image analysis depending on the acquisition media: Polarizing Microscope, Fluorescence Microscope and Scanning Electron Microscope (Secondary and Backscattered electrons modes)

Starting with the initial RGB images, processing has typically involved the conversion of these images to the HLS format (composed of components representing the hue, luminance and saturation). Techniques for enhancing contrast, noise removing and segmentation into binary images of the resulting grey level (intensity) images were also used.

Dealing with several images and each one having specific characteristics, the processing step has used a casuistic/task-oriented approach. For a thorough illustration of the image processing performed in this work two step-by-step examples of image processing and analysis are provided. Several operations were sequentially accomplished to perform the complete chain of image processing. Some of the most common operations applied in this study will then be presented in the following steps:

- **preprocessing:** grey level images (Figures 1a, 2a), more suitable to this dedicated application and representing only the intensity component (L), were obtained from the conversion of the initial colour RGB images to the HLS colour format, encoded as Hue, Lightness and Saturation (Gonzalez and Woods, 1993).

- **filtering:** different techniques have been used depending on the task/ purpose. *Opening* transforms (Coster and Chermant, 1985; Mauricio and Figueiredo, 2000; Serra, 1982) of the grey-scale images coming from the preprocessing, were, sometimes, performed for noise (small light details) removing and contrast enhancing (Figure 1b); *hole-fill* (Figure 2c), a classical, sophisticated and complex morphological algorithm has also been used, in some cases, on the background of binary images derived from the segmentation of grey level images, for removing small features (which are mostly noise in the image), and leaving a more statistically significant sampling of fracture system pattern and improving its detection in the cleaned final image (Figure 2d).
- **segmentation:** the grey level images (previously enhanced or not) were segmented into binary images for fracture patterns extraction and individualization from the background, performing either a *thresholding algorithm* (Figure 2b) or the so called morphological *black top-hat* transform (Figure 1c), to detect the dark area of the image. Another sophisticated algorithm combining a *thresholding* transformation with the *skiz* (the skeleton by zone of influence) of the back ground was also used to detect the fracture pattern in some particular images as shown in Figure 1d.
- **arithmetic** and **logical operations:** some of these operations were performed either on binary images during the processing chain procedure or to generate the final composed images (Figures 1e, 2d) obtained from the initial grey level images and the final refined and segmented binary images of fracture system patterns.

On the refined and segmented binary images previously derived from the image processing step, different types of parameters such as fracture length/ perimeter, area, density and orientation, were then obtained using basic formulae (Aires-Barros et al., 1991; Coster and Chermant, 1985; Figueiredo et al., 1993; Serra, 1982):

- The area measure (A (X)) returns, on a binary image (X), the number of non zero pixels in the image. If the image is calibrated, the area in pixels is scaled to give the area in the measurement unit (mm^2) associated with the image. The sample (A(S)) and fracture system pattern (A(fs)) areas were then estimated.
- The *perimeter* measure (L(X)) (mm) designates the length of the object boundary. It was estimated using the *Crofton* formula. For closed fracture system pattern, one pixel width, its *length* (*L(fs)*) was computed.
- The *rose of directions*, characterizing the directional structure of contours or line sets, is a powerful theoretical tool to point the major directions of an object (Figure 1f). It is the density of the contour length as a function of the tangent direction.

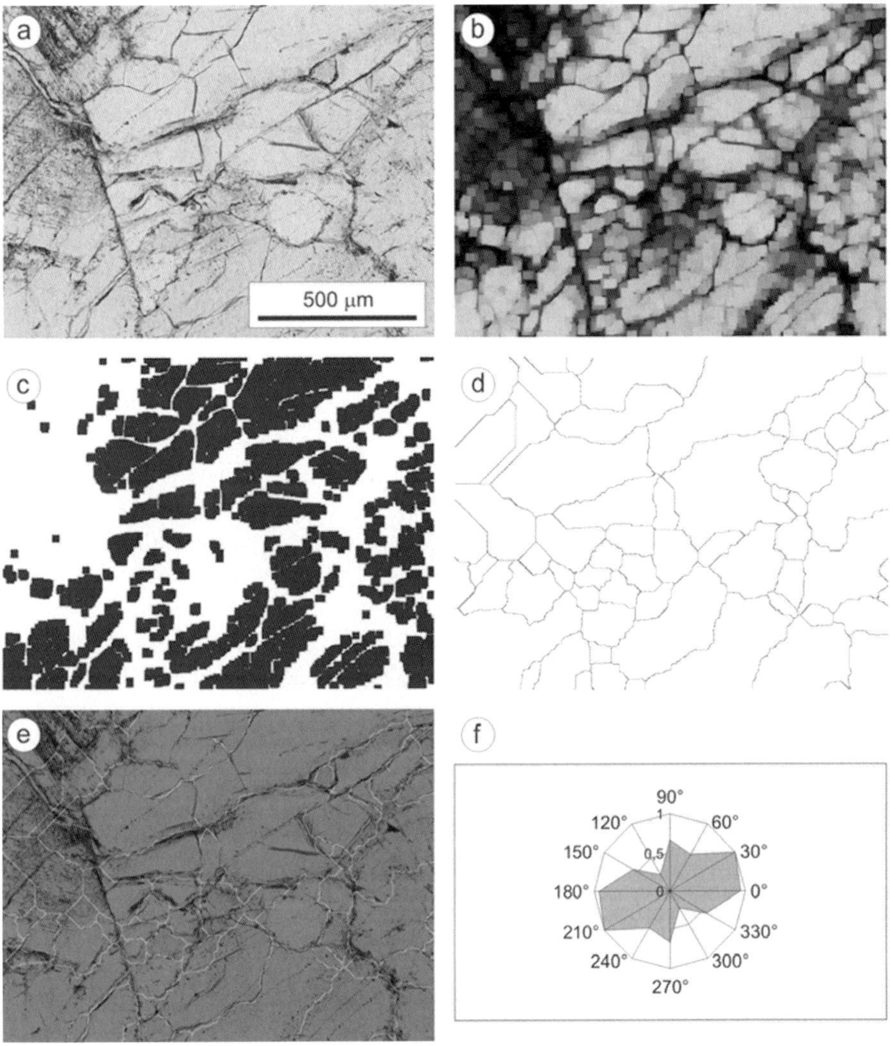

Figure 1. (a): Original image obtained with polarizing microscope (parallel nicols): grey level image; (b): Result of an opening performed on the image (a); (c): Binary image obtained by performing a "black top-hat" transform on the image (b); (d): Result of performing a "skiz" transform on the background of the image (c); (e) Composite image obtained by combining the original image a and the refined, cleaned and simplified binary image (d). The different fracture types could be identified in this image. (f): Rose of directions diagram for the whole fracture system pattern detected on the original image.

Based on some of these basic parameters additional ones were estimated, such as the roughness index (*RI (fs) = L(fs)/ A(fs)*) (useful for the distinction between *open* (RI< 1) and *closed* (RI=1) fracture system pattern, and where *L(fs)* and *A(fs)* are estimated in pixels) (Aires-Barros et al., 1994), the *specific surface area*, ($A_A(fs)$) (%), and the *specific perimeter/length* ($L_A(fs)$) (mm^{-1}).

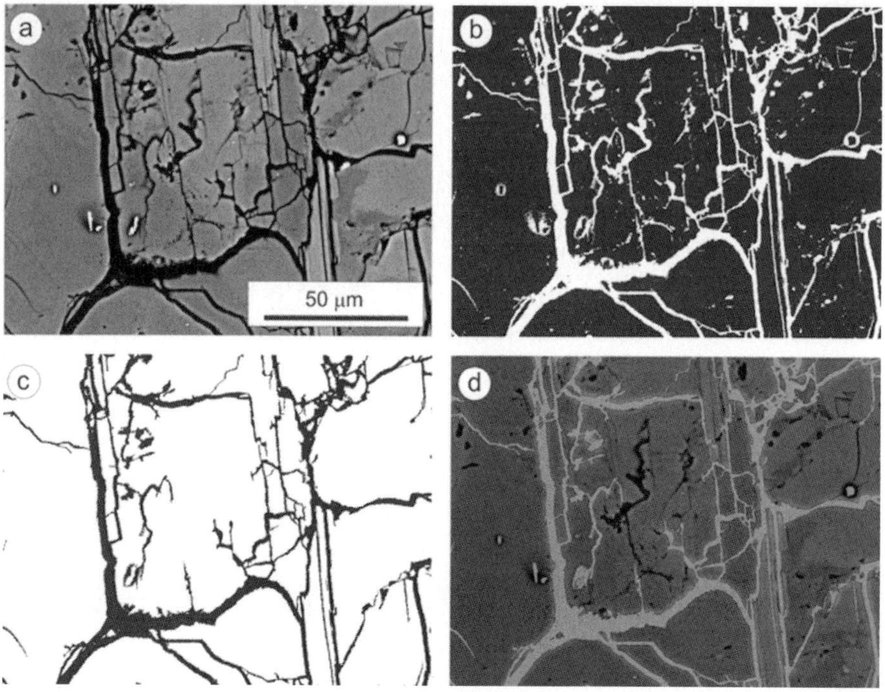

Figure 2. (a): Original image: grey level image; (b): Binary image obtained by thresholding of image (a); (c): Result of performing a "hole-fill" transform on the background of the image (b); (d): Composite image obtained by combining the original image (a) and the refined, cleaned and simplified binary image (c). The different fracture types can be identified in this image.

3. RESULTS

The samples of granite showed macroscopically in all cases a surface discolouration due to a fine deposit of ashes and a slight reddening affecting the first few millimetres of stone. No major fissures are detected macroscopically. The petrographic observations of the selected samples correspond with a fine to medium grained panalotriomorphic equigranular leucogranite. Principal minerals are quartz, plagioclase and potassium feldspar. Accessory minerals are biotite and apatite. Feldspar is generally sericitized and presents abundant pertitic textures.

The microscopic observation of thin sections of this granite shows that the outer portion of stone is much more fissured than the internal areas not affected directly by fire. The most noticeable fissures are sub-parallel to the burnt surface. Individual crystals present an intense internal fissuration in the external burnt area. This is especially noticeable in feldspars (both plagioclases and K-feldspars). Weak surfaces, such as exfoliation planes, altered areas (sericitized feldspars) or inclusions favour the fissuration.

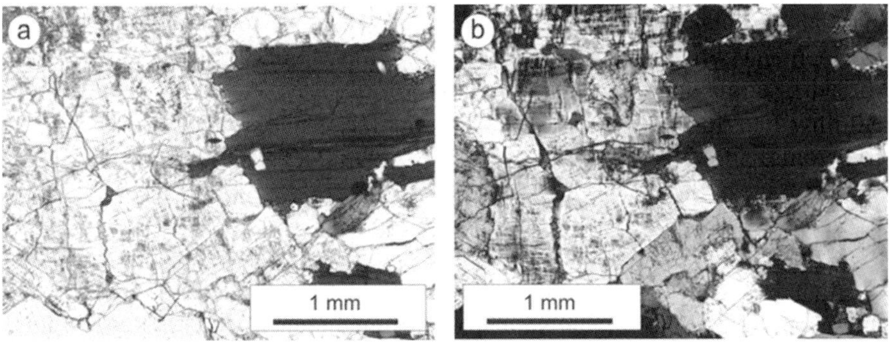

Figure 3. Photomicrographies of leucogranite samples from *Infante Don Luis* Palace. Parallel (a) and crossed (b) nicols.

Fluorescence Microscopy shows the existence of three types of fissures in the material: intergranular fissures (in the interface of two mineral grains), intragranular (breaking single mineral grains), and transgranular (a combination of the previous fissures). SEM images reveal that in the most external zones (0.5 cm deep from the surface) the three types of fissures are present (inter, intra and transgranular). Towards the interior of the stone sample (deeper than 0.5 cm from the surface), transgranular fissures disappear, and deeper, the intragranular ones disappear also (1 cm deep from the surface). There are no fire-induced fissures above 1.5 cm deep from the exposed surface of the samples. A set of analysis was then performed on a number of binary images, and statistics of some parameters and indexes on the measures performed are given in Table 1.

Table 1. Some basic statistics of the parameters measured on the Fracture System Pattern in 'fresh' and burnt samples

		A(S) (mm^2)	A(fs) (mm^2)	A_A(fs) (%)	L(fs) (mm)	L_A(fs) (mm^{-1})	RI(fs)
'Fresh' areas	**Minimum**	8.67	0.18	2.07	85.03	9.81	1.00
	Maximum	8.67	0.29	3.36	137.63	15.87	1.00
	Average	**8.67**	**0.24**	**2.71**	**111.33**	**12.84**	**1.00**
	St. Deviation	0.00	0.08	0.91	37.19	4.29	0.00

		A(S) (mm^2)	A(fs) (mm^2)	A_A(fs) (%)	L(fs) (mm)	L_A(fs) (mm^{-1})	RI(fs)
Burnt areas	**Minimum**	0.02	0.00	0.91	0.89	10.24	0.17
	Maximum	8.67	0.21	16.30	88.81	441.07	1.00
	Average	**0.98**	**0.04**	**8.60**	**23.00**	**123.39**	**0.52**
	St. Deviation	2.12	0.07	5.43	22.79	142.42	0.32

Symbols in Table 1 are as follows: A(S) is the area of the sample; A(fs) is the area of the Fracture System pattern; A_A(fs) is the specific surface area of Fracture System pattern; L(fs) is the perimeter/length of the Fracture System pattern; L_A(fs) is the specific perimeter/length of the Fracture System pattern and finally RI(fs) is the Roughness Index.

The calculated parameters (specific surface area A_A(fs), specific length L_A(fs) and roughness index RI(fs)) are the best for establishing the comparison between burnt and non-burnt areas of the studied stone: surface area in the burnt areas is about four times greater than in non-burnt areas. More evident is the increase in the specific length of fissures, which reaches values up to more than twenty times greater than the values obtained in non-burnt areas. Average roughness index decreases from the value RI = 1, which represent a totally 'closed' fracture system pattern to 0.52.

Regarding the orientation of fissures (Figure 4), burnt areas show a main direction of fissures parallel to the burnt surface. A second system of fissures appears perpendicularly or sub-perpendicularly in much smaller proportion. The closer to the burnt surface the more abundant are these perpendicular fissures.

The fissuration perpendicular to the surface has two origins; firstly the occurrence of fissures conjugated with the fissures parallel to the surface. Secondly, intergranular unions are more likely to open in the surface grains. The strain undergone by the minerals during a fire is proportional to the temperature distribution within the rock, which decrease exponentially inwards.

Relative proportion of inter-, intragranular fissures varies between burnt and non-burnt areas. Non-burnt areas present a homogeneous proportion of

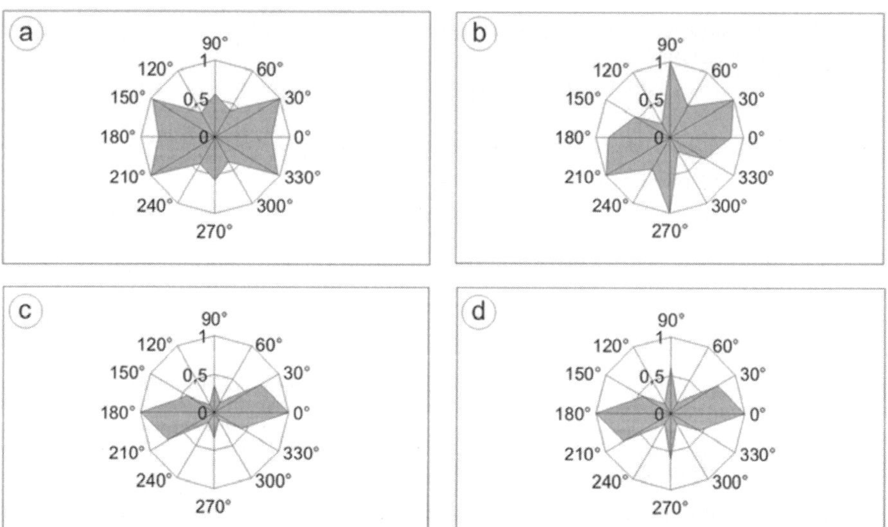

Figure 4. Roses of directions obtained from several images of granite from *Infante Don Luis* Palace. Angles are measured considering 0° the burnt surface.

the different types of fissures, though intergranular ones are slightly more abundant. The ratio intergranular/intragranular fissures is around 1.6. Burnt areas present an intense increase of the relative proportion of intragranular fissures. The values of the ratio intergranular/intragranular in this case is of the order of ~0.5.

4. CONCLUSIONS

Fire produces a noticeable mechanical decay in tough stones such as granite, as expected from other studies on natural stone materials. As it has been shown here, this decay is not only evident and measurable at the macro-scale but also at the micro-scale. As opposed to what is observed in granular stones, such as sandstones, in which predominant micro-effects generated by fire are related with mineralogical changes, in tough stones, such as the granite described in this paper, micro-effects of fire consist of the generation of new fissures and the growth of pre-existing ones. The process of fissuration of the studied burnt granites is circumscribed to the first 1.5 centimetres into the stone. Fire affects mainly the outer portion of rock, which even though it may not affect the bulk structural stability of the stone, it can have an outstanding impact on surface characteristics. Three types of fissures are found: intergranular, intragranular and transgranular. Transgranular fissures are sub-parallel to the material's surface and are found within the first half centimetre. In this outer area the temperature gradients are higher than in internal zones, both in terms of time and space, which promote a higher thermal stress enough for intragranular (and consequently transgranular) fissures to be formed. Transgranular fissures are the most important fissures to consider in decay processes such as spalling, as they form surfaces from which large fragments of the stone can detach. Inter and intragranular fissures can promote and enhance other decay processes such as granular disaggregation.

Transgranular, intragranular and intergranular fissures disappear progressively in this sequence as long one goes deeper following a longitudinal profile through the stone sections. The low conductivity of granite favours that the temperature gradients enough to promote fissuration during the fire event do not go deep within the stone and therefore no significant differences between burnt and un-burnt specimens are found deeper than 1.5 cm.

Three mechanisms of generation of fissures are observed: widening of pre-existing fissures, propagation of pre-existing fissures and nucleation of new fissures. The first mechanism to develop is the widening of pre-existing fissures, as it needs lower stress energy to be produced. Triple junctions of crystals are the most common areas where this mechanism takes place. New fissures develop mainly at the crystalline boundaries or are nucleated from crystalline 'defects', such as exfoliation planes, inclusions or previous micro-fissures.

Microscopic techniques and image processing and analysis techniques have been marked out as a highly suitable technique for studying micro-effects generated by fire in historic buildings, as far as they can be carried out in small samples, compatible with the size of samples that should be obtained from heritage buildings.

Best results are obtained with images from microscopic techniques which offer high contrast images or binary images, such as Scanning Electron Microscopy in Backscattered mode (SEM-BS) and Fluorescence Microscopy.

Parameters obtained from the image treatment are indicative of the state of 'micro-decay' and could be used to be extrapolated or correlated to the results obtained for the bulk stone decay.

The results obtained so far have demonstrated the usefulness of image processing and analysis techniques as a non-destructive and complementary tool to help establishing, in a qualitative and quantitative way, the fracture system pattern resulting from mechanical stresses induced by fire.

ACKNOWLEDGEMENTS

The authors want to acknowledge the assistance of the Luis Brú Electronic Microscopy Centre of the Universidad Complutense and also the technical support of E.M. Pérez-Monserrat during this work. The Spanish Ministry of Education and Science is also acknowledged for a Research Contract *Ramón y Cajal* (MAdeB).

This work was partially supported by FCT sub-project DECASTONE of the Centre of Petrology and Geochemistry, IST, Lisbon, Portugal and the project MATERNAS of the regional government of Madrid

REFERENCES

Aires-Barros, L. Maurício, A. and Figueiredo, C., 1991, Experimental correlation between alterability indexes obtained by laboratorial ageing tests and by image analysis, in: *Proceedings of the Int. Conf. on La Détérioration des Materiaux de Construction*, La Rochelle, France, pp. 199-208.

Aires-Barros, L. Maurício, A. and Figueiredo, C., 1994. Profilometer and image analysis applications to "in-situ" study of stone decay phenomena, in: *Proceedings of the 3rd Int. Symp. on the Conservation of Monuments of the Mediterranean Basin*, Venice, Italy, pp. 19-24.

Allison, R.J. and Goudie, A.S., 1994. The effects of fire on rock weathering: An experimental study, in: *Rock Weathering and Landform Evolution*, D.A. Robinson and R.B.G. Williams eds., Wiley, Chichester, pp. 41-56.

Chakrabarti, B., Yates, T. and Lewry, A., 1996. Effect of fire damage on natural stonework in buildings. Construction and Building Materials, 10(7): 539-544.

COST-C17, 2001. Memorandum of Understanding for the implementation of a European Concerted Research Action designated as COST C17 "Built Heritage: Fire Loss to Historic Buildings", pp. 17.

Coster, M. and Chermant, J.L., 1985. Précis d'Analyse d'Images. Éditions du Centre National de la Recherche Scientifique, Paris, pp. 512.

Ehling, A. and Kohler, W., 2000. Fire damaged Natural Building Stones, in *Proceedings of 6th Int. Congress on Aplied Mineralogy ICAM*, Göttingen, pp. 975-978.

Figueiredo, C., Mauricio, A. and Aires-Barros, L., 1993. Rose of directions algorithm: an improvement of Nachet Ns 1500 implementation, in *Proceedings of the 5th Portuguese Conf. on Pattern Recognition*, RecPad'93, Porto, pp. 227-231.

Gómez-Heras, M., Varas, M.J., Alvarez de Buergo, M. and Fort, R., 2004. Characterization of changes in matrix of sandstones affected by historical fires, in: *Proceedings of the 10th Int. Congress on Deterioration and Conservation of Stone*, D. Kwiatkowski and R. Löfvendahl, eds., Stockholm 2004, pp. 561-568.

Gómez-Heras, M., 2006. *Procesos y Formas de Deterioro Térmico en Piedra Natural del Patrimonio Arquitectónico*. Universidad Complutense de Madrid, Madrid, 339 pp.

Gonzalez, R.C. and Woods, R.E., 1993. *Digital Image Processing*. Addison-Wesley Publishing Company, New York, 716 pp.

Hajpál, M., 2002. Changes in Sandstones of Historical Monuments Exposed to Fire or High Temperature, *Fire Technology*, **38**(4):373-382.

Hajpál, M. and Török, A., 2004. Mineralogical and colour changes of quartz sandstones by heat, *Environmental Geology*, **46**:311-322.

Mauricio, A. and Figueiredo, C., 2000. Texture Analysis of grey-tone images by mathematical morphology: a non-destructive tool for the quantitative assesment of stone decay. *Mathematical Geology*, **32**(5):619-642.

Ortiz, P., Galán, E., Vázquez, M.A., Guerrero, M.A. and Ortiz, R., 2000. Comparación entre metodologías de levantamiento de mapas de alteración como herramienta de diagnosis, in: *Proceedings of the 5th Int. Symp. on the Protection and Conservation of the Cultural Heritage of the Mediterranean Cities*, Seville, Spain, pp. 188-191.

Serra, J., 1982. Image Analysis and Mathematical Morphology. Acad. Press Inc., London, pp. 610.

Zezza, F. 1989. Computerised analysis of stone decay in monuments, in: *Proceedings of the 1st Int. Symp. for the Conservation of Monuments in the Mediterranean Basin*, Italy, pp. 163-184.

Chapter 7.3

THE BEHAVIOUR OF NATURAL BUILDING STONES BY HEAT EFFECT

Mónika Hajpál
Laboratory of Building Physics, Budapest University of Technology and Economics, Bertalan Lajos u. 2., H-1111 Budapest, Hungary, hajpal@lab.egt.bme.hu

Abstract: This paper deals with the effect that fire or high temperature has on natural stone. A lot of historic buildings have suffered fires, either accidentally or from warfare. This topic had not received much attention till the latest time. In case of fires all attention was focused on protection of human lives and the damage to historic buildings was only addressed subsequently. Studies like the present one try to convince authorities to solve the problem for the conservation of burnt built stone heritage.

Key words: fire; natural stone; high temperature; petrological and petrophysical changes.

1. INTRODUCTION

Fortunately in the recent time the topic of the historical monuments (preservation, reconstruction, etc.) came more and more in the focus of the scientific interest[1,2,3]. Also the every-day-people are much more interested in these questions. This is also due to the fact that more attention has been continuously paid in all possible sources in newspapers and TV-programs and so on. It is known that also in previous centuries a lot of fire occurred, which caused significant cultural loss of built heritage. A number of ancient buildings have suffered from the effects of fire – only during 1992 as follows:

- Odd Fellow Palace, Copenhagen
- Proveant Garden, Copenhagen
- Christianborg Palace Church, Copenhagen
- Windsor Castle, Berkshire, England
- Redoutensal, Hofburg Palace, Vienna

These and other major incidents are well known, since they were published in the world press. Of course some other samples could be mentioned, like:

S.K. Kourkoulis (ed.), Fracture and Failure of Natural Building Stones, 439–445.

- Hampton Court, England (1986)
- Theatre "La Fenice", Venezia (1996)
- Cathedral of Torino "Sacra Sindone", Torino (1997)
- St. Michael Church, Budapest (1998)
- Church "dell'Assunta", Cavalese, Trento (2003)

The loss of historical monuments damaged by fire was the theme on some international conferences and other events (e.g. COST C17 Action, some intergovernmental bilateral project like Hungarian-Portuguese and Hungarian-Spanish).

The fire protection of the building of the built heritage hs not been worked out well enough. In many cases just the monument's nature makes the installation of efficient fire-fighting equipment more difficult or even impossible. However, the prevention of fire must have much higher priority for both the building and the contents of the building (e.g. furniture) since they are, also, part of the cultural heritage. It is inevitable that the normal fire fighting is concentrated on life safety, but the historical monuments need a special treatment, as well. In case of a monument, besides the direct damage, the fire causes an extra harm because one cannot actually substitute the injured parts. Special fire-fighting methods for the monuments would have to be worked out because in many cases the traditional way of fire fighting may cause such damages as the fire itself.

The measure of damage caused by fire depends on many factors. Among these there are some such material properties, which are hardly known yet. Since determining these properties and their changes due to high temperature is a complex problem, reaching detailed results demands lot of work and time.

2. MATERIALS AND METHODS

Initially the area of the present research project was the investigation of the behaviour of natural stones by heat effect[1-4]. On this field only few results existed. That's why at the beginning one dealt with sandstones due to their simple structure. For the time being the research is expanded for other stone types as well. In addition, coarse, freshwater and compact limestones, rhyolittuff are studied. Finally one should examine also marble and granite in the future. These are the stone types, which were the mostly used as building stones at historical monuments.

The signs of damage of former fires e.g. changes in colour, spalling, rounding off the corners, cracking, etc. were gathered, because they play a significant role in dating the stone elements or even the entire building. The injured stone material is exposed more intensively to the natural effects.

The effect of fire has been studied on cylinder-shaped specimens, which were drilled parallel and perpendicular to bedding. The specimens were heated in an oven at 6 different temperatures (150, 300, 450, 600, 750, 900 °C). The different test conditions were as follows: by room temperature (22 °C) air dry, water saturated and after 25 freezing cycle, burnt on different temperature air dry, water saturated and after 25 freezing cycle. The petrological (polarizing microscope, X-ray, SEM) and petrophysical (density, porosity, water adsorption, ultrasonic sound velocity, duroskop rebound, uniaxial compressive and indirect tensile strength, colour measurement) investigations have shown that heating changes the texture and mineral composition of various stone types. These changes influence the strength and durability of stone material.

3. INVESTIGATIONS AND RESULTS

The damage of natural stone building materials by heat effect was studied in terms of the physical phenomena involved. The heating produces not only colour changes at the stone materials but also influences their petrological and petro-physical properties, like inner texture, porosity, water adsorption, strength, hardness, weather resistance.

3.1 Petrological results

The mineralogical alteration stages[5] which could be associated to the heat analyses are the following: the minerals remain unaltered, formation of new mineral phases and disappearance of minerals.

Quartz and K-feldspar do not show significant alteration up to 900 °C, however the transformation of a-quartz to b-quartz associated with a volume increases. Another visible change in quartz and feldspar grains is the development of micro-cracks in grains and at grain boundaries above 600 °C.

Clay minerals and phyllosilicates are more sensitive to heat and show several transformations at elevated temperatures. The kaolinite structure collapses completely at around 550 °C. Illite-smectite mixed layer clay minerals are more stable than kaolinite. Illite can be still detected at 900 °C, although it looses the structural water (dehydroxylation) at 553 °C.

Besides mineralogical changes a colour change may also indicate the transformation of a clay mineral. For example glauconite will be orange (at 450 °C) and finally will become brownish red (at 900 °C). Another example is chlorite, which shows a colour change from green (at 22 °C) to yellow (at 900 °C) most probable due to the oxidation of iron (II) to iron (III).

In thin sections the first damages of calcite were detected at 450 °C, but major mineral transformation are only visible after heating at 600 °C. At 750 °C the collapse of the calcite structure is significant and easy to visualise under an optical microscope. At 900 °C the calcite and dolomite decompose to form CaO and MgO. Leaving the samples at room temperature induce the appearance of a new mineral phase, portlandite $Ca(OH)_2$, from the reaction that has been previously reported for limestones. The formation of portlandite is associated with volume increase and leads to the disintegration of heated cylindrical samples. In dolomite majority this kind of portlandite reaction was not observed, although the carbonate phase disappeared above 750 °C.

Goethite or jarosite (iron-bearing minerals) only show alteration at elevated temperatures. At 900 °C a new mineral phase, haematite, appears and both goethite and jarosite disappear. The heat resistant haematite is the final reaction product of any iron-bearing oxy-hydroxide at elevated temperatures.

3.2 Petrophysical results

Heat-related changes in mineralogy induce changes in physical properties of natural stones. In general, by increasing the temperature the bulk density decreases above 450 °C. Porosity has an opposite trend and the increasing will be remarkable also above 450 °C (Figure 1).

Figures 1-4 show the changes of some petrophysical properties as result of the heating. At heating on 450 °C and 600 °C some of the compact limestone samples are exploded and at elevated temperature (750, 900 °C) any samples decayed so it was impossible to use them. The coarse and freshwater limestone shows the portlandite reaction and also the disintegration of the limestone samples after higher heating temperature (600, 750, 900 °C) as well. The ultrasonic sound velocity is similar for all stone types: it decreases with increasing heating temperature (Figure 2).

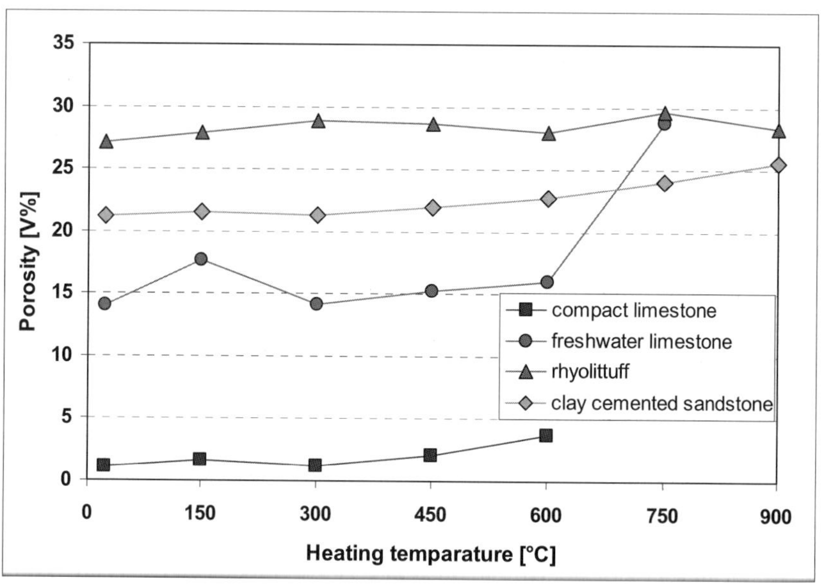

Figure 1. Porosity of different natural stone types as a function of heating temperature.

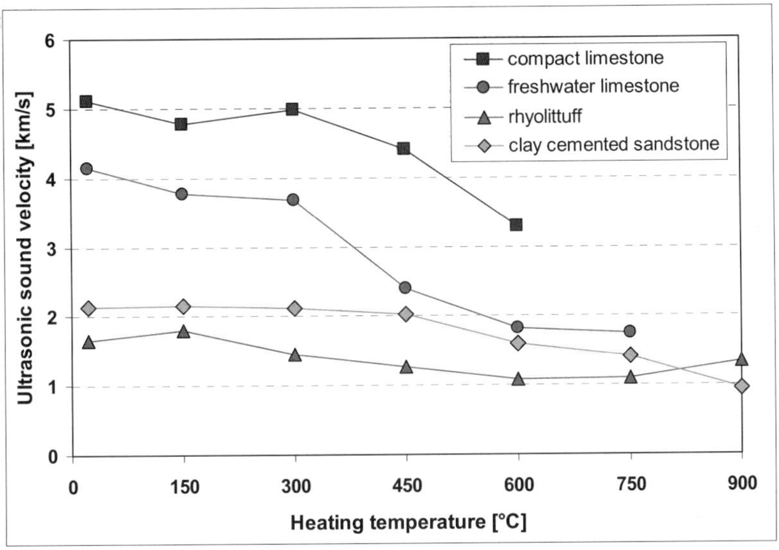

Figure 2. Ultrasound velocities for different natural stone types as a function of heating temperature.

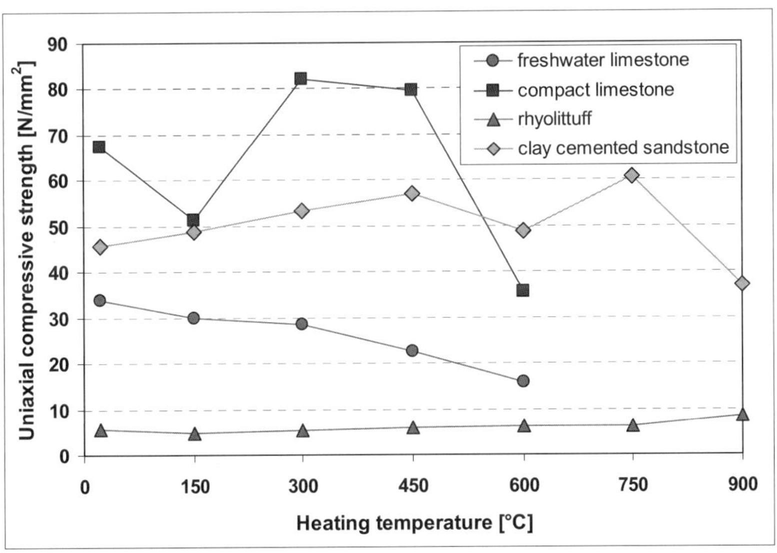

Figure 3. Uniaxial compressive strength of different natural stone types as a function of heating temperature.

The changes in indirect tensile strength and uniaxial compressive strength are not so uniform. The clay cemented sandstone and the rhyolittuff show an increase in uniaxial compressive strength with increasing the heating temperature. Limestones respond in a more sensitive manner to heat effect: they are characterised by a decrease in strength with increasing temperature (Figures 3, 4).

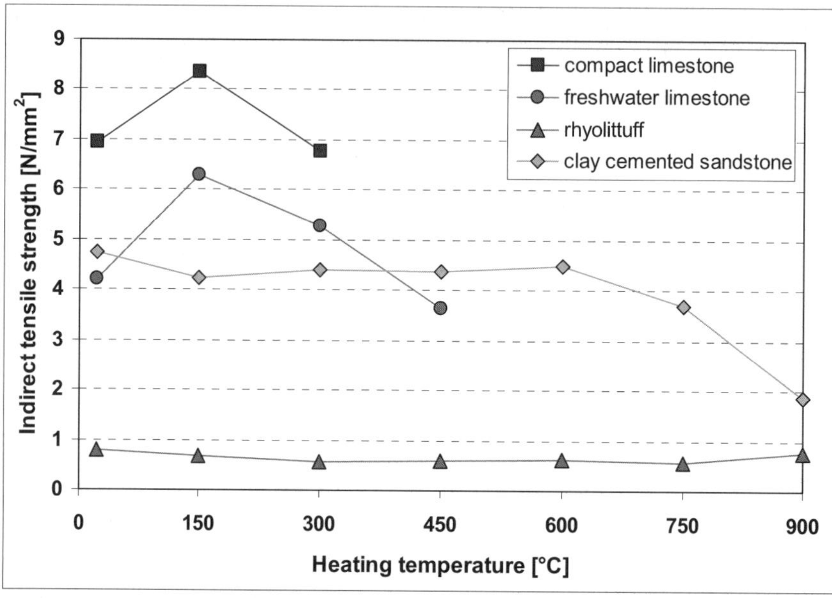

Figure 4. Indirect tensile strength of different natural stone types as a function of heating temperature.

After a fire, structures are exposed and thus weathering begins. In this study, some heat exposed specimens were water saturated before tested and also subjected to 25 freeze-thaw cycles.

All stone types show more or less colour changing due to the heating. For most of them the measured colour values show a positive shift from green towards red with increasing temperature, e.g. sandstones become more reddish after heating. This colour change is mostly attributed to the iron-bearing minerals.

4. CONCLUSIONS

The mineralogical composition and texture of natural stones significantly influence their resistance to fire and thermal changes. The heat resistance of different stone types depends on the type and amount of cementing material (grain/cement ratio), the grain size (fine, medium, coarse) and the grain to grain or matrix to grain contacts.

The investigated natural stone specimens showed textural and mineralogical changes during the laboratory heating tests carried out. The changes were mainly observed at elevated temperature. Some mineral transformations result to a volumetric increase, which generates thermal expansions cracks in stone. This effect can be responsible for the increase in porosity and the decrease in strength.

The clay containing stones are more resistant to heat as the calcareous types. The best properties are showed by the fine-grained matrix-rich sandstone and the rhyolittuff.

It is very important to continue this research, because the results could be applied in many fields. For example it was observed that the use of cold extinguishing water at the fire fighting is very harmful for the stone parts of the historical monuments and it can cause damages in the stone structure, which can lead to stability problems. The observations and the results of the present research contribute to the better knowledge of the mechanical properties of natural stones, something which is fundamental for conservation and restoration of the building stones. In addition, it can serve as a basis for the development of conserving materials as well as for structural calculations.

ACKNOWLEDGMENTS

Post-doctoral fellowship grant (No.: OTKA D45932), the Hungarian-Portuguese (No.: P-18/03) and the Hungarian-Spanish (No.: E-39/04) intergovernmental grants are appreciated.

REFERENCES

1. M. Hajpál, Feuerfestigkeitsuntersuchungen der Sandsteine von Baudenkmälern, *Diplomarbeit (In German with English abstract)*, (Budapest, 1995)
2. M. Hajpál, Investigation of burnt sandstone, in *Proceedings 1st International Conference on Fire Protection of Cultural Heritage*, (Thessaloniki, 2000), pp 349-361
3. M. Hajpál, Changes in sandstones of historical monuments exposed to fire or high temperature, *Fire Technology,* **38**(4), (Kluwer, 2002), pp 373-382
4. M. Hajpál, Changes in sandstones due to thermal effect, *PhD thesis (In Hungarian with English and German abstract)*, (Budapest University of Technology and Economics, Budapest, 2002)
5. Á. Török, M. Hajpál, Effect of temperature changes on the mineralogy and physical properties of sandstones. A laboratory study, *Restoration of Buildings and Monuments* **11**(4), 211-218 (2005)

Chapter 8: Petrography, Fabric and Properties

Chapter 8.1

FAILURE OF ANISOTROPIC MARBLE: THE PROCONNESIUM MARMOR OF ROMAN COLUMNS IN BRINDISI

Fulvio Zezza
Faculty of Architecture-DCA, University IUAV of Venice, 191 St. Croce, Venice, VE 30125, Italy

Abstract: The failure process of the roman columns located in Brindisi (Italy), where the ancient "Via Appia" ends, is outlined. In situ and in laboratory investigations have been performed to evaluate the choice of the restoration intervention starting from the foliation-fissuring relationship as long-term marble properties modification and the intrinsic characteristics of building material. The statistical-structural analysis of the fracturing rate has stated the systems of the neo-formation fractures and their distribution into the five components of the pedestal (*crepidoma, plinth, dado, cornice, orlo of the base and torus*). The geometry of the fractured elements and the marble decay as well as the mechanical characteristics of the marble blocks have been reconstructed through a non-destructive analysis of the pedestal and of the shaft. The inclusively collected data have allowed elaborating block-diagrams worked out in a semi-automatic form. The investigation has supplied the criteria both to evaluate the risk condition, advancing the shaft's remounting phase of the column, and to employ in an operative plan for the choice of the restoration intervention.

Key words: failure; fissuring; marble; non-destructive analysis; restoration.

1. INTRODUCTION

The terminal columns of the ancient Via Appia in Brindisi, standing in front of the port, are two (Figure 1). They represent an example of civil roman building in marble. One of the columns, named "cut off", is reduced at present to the pedestal after the ruin of its shaft in 1528; it is referred that the fall was caused by an earthquake but in the General Catalogue of Italian Earthquakes a sisma is mentioned in the previous year (1527). Vice versa the "entire" column reaches 19 meters in height.

S.K. Kourkoulis (ed.), Fracture and Failure of Natural Building Stones, 449–469.

Figure 1. The terminal columns of the Via Appia in Brindisi. On the right the pedestal of the column fell down in 1528.

Figure 2. The capital (Proconnesium marble) decorated with the busts of Jupiter, Neptune, Pallas and Mars and eight tritons.

The shaft of this column as well as the pedestal is of marble blocks; it supports a marble capital decorated with busts of Jupiter, Neptune, Pallas and Mars and by eight tritons (Figure 2). These sculptures, which belong to the Age of the Antonines, originate in the 1st or 2nd century A.D. The capital holds a dosseret which probably supported a statue. Owing to the fracturing rate and the decay of the pedestal, the shaft was dismantled on precaution in 1997.

Within the restoration project the strengthening of the pedestal is expected. The knowledge of the anisotropic characteristics of a rock is important to the designer of monument restoration. The perfect levelling of the support plane of the first shaft drum, and then of the whole shaft, on the torus requires really the preservation in the time of the flatness of such plane subjected to loads with oscillating pressure centre (wind effect on the shaft). It is likely that the dismounting of the shaft has altered the strain in the pedestal and in the bound dado owing to the variable geometry of the fractured elements (Figure 3). There are justified doubts that in the remounting phase of the shaft the application, although slow and progressive, of a load equal to eighty tons could produce mutual displacements by fissuring rate of the marble blocks.

Detailed analysis of the monument conservation to be applied in the context of the restoration project has been performed. The petrographic, chemical, mineralogical characteristics and the texture as well as the physical and mechanical properties of the marble have been examined. The anisotropy of the marble is linked to the structural characteristic (foliation) which makes evident the relationship with neo-formation fractures of the blocks of the pedestal as the statistical analysis confirms. Ultrasonic measurements performed through direct and indirect methods, and locally semi-direct, have determined the geometry of the fractured parts. The fabric of the marble and the dip of the foliation within the blocks have been taken into account in interpreting measures. Laboratory tests, performed on samples with different orientations in regard to the foliation planes, have analysed the relation between the compressive strength values and the ultrasonic velocities.

The approach to evaluate the failure/stone decay process of each component of the pedestal (*crepidoma, plinth, dado, cornice, orlo of the base and torus*) and the shaft is based on the correlation between the ultrasonic velocities and the mechanical properties expressed by index values related to mechanical stresses and long exposition (about two thousand years) to the agents of weathering.

2. THE BUILDING MATERIAL

The technique based on the cathodoluminescence phenomenon has cleared up that the Marmara Island is the source of the marble (Zezza F. et al., 1992) well-known as *Proconnesium marmor*. The blocks used to build the pedestal and the shaft are foliated; white and grey bands of some centimetres in thickness, bestow a very appreciable decorative effect to the employed metamorphic rock. The foliation undertakes different dips, according to the cutting and the laying of each blocks; it is almost horizontal or slightly bended along the shaft, while in the pedestal blocks it can reach high values of inclination up to the vertical.

Figure 3. The fractured pedestal which supports the shaft of the column (19 m high).

The petrographic analysis shows that the grey bands are formed by small crystals (mainly between 0.3 and 0.8 mm) while the light bands are composed by medium-large crystals (between 1.5-2.5 mm). The formers have a granoblastic structure tending towards the equigranular; the latters show a granoblastic heterogranular structure with calcitic crystals of irregular habit (Figure 4).

The volume weight values (Figure 4) do not show substantial difference between the grey and the white parts; the value is around 2.72 g/cm^3. There is, however, a notable weight reduction on the weathered external part of the marble. The absorption of water, determined by capillary rise method (Figure 4), indicates that the white bands, with greater crystal sizes and wider intercrystalline spaces than the grey bands, present absorption levels which are around double. However, once pitting has taken place, the grey marble reveals the highest absorption levels: in these cases the attained values are double of the ones of white marble and four times the ones of fresh grey marble. This tendency is confirmed by the data deduced by the total immersion water absorption test (Figure 4).

Thin section along the passage of white band (upper part) and grey part (lower part) of the marble

Natural volume weight values of the marble

sample	natural volume weight (gr/cm)
unweathered grey marble	2.72
unweathered white marble	2.73
weathered marble	2.50

X-ray diffractograms

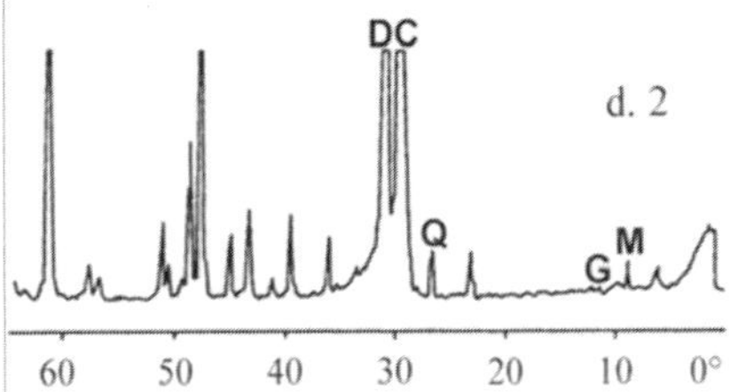

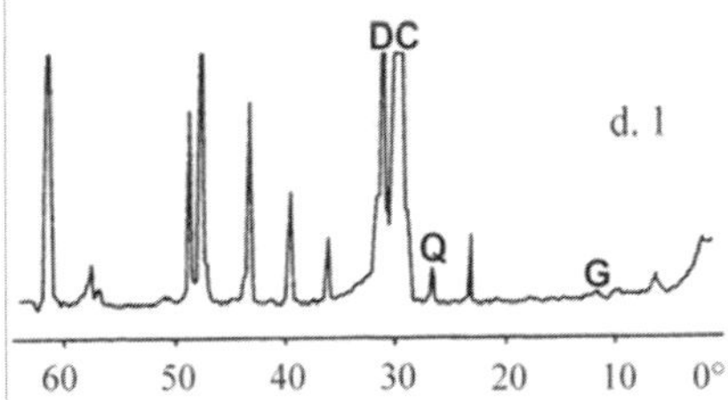

Water absorbtion by capillarity

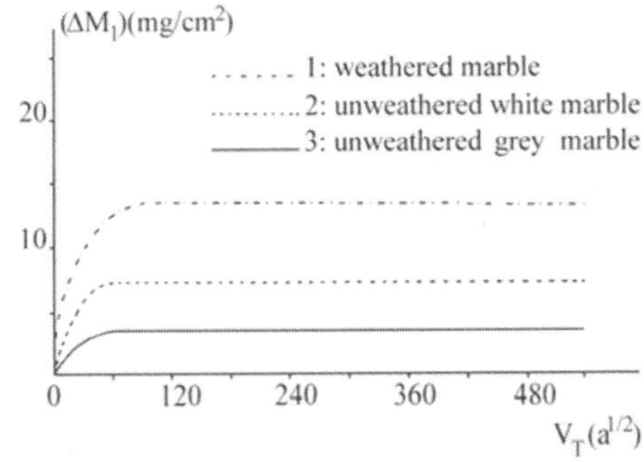

Water absorbtion by total immersion

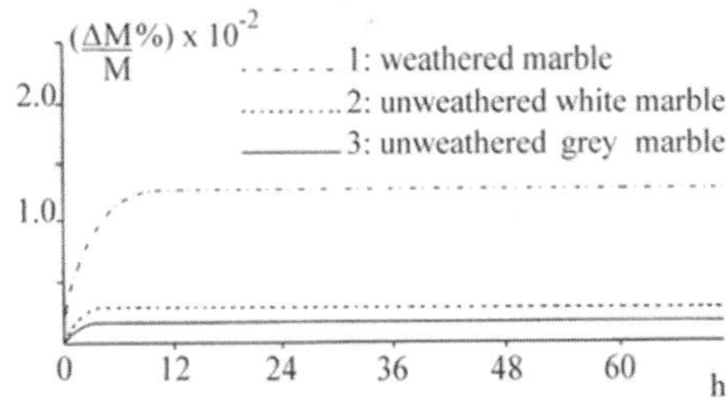

Concentration of soluble salts (1: white marble, 2: grey marble, 3: weathered grey marble)

Sample	Cond.	HCO_3^-	CO_3^-	Cl^-	NO_2^-	NO_3^-	SO_4^-	Na^+	NH_4^-	K^+	Mg^{++}	Ca^{++}
	(S/cm)	(mg/100$_{cc}$)										
1	51.1	2.684	0.360	0.0406	0.0029	0.0007	0.0598	0.0539	0.0087	0.0270	0.0380	0.833
2	68.0	4.271	0.180	0.1060	Ass.	0.0013	0.1481	0.1009	0.0079	0.0263	0.1594	0.93
3	92.4	3.783	0.660	0.0438	0.0048	0.0010	0.0819	0.0529	0.0126	0.0202	0.0222	1.601

Figure 4. Physical properties and petrographic, mineralogical and chemical composition of the Proconnesium marble (by F. Zezza et al., 1992).

At X-ray diffraction analysis, the grey and white bands show a paragenesis featured by calcite and dolomite with quartz (Figure 4.d1) enriched by muscovite in the grey bands (Figure 4.d2). The products of weathering, with particular reference to the soluble neo-formation minerals (sulphates, chlorides and nitrates), is shown in the table of Figure 4.

SEM and EDX examination of the weathered surfaces shows: a) the calcitic crystals are visibly corroded; the state of intercrystalline decohesion of the external surfaces is clearly visible, with consequent detachment of calcite flakes; b) the crystal size influences the development of pitting which tends to settle usually on grey sheets of small crystals (200-400 μm) and, less frequently, on larger crystals (500-700 μm); c) concentrations of carbonaceous particles, of about 20-40 μm in size, and of cubic sodium-chloride crystals (10-15 μm) are found in the intercrystalline spaces of the weathered surfaces; d) fungal growth (*Lecanora dispersa* and probably *Lecanora campestris*) along microfractures causes detachments of external surface flakes, affected by pitting (Figure 5).

Pitting phenomenon is common in many marble monuments. This type of decay is usually attributed to the action of lichens and fungus, even if someone thinks that it is due to localised corrosions by the action of water acidified by carbon dioxide which preferentially attacks rock surfaces with fine granulometry together with coarse veins of calcite crystals.

3. FOLIATION AND FISSURING

The anisotropy of the blocks is related to the primary (foliation) and secondary (fissuring) structures of the marble. The first ones are differently dipping in space and depend on the placement of the blocks of foliated marble. The second ones have an inclination varying from the horizontal to the vertical; these can be also orthogonal to the foliation planes (Figure 6). Nevertheless, the presence of fracturing joints, which are open along the dip of these planes, often corresponds to the different inclination of the foliation planes being closely related to them. Discontinuities are differently inclined and also have openings and spaces variable; the intersection of two or more discontinuities often causes the formation of wedges (Figure 7).

The lower component of the pedestal (*crepidoma*) is formed by two blocks. The first element ("crepidoma a") shows a foliation with sub-horizontal dip to which a discontinuity set is associated; there are other four joint systems differently inclined in space. The second element ("crepidoma b") has the foliation covered by a coating of superficial weathering; the most frequent discontinuities, horizontally arranged, are associated to other four prevalent systems. The second component (*plinth*) shows a sub-horizontal foliation to which a system of fissuring joints is associated; in addition to these discontinuities there are other four sets of fissuring joints with much varying dips. The most fractured side faces towards S where the highest concentration (about 60%) of discontinuities occurs. A well developed fracture line of S-W orientation, defines also an appreciable wedge. Some detachments are provoked by the fissuring.

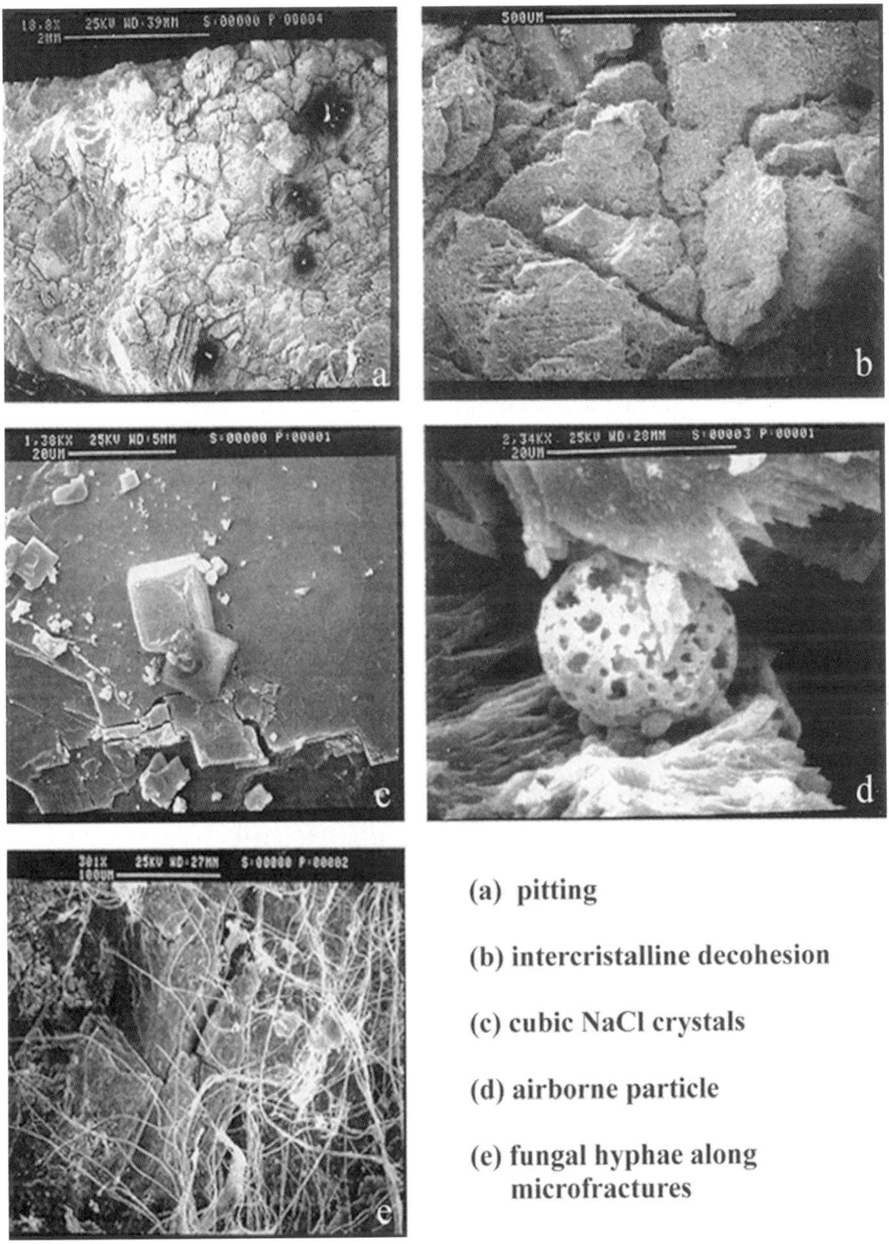

Figure 5. Weathered marble, by F. Zezza et al., 1992.

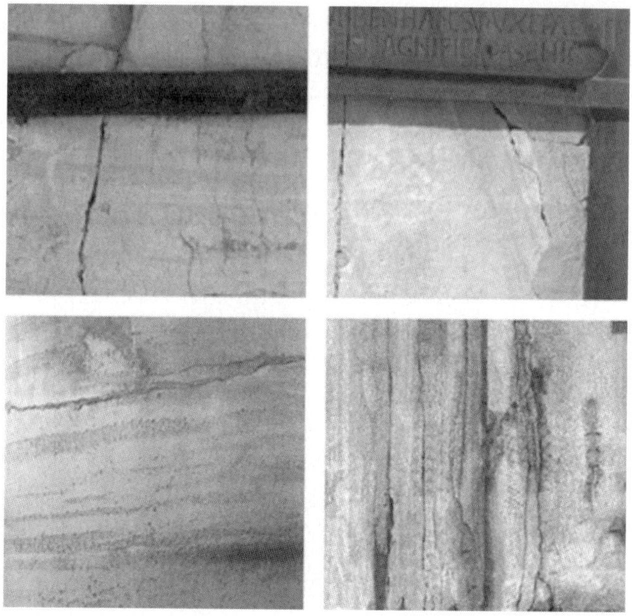

Figure 6. Relationship between fractures and foliation.

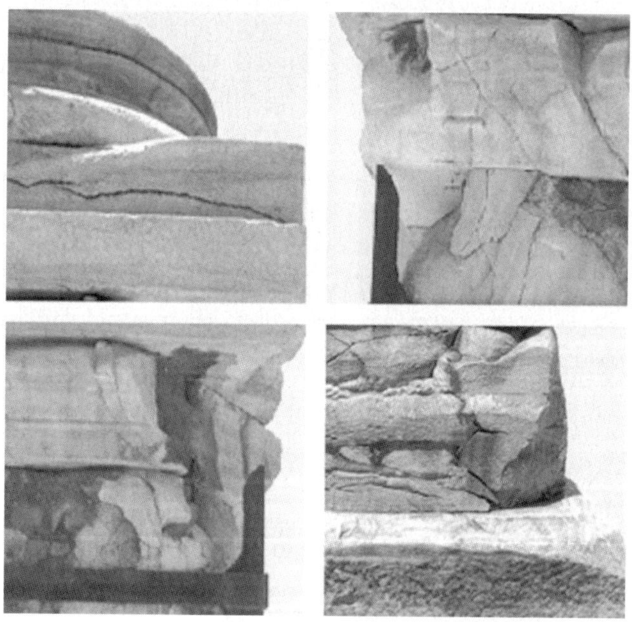

Figure 7. Fissuring within the marble blocks.

Two blocks compose the third component (*dado*). The first ("dado a") presents a foliation, both in vertical position and 60° dipping, strictly linked to the fractures of the block; in addition, the surface gives evidence of two joint systems; one is perpendicular to the foliation while the other is arranged in sub-horizontal position. The second block ("dado b") has a foliation of about 70°, dipping opposite to the foliation of "dado b". The systems of fissuring joints within this element are three; the first is related to foliation; the second is orthogonal to the first; the third lies sub-horizontally. This component is the most fractured among the five components of the pedestal and, besides, it is the only component which has in the past received preservation intervention (bounding). Indeed, the dado shows closed fractures, from 10 to 30 cm; the intersection of these with others, orthogonal, divides the dado in relatively small size fragments.

The fourth component (*cornice*) is characterized by a sub-horizontal foliation with the related fractures; in addition to these joints there are three more systems of fractures differently arranged. In this component the fractures normal to the foliation are also the most frequent; the detachments of the N-W corner of the block are due to them.

The fifth component (*orlo of the base and torus*) presents a sub-horizontal foliation which generates a set of discontinuities; there are other three systems of fractures differently oriented. In this component of the pedestal the fractures produce some wedges.

The drums of the shaft show a visible foliation disposed in horizontal or inclined (25°-30°) bands. The drum I, at the bottom of the column, presents the worse state of conservation; it is covered by mortar and it shows bounding traces. The other drums are generally not fractured, only the drums II and VII are intersected by notable fractures opened along the foliation planes. While the drum V puts in evidence fractures orthogonal to the foliation.

Also the marble of the capital shows a sub-horizontal foliation; there are no notable fractures altering the compactness of the block even if it shows small superficial detachments.

The data regarding the statistical-structural analysis are represented in the graphs of Figure 8 which provide evidence of only one representative plane for every discontinuity set, the foliation included, for a clear graphic representation. Every block forming the pedestal shows localized concentrations of fractures, differently settled into the pedestal components and determined by different dip of the foliation and eccentric loads, due to oscillating pressure centre. The ejection of parts in the marble components along their external zones is common. In addition to mechanical stresses, thermal variations represent a further fracturing cause linked to the loss of cohesion.

Within the fracture systems, the textural and structural properties of the crystalline lithotype influence in different ways the internal migration and crystallization of the soluble salt and originate different forms of stone decay. This suggests that the susceptibility to salt decay of marble basically depends on the "intrinsic parameters" of the rocks, such as pore system and texture. The decay evolution is regulated mainly by the textural features; the artefacts, worked in relation to the marble-anisotropy, show the deterioration

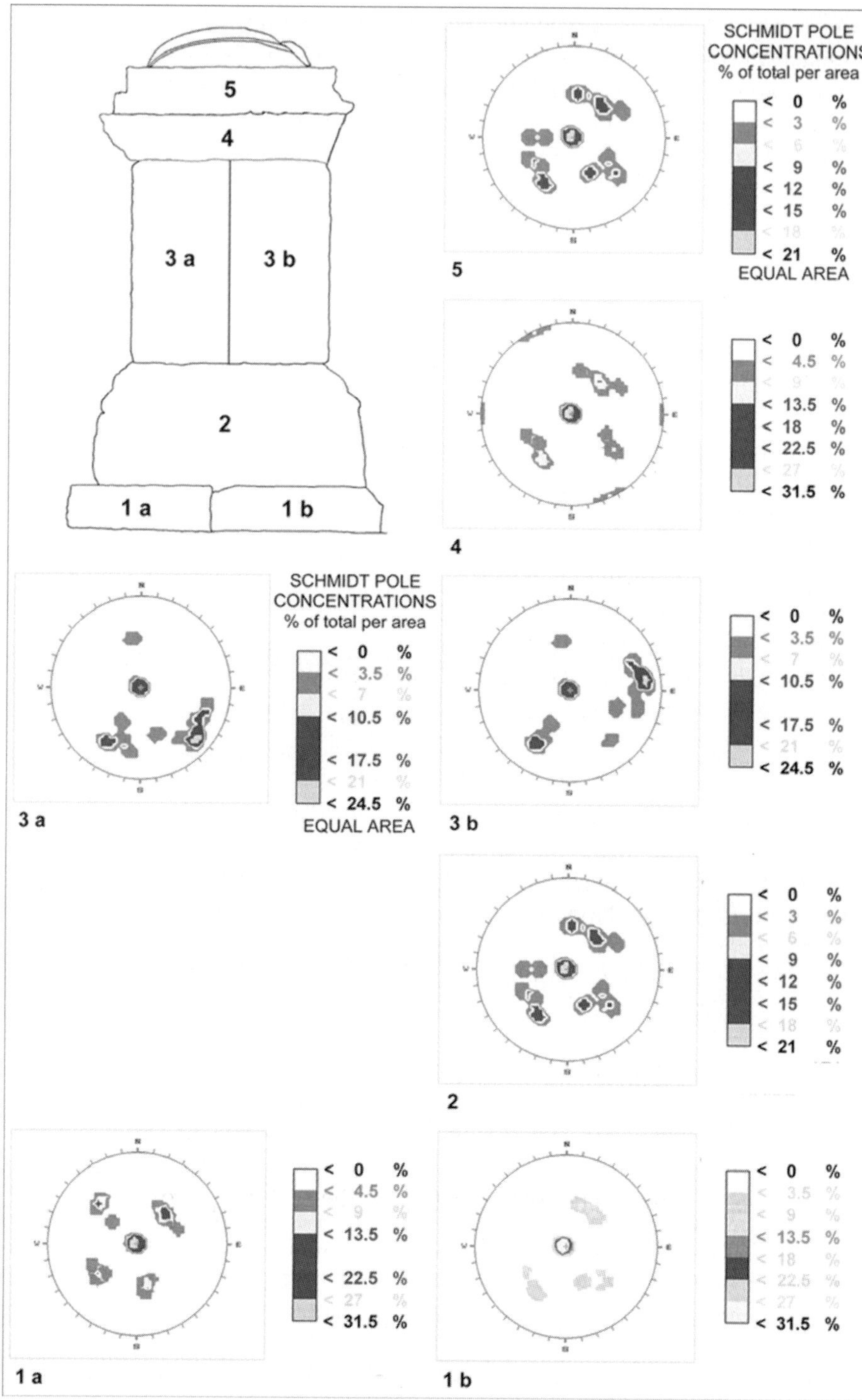

Figure 8. Fissuring rate of the pedestal components.

effects in parallel arrangement to the foliation lines, both for the early micro-cracks and granular disaggregations, as for the following large fracture and detachments. The textural features have influence on the kinetics of the decay process and show themselves directly proportional to superficial area, exposed to atmospheric attack and to loading stresses. Whatever the cause might be (physico-chemical, physico-mechanical, biochemical) of the marble decay process, the main effect is a weakening of marble tensile strength.

4. FAILURE

In the domain of the rock mechanics, failure is the process by which a body under stress loses cohesion and divides into two or more parts, commonly by means of a brittle fracture. In the field of monument conservation the stone decay form that manifests by means of the formation of fracture joints, which can entail the mutual displacement of the parts, is named *fissuring*. The systems of neo-formation fractures affecting the building material depend on planning defects (choice of material, structural errors) or extrinsic causes (occurrence of aggression from exterior). The fissuring can show an interrelation with the structural properties of the stone (foliation, lamination, bedding) or be independent of the building material structure. For a block-jointed system it is convenient to subdivide, according to Trollope and Brown (1966), the observed modes of failure as it follows: i) failure along continuous joints; ii) failure within the rock material; iii) composite failure involving rock material and joints. Where failure occurs along joints and through rock material laboratory tests have suggested that even with block-jointed materials some dilatation may accompany shear failure under low confining pressure. When the normal pressure is high enough to inhibit dilation composite failure develops. The failure which affects the roman columns in Brindisi belongs to the last mode of failure.

The ultrasonic pulses can be used as a non-destructive test to establish the fracturing state and the weathering degree of a stone. The velocity of longitudinal waves can, indeed, be correlated with the physical and mechanical characteristics (strength, Young's modulus) of a fractured marble. The direct method has demonstrated that the elastic longitudinal wave propagation in marbles is controlled to a large extent by: i) the specific petrofabric with V_p values relatively lower for direction of propagation parallel to c-axes maximum concentration of calcite crystal and higher velocity along a-axes concentration (Klima and Kulhanek, 1968; Kern, 1974); ii) the textural anisotropy (Zezza, 2002).

In the specific case, the values of propagation of velocity in the marble blocks are influenced by two factors: a) petrofabric of the marble; b) fissuring of the blocks and its state of decay.

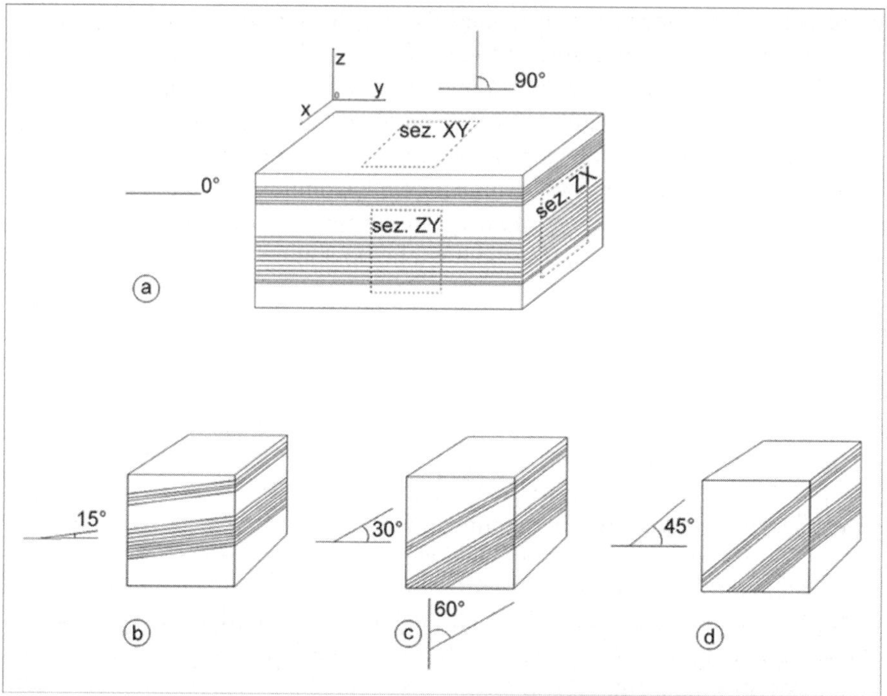

Figure 9. Ultrasonic measurements (direct method) parallel and perpendicular to the foliation (a) or with different values of inclination (b, c, d).

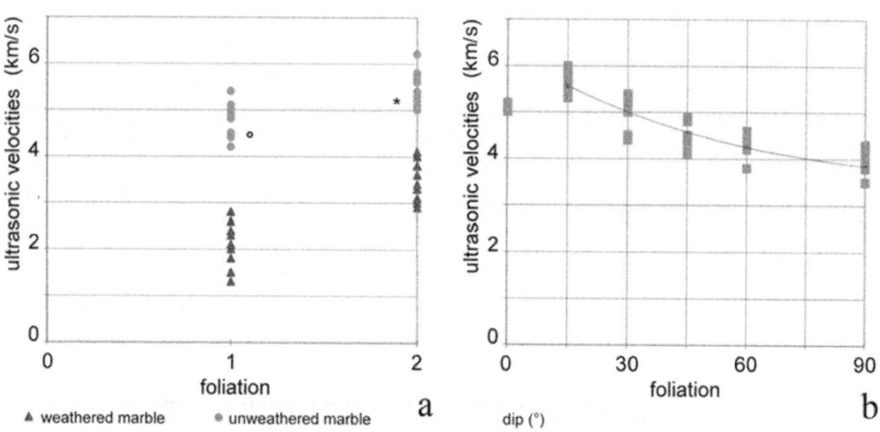

Figure 10. Comparison of the ultrasonic velocities between unweathered and weathered white (°) and grey (*) marble measured along the foliation (a) and the sloping bands (b).

To estimate the incidence of the two factors, petrographic analysis and ultrasonic measurements have been carried out respectively on thin sections analysed by optical microscope and on samples shaped in relation to the foliation planes (Figure 9). The thin sections have been examined according to parallel (XY section) or perpendicular (ZX section and ZY section) foliation planes. The section XY concerns a white band of marble. The heteroblastic structure shows crystals varying between 0.40 mm and 2 mm, with medium values around 1.20-1.40 mm, in size. The outlines of these crystals appear generally regular (idiomorphic); the isodiametric crystal does not show preferential elongations while the others reveal the axis of maximum elongation arranged alternately according to X and Y; therefore, directions of preferential development are not evident parallel to the foliation planes. The section ZX contains white and grey bands. The crystalloblastic structure of the grey bands is formed by crystals with generally regular outlines of smaller sizes (0.1-0.5 mm) than the ones of the white bands; apart of a low percentage (10%) of isodiametric crystals, the most part of the crystalline structure shows a preferential direction that coincides often with the X axis. The white bands are formed mainly (80%) by isodiametric crystals; the remaining percentage of them shows a direction of elongation corresponding to foliation planes. In the section ZY the crystals of the grey bands show a more pronounced elongation, according to Y axis, parallel to the foliation planes (about 65%-70%).

On the basis of petrographic properties the structural-textural anisotropy of the marble (size and orientation of the crystals, foliation, alternation of centimetric bands of different thickness) appears evident.

As regards the velocity of ultrasonic pulses both in parallel and orthogonal directions to the foliation and along the foliation planes, differently inclined, the values are indicated in Figure 10. There are appreciable differences for the measures of the medium values of velocity parallel to the foliation planes (Figure 10a); the type of analysed bands (4.8 km/s, white band; 5.6 km/s, grey band); the medium values of the ultrasonic velocity have appreciable decreases, respectively 2.0 km/s and 3.4 km/s, when the marble is weathered.

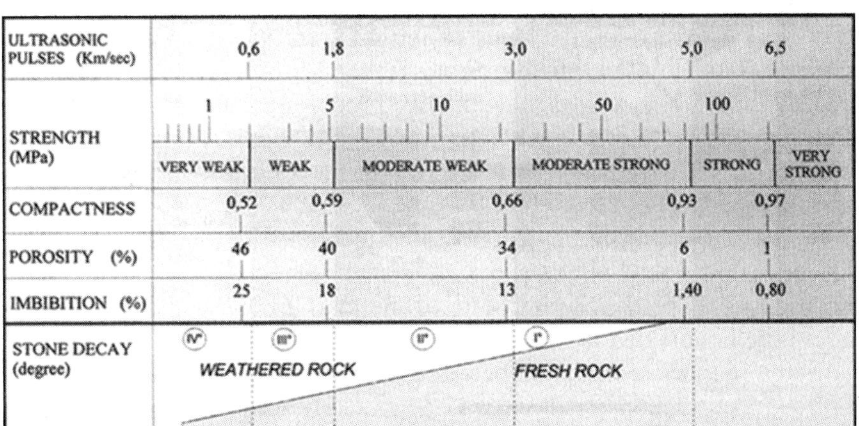

Figure 11. Ultrasonic velocities and degree of stone decay expressed by the average values of indices and physical-mechanical parameters (by F. Zezza, 1996, modified).

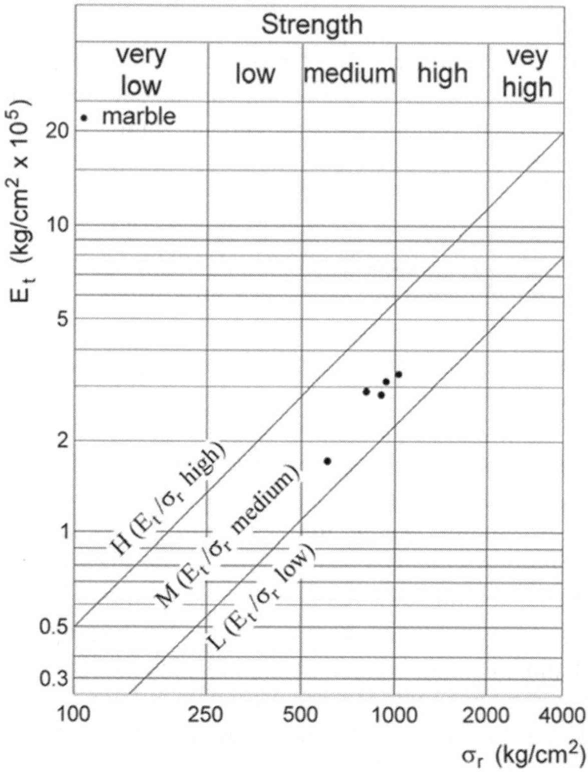

Figure 12. Mechanical classification of fresh Proconnesium marble based on uniaxial compressive strength and modulus ratio.

The inclination of the foliation influences appreciably the velocity of the elastic longitudinal waves. The graph of Figure 10b shows the change of velocity vs. the inclination of foliation. The values decrease between 0°-90° (15°: VL 5.4 km/s; 30°: VL 5.0 km/s; 45°: VL 4.6 km/s; 60°: VL 4.3 Km/s; 90°: VL 3,9 Km/s). On the basis of the different values of ultrasonic velocities according to the foliation planes, the index of anisotropy (AM%) has been defined which comes out directly proportional to the degree of the structural-textural anisotropy of the marble. The index of anisotropy values surveyed in the analysed marble lie between 17% and 30% and more frequently around 21%-22%.

As regards the fracturing degree, it is evident that it influences the spreading time of ultrasonic pulses and, consequently, the values of velocity, even conditioned by the weathering state of the marble. The employment of advanced methodologies in the field of non-destructive analysis (Zezza, 1996) allowed to define a correlation between the ultrasonic velocities, indices and physico-mechanical parameters of stone material (Figure 11). The characterization of the pedestal blocks is made by taking into account this type of correlation, also supported by laboratory tests. With regard to these tests, for example the uniaxial compressive strength on intact samples, *Proconnesium marmor* doesn't belong to marbles of the best quality; in fact, the measured values of resistance are of about 1088 kg/cm^2 (horizontal foliation), of 975 kg/cm^2 (vertical foliation) and decrease considerably to 520 kg/cm^2 when foliation is inclined at 60°. To these values there are corresponding values of Young's modulus respectively of 362000 kg/cm^2, 25000 kg/cm^2 and 173000 kg/cm^2. Therefore the *Proconnesium marmor* is, on the basis of uniaxial strength (C_o) and Young's modulus (E_t), of "medium" resistance (Figure 12) as the ultrasonic measurement, have valued.

The representations of the non-destructive analysis in situ show the position of the direct and semi-direct measurements on the south side of the pedestal (Figure 13) and the profile of the indirect and direct measurements carried out along the shaft (Figure 14).

A decay pattern, by way of example, is shown in Figure15 concerning the physico-mechanical classification with the textural, structural and morphological characteristics of a component (*dado*) of the pedestal. In Figure 16 the decay patterns for each component of the pedestal are represented whereby all the stressed elements (petrofabric, foliation, weathering, fissuring degree) contribute to assess the present state of conservation.

The application of the homeo-surface or direct method on the shaft has allowed to detect in situ the changes in physical conditions of decayed marble and the depth of weathered layers. The distance-time diagrams, which include the number of segments (each linear segment indicates the velocity value of every thin layer passed through) and their reciprocal relations, are representative of the buried structures related to weathering (F. Zezza, 1989). The section of Figure 17 demonstrates its state of conservation and provides significant evidence of a weathered surface layer varying from 1.4 to 3.5 cm in thickness. Weathered parts are well developed inside the drums (drums II and VII) where neo-formation fractures are opened along the foliation planes. The passage between the decayed and the fresh marble is irregular; the weathered layer, in addition, offers a structure formed frequently by three intervals, sometimes two and exceptionally four (dosseret). The three intervals structure prevails in the middle-lower part of the shaft while it seems to increase the thickness of the stone decay in the upper part. According to the recorded values of ultrasonic velocities, the weathered layers approximates from a geomechanical point of view a marble "moderately weak" to "moderately strong", while the underlying fresh marble is "strong" (VL = > 4.0 km/s).

5. RESULTS

The results of ultrasonic non-destructive analysis have established two typologies of weathering substantially different: the one is represented by the long-term process of weathering suffered by the *Proconnesium marmor*

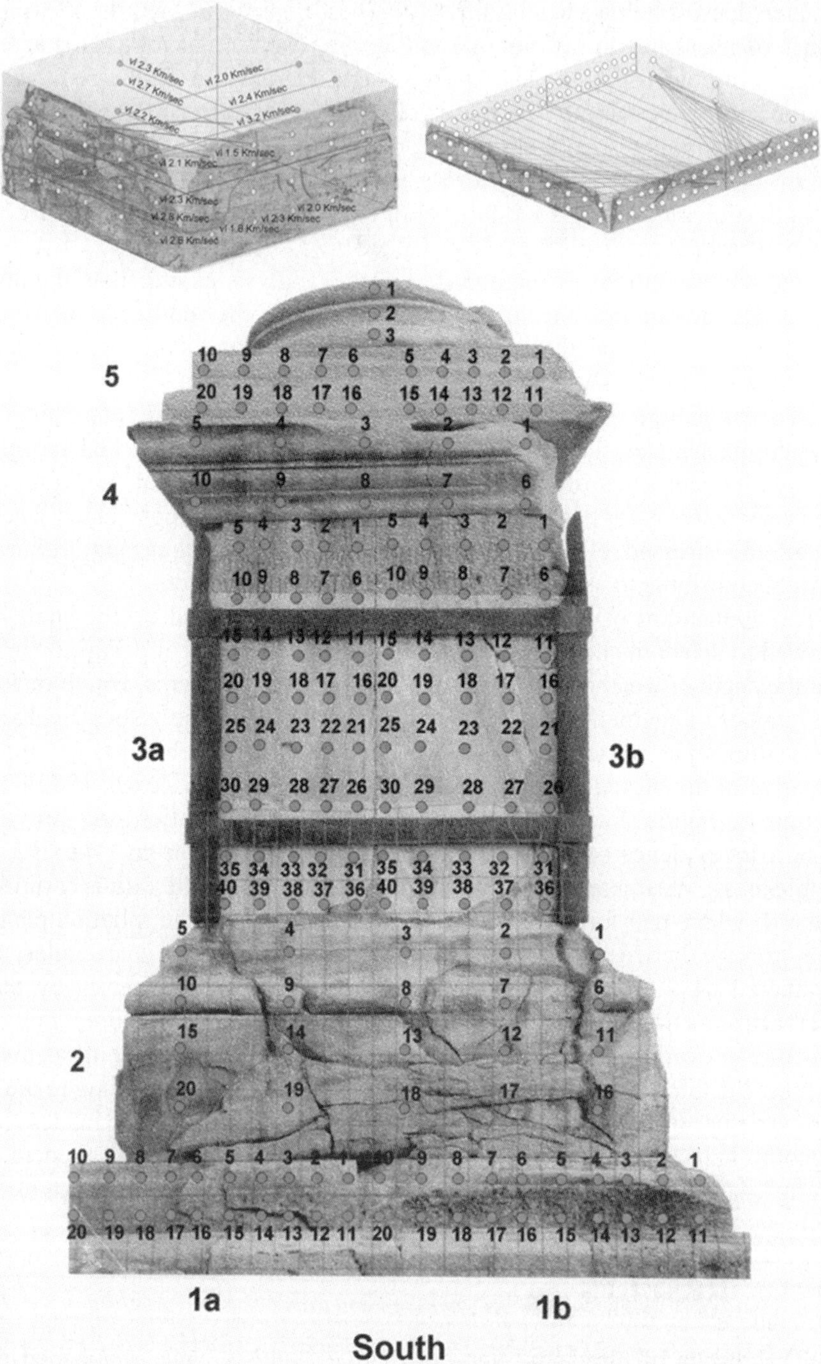

Figure 13. Location of the ultrasonic measurements (direct and semi-direct methods) on the south side of the pedestal.

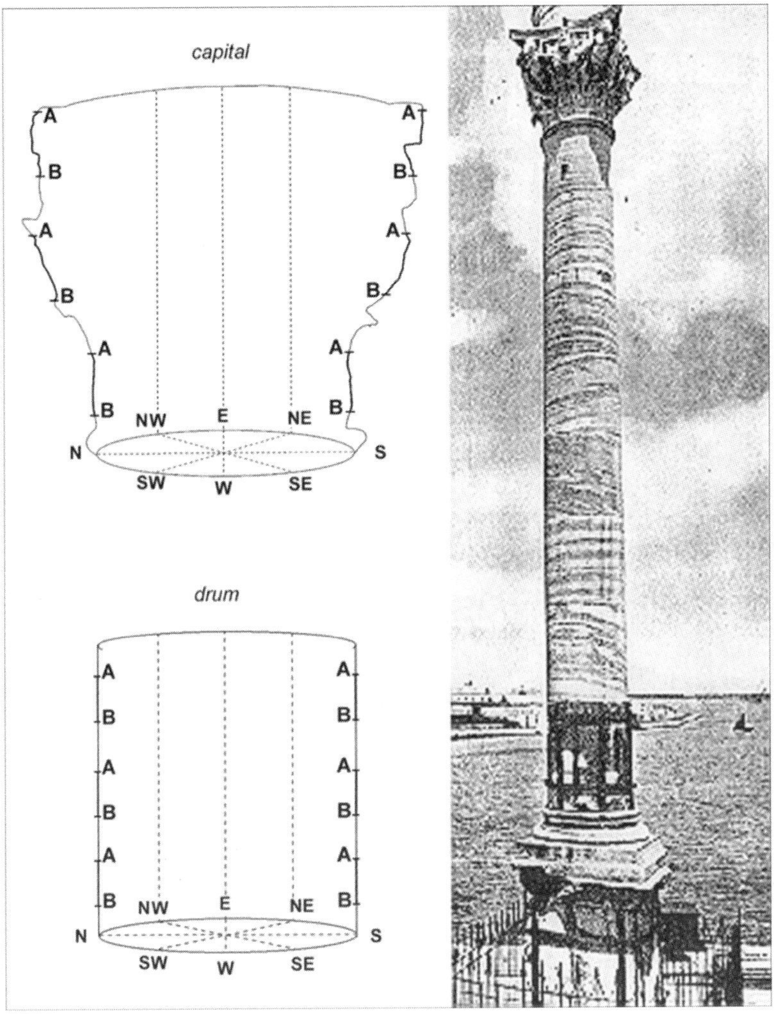

Figure 14. Plan of the ultrasonic measurement (direct and indirect methods) along the shaft and the capital of the column.

during about two thousand years of exposure to the decay agents; the other is linked to the fissuring state of the marble blocks, in particular, of the ones forming the pedestal of the column.

In reference to marble decay of the shaft, ultrasonic pulses allowed to point out that the sculptural parts (capital) or the shaped ones (drums and dosseret) suffered the effect of weathering agents to a maximum thickness of about 3.5 cm; therefore, the marble decay is confined to the surface. There are some exceptions affecting the drum I, very fractured, and the drums II and VII, intensively marked by neo-formation fractures along foliation planes. Suitable methodologies of restoration intervention, chosen on the basis of the acquired results, could control or solve the decay of the shaft.

Component: "dado", component of the pedestal with highest fracturing degree

***morphological characteristics**:* detachments as wedges at the corners
***structural characteristics**:* foliation (60° - 70°) sloping up to vertical
***ultrasonic velocities** (direct method):* values (2,5 – 3, 5 km/sec, locally 4,0 km/sec) increasing towards the inner parts
***mechanical classification**:* from "moderately weak" to "moderately strong"

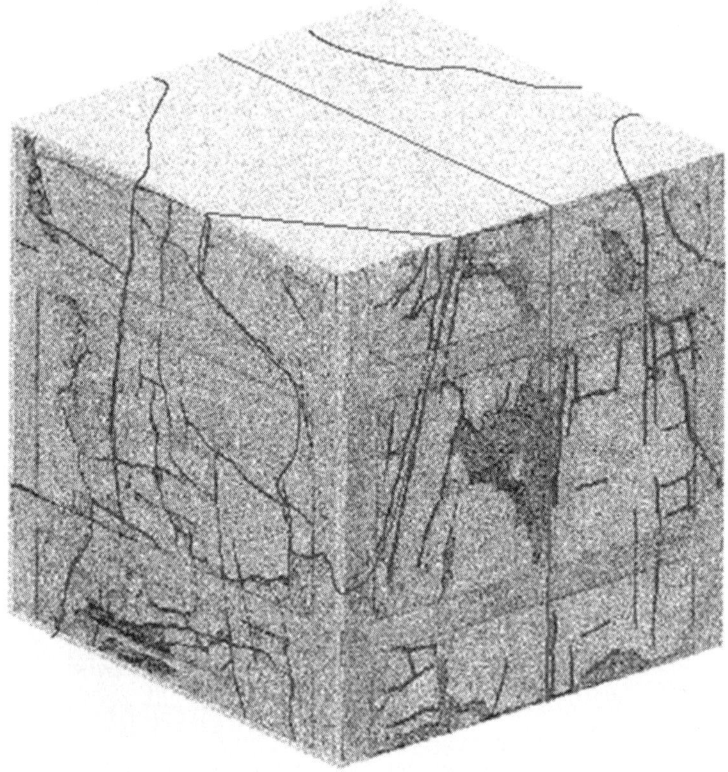

Figure 15. Form for a homogeneous collection of data, including the stone decay classification, used for the pedestal components.

Regarding the fissuring system of the pedestal of the "entire" Roman column, the statistical-structural analysis has shown that it is not different in comparison with that of "cut-off" column which fell down in 1528. Since the strength of *Proconnesium marmor* is not really very high, such an opening of joints can be facilitated along the foliation unfavourably arranged and the neo-formation fractures must be evaluated as a consequence of the irregular distribution of loads owing to the oscillations of the shaft exposed to the wind action. From this point of view the "faulty" fitting of the blocks could have caused the detachment that the ultrasonic pulses have pointed out.

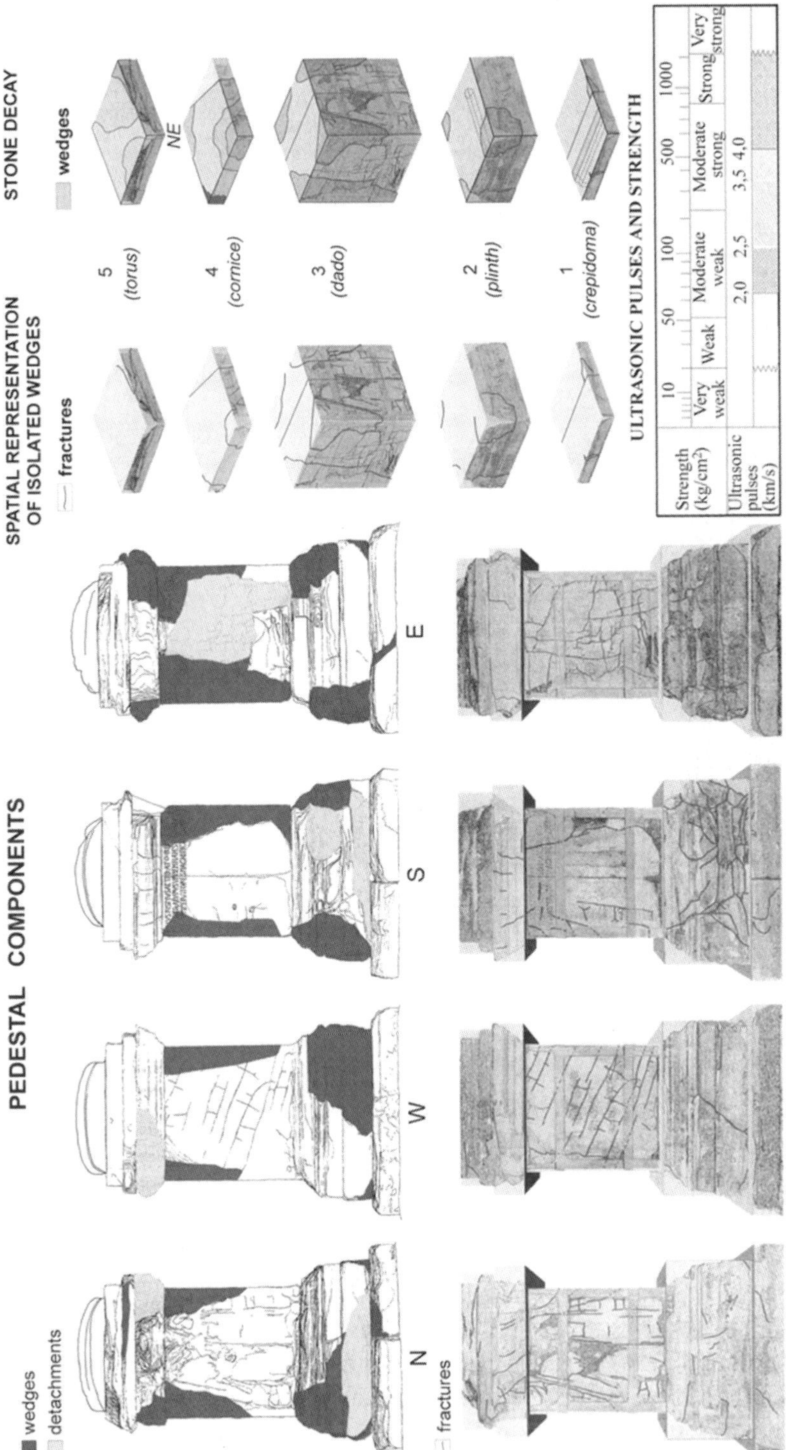

Figure 16. Failure and stone decay classification of the pedestal detected by ultrasonic measurements.

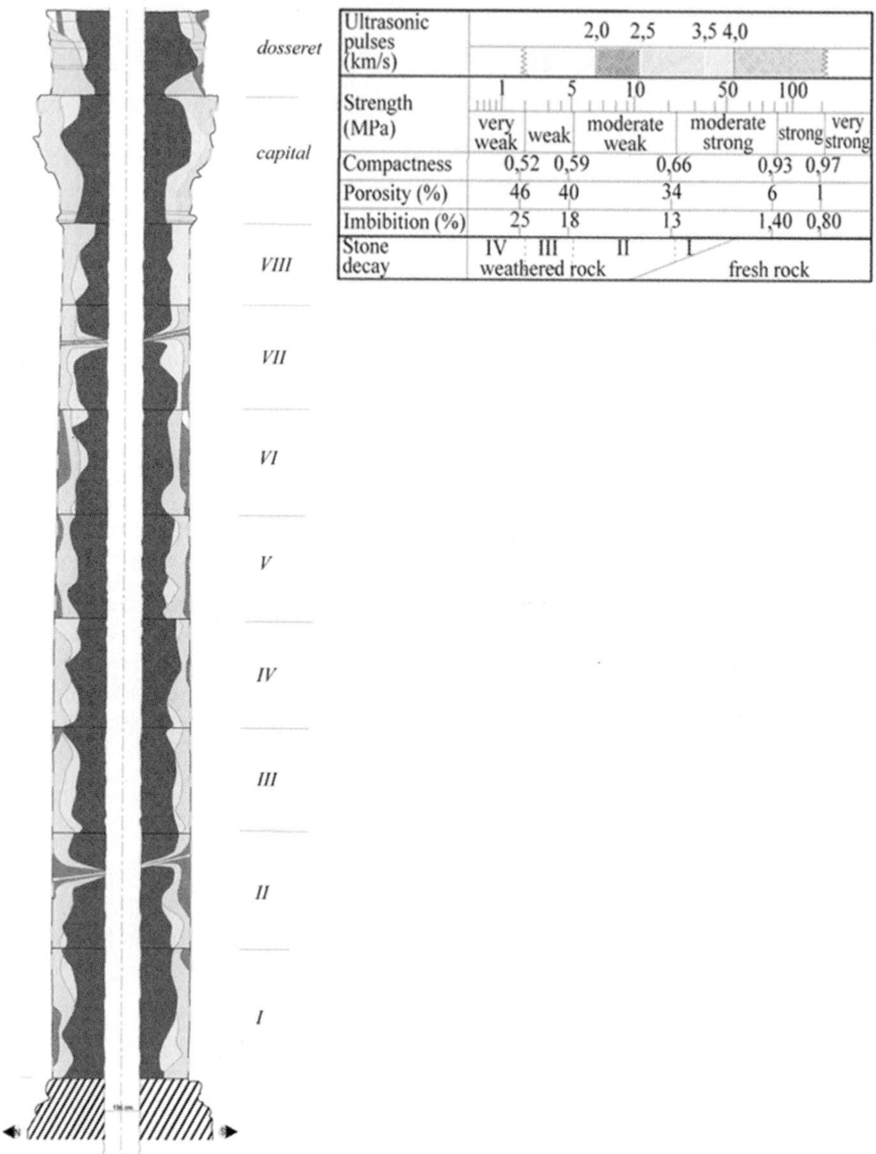

Figure 17. Stone decay classification of the column (shaft and capital) detected by ultrasonic measurements (indirect method).

Within the single blocks, in fact, detachments and wedges have such a position that it can not be mistaken for the one normally observed on failure in laboratory samples subjected to strength tests. Whereas, in the specific case, the edges and the tips marking perimetrally the lower and upper surfaces of the base blocks are the most damaged.

The non-destructive ultrasonic analysis has allowed classifying and characterizing from a geomechanical point of view the marble of the pedestal components. The indexes and parameters prove that the failure of the pedestal is very critical. The pedestal of the shaft which fell down in 1527 emphasizes that such a fissuring system characterizes a base reaching its breaking point in terms of supporting the considerable load transmitted by the shaft.

REFERENCES

Kern, H., 1974, Gefügeregelung und elastische anisotropic eines marmors, *Contr. Mineral. Petrol.*, **43:** 47-54.

Klima, K., Kulaner, O., 1968, Quantitative correlation between preferred orientation of grains and elastic anisotropy of marble, *IEEL Trans. Geosc. Electron.* ***GE-6*** (3), 139-143.

Trollope, D. H., Brown, E.T., 1966, Effective stress criteria of failure of rock masses, in *Proc. 1st International Congress of the International Society of Rock Mechanics*, **1** (3): 515-519.

Zezza, F., 1989, Computerized analysis of stone decay in monument, in *Proc. 1th International Symposium on Monument Conservation in the Mediterranean Basin*, 163-184.

Zezza, F., 1992, and Macrì F., Decrouez D., Barben V., Ramseyer K., Garcia-Rowe J., Saiz-Jimenez C., Origin, weathering and biological colonization of the marble terminal column on the Via Appia, Brindisi, Italy, in *Proc. 2nd International Symposium on Monument Conservation in the Mediterranean Basin*, 99-112.

Zezza, F., 1996, Decay patterns of weathered stones in marine environment, *European Commission Research Report*, **4:** 99-130.

Zezza, F., 2002, The susceptibility of marble to weathering and non-destructive controls on treated surfaces, in *Proc. 5th International Meeting for the Restoration of the Acropolis Monuments*, 628-638.

Chapter 8.2

PROVENANCE, DURABILITY AND DAMAGE ANALYSIS OF NATURAL BUILDING STONES BY MEANS OF PETROGRAPHICAL TECHNIQUES

Roland Dreesen, Peter Nielsen and David Lagrou
Flemish Institute for Technological Research, Materials Technology, Boeretang 200, B-2400 Mol, Belgium; roland.dreesen@vito.be, peter.nielsen@vito.be, david.lagrou@vito.be

Abstract: Several case studies taken from Belgium clearly illustrate the role and importance of optical microscopic investigation in provenance studies, quality control and damage analysis of both historical and modern natural building stones. The provenance of particular "exotic" historical building stones, such as the sandstone ashlars used in the former Roman town of Tongeren, has beeen unravelled thanks to a comparative petrographical analysis. Furthermore, the probable mechanism of frost-damaged and sulfate-attacked calcareous sandstone (the Balegem or Lede Stone) in the Beguinage of Mechelen, has been elucidated thanks to the use of high-quality thin sections and optical microscopic techniques. Finally, a combination of optical microscopic and back-scattering electron microscopic analysis, revealed the origin and processes leading to the un-aesthetic brown staining of modern limestone curbstones, the so-called bluestones.

Key words: natural building stones; petrography; provenance; frost damage; sulphate attack; staining; Belgium.

1. INTRODUCTION

This paper deals with a selection of case studies related to the provenance, durability and damage analysis of natural building stones in Belgium, where optical microscopy proved to be a powerful tool, especially in combination with more sophisticated tools such as automated image analysis and micro-analytical analysis (SEM and BSEM, with on-line WDS/EDS).

The mineralogical and textural composition of a building stone is like a fingerprint. Hence, it can be used in the accurate identification of "unknown",

S.K. Kourkoulis (ed.), Fracture and Failure of Natural Building Stones, 471–485.

strongly weathered or macroscopically unrecognizable historical building stones. Furthermore, the use of high-quality fluorescent epoxy impregnated thin sections is highly recommended, since it will allow to analyze the typology of important intrinsic physical parameters of building stones and cement-based building materials, such as the morphology and spatial distribution of the void system. For instance, the typology of micro-cracks is symptomatic of specific deterioration mechanisms such as swelling, shrinkage, frost-thaw cycles or even biodeterioration phenomena[1]. Moreover, the use of fluorescent light modes in microscopic thin section analysis will allow to assess changes in micro- and capillary porosity and to identify particular diagenetic phenomena s.l. that might have an impact on the durability of building materials[2].

2. PROVENANCE AND DURABILITY ASSESSMENT OF HISTORICAL BUILDING STONES

Over 35 different stone species have recently been inventoried in historical monuments of the province of Limburg (NE-Belgium) and described in an illustrated atlas in detail for the first time[3]. So far no reference documents existed on the market and the atlas has become an important reference tool for restoration architects, monument inspectors and building historians. Most inventoried indigenous stone species are sedimentary in origin; rare volcanic and low-grade metamorphic rock types have been imported from adjacent areas including Wallonia (S-Belgium), the Eifel area (W-Germany), and Champagne-Lorraine (N-France). Most indigenous building stones are no longer commercially available. This is due to a combination of factors: insufficient reserves, decreasing quality over the years, difficult workability, and - most importantly - as a result of competing imported materials. Therefore, historical monuments (especially their oldest parts such as Romanesque church towers) bear witnesses to the former richness in geological building materials in Flanders. Petrographical analysis has been used in order to discriminate between the different varieties of macroscopically analogous ferruginous, quartzarenitic, litharenitic and subarkosic sandstones. Moreover, thin section analysis was very helpful in assessing their provenance, weathering and bioreceptivity potential, in particular when the building stones were fine-grained or strongly weathered.

When possible, small chips of the weathered stone were sampled in situ on the monument. Comparative samples have also been taken from abandoned or active quarries or from standard collections stored at Belgian Universities or in the Belgian Geological Survey.

In a few cases, a petrographical investigation was the only way to trace the probable origin of particular natural building stones. This was the case for the "Roman" sandstone ashlars (Figure 1) commonly used in Roman and

Figure 1. Brownish red Roman sandstone (R), together with grey Carboniferous sandstones (C) and pale calcareous tufa (T). These are recycled ashlars from destroyed 2nd century Roman villas, actually used in the lower part of the Romanesque tower of the St-Martins church of Rutten, near Tongeren.

medieval monuments (the latter material being recycled) within the city and the surroundings of the historical (Roman) town of Tongeren (province of Limburg, north-eastern Belgium). A comparative petrographical study of samples from the sandstone ashlars and of samples taken from cored bore-holes in the deep subsurface of the Campine area, pointed to a probable common geological origin: Buntsandstein fluvial sandstones of Triassic age. Red and green varieties of the sandstones can be recognized, according to their stratigraphic position within the sedimentary suite. The latter corresponds to braided river and sheet floods deposits. The red-stained sandstones display typical iron oxide bearing clay coatings around the sand grains, related to diagenetic phenomena characteristic of the dry and warm continental paleoclimatic conditions during deposition[4]. The sandstones display a characteristic sublitharenitic to subarkosic composition with an important carbonate cement. The latter consists of idiomorphic ferroan carbonates (ferroan calcite, siderite). As a result of chemical weathering (dissolution) the ferroan carbonate rhombs dissolve and grade into iron-hydroxides. Further weathering produces a secondary (mouldic) porosity. The weathered Roman sandstone ashlars most probably represent a green variety of the Buntsandstein. Its provenance can tentatively be assigned to the surroundings of the Roman city of Trier in the Eifel area (NW-Germany). In thin section, the "Roman sandstone" ashlars often display completely weathered carbonate cements, whereby the relict goethite films perfectly mimic the outline of the original ferroan carbonate rhombs (Figure 2). A similar provenance has also been suggested for analogous sandstones encountered in Roman milestones, found near The Hague[5], on a mineralogical-geochemical basis. An acceptable re-

Figure 2. Micrograph of the Roman litharenitic sandstone, showing the weathered rhomb-shaped ferroan carbonate cement, now transformed into iron (hydr)oxides. Normal transmitted light. Non-dyed epoxy.

placement stone for this "Roman sandstone" includes a fine-grained olive-brown (kaki) sublitharenitic sandstone with carbonate cement, the Middle Triassic Udelfanger sandstone, that is actually quarried in the area of Trier.

As durability of historical building stones depends to a large extent on intrinsic parameters such as mineralogy and texture (e.g. porosity), petrographical analysis of thin sections is highly recommended as well. It helps to understand for example why macroscopically similar Belgian historical building stone types differ with respect to their weathering potential (especially frost- and sulfate-attack- resistance)[3].

The presence of frost-sensitive clay-rich lithic fragments and that of recrystallized clay and carbonate cements, explains for instance the low resistance of Carboniferous litharenitic or subgreywacke sandstones against physical and chemical weathering, as opposed to Upper Devonian (Famennian) micaceous subarkosic sandstones. Similarly, differences in pore morphology, including connected intergranular cavities versus mouldic porosity, might explain the differences in frost resistance of macroscopically very similar porous carbonate building stones, such as the Cretaceous Maastricht Limestone and the Palaeocene Lincent Limestone[3].

3. QUALITY ASSESSMENT IN RESTORATION

Microscopical analysis will also help the restoration architects in deciding whether or not frost-damaged and/or sulfate-attacked historical building stones, such as the paleo Cenozoic calcareous sandstones (commonly called white stones) need to be replaced. The latter calcareous sandstones, and more especially the Gobertange and Balegem Stones, are rather sensitive to frost and sulfate attack. Indeed, they commonly show scaling or exfoliation phenomena and they frequently develop black (soiled) gypsum crusts on the external surfaces that are not exposed to rainwater run-off.

The Balegem Stone (or Lede Stone) was certainly one of the most important historical building stones in Flanders during the 15th and 16th century and it has also been exported on a large scale to Zeeland and Holland (The Netherlands). From the 19th century onwards it has been replaced by the indigenous Gobertange white stone and more recently by a heterogeneous series of French and German limestones and sandstones. The Balegem Stone is still quarried on a small scale in Balegem (a village located 15 km south of Ghent, Belgium) for restoration purposes. It appears in outcrop as thin, decimetric calcareous sandstone beds within glauconite-bearing sands of the Eocene Lede Formation. Its characteristic pale bluish-grey color (fresh) soon turns into a yellowish-beige patina due to the oxidation of pyrite, glauconite and ferroan carbonate. Petrographically the Balegem stone can be classified as a fossiliferous glauconite-bearing calcareous sandstone with a carbonate content of 40-60% depending on the frequency of calcareous bioclasts. The microscopical texture varies from structureless, laminated or bioclastic packstone to grainstone fabrics, with varying degrees of bioturbation. The Balegem Stone displays an inhomogenous moldic (macro-)porosity, related to the dissolution of bioclasts (mostly shelly material) and also depending on the degree of carbonate cementation.

Samples of weathered (scaling off) Balegem stones were removed, from the frontage of the 17th century beguinage church in Mechelen (Figure 3) for further study (in 2003). The building stones of the frontage were mechanically cleaned (in order to remove the black soiled crusts and all loose scalings) prior to our sampling (Figure 3). Our microscopical study was essentially meant to detect the depth and intensity of weathering phenomena below the scaled external surface, in order to decide whether or not the stone should be replaced or treated with stone consolidants. Based on detailed microscopic observations the following scenario for the observed deterioration is proposed. Most of the optical microscopic observations have been subsequently corroborated by additional micro-analysis (SEM and BSEM).

Natural frost-thaw cycles have produced a conspicuous sub-horizontal micro-crack pattern in all of the studied samples. The latter micro-cracks run parallel to the external surface. These micro-cracks run both along grain boundaries and through the grains (Figure 4). The cracks occur until depths of 5 to 25 mm below the external surface and are responsible for the characteristic scaling of the building stone. An automated image analysis of micrographs of the samples clearly shows a highly increased porosity in the upper part of the investigated building stones (Figure 5). Macro-porosity (>1μ) values exceeding 30% have been measured in the topmost area below the external scaled-off surface. Normal background values for the porosity of the Balegem Stone are in the range of 2-7%[6].

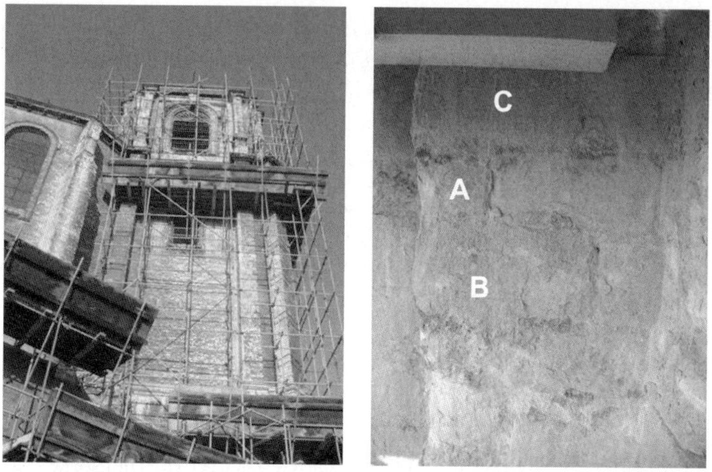

Figure 3. Left: Restoration works (2003) affecting the tower of the 17th century beguinage church in Mechelen, Belgium. Right: detail of the frontage with numbered blocks of Balegem stone (selected for petrographical analysis) still displaying scaling after removal of the black soiled gypsum crust.

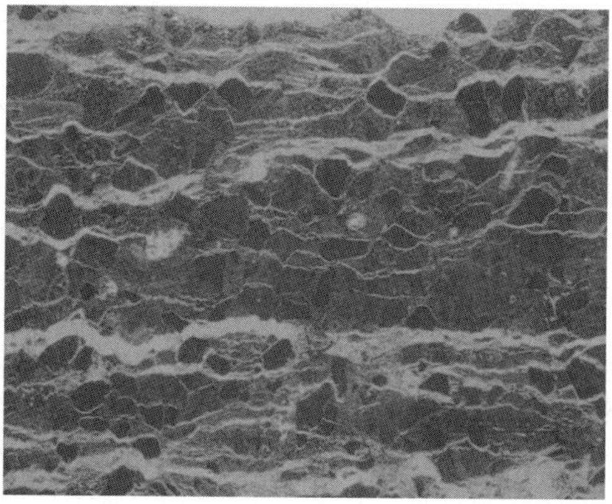

Figure 4. Conspicuous sub-horizontal micro-cracks affecting the Balegem Stone below the external surface. Micrograph taken in fluorescent transmitted light. Dark grains are sand grains. External surface is at the top.

The lobate structure of the glauconite grains most probably has facilitated their physical disintegration (Figures 6-7). Due to increased permeability of the frost-damaged area and the increased water circulation, some of the above physically disintegrated glauconite grains have been partly or completely dissolved (most likely depending on the clay-mineralogical maturity of the glauconite grains[7]) and replaced by fibrous or needle-like gypsum. In many of the surface-parallel cracks, gypsum is only present in direct contact with cross-cut glauconite grains, suggesting that the (dissolving) glauconite grains may have acted as favourable crystal nucleation sites for the precipitation of gypsum. This observation is important as it may explain why glauconite bear-

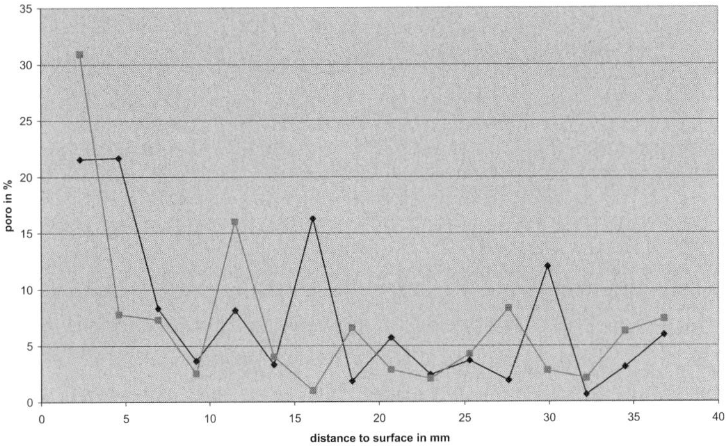

Figure 5. A graph depicting the depth evolution of macro-porosity as measured by automated image-analysis on the same thin section (two different runs located 50 mm from each other).

ing calcareous sandstones are more sensitive for sulfate attack than relatively pure limestones and calcareous sandstones without glauconite. The relationship between glauconite disintegration and gypsum precipitation needs to be further investigated. Furthermore, the fibrous or needle-like gypsum filling widened the cracks and triggered the formation of circum-granular cracks. Eventually, the disintegrated and completely oxidized glauconite grains weathered out in order to produce an important moldic porosity. This secondary porosity has been partially or completely filled with gypsum. Gypsum-filled ovoid secondary voids occur until depths of 30 mm below the external surface of the studied samples (Figures 8-9).

Our study revealed that gypsum is still present in the "sound" building stone sample, after cleaning and removal of the black soiled gypsum crusts, up to a few cm below the cleaned external surface. In order to guarantee the efficiency of the stone treatments (i.e. water repellents and consolidants), one should take measures in order to avoid water transport from the external surface into the core. Moreover the treatment should also allow for a sufficient vapor permeability so that the moisture inside the building stone can still evaporate.

4. UNRAVELING STAINING MECHANISMS

Optical microscopy of fluorescent epoxy impregnated thin sections and of polished slabs, especially in combination with BSEM, proved to be particularly useful in elucidating the mechanism leading to the un-aesthetically brownish staining of so-called "blue stone" limestones. The latter are virtually non-porous, hard, Paleozoic limestones, with a more or less pronounced natural bluish dark-grey color (due to their high organic carbon content).

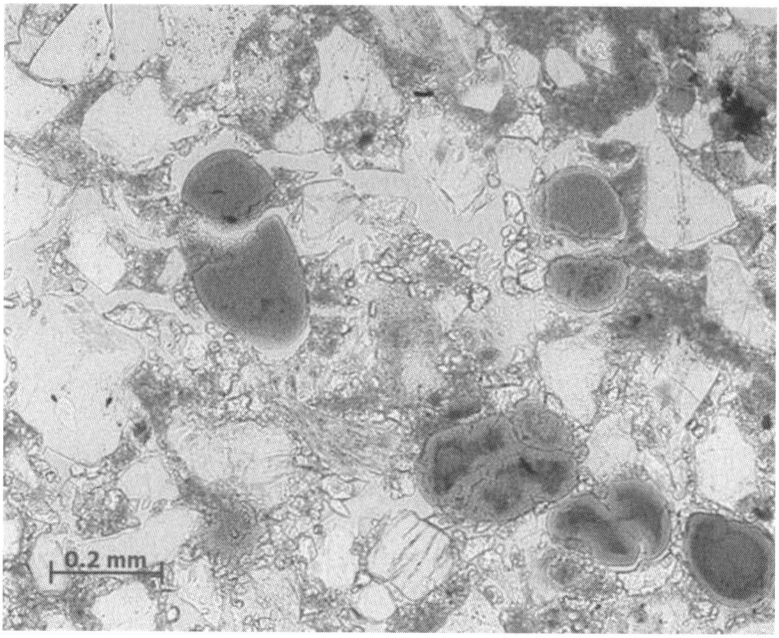

Figure 6. Frost-damaged Balegem stone. Micrograph displaying fracturing of lobate glauconite grains due to frost damage. Normal transmitted polarised light. Macro-porosity is highlighted by the use of a yellow dyed epoxy.

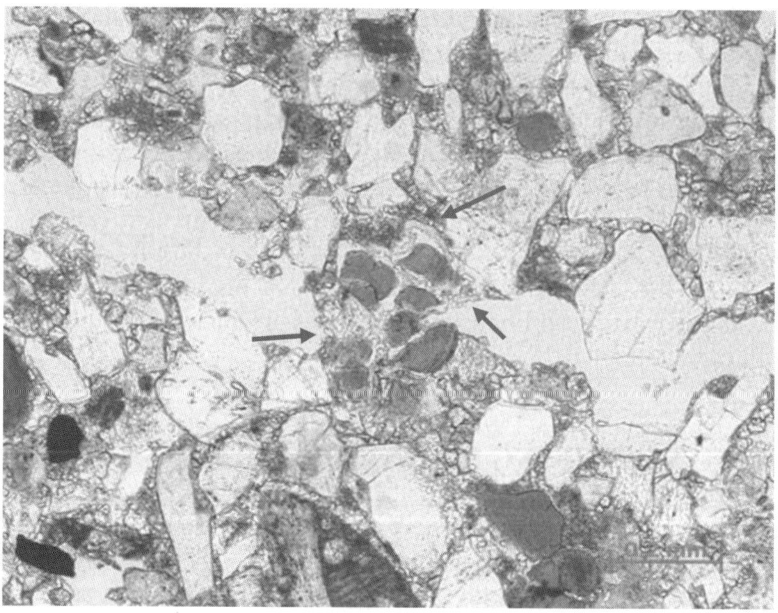

Figure 7. This picture shows a completely physically disintegrated former lobate glauconite grain. White mineral infilling (see arrows) is fibrous gypsum. Normal transmitted polarised light. Macro-porosity is enhanced by the use of a yellow dyed epoxy.

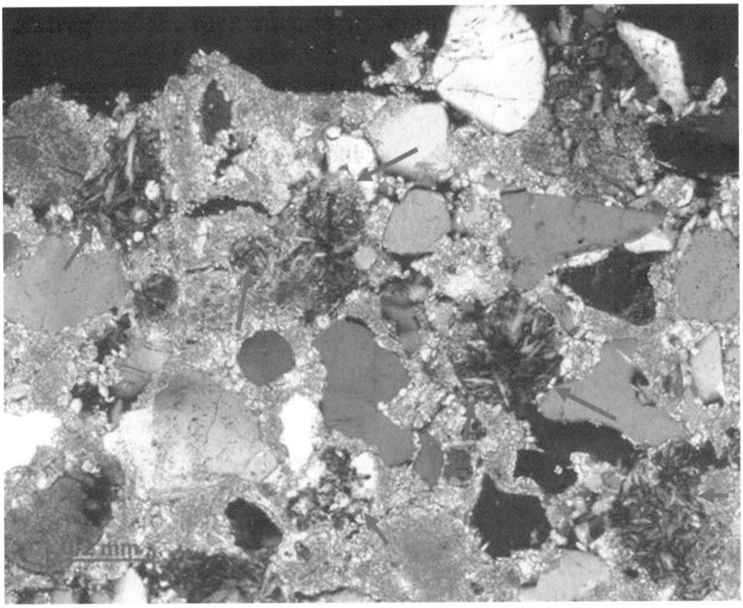

Figure 8. Frost-damaged Balegem stone displaying evidence for sulfate attack. Micrograph shows mouldic porosity completely obliterated by the infilling of fibrous gypsum crystals (arrows). The ovoid gypsum concentrations correspond to weathered and dissolved glauconite grains. External surface is at the top.

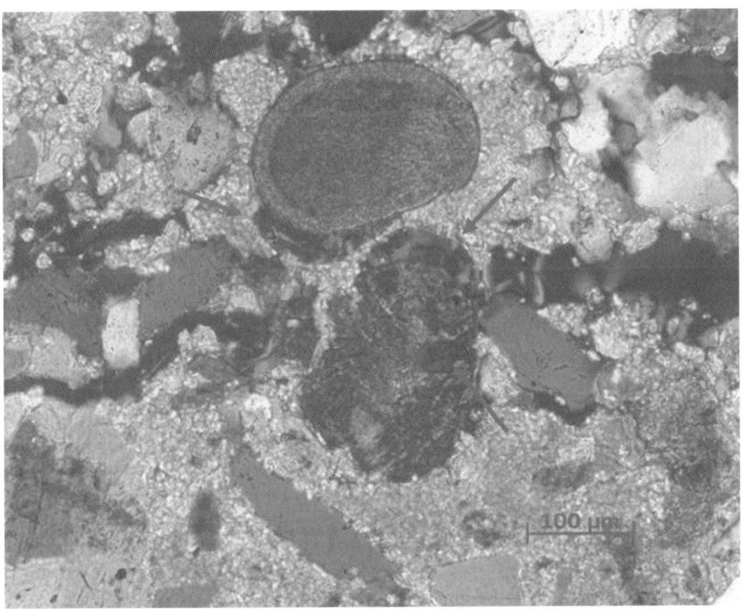

Figure 9. Micrograph showing the preferential precipitation of gypsum in direct vicinity of glauconite grains, both in the frost micro-cracks as in the circum-granular micro-cracks. Polarized transmitted light with crossed polarizers.

Two sets of blue limestone were studied: the traditional Lower Carboniferous Belgian blue limestone, known as "Petit Granit" and the Middle Cambrian Chinese blue limestone, actually being sold on the European market as "Chinese blue". The first is a crinoidal limestone (bioclastic wacke/packstone), the latter is a dolomitized oolitic limestone (oolitic grainstone). The samples studied are curbstones taken from building wharfs in Belgium. These limestones are frequently affected by pressure solution, producing numerous stylolites. The dolomitized parts of this type of limestones often display a brown staining (Figures 10-11), mostly in close connection with stylolites or joints.

It was commonly accepted by the field geologists that weathering of ferroan dolomite in the dolomitized limestone was responsible for the unaesthetic brown to yellow superficial staining of those Paleozoic carbonate rocks. This was explained by the fact that the concentration of ferrous iron in dolomite would be in excess of the amount required to saturate the calcite lattice. During dedolomitization, this excess of ferrous iron would be ejected and, under certain physico-chemical conditions be precipitated as colloidal ferric hydroxide that ages to iron-oxide[8]. Based on a detailed microscopic analysis of fluorescent-epoxy impregnated thin sections and on analysis of polished slabs under incident light conditions, a new scenario has been proposed by the authors to explain the brown staining mechanism[9].

In the Belgian blue stone a particular carbon- and pyrite-rich bioclastic wackestone facies, in combination with secondary dolomitization and the presence of a joint has triggered the weathering and subsequent staining. In the stone, the framboidal pyrite embedded in the dolomite crystals, and the general association of framboidal pyrite and dolomite suggest a genetic link (Figure 12). Dolomite, synthesized at surface temperatures under laboratory conditions in the presence of sulphate reducing bacteria, indicates that sulphate reducing bacteria may play a substantial role in catalyzing dolomite formation[10]. The link between dolomite precipitation and sulfate-reducing bacteria has also been reported from the Upper Tournaisian strata of Belgium[11]. The association of framboidal pyrite and dolomite may also explain the non-ferroan nature of the dolomite crystal lattice, as iron is preferentially incorporated in pyrite.

Meteoric weathering has resulted in the oxidation of framboidal pyrite, the dissolution of dolomite rhombs and the precipitation of iron (hydr)oxides in the intercrystalline dissolution voids (Figures 13-14), producing yellow-brown stains in the surrounding host rock (Figures 15-16). As all dolomite of the investigated dolomitized limestone samples is non-ferroan, weathering of ferroan dolomite alone cannot explain the un-aesthetic superficial staining of the limestones. Open stylolites and/or intercrystalline permeability channels related to the dolomitization of the host rock accelerated the weathering process by allowing meteoric water to penetrate the rock and oxidize the pyrite. Apparently, dolomitized blue limestone carbonates yielding enough framboidal pyrite are prone to this type of staining. The genetic link between secondary dolomitization and framboidal pyrite is particularly significant[9].

Figure 10. Brown staining of curbstones under atmospheric conditions: Cambrian Chinese oolitic limestone.

Figure 11. Brown staining of curbstones under atmospheric conditions: Lower Carboniferous Belgian bioclastic (crinoidal) limestone ("Petit Granit").

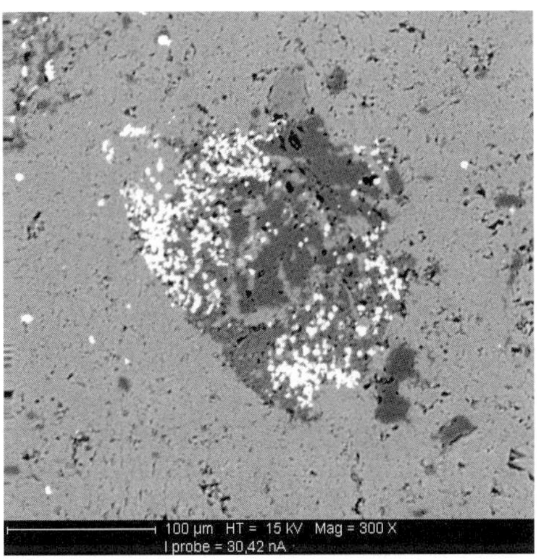

Figure 12. Belgian blue limestone. BSEM-image. Micrograph taken from the transition zone between the stained and the unstained parts of the limestone. Partly de-dolomitized dolomite crystal with numerous small ferric oxide/hydroxide particles (white). In this zone, the pyrite is often oxidized but the dolomites do not display dissolution features as in the stained areas.

Figure 13. Belgian blue limestone. BSEM-image, partly dissolved (black) and de-dolomitized planar dolomite crystals (see arrows). Micrograph taken within the stained part of the limestone, in the vicinity of the open stylolite.

Figure 14. Chinese blue stone. BSEM-image. Photomicrograph taken close to an open stylolite (located near upper left corner) showing dolomite crystals (dark grey) coated with ferric oxide/hydroxide (white). Locally planar cavities still exist (black). These cavities are most likely the result of dolomite dissolution (lower arrow), locally the cavities have probably been filled with calcite (upper arrow), which gives the impression that calcite has been dissolved. However, the planar outline of the cavities makes this very unlikely.

Figure 15. Chinese blue limestone. Micrograph taken under incident light conditions. Goethite (light grey) lining the dolomite rhombs and filling the inter-crystalline void system between the dolomite crystals. Area in the immediate neighbourhood of an open stylolite.

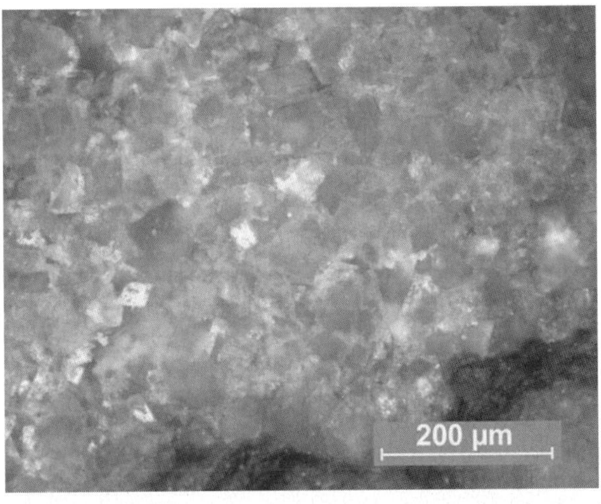

Figure 16. Chinese blue limestone. Same micrograph as Figure 15, taken under polarized incident light conditions, showing the conspicuous orange staining related to the goethite lining.

In the Chinese blue limestone the concentration of dolomitized patches near open stylolites with an increased pyrite content (concentrated after dissolving partially dolomitized micrite ooids) probably triggered the staining process. Element mapping of the dolomite crystals in the Chinese blue limestone indicates that the dolomites are non-ferroan in nature. In the vicinity of the open stylolites the alteration products of pyrite are precipitated as iron oxide/hydroxide in the intercrystalline dolomite pore space or in planar cavities occurring between dolomite crystals and the surrounding limestone host rock.

In the stained part of the Belgian blue limestone, pyrite has been completely oxidized to ferric or iron (hydr)oxide and dolomite crystals have been partially to completely dissolved resulting in the formation of significant moldic porosity. Element mapping of the dolomite crystals within the Belgian blue stone indicate that these dolomites are non-ferroan. Framboidal pyrite, or their alteration products, i.e. ferric or iron (hydr)oxide may however, occur embedded within the dolomite crystals. In the transition zone between stained and non-stained limestone areas, some pyrite has been oxidized to ferric or iron (hydr)oxide. Moldic porosity after dolomite as observed in the stained part of Belgian limestone, is however not observed. The dolomite crystals have been partially dedolomitized. Removed from the stained area, pyrite is still present while iron (hydr)oxides are scarce or not observed at all (Figures 17-18).

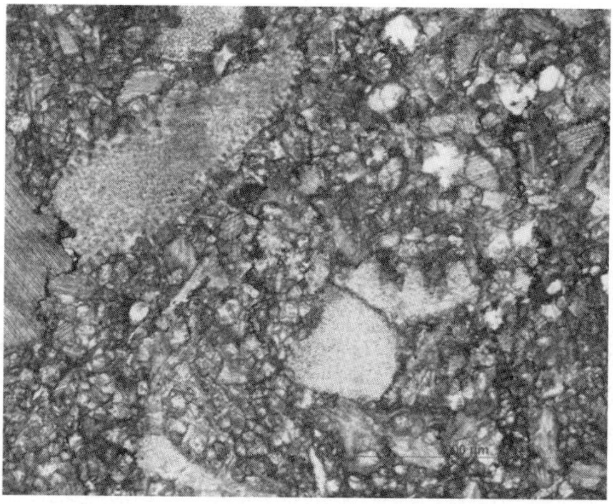

Figure 17. Dolomitized Lower Carboniferous Belgian crinoidal limestone ("Petit Granit"). Densely packed bioclasts (i.e. crinoid ossicles) corroded due to pressure solution. Thin section oriented perpendicular to the external weathered surface. Location about 1 cm below the surface. Moldic porosity is highlighted here by the yellow dyed epoxy. Normal transmitted light.

Figure 18. Dolomitized Belgian blue limestone. Compare with previous micrograph (same field of view, Figure 17). The moldic porosity (dissolution of de-dolomitized dolomite rhombs) is particularly conspicuous in fluorescent light conditions, as opposed to Figure 17.

5. CONCLUSIONS

Optical microscopy is a powerful investigation tool, especially when performed on good-quality thin sections, i.e. fluorescent epoxy impregnated thin sections. Moreover, the petrographical observations need to be corroborated by additional analysis such as automated image analysis, backscattering scanning electron microscopy (BSEM) and X-ray diffraction. Petrography can be successfully applied in unravelling the provenance of historical building stones, in assessing the durability of particular stone species, in monitoring the nature and extent of deterioration phenomena and in elucidating the mechanism responsible for un-aesthetically staining.

REFERENCES

1. R. Dreesen, O. Guillitte, J. Swings, and P.C. Nimis, in *Proceedings of the 7th Euroseminar on Microscopy Applied to Building materials*, edited by H. S. Petersen, J.A. Larbi and H.H.A. Janssen (1999), Delft University of Technology, Delft, pp. 267-276.
2. B. Laenen, R. Dreesen, and D. Van Rossem, in *Proceedings of the International Conference on Challenges of Concrete Construction. Sustainable Concrete Construction,* edited by R. K. Dhir, Th. D. Dyer and J.E. Halliday (2002), University of Dundee, Scotland, UK, Thomas Telford Publishing, London, pp. 335-354.
3. R. Dreesen and M. Dusar, Historical building stones in the province of Limburg (NE Belgium): role of petrography in provenance and durability assessment, *Materials Characterization* **53**, 273-287 (2004).
4. P. Bertier, R. Swennen, D. Lagrou, B. Laenen, and R. Kemps, Contrasting diagenesis of the Westphalian C & D sandstones of the Campine Basin (NE Belgium). *Sedimentology* (under review).
5. C.W. Dubelaar, 2000. Romeinse mijlpalen in het Wateringse veld. NITG-TNO Informatie, April 2000, 16-18.
6. R. Dreesen and C. Dingelstadt, Petrographical characteristics of building materials selected for CT-scanning, edited by M. De Cleene, Interactive physical weathering and bioreceptivity study of building stones, monitored by computerized X-ray tomography (CT) as a potential non-destructive research tool. *Protection and Conservation of the European Cultural Heritage, Research Report n°2*, European Commission Directorate-General XII for Science, Research and Development. (1995), pp.199-206
7. G.S. Odin, Green marine clays, *Developments in Sedimentology* **45** (Elsevier, Amsterdam, 1988).
8. B.D. Evamy, Dedolomitization and the development of rhombohedral pores in limestones, *J. Sed. Petrol.* **37**, 1204-1215 (1967).
9. R. Dreesen, P. Nielsen and D. Lagrou, in *Proceedings of the 10th Euroseminar on Microscopy Applied to Building Materials*, edited by J.J. Hughes, A.B. Leslie and J.A.Walsh (2005), University of Paisley, Scotland, UK, Extended abstracts and CD-ROM, pp. 25-26.
10. C. Vasconcelos, J.A. McKenzie, S. Bernasconi, D. Grujic, and A.J. Tien, Microbial mediation as a possible mechanism for natural dolomite formation at low temperatures. Nature **377**, 220-222 (1995).
11. P. Nielsen, R. Swennen, J.A.D. Dickson, A.E. Fallick, and E. Keppens, Spheroidal dolomites in a Visean karst system - bacterial induced origin? *Sedimentology* **44**, 177–195 (1997).

Chapter 8.3

INFLUENCE OF FABRIC ON THE PHYSICAL PROPERTIES OF LIMESTONES

Ákos Török
Budapest University of Technology and Economics, Department of Construction Materials and Engineering Geology, Stoczek u., H-1111 Budapest, Hungary, torokakos@mail.bme.hu

Abstract: The textural and physical properties of three Hungarian limestones; a compact, a porous and a travertine were studied in detail. Analytical methods included micro-textural analyses and laboratory testing of rock mechanical properties, such as density, water absorption, US sound velocity, UCS and indirect tensile strength. Cylindrical specimens were tested in air dry and water saturated conditions. Rock mechanical tests with combination of fabric analyses have shown that strength parameters depend not only on amount of effective porosity but also on the type of calcite cement and pore-size distribution. The porosity alone does not necessarily reflect the influence on water absorption on the properties of limestone.

Key words: limestone; travertine; microscopy; calcite cement; rock mechanics; water absorption; porosity; UCS.

1. INTRODUCTION

Limestone is one of the most common building and dimension stone. It has been used worldwide and monuments from the prehistoric times (eg. megaliths of Malta) are known to built from these carbonates. Although the primary mineralogical composition of various limestones is mostly calcite limestones show significant variations in properties and texture. From textural point of view spongy porous limestones to massive compact marble-like limestones are known[1]. The behaviour of these carbonates has been studied in many ways, especially focusing on physical properties[2] and the relationship between physical properties and different test conditions[3-4]. The present

S.K. Kourkoulis (ed.), Fracture and Failure of Natural Building Stones, 487–495.

study deals with three texturally very different Hungarian limestones. Petrographic and physical differences of a compact limestone, a porous limestone and a travertine are characterized by using microscopic analyses and rock mechanical laboratory tests. This study demonstrates that textural differences of calcitic rocks are also manifested in physical differences, especially when water is also present. The importance of calcite cement types and pore-size distribution is also emphasised.

2. METHODS AND MATERIALS

For laboratory analyses rock blocks were obtained from active quarries. Before sample preparation the limestone blocks were described in detail and were classified according to visual macroscopic properties. Thus from the three limestones seven textural types were identified and tested under laboratory conditions. From the limestone blocks cylindrical test specimens of 5 cm in diameter were drilled by using traditional coring techniques. Thin-sections were prepared by using resin impregnation to visualise textural and mineralogical properties. Physical and rock mechanical properties were determined under laboratory conditions for air dry and water saturated test specimens. Density properties, US sound velocities, UCS and indirect tensile strength (Brasilian tests) were measured on 175 cylindrical test specimens.

The three limestones studied represent wide ranges of origin and textures. The oldest one is Jurassic compact limestone from Gerecse Mountains Central Hungary. This red, fossiliferous limestone has been used as dimension stone from the Renaissance (e.g. Medieval royal palace of Visegrád, Hungary). It displays red nodules and both in use and in appearance it is very similar to the Amonitico Rosso of Verona.

The second set of samples represents one Miocene formation, with various lithologies. The white, yellowish porous limestones often contains ooids and thus it is similar to well-known dimension stones of UK and France such as the Great Oolite or Portland Limestone (UK) or Jaumont Limestone (France)[5]. This type has been used in several monuments of Hungary such as the House of Parliament, the Opera or the Basilica in Budapest. Similar types of limestone were used for the construction of several monuments in Vienna (St- Stephan's Church, The Opera, Palace of Schönbrunn)[6].

The third limestone is a travertine, which was formed from Pleistocene springs and thus it is of freshwater origin[7]. The travertine was already used by the Romans in Budapest and aqueducts and amphitheatres were constructed from this stone. The appearance and the usage are very similar to the classical Italian travertine of Tivoli (quarries are found near Rome)[8]. Travertine has been used in several monuments such as in the footings of Parliament and at Mathias Church in Budapest[9].

3. RESULTS

3.1 Macroscopic and microscopic properties

Red compact limestone has a mottled to nodular appearance (Figure 1). According to textural and X-ray diffractometry analyses besides calcite it contains clay minerals (illite) and hematite. The clay minerals and hematite accumulates in the darker red mottles. The microbioclastic wackestone fabric contains pelagic micro-fossils and fragments of ammonites (Figure 2a). The effective porosity is less than 0.5 % and is related to clayey styolitic seams.

Textural analyses have shown that porous limestone is divided into four different fabric categories: i) coarse grained bioclastic limestone (Figure 1c), ii) medium grained oolitic limestone, iii) fine grained oolitic limestone and iv) fine grained micritic limestone (Figure 1d). The first type exhibits a bioclastic ooidal grainstone texture with large (up to cm-size) shell fragments. Its pore system consists of large mm- to cm-size intergranular and smaller intragranular pores. Pore spaces are only partly occluded by sparitic calcite cements, the effective porosity is nearly 25%.

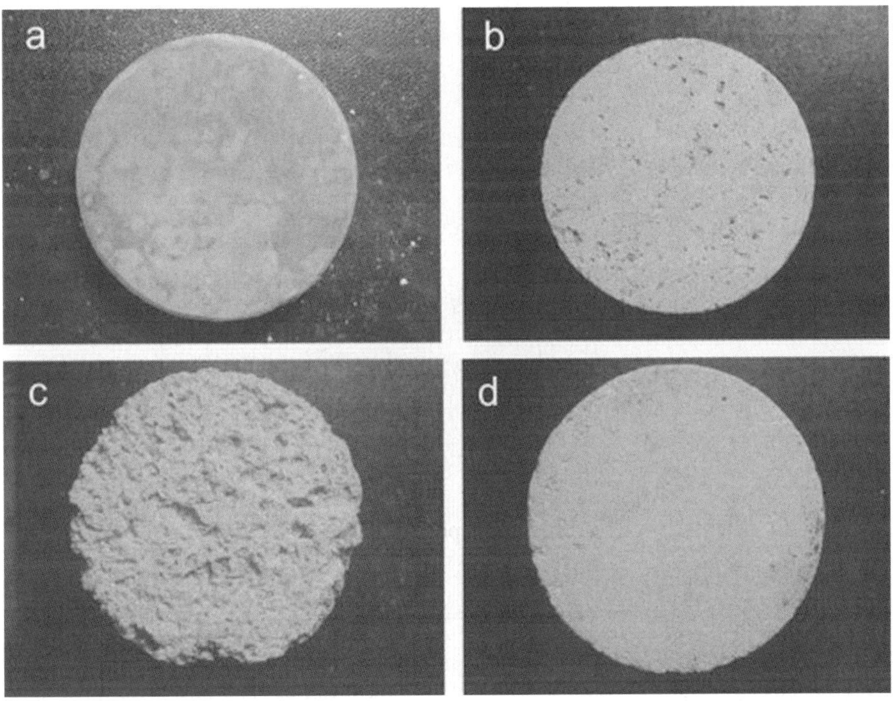

Figure 1. Macroscopic appearance of Hungarian limestones: (a) compact limestone; (b) travertine; (c) coarse grained bioclastic limestone; (d) fine grained micritic limestone.

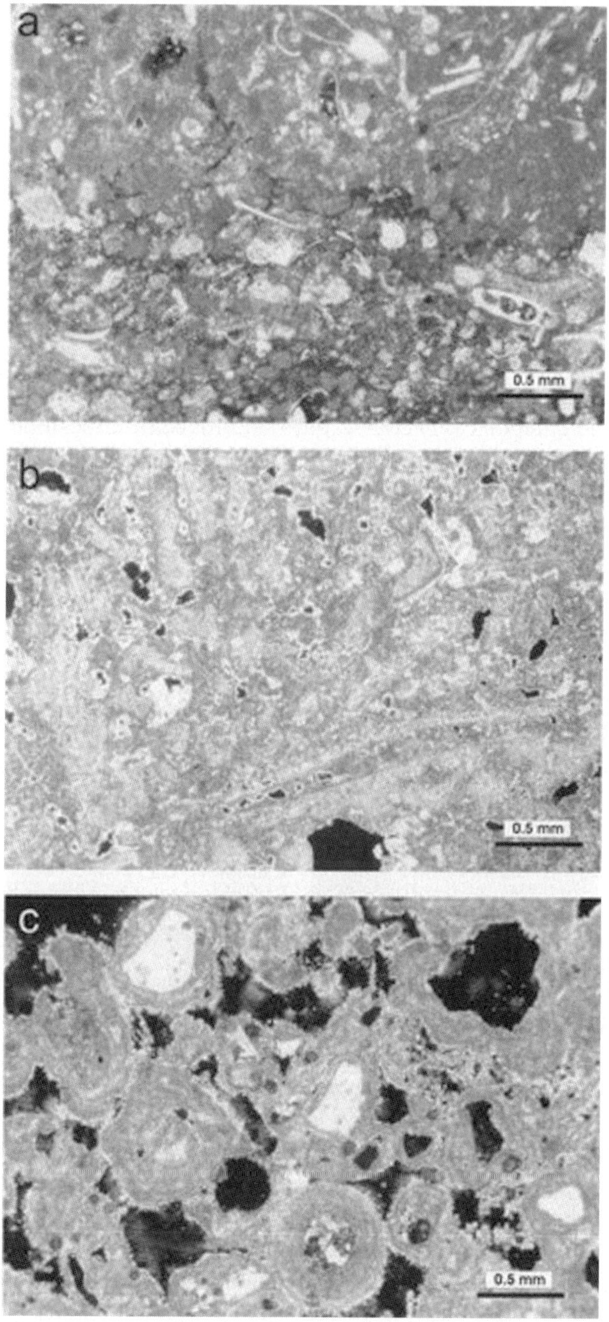

Figure 2. Microscopic image of Hungarian limestones showing various types of calcite cement and pore structures: (a) compact limestone with micritic calcite that contains small pelagic bioclasts and ferrous stylolites (lower part) showing no visible pores; (b) travertine with partly cemented pores and phytoclasts ; (c) oolitic limestone with minor cement between ooids and large irregular interconnected pores (black areas represent open pores).

Medium-grained oolitic limestone contains well rounded ooids of 0.1-0.8 mm in diameter (Figure 2c). The fabric is ooid grainstone[5]. Pores are mostly intergranular ones with an average pore size of 0.1 mm. Thin rim of sparitic calcite cement is often observed on pore walls. Its effective porosity can reach 28%.

Fine grained oolitic limestone is ooid-graintsone/packstone with smaller rounded carbonate particles ooids and peloids of 0.05-0.1 mm in size. The pores are also in the same order, but effective porosity exceeds 32%. Fine grained micritic limestone has a different fabric from the previous ones since small crystal-size calcite (i.e. micrite) is the dominant. Ooidal to bioclastic wackestones/packstones are the main fabric types, with very small matrix dependent pores providing an 8% of effective porosity.

Travertine is divided into two groups[10] based on fabric: laminated macro-porous travertine and massive travertine with small pores (Figure 1b). The first laminated-type, has large (up to cm-size), irregular pores that are parallel to bedding and often contain large calcified plant fragments. In thin section it is characterised as stromatolitic phytohermal to phytoclastic bound-stone. This type of travertine has an effective porosity of up to 10% which reflects the influence of a complex network of macroscopic intergranular and microscopic intragranular pores. The second travertine contains smaller but irregularly distributed pores, with effective porosity of 5%. Gastropods, smaller fragments of calcified reeds or oncoids comprise the main carbonate grains within the bioclastic wackestone to peloidal, oncoidal packstone fabric (Figure 2b).

3.2 Physical properties

Compact limestone has the highest density (2692 kg/m^3) which does not change significantly when the limestone is water saturated (2705 kg/m^3) (Table 1). The effective porosity is less than 1 % (0.5%) being the smallest from all analysed limestones (Figure 3). The maximum US velocity was measured in water saturated compact limestone with values of 5.2 km/sec. The compressive and tensile strength of compact limestone do not show significant changes when dry and water saturated samples are compared, 57.6 MPa to 41.7 MPa and 6.5 MPa to 5.6 MPa respectively (Table 1). Despite the low porosity the tensile strength of compact limestone is not the highest since according to tests massive travertine has a tensile strength of 6.9 MPa (Figure 4).

Porous limestone varieties have a wide range of density values starting from 1570 kg/m^3 and ending at 2356 kg/m^3. The trends in porosity are not necessarily similar to the densities since the highest porosity 33.1% was measured on fine oolitic limestone which has not the smallest dry density (1693 kg/m^3).

The porosities of the Miocene porous limestone are in between 8.5% and 33.1%. The smallest porosity value was recorded for samples of fine micritic limestone while the most porous type is the fine oolitic limestone with small but interconnected pores (Table 1).

Although the compressive strength of most Miocene limestones is below 10 MPa the very fine grained micritic limestone which has the smallest porosity (8.5%) shows a dry compressive strength of about 34 MPa, which drops to about 13 MPa when water saturated samples are tested. The US velocities, also, show some changes when dry and water saturated Miocene limestones are compared but this shift is opposite since air-dry US sound velocities are in the order of 1.6 to 3.5 km/sec while the water saturated ones are always greater (1.9 to 3.6 km/sec). It seems that the tensile strength of various Miocene limestones is in good correlation with porosity values (Figure 4).

The porosity of laminated travertine is nearly double than that of the massive one, 10.2 and 5.4% respectively (Figure 3). The densities are not significantly different (2251 kg/m^3 for the laminated, 2516 kg/m^3 for the massive one). The strength parameters reflect better the textural differences since the dry compressive strength of laminated travertine is 31.2 MPa, whereas strength value for massive travertine is 89.1 MPa in average. The compressive strength does not show marked change when the test results of water saturated and dry travertine is compared. Nevertheless, indirect tensile strength slightly decreases with water-saturation. The relationship between porosity and tensile strength is not so clear for travertine since massive travertine represents a minor anomaly from the general trend line (Figure 4).

Table 1. Physical properties of Hungarian limestones

Type	Porosity [V%]	Density dry [kg/m^3]	Density water sat. [kg/m^3]	Compr. strength dry [MPa]	Compr. strength wat.sat. [MPa]	Tensile strength dry [MPa]	Tensile strength wat.sat. [MPa]
Compact limestone	0.5	2692	2705	57.6	41.8	6.5	5.6
Coarse bioclastic	24.3	1570	1725	4.3	3.5	1.1	0.7
Medium-grained oolitic	28.7	1573	1802	5.3	4.4	1.0	0.4
Fine oolitic limestone	33.1	1693	1908	6.7	3.8	0.5	0.3
Fine micritic, small pores	8.5	2356	2470	33.6	13.1	2.6	1.0
Laminated travertine	10.2	2251	2341	31.2	31.9	4.7	3.0
Massive travertine	5.4	2516	2566	89.1	87.8	6.9	6.5

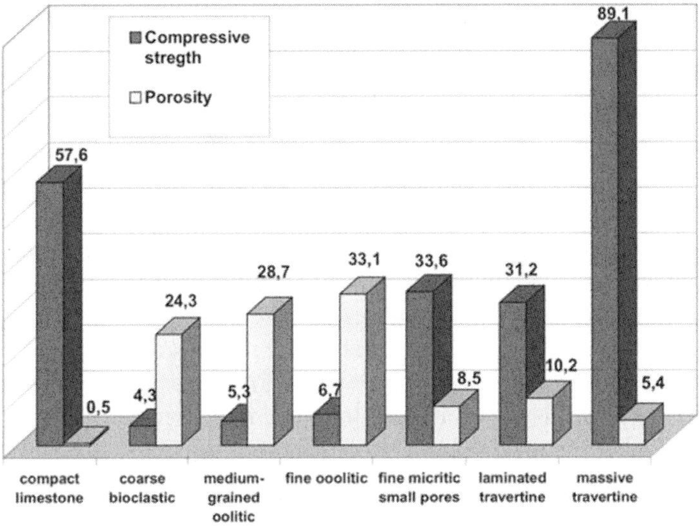

Figure 3. Compressive strength (MPa) and porosity (%) of limestones with different fabric.

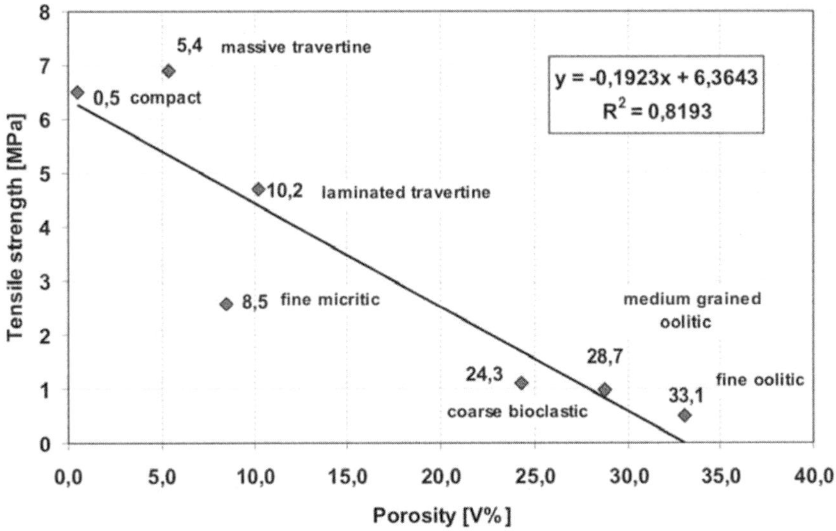

Figure 4. Correlation between tensile strength and porosity of limestones (porosity is shown in numbers).

4. DISCUSSION

The density of limestones varies between 1570 kg/m^3 (coarse bioclastic limestone) and 2692 kg/m^3 (compact limestone), but it is not in direct correlation with porosity values. The discrepancies are observed for very porous limestones. US sound velocities reflect the compressive strength values.

There is a relationship between tensile strength and porosity of limestones[11] although it seems that at some limestone types (fine micritic with small pores and laminated travertine) the correlation between porosity and tensile strength is less evident (Figure 4). This feature is related to the various calcite cementation and microscopic fabric of the limestones. The fine micritic limestone has micritic cement (1-5 μm size calcite crystals) with very small pores and thus its strength values drop below the trend line, while laminated travertine has coarser micro-sparitic cements (15-30 μm) which cause a shift of values above the trend line.

The tensile strength values of limestones with high porosity show a drastic drop when air dry and water saturated test results are compared (medium oolitic to 42%, fine micritic with pores to 37%) (Table 1). Massive travertine with higher porosity is less sensitive to water than compact limestone since its tensile strength from 6.9 MPa decreases to 6.5 MPa, while for compact limestone this change is 6.5 MPa to 5.6 MPa.

For the compressive strength the porosity and strength values are more scattered. Massive travertine with its effective porosity of 5.3 % has a uniaxial compressive strength (UCS) of 89.1 MPa while compact limestone with lower effective porosity (0.5 %) displays a smaller UCS (57.6 MPa) (Figure 3).

5. CONCLUSIONS

The porosity of limestones provides preliminary information on strength values but the correlation between porosity and tensile strength depends on the type of calcite cement, i.e. the fabric of limestone.

Densities not necessarily indicate the porosity of limestone; discrepancies are mostly observed in porous limestones.

Porous limestones when cemented by fine grained micritic calcite (μm-size crystals) are more sensitive to water than those which are cemented by micro-sparitic calcite (tens of μm).

Significant variations in strength and porosity are observed even within one type of limestone depending on the texture.

Rock mechanical tests with combination of fabric analyses have shown that strength parameters depend not only on the amount of effective porosity but also on the type of calcite cement. Consequently detailed micro-fabric analyses provide valuable information on the durability and strength of limestones.

ACKNOWLEDGEMENTS

This work was partly financed by the Bolyai János research grant (BO/233/04) and by the Hungarian Science Found (OTKA, K 63399). The support of Hungarian-Portuguese (P-18/03), Hungarian-Spanish (E-39/04) and Hungarian-Czech intergovernmental grants are also appreciated. In laboratory analyses E. L. Árpás, Gy. Emszt and E. Horthy provided help.

REFERENCES

1. M.E. Tucker and V.P. Wright, *Carbonate sedimentology*, (Blackwell, Oxford, 1990)
2. F.G Bell, Durability of carbonate rock as a building stone with comments on its preservation. *Environmental Geology*, **21**, 187-200. (1993).
3. G. Tsimbaos, N. Sabatakakis, Considerations on strength of intact sedimentary rocks, *Engineering Geology*, **72**, 261-273 (2004).
4. B. Vásárhelyi, Statistical analysis of the influence of water content on the strength of Miocene limestone, *Rock Mechanics and Rock Engineering*, **38**(1), 69-76 (2005).
5. Á. Török, Oolitic limestone in polluted atmospheric environment in Budapest: weathering phenomena and alterations in physical properties, in: *Natural Stones, Weathering Phenomena, Conservation Strategies and Case Studies*, edited by S. Siegesmund, T.S. Weiss and A. Vollbrecht (Geological Society, London, Special Publications 205, 2002), pp. 363-379.
6. Á. Török, N. Rozgonyi, R. Prikryl, J. Prikrylová, Lethakalk: the ornamental and building stone of Central Europe, an overview, in: *Dimension stone*. Edited by R. Prikryl (Balkema, Rotterdam, 2004), pp. 89-93.
7. A. Pentecost, The Quaternary travertine deposits of Europe and Asia minor, *Quaternary Science Reviews*, **14**, 1005-1028 (1995).
8. I. Sindraba, K.C. Normandin, G. Cultrone, M.J. Scheffler, Climatological and regional weathering of Roman travertine, in: *Architectural and sculptural stone in cultural landscape* edited by R. Prikryl and P. Siegel (The Karolinum Press, Prague, 2004), pp. 211-228.
9. Á. Török, Comparison of the Processes of Decay of Two Limestones in a Polluted Urban Environment. in: *Stone Deterioration in Polluted Urban Environments*, edited by D.J. Mitchell, D. E. Searle, (Science Publishers Inc., Enfield, 2004), pp. 73-92.
10. Á. Török, Facies analysis and genetic interpretation of travertine. Buda Vár-hegy, Hungary. *Acta Geologica Hungarica*, **46**(2), 177-193 (2003).
11. D. Benavente, M.A. García del Cura, R. Fort, S. Ordónez, Durability estimation of porous building stones from pore structure and strength, *Engineering Geology* **74**, 113-127 (2004).

Chapter 8.4

DISCRIMINATION OF GREEK MARBLES BY TRACE-, ISOTOPE- AND MINERALOGICAL ANALYSIS

Vassilis Perdikatsis[1], Konstantinos Kritsotakis[2], Theodoros Markopoulos[1] and Konstantinos Laskaridis[3]

[1]Department of Mineral Resources Engineering, Technical University of Crete, 73100 Chania, Greece, vperdik@mred.tuc.gr; [2]Institut fuer Geowissenschasften, Saar str. 21, Johanes Gutenberg Universitaet, 55099 Mainz, Deutschland; [3]IGME - Lithos Laboratory, 1st km Markopoulou Avenue, 190 02 Peanea, Attiki

Abstract: Petrographically marbles are crystalline carbonate rocks with main mineral phases Calcite ($CaCO_3$), Calcitic marbles or dolomite ($CaMg(CO_3)_2$), dolomitic marbles. The commercial and archaeological interest for the characterization of provenance of marbles from specific marble quarries is great. Within this aim several studies have been carried out in the past, which referred to the petrographic characterization as well as to trace element analysis. In the present work a farther attempt is made for the discrimination of the origin of marbles, based on their mineralogical characterization and analysis of major and trace elements as well as isotopes analyses. In particular, 8 major elements (Si, Al, Fe, Mn, Mg, K, Ca, Na), 27 trace elements: (Li, Be, B, Sc, B, Cr, Co, Ni, Cu, Zn, Ga, Ge, Rb, Sr, Y, Zr, Nb, Sn Sb, Cs, Ba, Hf, Pb, Bi, Th, U), 14 REE (Rare Earth Elements): (La, Ce, Pr, Nd, Sm, Eu, Tb, Gd, Dy, Ho, Er, Tm, Yb, Lu) and the ratio of isotopes: Pb, Sr und Rb (204/206Pb, 207/206Pb, 208/206Pb, 207/204Pb; (85/87Rb); (84/86Sr, 87/86Sr, 88/86Sr87/86Sr) have been measured. Eleven different Greek marbles have been analyzed from various marble regions. The study of the analytical results shows that the discrimination of marbles is possible based on the trace elements in combination with the isotope analysis. Binary and ternary chemographic diagrams like 207Pb/206 Pb-208Pb/206Pb, 87Rb/86Sr-206Pb/204Pb, Ti-B-Be, (Sn+Pb)-(V+Cr)-(Sc+Y) and the ratio 147Sm/146Nd contribute to, more or less, a clear group discrimination of the Greek marbles. This discrimination is significantly improved involving the mineralogical analysis combined with Image Analysis Techniques. By the ternary diagrams Pr-Nd-Sm, Tb-Gd-Dy and Tm-Yb-Lu a discrete and significant grouping of all examined Greek marbles is achieved, which could be probably used for the discrimination of the Greek marbles from other localities of the Mediterranean area.

Key words: marbles; trace elements; isotope analyses; X-ray diffraction; image analysis.

S.K. Kourkoulis (ed.), Fracture and Failure of Natural Building Stones, 497–515.

1. INTRODUCTION

Marble is the most important decorative building stone in the human history. Buildings of marbles, predominantly from the Hellenistic, Hellenistic - Roman and later period, can be regarded as indicators for the culture-historical development of the European settlement area. The provenance allocation of the marbles, which were used for monuments, sculptures, buildings and other historical culture objects, is very important for the archaeological study and give interesting information about the culture and the trade relations within Europe and the neighboring regions. Even for modern buildings of marbles, if restoration is required, the knowledge of the provenance of the marbles is very important.

The archaeological interest for the regional allocation of marble artifacts at certain marble occurrences of the Mediterranean area began relatively early. By the material allocation of marble artifacts to certain delivery areas (quarries) possible historical cultural and trade relations could be archaeologically reconstructed within Europe. As it was shown in the case of ceramic artifacts, (Kritsotakis, 1994; Schneider and Rother, 1991), it should be possible, due to mineralogical and geochemical criteria, to locate historical marble objects with probably good safety. This can be done by comparative mineralogical-geochemical investigations with "groups of reference marbles" from different historical settlement regions of Europe. In addition marble fragments, belonging together as well as replicates could be more easily recognizable with the help of the groups of reference marbles.

A condition for a Data Bank of groups of marble references is a systematic sampling of historical and recent marble occurrences within the regions of interest. Furthermore it is necessary that from each sampled marble occurrence (quarry) a large number of samples is collected, representative of the occurrence, and examined mineralogically as well as geochemically.

Lepsius (1890) described marble occurrences and artifacts from different regions of Greece in detail and tried to differentiate chemically the Penteli marble from the cycladic marbles. He compared thereby the Fe_2O_3 contents of these marbles and found for the Penteli marble a Fe_2O_3 content of 0.2 %, while in the cycladic marbles no Fe_2O_3 was detected, obviously because at that time the Fe_2O_3 content laid below the analytical detection limit.

The first scientifically documented publications, which were concerned comprehensively with the geostructural and stratigraphical situation of the marble occurrences in the Aegean area, appeared during the late ’40s. Marinos (1948) examined the Aegean marbles petrographically and showed that their texture, crystalinity, twinning, weathering and roughness as well as some further mechanical properties depend on the geostructural position of the marbles. The more deeply their geotectonic position is, the larger, more uniform and clear (high P-T conditions) their calcite crystals are, and therefore they are suitable for sculpture work.

Martin (1965) and Papagiorgakis (1963, 1967) described in detail the geotectonic and petrographic environment of the marbles, which were found in antique Greek architecture, while Dworakowska (1975) gave a broad overview of the localities of some historical quarries. The classical marble occurrences of Asia Minor were described likewise in detail by Monna and Pensabene (1977). In the meantime a large number of works were concerned with the mineralogical petrographic and/or geochemical characterization (main and trace elements, carbon-, oxygen and Sr-Isotopes) of marble artifacts and marble occurrences from the Mediterranean area. A large part of these contributions is coherently discussed in the Monograph by Kempe and Harvey (1983), while in the Monograph by Herz and Waelkes (1988) many relevant research papers are presented in detail. Interesting abstracts with the same topic are also found in the proceedings of ASMOSIA. Two interesting contributions of Moens et al. (1987, 1988) are concerned with the investigation and characterisation of the Asia Minor marble occurrences.

Lately for the characterisation of the Mediterranean marbles and/or marble artifacts spectroscopic methods were used. Thus, the cathode luminescence was used in connection with petrographic-geochemical investigations (Barbin et al., 1992a; Schmid et al., 1999) as well as the electron paramagnetic resonance spectroscopy (Baietto et al., 1999) for the discrimination of white marbles of the cycladic islands and the Greek mainland (Vakulis et al., 2000).

Studying the literature, it was shown that most investigations considered only a few regional marble occurrences of the Mediterranean area. Besides, the number of examined samples cannot be regarded as representative. Furthermore the obtained results for the same marble, with different research methods, are partly contradictory and in themselves not consistent.

Petrographic and textural studies, (Weiss, 1954; Herz, 1955b; Renfrew and Peacey, 1968; Germann, 1978), supplied partial useful results but the investigations are very time consuming, partly subjective, not always reproducible and require a large experience from the scientist. Although the numerous chemical analyses of main- and trace- elements - of marbles from different Mediterranean Localities, (Rybach and Nissen, 1965; Renfrew and Peacey, 1968; Andreae et al., 1972; Conforto et al., 1975; Germann, 1978; Lazzarini et al., 1980), are well reproducible, they brought no satisfying significant discriminations. The reason for this lies obviously in involving only few main- and trace-elements for the discrimination.

In recent time an attempt was often undertaken to discriminate Hellenistic and Roman marble artifacts by involving the stable isotopes ($^{13}C/^{12}C$, $^{18}O/^{16}O$), Lazarini (1995, 2000, 2002), and the Sr isotope ratios (Craig and Craig, 1972; Manfra et al., 1975; Herz and Wenner, 1978; Coleman and Walker, 1979; Gale, 1979; Gast et al., 1979; Germann et al., 1980; Herz, 1987). Although at the beginning a very good discrimination of some Greek marble occurrences could be achieved, by the data of Craig and Craig (1972), the discrimination of the groups became more unclear by plotting data from further localities into this diagram, and therefore is useless as a group discrimination tool.

The stated difficulties arouse the impression that the Mediterranean marble occurrences are neither with petrographic nor with chemical methods reliably discriminable. It should however be possible, due to geochemical considerations, by involving additional trace elements, e.g. the Rare Earth Elements (REE), as well as the Rb-Sr and Pb-Isotopes to receive more effective discrimination for these marbles. Today it is possible, by the means of solution and Laser Ablation - ICP MS, to receive very fast precise analytical data for a large number of elements.

In the context of a pilot project, between the Institute for Geology and Mineral Exploration (IGME) in Athens, the Technical University of Crete (Mineral resources department) and the Institute fuer Geowissenschasften, Universitaet Mainz, only a limited number of marble samples from different occurrences in Greece was analyzed for their chemical and isotope composition. In this project as many as possible chemical elements and isotope rations from different marble occurrences have been analyzed, to test whether it is at all possible to achieve a clear discrimination from each other. In case of a clear discrimination of the marble occurrences, the project should be extended to the systematic investigation of all well-known historical and recent marble occurrences of Greece. The data obtained and chemographic discriminants are to be summarized afterwards in the form of a geochemical Atlas. These data should serve to the characterization of marble artifacts regarding their origin. In a long-term project it is intended to work on all well-known marble occurrences of the Mediterranean area.

2. MARBLE: DEFINITION AND MINERALOGY

Petrographically marbles are crystalline carbonate rocks with main mineral phases Calcite ($CaCO_3$), Calcitic marble or dolomite ($CaMg(CO_3)_2$), dolomitic marble. Contrary to this petrographic definition in the decorative stone industry, marbles are, generally, all easily machinable rocks, i.e. sharpen and polishing. Therefore in the stone-processing industry not metamorphic rocks and fine-grained limestones are also called "marbles".

Marbles are from the petrographic point of view, recrystallized carbonate rocks, named also crystalline limestones, where the recrystallisation takes place via the thermal stress of sedimentary carbonates. Pure limestones are converted by thermal metamorphose to fine up to coarse crystalline Calcitic marbles. From silica and/or clay containing carbonate rocks, depending on metamorphose degree, additionally mineral phases can occur as accessory minerals in the marble, like: Quartz, Chlorite, Epidote, Zoisite, amphibole, Clinopyroxene, garnet, plagioclase, light and dark mica. From these, marble designations result, like Zoisit-, or Chlorite-marble.

Table 1. Analysed marbles from different localities in Greece.

Sample Number	Locality	Mineralogical characterization
M 1	Crystalline, Thassos	Calcitic marble
M 2	White, Prinos, Thassos	Dolomitic marble
M 3	Yellowish-white, Platanotopos, Kavala	Dolomitic marble
M 4	Black, Vathylakos, Drama	Calcitic marble
M 5	White, Volakas, Drama	Dolomitic marble
M 6	White, Menikion, Serres	Calcitic marble
M 7	Roditis, Kozani	Calcitic marble
M 8	Grey, Tranovaltos, Kozani	Calcitic marble
M 9	White, Tranovaltos, Kozani	Calcitic marble
M 10	Crystalline, Naxos	Calcitic marble
M 11	White, Dionyssos, Penteli (Attica)	Calcitic marble

The frequently arising clay parts in the primary limestones are converted by the metamorphic processes to mineral associations of the green and/or Amphibole fazies and form the well-known banded marbles, or by syn- or post-metamorphic tectonic processes, irregularly distributed, darkly colored engagements. Such a different colour veining, lend a decorative effect, particularly if, as in some crystalline limestones, the differences in color are intensive. The color of crystalline limestones is not definable, since nearly all color nuances and color combinations can occur, from pure white up to deeply black varieties. The frequently occurring colors form a white to light-grey carbonate matrix with gray-green to black veins.

The structure of the marbles is massive. Banded types arise only if the marbles contain larger quantities of phyllosilicates or needleform minerals (mica, graphite, needleform amphiboles). In consequence of the crystallization friendliness of the carbonate minerals, granoplastische (coarse grained) Calcitic and dolomitic structures arise in tectonically strongly stressed areas, which show a close interconnection of the crystallites, where the pore area of the marbles becomes extraordinarily small (up to 0.01 volume per cent). The grain structure analysis usually shows no uniformed orientation of the carbonate minerals.

Marbles occur in all crystalline areas, where they are most strongly represented in low metamorphic series.

3. SAMPLING AND ANALYTICAL TECHNIQUES

For the present investigation marble samples from 11 different marble occurrences from Greece were analysed. In Table 1 the localities and the mineralogical characterisation of the samples are given.

The geographic localities of the occurrences indicated in Table 1 are registered in the map of Greece (Figure 1) as well as the localities of well-known marble occurrences. With the exception of Penteli (Attica), Thassos and Naxos, the examined marble occurrences originate from north Greece.

Table 1. Analysed marbles from different localities in Greece.

Sample Number	Locality	Mineralogical characterization
M 1	Crystalline, Thassos	Calcitic marble
M 2	White, Prinos, Thassos	Dolomitc marble
M 3	Yellowish-white, Platanotopos, Kavala	Dolomitic marble
M 4	Black, Vathylakos, Drama	Calcitic marble
M 5	White, Volakas, Drama	Dolomitc marble
M 6	White, Menikion, Serres	Calcitic marble
M 7	Roditis, Kozani	Calcitic marble
M 8	Grey, Tranovaltos, Kozani	Calcitic marble
M 9	White, Tranovaltos, Kozani	Calcitic marble
M 10	Crystalline, Naxos	Calcitic marble
M 11	White, Dionyssos, Penteli (Attica)	Calcitic marble

The mineralogical analysis was performed by X-ray powder diffraction techniques. A Siemens D500 diffractometer was used with copper radiation and graphite monochromator. The qualitative evaluation was carried out by the Diffrac Plus software and the PDF (Powder Diffraction File) data base. The quantitative analysis was performed by Rietveld refinement technique. This method gives the most accurate quantitative analysis, which can be achieved by X-ray diffraction analysis.

Figure 1. Distribution of the most important marble occurrences in Greece.

Thin sections were prepared and photos were taken with a digital photo camera. The photos were taken from a stereo transmission microscope, which was converted to a simple polarizing microscope, adding two commercial polarizing filters. In this way the optical field was >1 cm, instead of a few millimeters in a normal polarizing microscope, even in low magnification. This arrangement allows to measure coarse crystalline marbles, with grains >2 mm. The grain size of the analysed marbles was determined by image processing techniques followed by statistical evaluation.

For the chemical analysis, about 10 g of each sample were grounded, in an agate mortar, and twice 0.1 g sample material were dissolved with 10 ml 1n HNO_3 ultrapure in Teflon bottle. Not dissolved portions were filtered and the solutions were diluted with ultrapure H_2O into a Teflon bottle to a final dilution 1000. With these solutions all further analytical work was carried out. The sample solutions were analyzed with a PQ3 S Quadrupole ICP MS of the company VG-Instruments. For the analyses the element In with a concentration of 10 ppb was used as internal standard. The optimization work of the system took place with the necessary reference solutions. The reproducibility and accuracy of the measuring data were tested with the help of synthetic solutions and international rock and isotope reference standards. Isobar mass interferences were corrected by optimized correction factors as far as possible.

Quantitatively 8 main elements (Si, Al, Fe, Mn, Mg, K, Ca, Na) and 27 trace elements: (Li, Be B, Sc, V, Cr, Co, Ni, Cu, Zn, Ga, Ge, Rb, Sr, Y, Zr, Nb, Sn, Sb, Cs, Ba, Hf, Ta, Pb, Bi, Th, U), 14 REE: (La, Ce, Pr, Nd, Sm, Eu, Tb, Gd, Dy, Ho, Er, Tm, Yb, Lu) and the Pb-, Sr- and Rb- isotope ratios: (204/206Pb, 207/206Pb, 208/206Pb, 207/204Pb, 208/204 Pb); (85/87Rb); (84/86Sr, 87/86Sr, 88/86Sr, 87/86Sr) were measured.

4. RESULTS AND DISCUSSION

The discussion and presentation of the extensive analytical data, will be done in form of binary and ternary chemographic diagrams. From a large number of diagrams, there are presented and discussed only those which contribute to more or less clear group discrimination (The analytical data are available on request). The mineralogical composition and the cristallinity of the marbles are given in Table 2. The physical properties and the chemical composition of the examined marbles are given in Table 3 and Table 4.

Based on Table 2, all examined marbles consist mainly of the minerals calcite, dolomite and quartz, with some trace of muscovite in samples M6, M7 and M11. The marbles M2, M3 and M5 are nearly pure dolomitic, the M3-Marble contains 3.5% Calcite while M2 and M5 contain only ~1% calcite. The rest of the marbles are calcitic, which, besides quartz, contain also small amount of dolomite (~3 %).

Table 2. Mineralogical composition, crystallinity and grain size of the marbles samples

Marble Sample	Calcite %	Dolomite %	Muscovite %	Quartz %	Crystallinity	Mean Grain Size (μm)	Std. Ddev.
M1 Kristalin Thassos	100	-	-	<0.1	coarsely crystalline	891	437
M2 White, Prinos Thassos	1.5	98.4	-	<0.1	coarsely crystalline	983	366
M3 Yellowish Platanotopos/Kavala	3.5	96.4	-	<0.1	Medium crystalline	178	69
M4 Black, Vathylakos/ Drama	99.8	-	-	0.2	coarsely crystalline	509	163
M5 White, Volakas/Drama	1	99	-	<0.1	Medium crystalline	190	53
M6 White Menikion/Serres	99.5	0.5	Trace	<0.1	coarsely crystalline	522	200
M7 Roditis/Kozani	97.6	2	Trace	0.4	coarsely crystalline	501	153
M8 Gray, Tranovaltos/ Kozani	99	0.5	-	<0.1	Medium crystalline	171	58
M9 White,Tranovaltos /Kozani	96.5	3	-	<0.1	Medium crystalline	240	72
M10 Kristalin, Naxos	99	0.5	-	0.5	coarsely crystalline	1028	506
M11 White Dionyssos/Penteli	98	0.5	trace	2	Medium crystalline	289	85

Although the grain sizes of the studied marbles have a relative big standard deviation, we can recognize differences between the different marble localities. Samples M1, M2 (Thassos) and M10 (Naxos) are characterized as coarsely crystalline while the rest of the marbles are medium crystalline. Among the medium crystalline marbles, the marbles M3 (Kavala), M8 (Gray/Kozani), M9 (White/Kozani) and M11 (Dionyssos) have the lower crystallinity (microscopical pictures and grain size distribution are given in the Appendix).

Combining the mean grain size of the marbles with the physical properties, a relation could be recognized between grain size and compressive strength. The marbles M3, M5 and M8 have the smaller grain sizes and highest compressive strength. The same relation is observed also after freezing and thaw. The concentrations of the elements Si, Al, Fe, Mn, Mg, K, Ca, Na are quite low in all marbles (Table 3). A clear discrimination of these marbles is not easy due to their mineral composition and/or their main element concentrations. It can be differentiated between calcitic and dolomitic marbles, but a clear differentiation within the group of either the calcitic or the dolomitic marbles is not easy too.

Using the Fe-concentration in Table 3 a simple discrimination of the examined marbles is possible in marbles with Fe_2O_3 <0.1% or Fe_2O_3 >0.1%.

Table 3. Physical properties of the marble samples

	Apparent Specific Weight (DIN 52102)	Porosity and Water Acces. (DIN 52102)	Absorption Coefficient (DIN 52103)	Modulus of elasticity (DIN 1048, part 5)	Compressive Strength (dry) (DIN 52105)	Modulus of Rupture (dry) (DIN 52112)	Compressive Strength after freezing and thaw (DIN 52104) and (DIN 52105)	Abrasion Resistance (DIN 52108)	Impact Strength (UNI-U 32.07.248.0)
	Kg/m^3	% vol.	%wt.	GPa	MPa	MPa	MPa	mm	cm
M1	2714	0.14	0.05	18	60	9	64	2.20	38
M2	2846	0.28	0.1	21	80	10	103	2.49	40
M3	2850	0.60	0.21	42	120	11	81.2	2.06	29
M4	2711	0.21	0.08	23	89	18	72	2.26	41
M5	2825	0.54	0.19	35	139	10	103	2.20	59
M6	2712	0.17	0.06	34	77	24	86	2.35	46
M7	2732	0.19	0.07	25	63	20	81	2.28	33
M8	2726	0.29	0.11	24	114	24	104	2.3	53
M9	2719	0.28	0.1	25	90	9	84	2.68	65
M10	2710	n.d	0.09	35.5	89	12.8	n.d	8.56	n.d
M11	2717	n.d	0.11	57.2	111	19.2	n.d	6.7	n.d

Table 4. Concentrations of the main elements in the marble samples

	M 1	M 2	M 3	M 4	M 5	M 6	M 7	M 8	M 9	M 10	M 11
CaO	53.50	30.40	30.0	54.50	30.60	54.00	53.20	53.60	53.30	55.60	54.80
MgO	1.92	21.90	20.8	0.66	21.50	0.7	2.60	0.84	0.84	0.50	1.55
SiO_2	0.50	0.60	1.2	0.20	0.20	0.6	0.65	1.00	0.80	0.07	1.10
Fe_2O_3	0.05	0.07	0.20	<0.05	<0.05	0.08	0.10	0.10	0.15	0.14	0.14
Al_2O_3	0.11	0.08	0.16	<0.05	<0.05	0.04	0.20	0.15	0.20	0.02	0.20
K_2O	<0.01	0.01	0.01	<0.01	<0.01	0.03	0.03	0.02	0.03	0.02	0.09
Na_2O	<0.01	<0.01	0.04	0.01	0.01	0.01	0.01	<0.01	<0.01	0.04	0.04
MnO	0.01	0.01	0.02	<0.01	<0.01	0.01	0.02	0.02	0.02	0.02	0.02
LOI	43.44	46.93	46.1	43.60	47.60	43.50	43.20	43.00	42.80	43.00	43.05

The yellowish white marble from Platanotopos/Kavala (M3) is characterized by a high Fe-concentration. The yellowish colour of this marble is obviously caused by its high iron content. Also the light reddish colour of the Roditis/Kozani (M7) marble is due to its relatively high Fe-concentration. The Calcitic marble from Dionyssos/Penteli (M11) is likewise iron-rich. Of the examined marbles only the marbles from Vathylakos (calcitic marble) and Volakas (dolomitic marble) near Drama (M4, M5) differ, because they are iron free.

Within the dolomitic marbles the calcite content could possibly serve as discrimination criterion, because, as seen from Table 2, the marble from Platanotopos/Kavala (M3) contains some calcite, higher than in the two other dolomitic marbles (M2, M5) but differs significantly in grain size from M2 (Thassos). However, this is not a safe criterion, because we do not know, the calcite content and the grain distribution in other dolomites, which were not analyzed here. The dolomitic marbles from Prinos/Thassos (M2) and Volakas/ Drama (M5) are not differentiated mineralogically from each other but are clearly differentiated in terms of grain size distribution. The discrimination effectiveness is improved, if apart from the mineralogical analysis, additional criteria are involved, like petrographic (crystalinity, crystall form, crystal/ grain size, crystal orientation) and macroscopical (color, color distribution).

For the discrimination of historical groups of marble objects it is shown that binary and ternary element correlations are particularly well suited. With such representations the concentrations of two or three elements (object-characteristic properties) are correlated to each other. All objects with more or less similar concentrations form discrete point clusters, which are characteristic for the respective group of objects. For a clear and reliable discrimination of the examined marbles several binary and ternary trace element correlations were examined for their discrimination power. The ternary representations Be-B-Ti, (Figure 2) and (Sc+Y)-(V+Cr)-(Sn+Pb), (Figure 3) supplied the best separation.

In the ternary diagram Be-B-Ti the localities and the occurrences within the same locality are quite well separated. In this way, the two samples of Thassos (M1, M2) and the two samples of Drama (M4, M5) are clearly discriminated. On the other hand, the three marble samples of Kozani (M7, M8, M9) though separated, lie not far from each other. The marbles of Naxos (M10) and Dionyssos (M11) are separated too, but the distance of the representing points is not large enough for a reliable discrimination. Additionally, if the grain size distribution is taken into consideration, a better discrimination will be achieved. Furthermore the points of the marble from Serres (M6) and the Roditis Kozani (M7) marbles lie quite close and an overlapping of their ranges is expected.

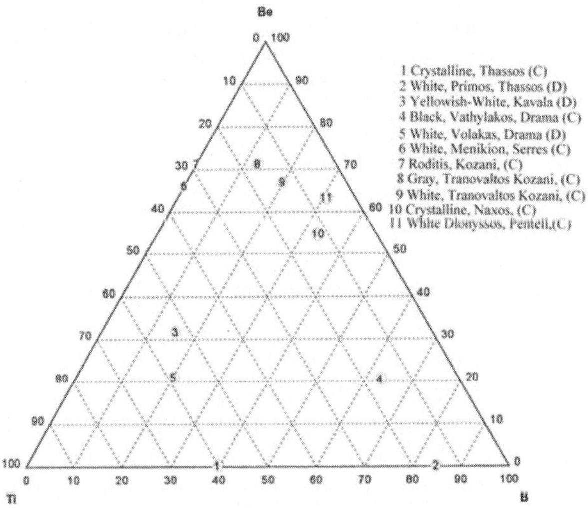

Figure 2. Discrimination of the marbles from Greece in the ternary diagram Be-B-Ti.

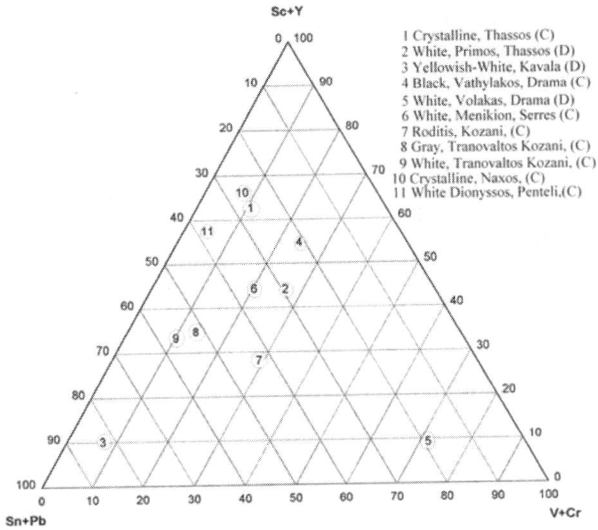

Figure 3. Discrimination of the marbles from Greece in the ternary diagram (Sc+Y)-(V+Cr)-(Sn+Pb).

By the ternary diagram (Sc+Y)-(V+Cr)-(Sn+Pb) (Figure 3) a discrimination of the examined marble occurrences also takes place. In this case the separation distances of the representing points are not so large, so certain partial overlapping of the representing ranges can occur. In the above diagram the two marble samples from Thassos (M1, M2) are clearly separated, as well as the two marbles from Drama (M4, M5). As for the three samples from Kozani the samples M8 and M9 are well together while the sample M7 is separated from the two others, with significantly bigger grain size. Here the point of representing the marble from Naxos (M10) lies quite close to the point representing the marble from Thassos (M1), a fact which is not so favourable but the Naxos marble (M10) is coarser then the Thassos marble (M1). The Rare Earth Elements (REE) were also quantitative analysed and their concentrations were tested for a discrimination.

In Figure 4 the geochemical distribution of the REE in the examined marbles is displayed. The measured concentrations are normalized on the average concentrations of these elements in the Upper Earth Crust. Although such distribution figures do not appear suitable for group discrimination, they supply some geochemically important information. The marbles are relative to the upper earth's crust at REE depleted (Figure 4). It means that the concentrations for all REE in the marbles lie lower than the concentrations of these elements in the Upper Earth Crust. Furthermore the heavy REE are more strongly enriched than the light REE in the marbles, which corresponds to the geochemical behavior of the limestones.

The marbles M4 (Vathylakos/Drama), M1 (crystalline/Thassos), M10 (crystalline/Naxos) and M11 (Dionyssos/Athens) show characteristic distri-

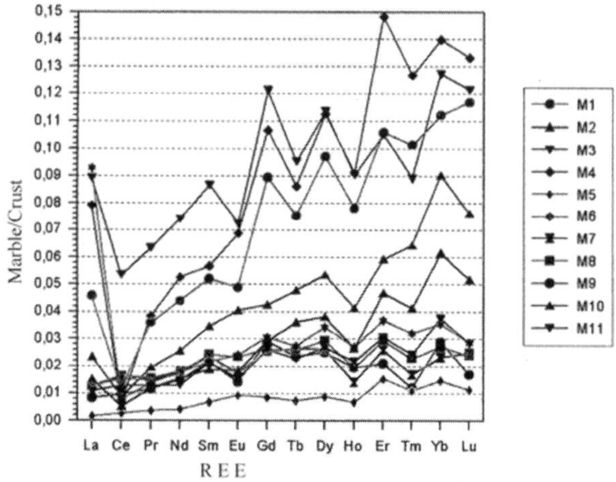

Figure 4. REE distribution of the analyzed marble occurrences from Greece.

bution. These four marbles show a more or less Ce-anomaly and have, with the exception of Ce, higher significant REE concentrations than the remaining marbles. The dolomitic marble of Volakas (M5) has the lowest REE-concentrations. Generally speaking, it can be said that a discrimination of marble occurrences is not to be expected with the help of the REE-distribution.

Using the $^{147}Sm/^{146}Nd$ isotope ratio we achieve a quite good separation of the individual marble samples (Figure 5). The marbles from Thassos (M1, M2) are separated from the others and lie well together. The three samples from Kozani (M7, M8, M9) are split into two groups. The two Calcitic ones, the grey (M8)/ and the white (M9)/(Tranovaltos), lie well together while the reddish Calcitic marble (M7)/(Kozani) separates clearly from the previous two.

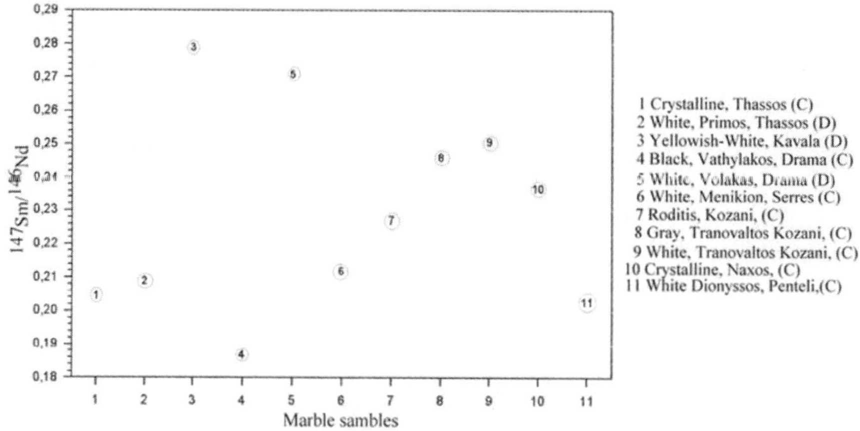

Figure 5. Plotting of the analysed marble occurrences as function of the $^{147}Sm/^{146}Nd$ isotope ratio.

Also the two marble samples from Drama, black Calcitic marble from Vathylakos (M4) and dolomitic marble from Volakas (M5), have a clear separation. The separation of the three specimens from Serres (M3), Naxos (M10) and Dionyssos (M11) is likewise very good.

By the discussed chemographic representation, the discrimination of historical marble objects and their allocation to regional stores seems to be quite possible. The discrimination power of this diagram can be regarded as safe only if it is confirmed without overlapping by additionally analyzed marble occurrences.

Besides, on the search for an efficient discrimination criterion for historical marble objects an attempt was made by using different isotope ratio plots. Two plots (208Pb/206Pb-207Pb/206Pb and 206Pb/204Pb-87Rb 86Sr) which give relatively reasonable results are presented and discussed here. Although the representing points of the analysed marble occurrences are separated for the Pb-Isotopes reasonably well (Figure 6), the separation distances are not large enough, in order to avoid an overlapping with further measuring occurrences. The two marbles from Drama (M4, M5) are also clearly separated by the isotopes and the same applies to the marbles from Thassos (M1, M2). The marbles from Kozani (M7, M6, M9) are also separated in a similar way like the trace elements. The two samples from Tranovaltos/Kozani (M8, M9) have the same isotope ratio, therefore overlap but they differ in their dolomite content and also in grain size, while the sample (M7)/(Roditis) possess a strongly deviating Pb-isotope ratio. The specimens from Kavala (M3), Naxos (M10) and Dionyssos (M11) are also separated clearly by means of the Pb-isotope ratios.

A better separation efficiency of the examined Greek marbles represents the (^{204}Pb/^{206}Pb)- (^{87}Rb/^{86}Sr)- isotope ratios of the samples (Figure 7).

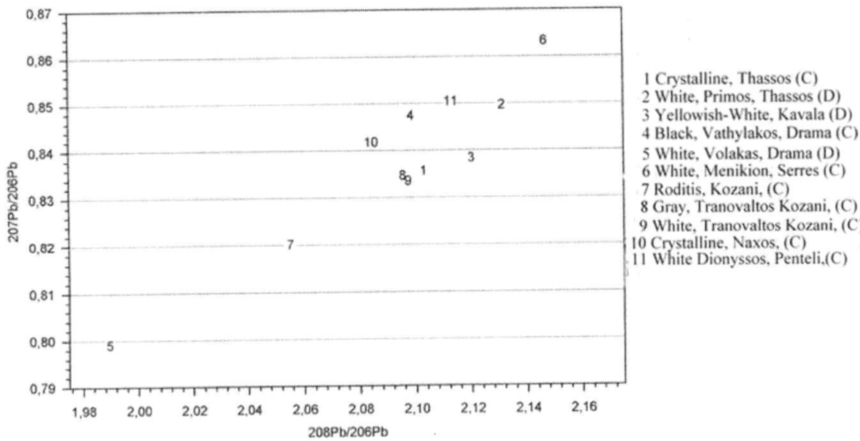

Figure 6. Discrimination of the analysed Greek marble occurrences by means of the Pb-Isotopes.

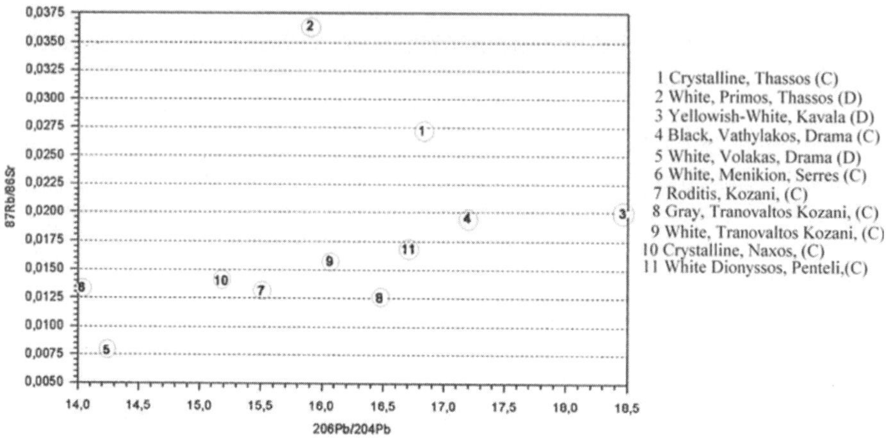

Figure 7. Discrimination of the examined Greek marble occurrences by the means of $^{206}Pb/^{204}Pb$)- ($^{87}Rb/^{86}Sr$)- isotope ratios.

By this representation the different marbles are also separated, but with larger separation distances. The discrimination conditions within the localities remain the same. Here the Thassos marbles (M1, M2) are likewise separated as well as the marbles from Drama (M4, M5). For the marbles from Kozani (M7, M8, M9) the separation is also clearly recognizable. The discrimination of the marbles from Kavala (M3), Naxos (M10) and Dionyssos (M11) is also very good in this representation.

5. CONCLUSIONS, REMARKS

The discussion of the results showed that a more or less clear discrimination, of historical and recent marble objects, can be achieved by using petrological and mineralogical analysis, grain size analysis, trace elements as well as selected isotope ratios. The petrologic and the quantitative mineralogical analysis can classify the marbles regarding their calcitic or dolomitic character. The grain size analysis is a further factor for the characterization of a marble quarry. Trace elements as well as selected isotope analysis can lead to a clear discrimination especially in combination with the above analytical techniques. However, in order to be able to make statistically safe statements regarding the discrimination of the chemographic representations discussed above, substantially more marble samples have to be analyzed, from everyone of the sampled localities. In addition, marble samples from further localities should be analyzed and represented chemographically, in order to confirm the discrimination efficiency of these diagrams.

Furthermore, it is to be examined whether marble occurrences from historically important regions of the Mediterranean area, like Italy, Greece and

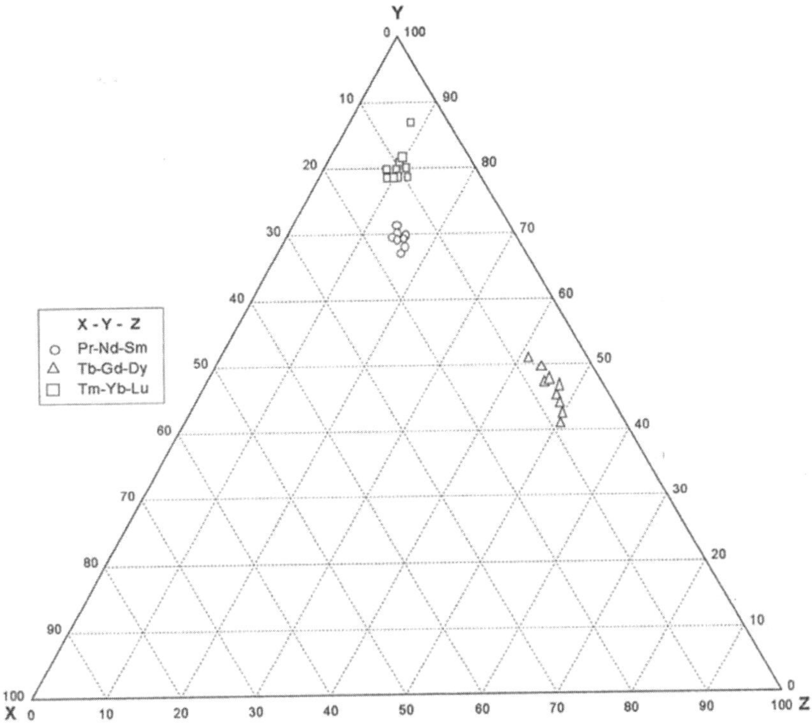

Figure 8. Clusters of the Greek marbles by three different ternary element correlations.

Asia Minor, are separated geochemically. It seems, according to this study, that certain ternary trace element correlations lead to extremely significant groupings of the marbles of Greece (Figure 8). In this ternary diagram, for all examined marble samples, three different REE-correlations were used and graphically projected: Pr-Nd-Sm, Tb-Gd-Dy and Tm-Yb-Lu. It is recognized from this illustration, that all three element correlations gave discrete and significant grouping for the examined Greek marbles.

For the archaeological research it would be therefore interesting to find out, whether marble occurrences from different delivery localities of Mediterranean area, like Italy and Asia Minor, give closed cluster areas, significant and different, as these of Greece. In this case one could use this diagram (Figure 8) to determine and verify first the global delivery region of the marble objects and afterwards by fine solvent diagrams (Figures 5 and 7) to find out the marble queries within the well known large region.

The results of the pilot research project presented here rely only on a small number of samples and are considered therefore as "provisional" and not “representative”. In order to achieve more founded research results, a research project with appropriate financial and personnel resources is to be organized.

REFERENCES

Andreae, B., Oehlschlegel, G., and Weber, K., 1972, Joining the fragments of a Meleagersarkophags in Frankfurt and Kassel, *Yearbook German Archaeological Institut.* **87**: 388-432.

Baietto, V., Villieneuve, G., Schvoerer, M., Bechtel, F., and Heart, N., 1999, Investigation of electron paramagnetic resonance peak into some powdered Greek white marbles, *Archaeometry* **41**(2)**:** 253-265.

Barbin, V., Ramseyer, K., Decrouez, D, Bruns, S. J., Chamay, J., and Maier, J.-l., 1992, Cathodenluminenscence of white marbles: on overview, *Archaeometry* **34:**175-183.

Bruno, M., Lazzarini, L., Soligo, M., Turi B., and Varti-Matarangas, M., 1997, in *Proceedings of the First International Conference on the Archaeology of Paros and the Cyclades*, Paros, 2-5 October 1997.

Coleman, M., and Walker, S., 1979, Stable of isotopes identification of Greek and Turkish of marbles, *Archaeometry* **21:**107-112.

Conforto, L., Felici, M., Monna, D., Serva, L. and Taddeucci, A., 1975, A preliminary evaluation of chemical data (trace of element) from classical marble quarries into the Mediterranean, *Archaeometry* **17:**201-213.

Craig, H., and Craig, V., 1972, Greek of marbles: determination of provenance by isotopic analysis, *Science* **176:**401-403.

Dworakowska, A., 1975, *Quarries of Ancient Greece*, Polish Academy of Sciences, Worclaw.

Gale, N.H., 1979, in *Proceedings of the International Symposium on Archaeometry and Archaeological Prospection*, London, Abstracts 18, 18.

Gast, R., Germann, K., and Eilert, E., 1979, Petrographic and geochemical investigations for the regulation of marbles of brightistic grave places Thessaliens. "La Thessalie, Collection de la Maison de l'Orient Mediterranéen 6, *Série archéologique* **5:**51-62.

Germann, K., 1978, Problems and possibilities, geoscientific certificates of origin for marbles.in: *Minerali raw materials as culture-historical source of information*, edited by H. W. Hennicke, Association of German email specialists. Hagen, 173-190.

Germann, K., Woodman, G., and Winkler, F.J., 1980, Determination of marble provenance: limits of isotopic analysis, *Archaeometry* **22:**99-106.

Gorgoni, C., Lazzarini, L., Pallente, P., and Turi, B., 2002, in *Asmosia 5*, Achetype Publications London.

Herz, N., 1955b, Petrofabrics and classical archaeology, *American journal of Science* **253:**299-305.

Herz, N. and Wenner, D. B., Tracing the Origins of Marble, Archeology **34**: 14-21 (1981).

Herz, N., 1987, Carbon and oxygen isotopic ratios: A data cousin for classical Greek and novel marble, *Archaeometry* **29:**35-43.

Herz, N., and Waelkens, M., Eds., 1988, *Classical marble: geochemistry, technology trade.* NATO ASI Ser. E, 153, Kluwer Academic Publishers, Dordrecht.

Kempe, D.R.C., 1983, in: *The petrology of archaeological artefacts*, 80-153. edited by D. R. C. Kempe, and Harvey, A.P., pp. 374. Clarendon press, Oxford 1983.

Kritsotakis, K., 1994, Mineralogical and geochemical characterisation of the praehistorischen ceramic(s) of Hellbrunnerberg with Salzburg, *Jahrb. RGZM* **41:**145-177.

Lazzarini, L., Moschini, G. and Stievano, B.M., 1980a, A contribution ton the identification of Italian, Greek and Anatolian of marbles through A petrological study and the evaluation of approx./SR ratio, *Archaeometry* **22:**173-182.

Lazzarini, L., Masi, U., and Tucci, P., 1995. *Asmosia III Athens: Petrrographic and geochemical features of the Carystian marble, "Cipollino verde", from the ancient quarries of Southern Euboea (Greece).*

Lazzarini, L., Pensabene, P., and Turi, B., 1995, *Asmosia IV. Isotopic and petrographic characterization of marmor Lesbium, Island of Lesbos, Greece.*

Lazzarini, L, and Cancelliere, S., 2000, Characterization of the white marble of two unpublished ancient Roman quarries on the Islands of Fourni and Skyros (Greece), *Per. Mineral.* **69**(1)**:**49-62.

Lepsius, G.R., 1890, *Greek marble studies*, Papers in Prussian Academy of Sciences to Berlin, Philosophy-historical class, 1-153.

Monna, D., and Pensabene, P., 1977, *Marmi dell'Asia Minore. Consiglio Nazionale della Ricerche*, Rome.

Manfra, L., Masi, A., and Turi, B., 1975, Carbon and oxygen of isotopes ratios of marbles from some ancient quarries of western Anatolia and their archaeological significance *Archaeometry* **17:**215-221.

Marinos, G.P., 1984, Notes on the structure of Greeks of marbles, *American Journal of Science* **246:**386-389.

Martin, R., 1965, *Manuel D ' Architecture Grecque, Volume 1: Matériaux et Tec*hniques. A. and J. Picard, Paris.

Moens, L, Roos, P., and De Rudder, J., 1987, Identification of archaeologically interesting white marbles by instrument valley neutron activation analysis (in a) and petrography: comparison between samples from Afyon to the Usak (Turkey), *J. Trace Microprobe of techniques*, **5**(2/3)**:**101-114.

Moens, L, Roos, P., De Rudder, J., Hoste, J., De Paepe, P., Van Hende, J., Marechal, R. and Waelkens, M., 1988, White marble from Italy and Turkey: on archaeometric study based on minor and of trace element analysis and petrography, *J. Radioanalytical Nuclear Chemistry*, **123**(1)**:**333-348.

Papageorgakis, J.E., 1963, The historical marble quarries of Thessalien, *Praktikà Akadimias Athinon* **38:**564-572 (in Greek).

Papageorgakis, J.E., 1967, The rocks of Greece usable in the marble industry, *Annales Géologiques of the Pays Hélleniques* **18:**193-270.

Renfrew, C. and Peacey, J.S., 1968, Aegean marble, A petrological study, *Annual of the British School RK of Athens* **63:**45-66.

Rybach, L., and Nissen, H.-U., 1965, Neutron activation of Mn and well traces into marbles worked by the ancient Greeks. In: *Radiochemical methods of analysis, Volume I*, Internationally Atomic Energy Commission, Vienna, 105-117.

Schmid, J., Ambuehl, M., Decrouez, D., Mueller, S., and Ramseyer, K., 1999. Quantitative fabric analysis approach ton the discrimination of white marbles, *Archaeometry* **41:**239-252.

Schneider, G. and Rother, A., 1991, Chemical-mineralogical investigations of people-migration-temporal ceramic(s) of rounding mountain. In, the round mountain of Urach IX., *Heidelberger Acad. Wiss. Schr. Bd.* **15:**190-223.

Vakulis, Th., Saatsoglu Paliadeli, Ch., and Maniatis, Y., 2000, in: *Abstracts of the VIth internationally Conference ASMOSIA 2000*, Venice, June 15-18, 2000.

White, L.E., 1954, Fabric analysis of some Greek of marbles and its applications tons archaeology, *American journal of Science* **252:**641-662.

APPENDIX

Microscopic picture of the marble samples and their grain size distribution.

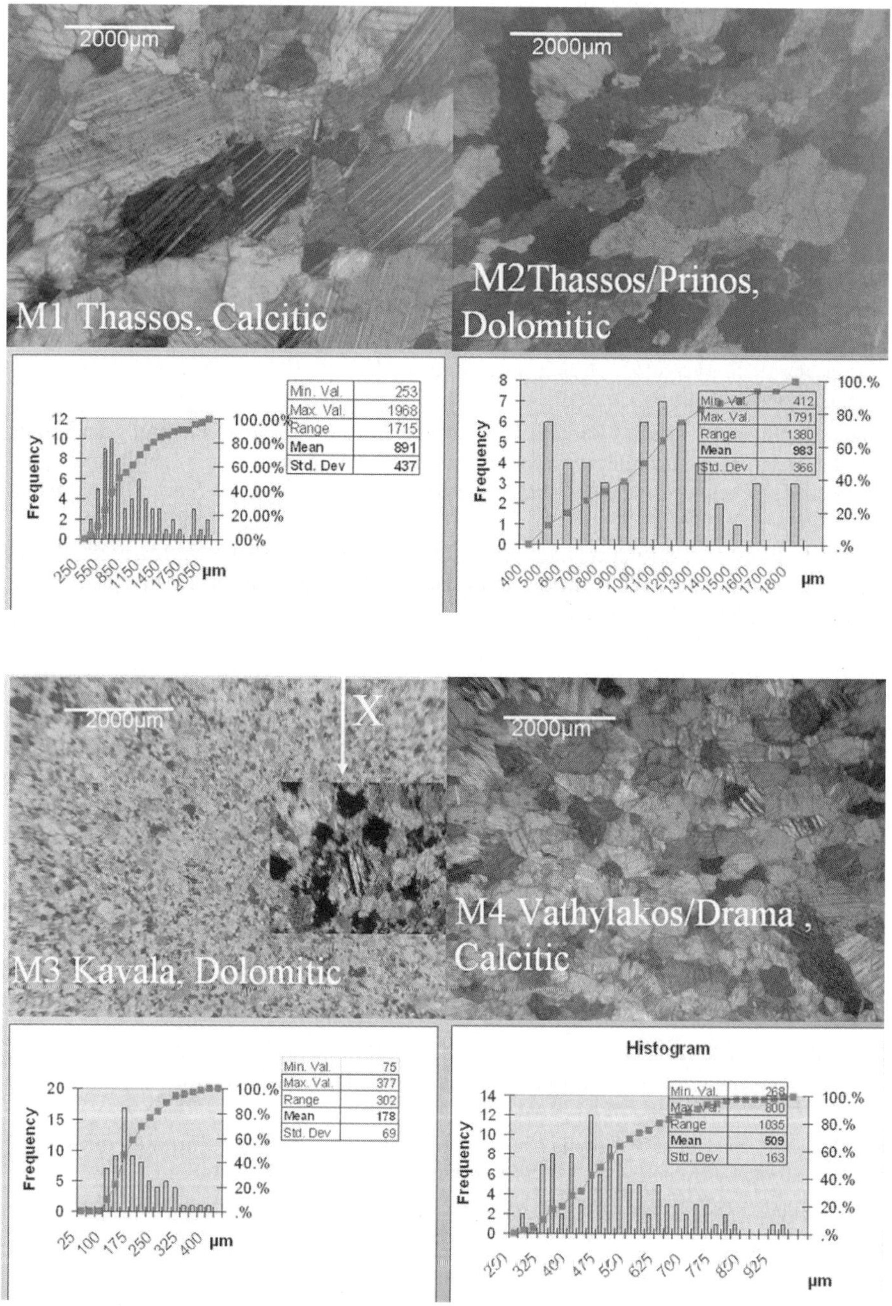

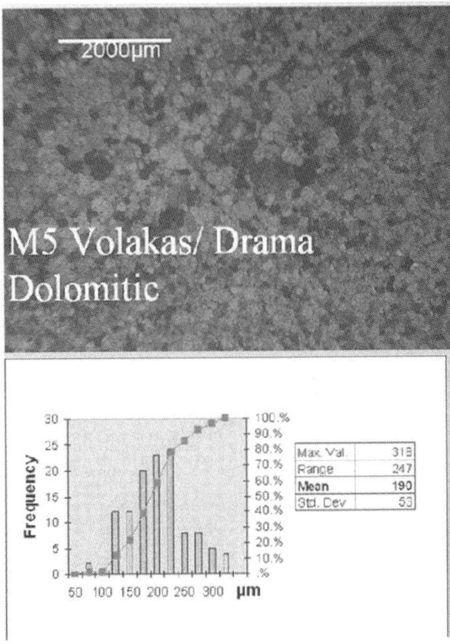
2000µm
M5 Volakas/ Drama
Dolomitic
Frequency
µm
Max. Val. 318
Range 247
Mean 190
Std. Dev 53

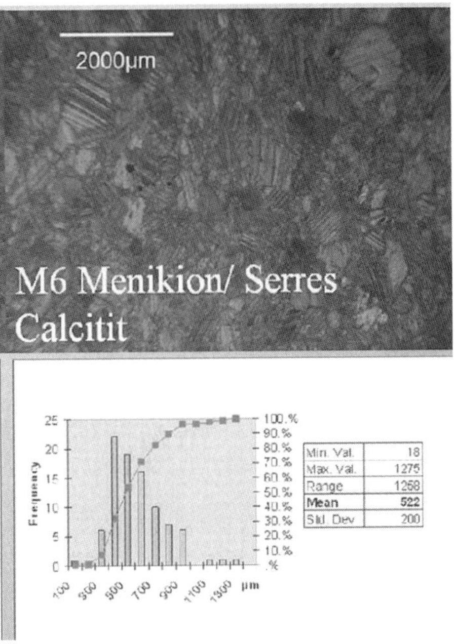
2000µm
M6 Menikion/ Serres
Calcitit
Frequency
µm
Min. Val. 18
Max. Val. 1275
Range 1258
Mean 522
Std. Dev 200

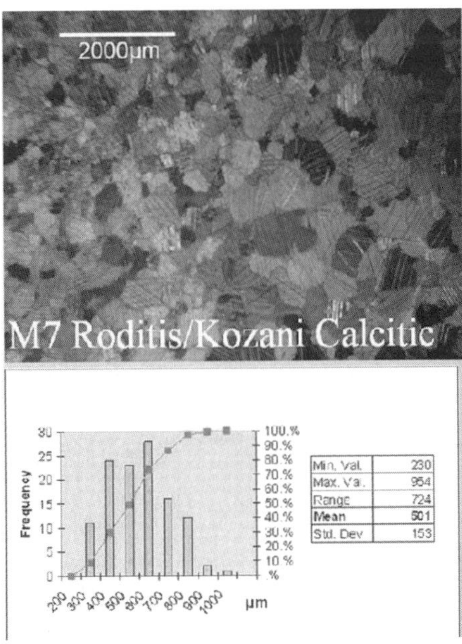
2000µm
M7 Roditis/Kozani Calcitic
Frequency
µm
Min. Val. 230
Max. Val. 954
Range 724
Mean 501
Std. Dev 153

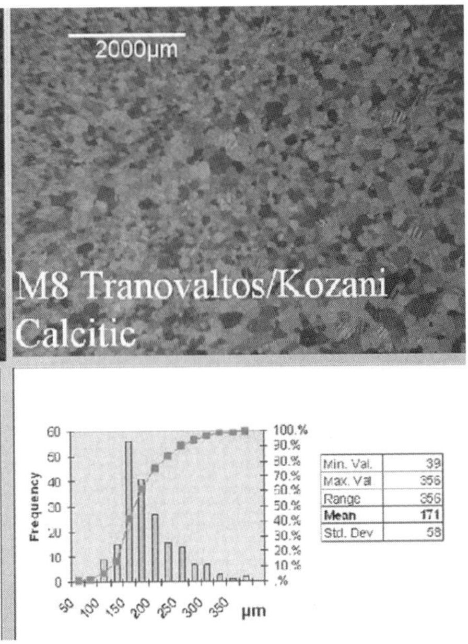
2000µm
M8 Tranovaltos/Kozani
Calcitic
Frequency
µm
Min. Val. 39
Max. Val 356
Range 356
Mean 171
Std. Dev 58

Chapter 9: Surface Treatment

Chapter 9.1

A HIGH-RESOLUTION VIEW AT WATER REPELLENTS AND CONSOLIDANTS: CRITICAL REVIEW AND RECENT DEVELOPMENTS

Veerle Cnudde[1], Manuel Dierick[2], Bert Masschaele[2] and Patric JS. Jacobs[1]
[1]Dept. of Geology and Soil Science, Ghent University, Krijgslaan 281/S8, Ghent; [2]Dept. of Subatomic and Radiation Physics, Ghent University, Proeftuinstraat 86, Ghent, Belgium

Abstract: The growth in the renovation and maintenance activities of old buildings has largely contributed to the success of water repellents and consolidants. The greatest difficulty however, is the selection of the most appropriate one out of this wide range of products. Although certain products can work splendidly on a certain stone type, they can be harmful to another, especially when they alter important petrophysical material characteristics. Therefore combined research is necessary on the properties of the conservation products in combination with the determination of some important technical and moisture transfer parameters of the stone material itself. Different traditional research techniques exist in order to determine these technical and moisture transfer properties. These traditional techniques, in combination with advanced 3D visualization techniques, like X-ray and high-speed neutron tomography, turn out to offer important additional information on both the stone material and on the transfer method of the conservation products inside the stone material. In this paper, results of traditional research in combination with data obtained from X-ray micro-CT and high-speed neutron tomography is given, as well as a comment on the recent developments in material research by means of tomography.

Key words: water repellent; consolidant; X-ray micro-CT; natural building stone; high-speed neutron tomography.

1. INTRODUCTION

Since sooner or later all natural building stones have to deal with weathering, many conservation products were put on the market in order to reduce the decay rate and to strengthen decayed stone. Although many products are available, it remains difficult to select the most appropriate product for a specific type of stone. Since water repellents and consolidants, two very important conservation products, can alter some important petrophysical material

S.K. Kourkoulis (ed.), Fracture and Failure of Natural Building Stones, 519–540.

parameters, their use remains an intervention with an uncertainty factor. Depending on the internal structure of the stone material, the impregnation and localization of the conservation products inside the stone material will vary. Since the efficiency of consolidants and water repellence is influenced by their impregnation depth and localization inside the stone, it remains important these parameters to be determined in combination with the structural properties of the stone material. Previous studies by Cnudde et al. (2004), Cnudde & Jacobs (2004a) and Cnudde et al. (submitted, a) proved that conservation products like water repellents and consolidants can be located with X-ray micro-CT inside the stone material, while Dierick et al. (2005) and Masschaele et al. (2004a, 2004b, and submitted) demonstrated that high-speed neutron tomography can be applied as well for the visualization of conservation products and other fluid flow.

Since most traditional research techniques focus only on the determination of one parameter at a time, while many parameters should be considered instead, a combined research using both the non-destructive visualization techniques and the traditional research techniques was performed.

In this paper the influence of two conservation products, a siloxane based water repellent and an ethylsilicate based consolidant, were tested with traditional research techniques in combination with results derived from X-ray CT and high-speed neutron tomography. Additionally, recent developments in material research by tomography are considered.

2. MATERIAL

In this study, two types of natural building stones were selected based on their high porosity and their pure (mono-)mineralogical composition: a highly porous bioclastic limestone from Maastricht (Maastrichtian, Upper Cretaceous) and a quartz arenite of Upper-Landenian age (Paleocene, Paleogene), known as the Bray sandstone. Very typical for the Maastricht limestone is its high porosity (up to 53%). Depending on the cementation degree, the Bray sandstone has more sandy varieties with a higher porosity (up to 27%) and more quartzitic varieties with lower porosity (min. 3.5%).

3. CONSERVATION PRODUCTS

3.1 Water repellents

A description of some water repellents and their properties is critically reviewed by González (2000), WTCB (2002), De Witte (2003), Young et al. (2003), and others. Besides silanes, siloxanes, and silicones, organic metal combinations and perfluoropolyethers were considered as water repellent products. Currently more than 90% of the applied water repellents contain siloxanes (WTCB, 2002), while polymer silicones are almost no longer used for the water repellent treatment of building walls. Organic metal combinations, exceptionally used as water repellents (WTCB, 2002) are based on aluminiumstearate or butyltitanate, while perfluoropolyethers are polymeric chains built upon just three elements: carbon, oxygen, and fluorine.

For this study Hydro10, an oligomer siloxane (10 vol-% in white spirit) with a small quantity of trifunctional monomers, was selected as water repellent, with an experimental dry weight of 8% and a density of 0.77 g/cm^3 at 20 °C before polymerization. After polymerization, when the polymer was composed out of $[(R)_2\text{-Si-O-}]$ as repeating units, it increased to 1.07 g/cm^3.

3.2 Consolidants

Many consolidants are available on the market. For the inorganic based consolidants, siliceous consolidants (alkali silicates, silicofluorides or fluosilicates) and alkaline earth hydroxides can be considered. Their reaction occurs mainly on the surface layers of the material, as a result of the consolidants poor penetration (Clifton, 1980). In general organic-based materials, like acrylic consolidants, vinyl polymers, epoxies, polyurethane, and polyester resin, have a good adhesion to the substrate and are good at taking up dimensional changes in stone (such as thermal expansion and contraction). Some disadvantages are their sensitivity to unstable environmental conditions and their vulnerability to heat or UV light. Usually the penetration depth, which is often low, depends greatly on the ability of the solvent to carry the consolidant into the stone and the percentage of moisture in the stone. Alkoxysilanes are regarded by many stone conservators as being among the most promising stone consolidating materials, due to their abilities to penetrate deeply into porous stone and by the fact that their polymerization can be delayed until deep penetration has been achieved (Clifton, 1980). The alkoxysilanes can be diluted with solvents to reduce their viscosities and to influence their penetration. By the production of the silica end product there is a definite consolidating effect and many silane-based products seem to increase the strength (flexural, compressive, tensile, etc.) of damaged stone (Garrod, 2001). Unfortunately, there is some colour change with most types of silanes, although studies show that this usually lessens after about 18 months. More information on types of stone consolidating materials, their performances, and uses are critically reviewed by Clifton (1980), Price (1996), Garrod (2001), Young et al. (2003) and Moropoulou & Farmakalidis (2003).

In this study SH75 and SH100, both ethylorthosilicate based consolidants, were used. While SH75 is diluted in methylethylketon (25%), SH100 is solvent free. The products are a combination of pure ethyl silicate and some oligomers, which are low molecular prepolymers that are formed by controlled pre-condensation of ethylorthosilicate, while a catalyst (dibutyltin-laurate) is added to control the reaction speed. The basic elements used for SH100 have a different composition (amount of ethylorthosilicate oligomers) compared to those used to prepare the SH75. The experimental dry weight of SH75 and SH100 was respectively 32% and 52%. The density of SH75 and SH100 before polymerization was respectively 0.93 g/cm³ and 0.95 g/cm³ at 20 °C, while after polymerization it attained 1.67 g/cm³ for both products.

4. X-RAY TOMOGRAPHY

X-ray computed tomography provides three-dimensional images of the internal structure of a scanned sample, non-destructively, by calculating the X-ray attenuation within objects. To obtain the 3D information, an X-ray tomograph first records a high number of 2D radiographs of the sample, taken from different angles by rotating the sample relatively to the X-ray source-detector system (Figure 1).

Each radiograph records the intensity (I) of X-ray photons after passing through the object.

$$I = I_0 \cdot e^{-\int \mu(s) \cdot \rho(s) \cdot ds} \qquad (1)$$

Lambert-Beer's law (Eq. 1), a basic equation for attenuation of a mono-energetic beam through a homogeneous material, relates the recorded intensity (I) with the initial X-ray intensity (I_0) of X-ray photons and the linear attenuation coefficient $\mu(s).\rho(s)$ of the object being studied, with $\int ds$ the length of the X-ray path through the material. The linear attenuation co-

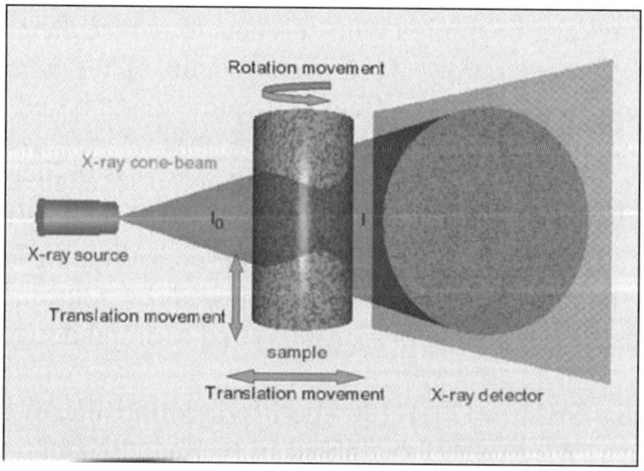

Figure 1. Schematic set-up of a micro-CT system.

efficient μ is a measure for attenuation per unit distance and depends on both density and atomic number. A radiograph therefore contains the absorption information inside the three-dimensional object, integrated along the line of sight. For a single CT scan hundreds of radiographs are acquired from different rotation angles between 0° and 360°. After the data collection, a Feldkamp-algorithm (Feldkamp et al., 1984) is used to calculate the μ value for each position in the sample. For more information concerning X-ray CT, Jacobs et al. (1995), Carlson et al. (1999), Ketcham & Carlson (2001) and Cnudde et al. (in press) are referred.

The major advantage of X-ray tomography is the fact that the internal structure of a scanned object can be made visible, without any sample preparation, in a relatively short time and completely non-destructively. 3D quantification of this internal structure is possible, and changes due to treatment can be monitored. Spatial resolution below one micron can be achieved, though only for small samples. As a rule of thumb one can state that the spatial resolution is of the order of 1/1000th of the sample size. The major disadvantage (in a geological context) is the fact that water and solvents containing elements with a low atomic number are hardly visible when using X-ray CT. To visualize these products, doping agents often need to be added in order to obtain a higher X-ray attenuation. Too high concentrations of attenuating material can result in star artifacts in the calculated cross-sections, reducing the overall image quality.

5. HIGH-SPEED NEUTRON TOMOGRAPHY

Besides X-rays there are other types of radiation that can be used to investigate the inner structure of samples. One of them is a neutron beam. Neutrons have fundamentally different absorption characteristics compared to X-rays. Whereas X-rays interact with the electron cloud around an atom, neutrons interact with the atom's nucleus. Neutrons are very sensitive to water or other hydrogen rich compounds, resulting in strong absorption. On the other hand, neutrons can penetrate easily through metals, such as aluminum or even lead, in contrast to X-rays. Neutron tomography is used for both industrial non-destructive testing and scientific applications, such as imaging biological samples or visualizing hydrogen distribution in fuel cells.

The strong neutron absorption of hydrogen, combined with the relatively low absorption of rock-type materials, makes neutron tomography ideally suited for the study of water penetration in rocks. Without the aid of doping agents one can visualize the conservation agents and the water distribution in a rock in 3D. The strongest available neutron sources make it even possible to monitor the fluid distribution in quasi real-time in 3 dimensions, with a time step of the order of 10 seconds and a spatial resolution of the order of 0.6 mm. This is called dynamic neutron tomography, and has been applied successfully to visualize the actual penetration of conservation products themselves, as well as their resulting effectiveness when the sample is exposed to water.

Neutron tomography has some disadvantages too. Neutron beams are available at nuclear research reactors and so called spallation sources. Both are very large scale facilities. Neutron tomography therefore remains mainly limited to research environments. Another disadvantage is the spatial resolution, which is of the order of 250 μm under good circumstances. Still, neutron tomography offers unique opportunities where X-rays fail.

6. RESULTS

A brief summary of the influence of the selected conservation products, Hydro10, SH100 and SH75, on certain material properties are given below. Traditional research techniques were used as well as the non-destructive X-ray micro-CT and high-speed neutron tomography.

6.1 Compressive strength

One of the technical properties that can be influenced by consolidation is the compressive strength. The uniaxial compressive strength R (N/mm^2) was determined with the loading normal to their anisotropy, according to the European standard EN 1926 (1999) on Maastricht limestone samples before and after consolidation with the consolidants SH75 and SH100, by either spray flow or total absorption. The compressive strength values are presented in Table 1. Statistically the amount of absorbed product per area by the Maastricht limestone compared to the compression resistance was tested. From these calculations it could be concluded that for the Maastricht limestone, treated with SH100, 52% of the variance in the compressive resistance data can be predicted from the amount of absorbed product/area (g/m^2), while this was only 26% for samples treated with SH75. The results indicate that for the consolidation of Maastricht limestone, SH100 has a higher influence on the compressive resistance than SH75.

Table 1. Overview compressive strength results Maastricht limestone.

Compressive resistance (N/mm^2)	Min.	Average	Max.
Untreated	2.1	3.2	4.6
SH75 spray flow	3.2	3.7	4.2
SH75 total absorption	2.7	3.9	6.4
SH100 spray flow	2.4	3.2	5.2
SH100 total absorption	3.4	4.5	5.5

6.2 Drilling resistance

Since consolidation should influence the drilling resistance, this property was determined by a portable drilling resistance measurement system (DRMS) on untreated and consolidated Maastricht limestone and Bray sandstone samples. During the experiment, holes of 5 mm diameter were drilled on the stone surface for a depth of 1 cm, while measuring the force (N) as a function of the penetration depth (Tiano, 2003). The apparatus was operated with a rotational speed of 600 rpm and a penetration rate of 10 mm/min. Maastricht limestone turned out to be extremely soft, while Bray sandstone ranges from extremely soft (<1 N), to extremely hard (>100 N). In general a firm link between porosity values and drilling resistance was found. Although consolidation with SH100 increases, in general, the drilling resistance from the upper layer up to 6 mm deep, it often remains difficult to determine the exact reason for the increasing resistance, since local heterogeneities sometimes have a larger impact on this property than consolidation. SH75 turned out to have a much lower influence on the drilling resistance than SH100 for both Bray sandstone and Maastricht limestone.

6.3 Colour measurements

Colour changes due to treatment with conservation products can be important. A spectrophotometer (X-Rite SP60) was used to determine a possible colour deviation between untreated and treated samples. After measuring the spectral curves of a large distribution of Maastricht limestone and Bray sandstone samples, the colour was mapped into a colour space. To determine colour changes, ΔE^*_{ab} was introduced (WTCB, 2002; Berger-Schunn, 1994):

$$\Delta E^*_{ab} = (\Delta L^{*2} + \Delta a^{*2} + \Delta b^{*2})^{1/2}$$

with the L*-axis, known as the lightness and extending from black (L=0) to white (L=100), and the coordinates a* and b*, representing respectively redness-greenness and yellowness-blueness. The results of the colour measurements are presented in Table 2. Depending on the stone type different colour changes are observed but, in general, treating Bray sandstone and Maastricht limestone with the consolidants SH100 and SH75 and water repellent Hydro10 lowers the average L, while the a and b value rises, resulting in a more red and more yellow colour.

6.4 Contact angle

The contact angle (θ), resulting from the balance among the three surface tensions acting at the interface between liquid, solid and gas, is an important parameter due to its influence on the capillarity of a stone. Contact angle changes due to treatment with conservation products were obtained by an optical contact angle device (OCA 20). Additionally the impregnation time of 3 μl ultra-pure water droplets was determined on the stone surfaces before and after treatment. For Maastricht limestone and the highly porous Bray sandstone samples, the contact angle in equilibrium was impossible to be determined, since the water droplet was absorbed in less than 20 ms. The impregnation speed for 3 μl water in the Bray sandstone was closely related to the porosity and ranged between 76 ms and 166662 ms, with an average of 14361 ms. For all untreated Bray sandstone samples, even those with the lowest porosity, the contact angle θ was smaller than 90° (hydrophile), while after treating the samples by spray flow with either Hydro10, SH100 or SH75, their contact angle and impregnation speed increased (Table 3).

Samples treated with Hydro10 turned out to have a very high contact angle, as well as the consolidated ones (Table 3). Due to consolidation the porosity and pore size distribution at the examined surface will change and consequently the contact angle also, but the high contact angle results for the consolidated Maastricht limestone can not be explained only by porosity changes. More likely, the high contact angles of the consolidated limestone samples are due to uncompleted polymerisation of the consolidants. After more than 7 weeks it seemed that the limestone samples treated with SH75 were still having a water repellent effect, while on the consolidated Bray sandstone samples the hydrophobic effect was reducing.

Table 2. Results of the colour measurements on Bray sandstone and Maastricht limestone, before and after treatment.

Bray sandstone				
Average value	L	A	B	ΔE^*_{ab}
Untreated	74.2 (± 6.7)	3.7 (± 2.0)	12.9 (± 4.6)	
Treated with SH75	69.1 (± 3.9)	3.8 (± 1.5)	14.7 (± 3.8)	5
Treated with SH100	68.3 (± 3.5)	4.1 (± 1.5)	15.7 (± 4.1)	7
Treated with Hydro10	71.5 (± 6.9)	3.7 (± 1.4)	14.1 (± 4.7)	3
Maastricht limestone				
Untreated	79.0 (± 1.7)	4.2 (± 0.7)	24.4 (± 2.4)	
Treated with SH75	76.4 (± 1.1)	5.0 (± 0.4)	26.8 (± 1.4)	4
Treated with SH100	76.5(± 1.2)	5.1 (± 0.5)	26.8 (± 1.5)	4
Treated with Hydro10	75.2 (± 0.9)	5.9 (± 0.4)	29.8 (± 1.1)	7

Table 3. Overview of the average contact angle data and impregnation time for Bray sandstone and Maastricht limestone treated with Hydro10, SH100 or SH75.

Average Contact angle (°)			
	Hydro10	SH100	SH75
Sandstone	124 (±4)	107 (±16)	104 (±13)
Limestone	130 (±3)	118 (±4)	124 (±3)
Average impregnation time (s)			
Sandstone	1339 (±212)	928 (±412)	635 (±521)
Limestone	1865 (±429)	947 (±800)	1276 (±609)

6.5 Capillarity

Following the European standard EN 1925, the water absorption coefficient by capillarity C ($g/m^2s^{1/2}$), the free porosity N_{cap} (%), the water absorption by capillarity (%) and diffusion (%) was determined for several Maastricht limestone and Bray sandstone samples before and after treatment with either Hydro10 (Table 4), SH100 or SH75.

The results confirm that the application of Hydro10 has a major influence on the capillarity of the samples, since for those samples almost no capillarity occurs (Table 4, Figure 2), even for the Maastricht limestone with its high porosity (Table 4). Only due to diffusion some water is absorbed by these stone samples. The first hour samples treated with SH75 tend to act water repelling, while afterwards the capillarity rises (Figure 3). In general, it seems that treatment with the consolidants SH100 and SH75 tends to reduce the water absorption by capillarity, while the importance of diffusion rises.

Table 4. Overview of capillarity data for Bray sandstone and Maastricht limestone before and after treatment with water repellent Hydro10: (1) free porosity (%), (2) capillarity coefficient ($(g/m^2)/s^{1/2}$), (3) water absorption by capillarity (%), (4) water absorption by diffusion (%).

	(1)	(2)	(3)	(4)	(1)	(2)	(3)	(4)
	Sandstone				Sandstone + Hydro10			
Average	9.0	252.4	4.4	1.6	0.0	1.4	0.0	2.6
std. Dev	7.8	283.2	3.9	2.2	0.0	1.3	0.0	2.9
	Limestone				Limestone + Hydro10			
Average	39.6	2370.6	31.7	4.3	0.2	243.7	0.2	23.1
std. Dev	0.7	243.4	0.7	0.7	0.1	217.6	0.1	2.9

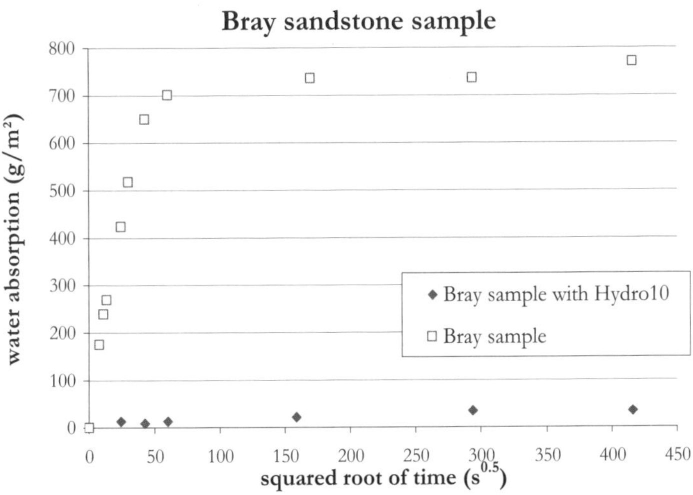

Figure 2. Water absorption by capillarity for a Bray sandstone sample before and after treatment with Hydro10.

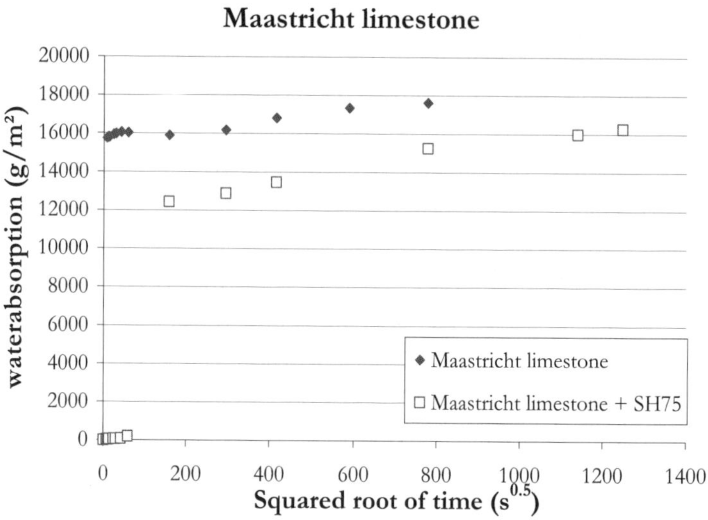

Figure 3. Water absorption by capillarity for Maastricht limestone sample before and after treatment with SH75.

6.6 Drying rate

After 24 hours of full immersion in water, the drying rate was determined on Bray sandstone and Maastricht limestone samples before and after treatment with SH75, SH100 or Hydro10. Treatment with Hydro10 resulted for both Bray sandstone and Maastricht limestone into a reduction of the total amount of absorbed water, while the drying time was multiplied by a factor 3 for the Bray sandstone and by a factor 2 for the Maastricht limestone (Table 5). The weight loss during the first phase was reduced by a factor 7 for the Bray sandstone and by a factor 2 for the Maastricht limestone, due to Hydro10 treatment, while the speed of the second drying phase was increased for both stone types.

The drying time for the consolidated sandstone and consolidated Maastricht limestone was respectively increased with a factor 2.5 and a factor 2. There is sufficient evidence to suggest a difference between the percentage water absorption, the total drying time and the drying speed in the first and second phase due to treatment with Hydro10 and SH75 on both stone types. For Bray sandstone and Maastricht limestone treated with SH100, the percentage of water absorption and the drying rate during the second phase proves no sufficient evidence to suggest a difference between the untreated and consolidated samples, while the total drying time and the drying speed during the first phase illustrates a systematic difference (Table 5). In general, it can be stated that treatment with Hydro10, SH100 or SH75 significantly changes the drying time of the samples, even if they lower the amount of absorbed water.

Table 5. Overview of data obtained during the drying rate experiment for untreated Bray sandstone and Maastricht limestone samples and for treated samples with either Hydro10, SH100 or SH75; (1) water-absorption (%), (2) (weight loss/area)/time 1st phase (g/m².h), (3) (weight loss/area)/time 2nd phase (g/m².h); between brackets the standard deviation value is given.

	(1)	(2)	(3)	(1)	(2)	(3)
	Sandstone			Sandstone + Hydro10		
Average	4.80 (3.67)	110.52 (15.15)	1.76 (1.21)	3.79 (2.82)	15.44 (7.92)	2.29 (1.29)
	Sandstone			Sandstone + SH100		
Average	3.72 (2.97)	110.72 (24.04)	1.45 (2.00)	3.63 (3.00)	48.18 (36.64)	0.88 (0.76)
	Sandstone			Sandstone + SH75		
Average	3.63 (2.63)	109.30 (19.58)	1.70 (1.70)	3.67 (2.69)	52.91 (17.71)	0.76 (0.52)
	Limestone			Limestone + Hydro10		
Average	35.96 (1.17)	88.10 (25.89)	4.22 (2.26)	23.09 (2.90)	44.69 (18.85)	11.32 (2.27)
	Limestone			Limestone + SH100		
Average	34.76 (0.62)	86.43 (20.67)	6.57 (3.37)	34.49 (2.04)	111.46 (21.82)	5.73 (3.32)
	Limestone			Limestone + SH75		
Average	35.50 (0.96)	79.18 (23.94)	5.29 (3.16)	29.98 (4.54)	81.92 (20.45)	2.54 (1.15)

6.7 Water vapour permeability

The water vapour permeability coefficient δ (kg/m.s.Pa) was determined before and after treatment with either Hydro10, SH100 or SH75 on both Bray sandstone and Maastricht limestone (Table 6), by using the "cup test" method, according to the European standard EN ISO 12572 (2001).

Table 6. Overview of permeability coefficient for Maastricht limestone and Bray sandstone, before and after treatment

Sample	permeability coefficient δ (kg/m.s.Pa)		
	Mean (std.)	Min	Max
Maastricht limestone: no treatment	0.00056 (0.00006)	0.00049	0.00062
Maastricht limestone + Hydro10	0.00046 (0.00003)	0.00043	0.00050
Maastricht limestone: no treatment	0.00059 (0.00006)	0.00047	0.00068
Maastricht limestone + SH100	0.00042 (0.00006)	0.00036	0.00048
Maastricht limestone: no treatment	0.00056 (0.00007)	0.00048	0.00069
Maastricht limestone + SH75	0.00040 (0.00003)	0.00037	0.00057
Bray sandstone: no treatment	0.00019 (0.00011)	0.00009	0.00034
Bray sandstone + Hydro10	0.00011 (0.00009)	0.00003	0.00021
Bray sandstone: no treatment	0.00025 (0.00017)	0.00010	0.00053
Bray sandstone + SH100	0.00020 (0.00018)	0.00006	0.00050
Bray sandstone: no treatment	0.00022 (0.00018)	0.00007	0.00063
Bray sandstone + SH75	0.00009 (0.00006)	0.00002	0.00016

Based on the permeability measurement results (Table 6) it seems that the water vapour permeability is reduced by a factor 1.2 for Maastricht limestone and by a factor 1.7 for Bray sandstone due to treatment with Hydro10. This indicates that although diffusion of the water vapour molecules occurs in the hydrophobic treated pores, this diffusion has decreased. From the same experiment, it seems that both SH100 and SH75 are reducing the water vapour permeability, but consolidation with SH75, reduces the water vapour permeability the most, by a factor 1.4 for the Maastricht limestone and a factor 2.4 for the Bray sandstone.

6.8 Water absorption under low pressure

The efficiency of a water repellent can be tested by water absorption measurements with a Karsten tube (Commission 25-PEM, 1980). This water tube measures the water absorption coefficient (WAC), which is the amount of water absorbed between the 5th and 10th minute, while the hydrostatic pressure on the exposed surface is determined by the height of the water level in the Karsten tube. When the water level is 98 mm, this corresponds to a wind speed of 40 m/s, perpendicular to the test surface (WTCB, 2002).

The experiment was performed before and after treatment with Hydro10, SH100 or SH75. Before treatment, the WAC was impossible to determine for Maastricht limestone, which absorbed the water in such a fast way that the Karsten tube was impossible to fill. For the Bray sandstone a relation between the porosity and the WAC was observed. All Bray sandstone and Maastricht limestone samples treated with Hydro10 turned out to absorb no water during the Karsten pipe experiment. For the Bray sandstone samples treated with SH100 and SH75, an overall reduction of the WAC was observed, except for those treated with SH75 and an open porosity of more than 16%. The average WAC remained rather similar for the Maastricht limestone treated with SH75 and SH100.

6.9 Porosity and pore size distribution

Several methods, each with its own advantages, characteristic limitations and disadvantages, exist in order to determine porosity. Besides water absorption under vacuum and mercury intrusion porosimetry (MIP), X-ray micro-CT can be used as an extra research tool for the determination of porosity and pore size distribution based on the information of the reconstructed 3D images. As described by Cnudde & Jacobs (2004a) total porosity, partial porosity and pore size distribution, based on the maximum inscribed diameter can be determined in 3D with the home-made 3D software µCTanalySIS. Although this data can be used to relatively compare data from different samples, it has to be interpreted with care due to discretization and resolution limitations. As described by Cnudde et al (submitted b) it remains difficult to compare the X-ray micro-CT data and porosity data from other techniques due to their different approach and physical basis. Even if there appears to be a correlation between different techniques, this does not confirm that both methods are accurate and unbiased, although it is highly probable. For validating the problem, often the "change of support" causes a problem, which is very critical in heterogeneous materials.

The pore size distribution of Maastricht limestone and Bray sandstone is a very important aspect in conservation research. Porosity and pore size distribution have not only an impact on capillarity, water vapour transport, contact angle and drying rate but also on the impregnation of the products applied on the stone. Due to local heterogeneities, the porosity of the Bray sandstone can strongly decrease which consequently effects the penetration of conservation products and their transport inside the stone. When the polymer size of the conservation product is larger than the pore size, pores will remain untreated. Mostly the transition of untreated to treated pores will be gradual over a broad pore size range due to the statistical character of the polymerization process (Carmeliet et al, 2002). In coarse pores the polymer network can fully develop, but these polymer structures can also clog the pores, leading to a reduction of the connectivity within the pore structure. The pore section reduction due to the development of the polymer network in the pore space will depend on the pore size. When the polymer size equals or exceeds the pore size of the stone material, pore clodding will occur, while otherwise the polymer will result in a reduction of the pore size radius (Carmeliet et al., 2002). Clodding and pore size reduction will exert an influence on some very important rock characteristics, including capillarity and water vapour permeability.

When water repellents are applied, it is important that the whole treated region is totally water repellent to prevent incoming water being captured behind the treated water repellent layer. At this point the pore size distribution and the pore structure of the stone are of great importance, since for pores larger than 0.3 mm, the water repellent will not work sufficiently enough to repel water (WTCB, 2002; De Witte, 2003). From MIP results, Maastricht limestone and Bray sandstone turned out to consist respectively of 0.9% and 2.3% pores larger than 300 μm. The smallest detected pore diameter for both Maastricht limestone and Bray sandstone was 4.5 nm. For Maastricht limestone and Bray sandstone respectively 76% and 86% of the pores have a diameter between 1 μm and 96 μm. Although these data give a good indication of the pore size distribution, it is important to keep in mind that MIP has the tendency to overestimate the smaller pores, which might lead to erroneous conclusions on their number. The information of the μCTanalySIS Maastricht limestone and Bray sandstone indicated that both stone types were having pores with a maximum inscribed diameter of 292 μm, also indicating that water repellent treatment for both stone types could be a risky intervention. The pore size distribution information reveals that both Bray sandstone and Maastricht limestone have a small percentage of pores that are of the size for which the water repellency is not working efficiently any more. Derived from the MIP data it can be presumed that the ethylorthosilicate molecules and the polymers of the Hydro10 can impregnate the smallest pores of both Bray sandstone and Maastricht limestone. Since SH75 contains beside ethylorthosilicate molecules a small fraction of oligomers also, it can thus be assumed that these oligomers will not impregnate the smallest pores of both types of natural building stones and pore clodding is possible.

The relation between the capillarity of the stone and the working capability of the water repellent is very important. Based on the capillarity results, it can be stated that the hydrophobic treatment works perfectly for the Bray sandstone, since it reduces the capillarity coefficient by a factor of 180, while this is only by a factor 10 for the Maastricht limestone (Table 4). The large standard deviation of the capillarity coefficient for the untreated Bray sandstone is due to its porosity heterogeneity. The large standard deviation of the treated Maastricht limestone however, indicates that certain samples were treated more effectively than others. This phenomenon can probably be explained by occasionally occurring large pores, which prevent the full action of the water repellent product. Due to the hydrophobic treatment, for both Bray sandstone and Maastricht limestone, the free porosity and the water absorption by capillarity is reduced extremely, while the water absorption by diffusion is raised. It is expected that the water repellent samples will built up water droplets in their pore structure slower than untreated ones. This will consequently result in a higher water vapour resistance, than in the untreated samples where many droplets will form water islands. Due to these water islands, the water vapour permeability, attributed to microscopic liquid flow transport occurring in adsorbed liquid water sites and water filled regions due to capillary condensation, increases (Carmeliet et al., 2002). Since the Maastricht limestone has a higher pore interconnectivity than the Bray sandstone, the overall diffusion transport is facilitated, resulting in a higher water vapour permeability compared to the Bray sandstone.

Based on the water vapour permeability results for the untreated and treated samples (Table 6) a reduction of the water vapour permeability by a factor 1.2 for the Maastricht limestone and by a factor 1.7 for the Bray sandstone due to treatment with Hydro10 is observed. This indicates that although diffusion of the water vapour molecules occurs in the hydrophobic treated pores, this diffusion is reduced. From the same experiment for the consolidated samples, it seems that consolidation with SH75, reduces the water vapour permeability more than SH100, that is by a factor 1.4, for the Maastricht limestone and a factor 2.4 for Bray sandstone. The reduction of the water vapour permeability can be possibly explained by the reduction of the pore diameters, due to the presence of the polymer networks on the pore walls, due to a shift towards less effective water vapour transfusion transfer mechanisms (Carmeliet et al., 2002), and due to a decrease of the pore connectivity. Since in general the average pore size of the Maastricht limestone is larger than the Bray sandstone, the chance to obtain pore clodding in the Bray sandstone due to treatment is higher. This probably explains why the water vapour permeability is reduced more in the consolidated Bray sandstone. It is known that the basic elements of SH100 have a different composition (amount of ethylorthosilicate oligomers) than those used in SH75 and that SH75 is more viscous than SH100 (Cnudde et al., submitted a). This information can help to explain why the water vapour permeability is more reduced in the Bray sandstone samples treated with SH75 than SH100. One of the reasons is that SH75 will probably block more pores for water vapour diffusion than SH100. Since Maastricht limestone has on average larger pores than Bray sandstone, its amount of blocked pores is smaller. Other observations that need to be considered, are the contact angle measurements illustrating that samples treated with SH75 continued to have a large water repellency after 7 weeks, and capillarity measurements, demonstrating an initially water repellency for those samples. Since it is expected that Hydro10 does not block the pores like the consolidants, the water vapour permeability reduction should mainly be explained by a reduction of the effective water vapour diffusion transfer mechanism.

6.10 Localization of water repellents and consolidants

When consolidation is planned, each stone structure should be considered as a unique problem. There is a long way between the start of the consolidation, where the consolidant is a fluid, and the end, where it turns out to be solid. The penetration of the consolidant will depend on factors like the viscosity of the liquid and the size of the pores. Young et al. (2003) mention that a treatment should penetrate the stone to at least 25 mm or to the depth of stone deterioration. This should result in a gradual transition in the thermal and mechanical properties from the treated exterior surface to the inner layer of the untreated stone. Since the viscosity of SH75 is larger than that of SH100, its penetration inside the Bray sandstone will probably be more easily obstructed in the low porosity areas than for SH100. As mentioned by Cnudde et al. (submitted a) the consolidants SH100 and SH75, and the water repellent Hydro10 can be visualized inside the stone by means of X-ray micro-CT, due to the doping of the original products with a higher attenuating material (Cnudde et al., 2005). On both SEM and reconstructed X-ray micro-CT images the consolidants SH100 and SH75 could be found over a depth of more than 1 cm, which was the limited sample size of the X-ray micro-CT system, in both Bray sandstone and Maastricht limestone. Micro-CT images also revealed that the water repellent Hydro10 could be found over a similar depth in both stone types, but depending on local heterogeneities inside the stone material, could be spread irregularly over the sample (Figure 4).

Figure 4. X-ray micro-CT reconstructions indicating a heterogeneous spreading of doped Hydro10 inside the Bray sandstone.

The X-ray micro-CT images in combination with μCTanalySIS data revealed that even in a sample with a surface area of 25 cm^2, a product can penetrate and spread itself in a very different way. Fluorescent microscopical research revealed an overall pore size reduction over the entire sample and no pores turned out to be completely blocked by the consolidant, which was mainly due to shrinkage of the silica-gel. For the samples treated with the water repellent, pores indeed turned out to be blocked by the water repellent, since often no resin applied under vacuum could fill the pores.

Preliminary experiments of high-speed neutron tomography for the visualization of water, water repellents and consolidants inside natural building stones demonstrated that this technique is an ideal tool for the investigation of hydrogen-rich fluids in porous media (Masschaele et al., 2004a, 2004b; Dierick et al., 2004). One of the experiments consisted of the visualization of water flow inside a Bray sandstone sample, with locally strong reduced porosity due to high amounts of goethite (Figure 5). High-speed neutron tomography visualized the water release on the top surface and its fast migration into the more porous volume. The migration of the fluid front could be followed throughout the stone volume as a function of time. When the progress of the fluid front was slowed down intensively, more water was added on the surface. This secondary front quickly propagated through the already wetted part and is illustrated in Figure 5. This experiment confirms data from previous experiments, suggesting that the water will preferentially penetrate the volume with low goethite amounts and consequently higher porosity.

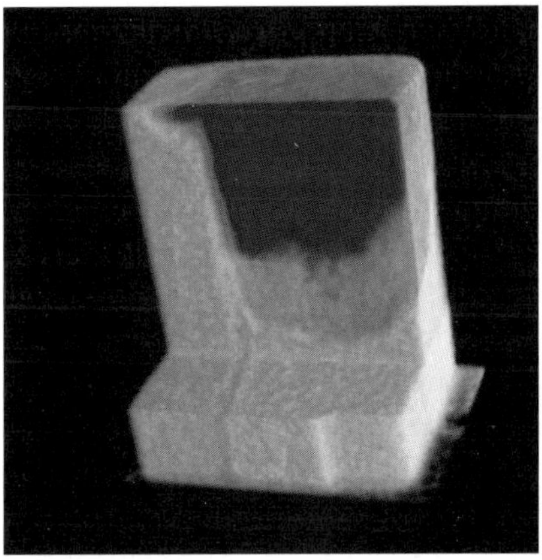

Figure 5. 3D reconstruction of layered Bray sandstone sample. The left part contains the higher amount of goethite, compared to the right part. The dark blue volume represents the secondary fluid front, compared to the first absorbed water (light blue).

X-ray micro-CT and optical microscopy detected a strong porosity reduction up to almost zero in the areas containing high goethite amounts. The high-speed neutron tomography indicates almost no water penetration in these areas, which was confirmed by the Karsten tube test. Additionally the impregnation time of a water droplet during the contact angle experiment and the total drying time turned out to be much higher for samples containing high amounts of goethite. The presence of goethite will also influence the penetration depth and spreading of a conservation product. Based on the X-ray micro-CT results, it seems that most of the product is blocked by the areas with low porosity and high goethite concentration, indicating that the migration and distribution of conservation products strongly depends on the local structure inside the natural building stones. High-speed thermal neutron tomography experiments for the visualization of consolidants and water repellents, inside natural building stones, turned out to be positive (Dierick et al., 2004; Masschaele et al., submitted). Figure 6 demonstrates the impregnation of the consolidant SH100 inside a Maastricht limestone.

Figure 6. 3D reconstruction of Maastricht limestone treated with SH100.

6.11 Salt visualization

Previous experiments by Cnudde & Jacobs (2004b) demonstrated that salts can be visualized by X-ray micro-CT. In order to study the effect of a water repellent product on salt migration, a salt weathering experiment was set up. In this experiment untreated Bray sandstone samples and samples treated with the water repellent Hydro10 containing 33% 3-bromopropyltrimethoxysilane were scanned with X-ray micro-CT before and after several salt weathering cycles. For the salt weathering test the European Standard EN 12370 was used as guidance, without having the intension to determine the salt weathering resistance of the samples literally. The untreated samples revealed a clear efflorescence and subflorescence on the micro-CT images. Due to the presence of the salt accumulation in the upper layer of the untreated Bray sandstone sample, a porosity reduction of more than 63% was noticed with the μCTanalySIS software. In general, based on the micro-CT images derived after each salt weathering cycle, for the untreated samples it could be concluded that the amount of salt accumulation inside the pores increased with further weathering cycles. Micro-CT images revealed that the high attenuating doped Hydro10 was primarily spread rather irregularly inside the Bray sandstone samples. Figure 7 indicates that salt is accumulated inside the stone sample in the regions just underneath higher concentrations of doped Hydro10. Although in the reconstructed cross-sections, sometimes almost no pores can be found locally, in reality this will probably be an exaggeration due to the combination of the high attenuation of bromine from the doping product, the partial volume effect and the resolution limitation. The porosity reduction due to the salt accumulation on the other hand is not exaggerated, since no doping agent was added. No efflorescence could be found on the sample treated with doped water repellent. Based on the micro-CT images it can be concluded that for

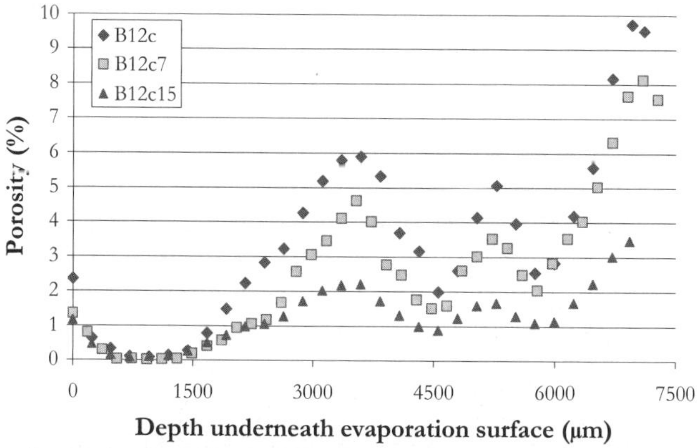

Figure 7. Porosity results determined with μCTanalySIS on reconstructions of sample B12 scanned before and after 7 and 15 cycles of saltweathering.

the untreated sample, the salt is mainly present in the upper 0.5 mm, while for the samples treated with water repellent, the salt accumulation was spread over the entire sample, with no high concentration in the upper surface. This indicates that the combination of local porosity reduction and water repellent product is blocking the salt migration transport, preventing evaporation of water in the upper surface, in combination with efflorescence.

7. RECENT DEVELOPMENTS AND FUTURE PROSPECTS IN MATERIAL RESEARCH

High resolution X-ray tomographs are limited in spatial resolution mainly by the dimensions of the X-ray focal spot from which they originate. In recent years X-ray tubes became available with focal spot sizes down to 1 micron or less. This opens the way for a whole new range of applications where high spatial resolution is required. Especially for geological purposes this bridges the gap between the available macroscopic investigation techniques and ultra high resolution techniques such as scanning electron microscopy. A much more detailed view of the porosity of rocks can be visualized and quantified. Such tomography setups require much more consideration in terms of alignment and sample control.

Dynamic Neutron Tomography is a brand new technique and is gaining much attention amongst research groups. There is still much room for improvement, both in terms of spatial and temporal resolution as reported in Dierick et al. (2005) and Masschaele et al. (2004). Currently the aim is to achieve a time resolution of the order of 3 seconds, combined with a spatial resolution of the order of 250 μm. This requires the development of a tomography setup that is sufficiently sensitive to allow extremely short exposure times. It also requires the development of dedicated software in order to process the enormous amounts of data that are recorded. Reconstruction of tomographic data is a time-consuming task and therefore requires a high performance software tool such as Octopus (www.xraylab.com).

8. CONCLUSION

The water repellent Hydro10 and the consolidants SH100 and SH75 tend to alter some characteristic stone parameters. The water repellent Hydro10 influences colour, impregnation speed and contact angle of water, capillarity of the stone, drying rate, water vapour permeability and water absorption under low pressure. Both SH75 and SH100 influence properties like compressive strength, drilling resistance, colour, impregnation speed and contact angle of water, capillarity, drying rate, water vapour permeability and water absorption under low pressure.

To visualize the water repellents Hydro10 and the consolidants SH75 and SH100 with X-ray micro-CT, the products need to by mixed with a higher attenuating material (e.g. 3-bromopropyltrimethoxysilane). X-ray micro-CT enables to locate conservation products in relation with local heterogeneities inside the natural building stones. The influence on a water repellent treatment on the salt migration inside a sandstone sample was monitored and revealed salt accumulation heterogeneously spread inside the sample, depending on the location of the water repellent and on local porosity reduction.

In general it seems that the hydrofuge sometimes is completely closing off pores for further fluid transport. This can also partially explain the water vapour permeability reduction and the reduction in water absorption after full impregnation. Based on the obtained information it also seems that the blocking by the water repellent is in close relation with the total porosity and the pore size distribution of the sample, since for Maastricht limestone no pores tend to be closed from the original pore interconnectivity. The pore blocking almost never occurred in the highly porous Maastricht limestone, while it often did in the Bray sandstone, with its average porosity between 5 and 25%. Since most of the natural building stones have a pore structure similar to that of the Bray sandstone, the pore blocking due to water repellent treatment will more frequently be present than assumed.

Additionally, high-speed neutron tomography turns out to be extremely important in the research on fluid flow in stone material. The resolution and the more practical fact that neutron sources are less accessible induces some limitations, but the fact that water, water repellents and consolidants can be visualized inside larger samples, without any doping is of high importance in the research domain of fluid flow inside natural building stones.

REFERENCES

Berger-Schunn, A., 1994, Practical Colour Measurement: A primer for the beginner, a reminder for the expert.

Carlson, W.D., Denison, C., and Ketcham, R.A., 1999, High-Resolution X-ray computed tomography as a tool for visualization and quantitative analysis of igneous textures in three dimensions, Electronic Geoscience **4**:3.

Carmeliet J., Houvenaghel, G., Van Schijndel, J., and Roels, S., 2002, Moisture phenomena in hydrofobic porous building material Part1: Measurements and physical interpretations, Int. Zeitschrift für Bauinstandsetzen und Baudenkmalpflege, Jahrgang **8**, Heft 2/3:165-183.

Clifton J.A., 1980, "Stone consolidation materials, A status report". U.S. Department of commerce/National Bureau of Standards, Technical Note. N° 1118, V-1980.

Cnudde, V., Cnudde, J.P., Dupuis, C., and Jacobs, P.J.S., 2004, X-ray micro-CT used for the localization of water repellents and consolidants inside natural building stones, Materials Characterization **53**:259 271.

Cnudde, V., Dubruel, P., De Winne, K., De Witte, I., Masschaele, B., Jacobs, P., and Schacht, E. Conservation products inside building stones, Engineering Geology, submitted a.

Cnudde, V., Cwirzen, A., Masschaele, B., and Jacobs, P.JS., Porosity and microstructure characterization of building stones and concretes, Engineering Geology, submitted b.

Cnudde, V., and Jacobs, P., 2004a, Preliminary results of X-ray micro-tomography applied in conservation and restoration of natural building stones, X-ray CT for Geomaterials: Soils, Concrete, Rocks, Otani & Obara (eds.), pp. 363-371.

Cnudde, V., and Jacobs, P.J.S., 2004b, Monitoring of weathering and conservation of building materials through non-destructive X-ray computed microtomography, Environmental Geology, Special Issue: Stone Decay Hazards **46**(3-4):477-485.

Cnudde, V., Masschaele, B., Dierick, M., Vlassenbroeck, J., Van Hoorebeke, L., and Jacobs, P., Recent progress in X-ray CT as a geosciences tool, Applied geochemistry, in press.

Commission 25-PEM Protéction et erosion des Monuments, 1980, Recommended tests to measure the deterioration of stone and to asses the effectiveness of treatment methods, Test No. II.4, Water absorption under low pressure, Materials and structures **13**(75):201-205.

De Witte, E., 2003, Hydrofoberen van natuursteen - herstelmortels - Anti-graffiti, Presented at the Renovation Course. session 3, Sint-Lieven Hogeschool, Belgium.

Dierick, M., Vlassenbroeck, J., Masschaele, B., Cnudde, V., Van Hoorebeke, L., and Hillenbach A., 2005, High-speed neutron tomography of dynamic processes, Nuclear instruments & methods in physics research section A, accelerators, spectrometers, detectors and associated equipment **542**(1-3):296-301.

EN ISO 12572, june 2001, Hygrothermal performance of building materials and products – determination of water vapour transmission properties.

European Committee for Standardization, 1999, European standard EN1926. Natural stone test methods. Determination of compressive strength, Brussels, CEN.

European standard EN 1925, 1999, Natural stone test methods – Determination of water absorption coefficient by capillarity, Brussels, CEN.

Feldkamp, L.A., Davis, L.C., and Kress, J.K., 1984, Practical cone-beam algorithm. Journal of the Optical Society of America. **1**:612–619.

Garrod, E., 2001, Stone Consolidation - halts decay and prolongs life, Reproduced from The Building Conservation Directory 2001, http://www.buildingconservation.com.

González, R.F., López de Azcona, M.C., Mingarro Martin, F., Alvarez de Buergo Ballester, M., and Rodriguez Blanco, J., 2000, A comparative study of the effeciency of siloxanes, methacrylates and microwaxes-based treatments applied to the stone materials of the Royal Palace of Madrid, Spain, 9th International Congress on Deterioration and Conservation of Stone. Venice 19-24 June 2000, pp. 235-243.

Jacobs, P., Sevens, E., and Kunnen, M., 1995, Principles of computerised X-ray tomography and applications to building materials, The Science of the Total Environment **167**:161-170.

Ketcham, R.A., and Carlson, W.D., 2001, Acquisition, optimization and interpretation of X-ray computed tomographic imagery: applications to the geosciences, Computers & Geosciences, **27**(4):381-400.

Masschaele, B., Dierick, M., Cnudde, V., Van Hoorebeke, L., Delputte, S., Gildemeister, A., Gaehler, R., and Hillenbach, A., 2004a, High-speed thermal neutron tomography for the visualization of water repellents, consolidants and water uptake in sand and lime stones, Radiation physics and chemistry **71**(3-4):807-808.

Masschaele, B., Dierick, M., Van Hoorebeke, L., Cnudde, V., and Jacobs, P., 2004b, The use of neutrons and monochromatic X-rays for non-destructive testing in geological materials, Environment Geology, Special Issue: Stone Decay Hazards **46**(3-4):486-492.

Masschaele, B., Cnudde, V., Jacobs, P., Dierick, M., and Van Hoorebeke, L, Use of high-speed neutron tomography for the direct 3D visualization of water repellents and consolidants in limestone and sandstone samples, Engineering geology, submitted.

Moropoulou, A., and Farmakalidis, L., 2003, Recommendations for Compatible Consolidant Materials. ITECOM Advanced study course: Innovative technologies and materials for the conservation of monuments 8 – 20 December 2003.

Price, C. A., 1996, Stone Conservation. An Overview of Current Research, The Paul Getty Trust., USA.

Tiano, P., Biodegradation of cultural heritage:, http://www.arcchip.cz/w09/w09_tiano.pdf.

WTCB, 2002, Technische Voorlichting 224: Waterwerende oppervlakte-behandeling, Juni 2002.

Young, M.E., Cordiner, P., and Murray, M., 2003, Chemical consolidants and water repellents for sandstones in Scotland, Historic Scotland Research Report 2003, pp. 226.

Chapter 9.2

A CRITICAL APPROACH TO SURFACE AND POROUS STONE ANALYSIS METHODS

M. Brugnara, C. Della Volpe, D. Maniglio and S. Siboni
Dept. of Engineering Materials and Industrial Technologies, University of Trento, Via Mesiano 77, 38050 Trento, Italy

Abstract: The aim of this paper is to give a correct interpretation to contact angles values that are obtained on porous surfaces by applying different methods. First a brief description of the physical meaning of the contact angle value is given, then a new methodology for the contact angle analysis by using the Wilhelmy balance is presented. Finally some limitations in the use of the Washburn's equation are shown.

Key words: wettability; porous materials; contact angle; Washburn's equation.

1. INTRODUCTION

It is well known that protection operations have the purpose to slow down or to make less probable the inevitable transformation processes every monument suffers and that are provoked by the aggressive environment where the work of art lives. Particularly, with the protection one wants to operate both on the alteration causes, correlated to the environmental factors and on the caused processes. Regarding the protective agents, today the most frequent problem consists in the use of new chemical products, which are not sufficiently experimented and whose effects are not known in a middle and long term. At the base of a balanced intervention it is therefore important to know both the materials and the techniques very well. Insofar it is necessary to analyze in every single case the most opportune treatment, appraising advantages, disadvantages, risks and probabilities to achieve a positive result, also effecting appropriate laboratory tests on the material-produced system.

S.K. Kourkoulis (ed.), Fracture and Failure of Natural Building Stones, 541–551.

The surface and porosity analysis to obtain characteristics of the stone as wettability and water absorption ability have a very long tradition. In fact the theories and the equations on which some common methods are based have been discovered even in the XIX century. Here reference is made to the application of Young's and Laplace's equations[1,2] and to the Washburn's equation[3] in the sessile method or in the Wilhelmy[4] technique for the determination of the contact angle and in the liquid absorption measurements for the determination of the stone porosity.

In the case of the stone protection by applying polymeric agents the evaluation of wettability modification or water absorption have an important practical and theoretical role. Notwithstanding this long tradition of experimental and theoretical results there is a long debate about the validity of the results and even about the exact meaning of the experimental results[5].

In the present paper it will be briefly shown how the evaluation of surface stone wettability and porosity may be considered in the light of recent experimental results. Initially attention is focused on the static angles and it is explained why these values do not well correlate with the liquid absorption. It appears that the wettability properties may be easily modified without decreasing the effective liquid absorption, while the real trouble is hidden in the real meaning of static contact angle. Only the use of the advancing and receding angles may overcome this misunderstanding; on a rough and heterogeneous material these angles may be attributed to the effect of both of these properties and so it is not correct to correlate the high advancing contact angles with the chemical composition of the material itself. When a stone is protected against the water absorption through the use of a polymeric agent in some way distributed on its exposed face the presence of the polymer induce an immediate increase of the advancing angles, due to its reduced wettability, while the portion of stone not covered allows in any case a significant absorption of water. Only when the increase in the concentration or in the final mass of the deposited polymer reduces the non-covered stone surface the absorption of water is really reduced; this phenomenon is followed by using the receding angle, which is generally connected with the most easily wettable portion of the surface.

This kind of measurement can be led using a Wilhelmy balance but the absorption of the test liquid often makes this evaluation difficult and introduces a systematic error in the final values of the contact angles. In this case a mathematical model of the absorption has been developed, which uses a Washburn-like approximation to calculate from the absorbed liquid mass and the interval time of the experiment, the speed of the imbibition; introducing this adjustment in the original data the experiment is corrected in an acceptable way. Moreover if the experiment is made on a porous and so on a reasonably rough material, it can be performed in a perfectly wettable liquid to recalculate the Wenzel ratio of the material; as a final result the contact angle of the wet solid is easily obtained.

Finally the mechanism of the absorption itself, described by the Washburn equation appears not well related to the size pores obtained by other methods and does not give values of the contact angles in agreement with those obtained through direct surface analysis.

2. MATERIALS AND METHODS

Two different porous materials were analyzed: the Noto Stone, an organogenic calcarenite coming from the caves of Palazzolo (Siracusa, Italy), largely employed in the local historical buildings and a ceramic material having a controlled porosity.

Specimens for the absorption and contact angle measurements had a size of 10x3x30 mm while ceramic samples of sizes 5x2.5x30, 10x2.5x30, 25x4.5x30 and 25x7.6x30 mm were used for the study of the dependence of the absorption parameters on the geometrical shape.

3. THE PHYSICAL MEANING OF THE CONTACT ANGLE

In the field of monument protection the use of contact angle has a long tradition[6-8]; notwithstanding this tradition it is often difficult to accept the approach which is used in papers or even in procedures proposed by official norms; in fact the actual application of the contact angle may be shortly expressed as the use of the static or of advancing angles on surfaces which are always heterogeneous and rough, i.e. non-ideal. Obviously this can be performed if and only if the intrinsic limits of this approach are known and explicitly considered.

These limits may be easily conceived referring to the concept shown in Figure 1 in which one can compare the profile of surface free energy of an ideal and of a real interface versus the contact angle. The first one presents only one minimum, corresponding to the Young angle, while the second one shows many minima in a wide interval of values; each one of these minima being a metastable minimum, produced by the roughness or heterogeneity, which may be found as the result of a contact angle measurement in different experimental conditions.

In a given experimental condition, the highest one corresponds to the advancing angle and the lowest one to the receding angle; note that the curve is not necessarily symmetric; however it is possible, using modern experimental methods as VIECA[9,10], to evaluate the angle corresponding to the energy minimum, e.g. the Wenzel angle for the rough surfaces; this angle does not necessarily corresponds to the Young angle.

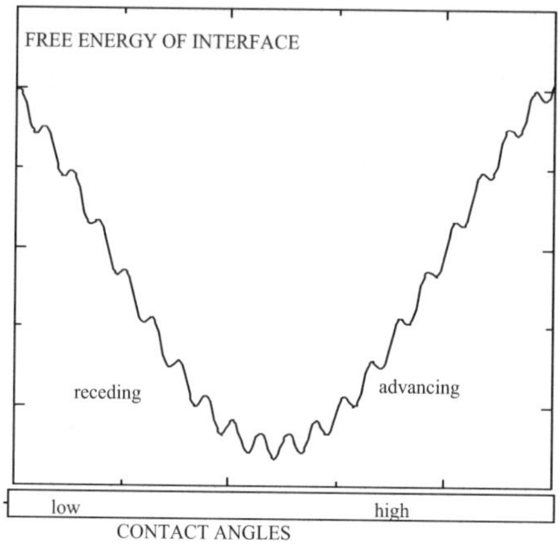

Figure 1. The surface free energy of the triphase system versus the experimental contact angle; the shape of the local minima is important in its effect on the real mobility of the meniscus.

The general correlation among all the angles may be captured by Figure 2 for a moderately rough surface. It appears clear that the current use of the static or advancing angle is often misleading. In order to obtain practical indications from the experimental results one should be fully aware of the real meaning of the angle which is measured.

The surfaces evaluated in the case of monument protection are among the most complex ones, because they may be rough, heterogeneous and porous. The effect of porosity will be analyzed in the following using a new mathematical model; the effect of heterogeneity has been already analyzed in literature; on a flat, but heterogeneous surface, one will obtain the so-called "hysteresis graph"[13] which allows to assign the advancing angle to the lowest energy portion of the surface and the receding one to the highest energy portion. The correlation is not linear, as clearly evidenced in Figure 3: a low percentage of hydrophobic surface can increase the advancing contact angle.

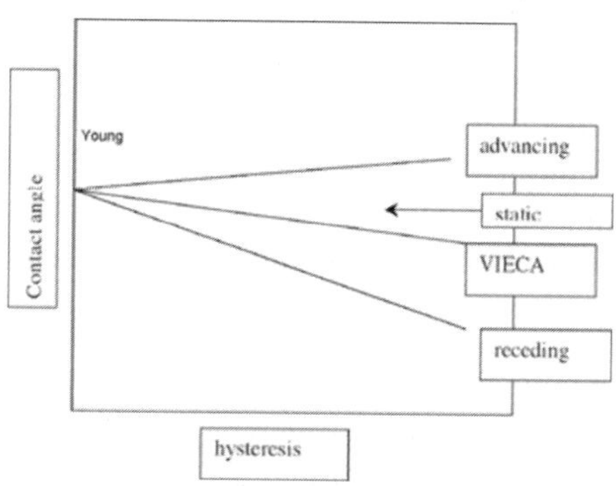

Figure 2. The variation of advancing, receding and VIECA contact angles with the surface roughness, which is generally correlated with a hysteresis. The Young angle may be only obtained by extrapolation[11, 12].

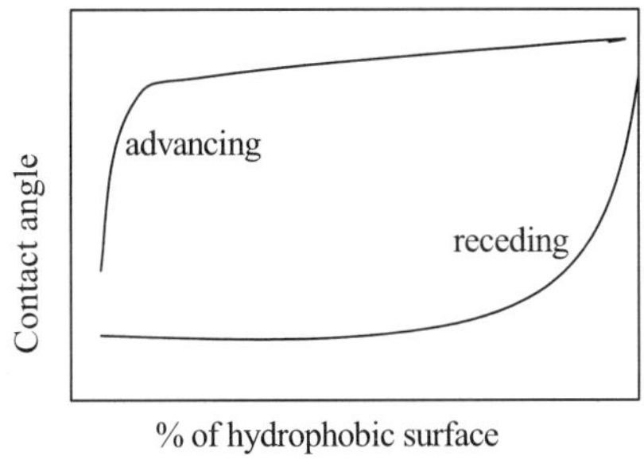

Figure 3. The Cassie Baxter Contact Angle and the hysteresis on a heterogeneous surface.

The conclusion may be that on a moderately rough surface, even neglecting the effect of roughness, the advancing angle is monitoring the lowest energy portion of a surface; as a consequence the use of the contact angle to check the efficacy of a polymer used as protective agent is misleading; its increase may simply confirm the mere presence of the protective and not its efficacy; a more efficient way to check the efficacy of the protective is to follow the evolution of the receding angle which increases with the percentage of the protective agent as shown in Figure 4.

A second point to stress is that it is common to find very high advancing angles on protected surfaces, often higher than the values characteristic of the flat protective agent alone. This experimental finding may be strongly misleading, inducing to accept the conclusion of the enormous efficacy of the used material, a fact which is difficult to reconcile with the contemporaneous water absorption; water cannot be absorbed when the contact angle is above 90°. The reason of this is in the eventual formation of a "composite" surface (today named "superhydrophobic"), already revealed by the traditional lite-

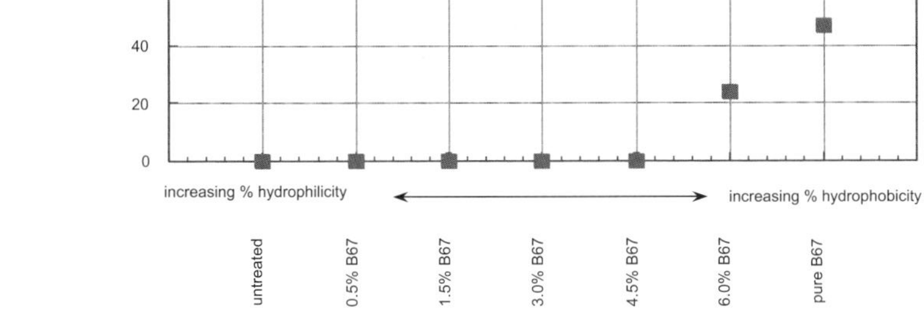

Figure 4. The advancing and receding contact angles of water on Noto stone protected with solution of Paraloid B67 of increasing concentration.

rature, which may transform a strongly rough surface in a heterogeneous one, using the air captured in pores as well as in the case of aquatic birds feathers. This situation is valid on the protected portion of the surface only, and for this reason appears as strongly metastable.

A last point to mention is the consequence of time-ageing of the protective agents; the effect of UV radiation and weathering may induce the formation of newly chemical functions, oxidizing the protective and altering its continuity; the final effects are very complex and difficult to understand in terms of surface properties only. A general decrease of the contact angles, may reveal this alteration, but it is necessary to use other techniques to understand the situation.

4. MODELING THE WILHELMY EXPERIMENT ON A POROUS MATERIAL

In order to correctly use the Wilhelmy experiments on porous samples it is necessary to model the mechanism of liquid absorption. The mass of the porous sample at the beginning and at the end of a typical Wilhelmy experiment are not the same, as also the slope of the force/immersion depth curve does not correspond to the buoyancy coefficient $-\rho gA$ expected for a sample with constant cross-sectional area A immersed in a liquid of density ρ. Therefore, appropriate corrections are needed.

The main idea of the model is that liquid penetration follows a Washburn-like law and defines empirically a mean absorption coefficient α as follows:

$$\alpha = (m_B - m_A)/\sqrt{t_B - t_A}\,,$$

where m_A and m_B are the initial and final mass of the sample for a typical Wilhelmy run of duration t_B-t_A, the dependence on the square root of time being just suggested by Washburn law. a can be easily calculated from the experimental data. At any time t larger than t_A, which corresponds to the zero depth of immersion, the total force measured by the microbalance will be corrected by subtracting the mass of the absorbed liquid:

$$F_{corr} = F_{original} - \alpha\sqrt{t - t_A} \tag{1}$$

It is understood that whenever full absorption takes place in a time shorter than the duration of a Wilhelmy cycle, the above correction must be referred only to the absorption time interval, by introducing a constant weight correction in the rest of the measurement. In any case the liquid advances on a solid surface already wetted, owing to the absorption process, and this could be a serious limit of the model when the contact angle is nonzero.

Another important effect to take into account in the analysis of a Wilhelmy experiment on a porous sample is the liquid evaporation from the solid surface. This phenomenon can be reasonably modelled by assuming that evaporation takes place at a constant rate per unit time and area and thus expressing the mass loss of the sample due to evaporation by the formula

$$\Delta m_{evap} = \int_A^B \beta \cdot S(t)dt \tag{2}$$

where S(t) denotes the evaporation area as a function of time and β is a mean evaporation coefficient per unit time and area. Such a coefficient can be empirically calculated by prolonging the final stage of the Wilhelmy measurement, when the sample is completely withdrawn form the liquid and the evaporation area is constant, so that the sample mass decreases linearly with time. S(t) typically increases with the height of the liquid within the sample and therefore with the square root of time, owing to the absorption phenomenon. Such a conjecture can be experimentally verified by checking (through a set of digital images) that the height of the liquid in the sample follows a square root law versus time and subtracting the height of the liquid meniscus. A height of the liquid less than the meniscus or negative must be regarded as a zero contribution to evaporation. A law of the form $S(t)=\sigma(t-t_A)^{1/2}$ is in good agreement with the experiment and introduces a new empirical constant s which can be interpreted as the evaporating area one second after the ZDOI time t_A. The sum of the masses evaporated per second, β.S(t), provides the total amount of evaporated liquid, which is then summed up to the amount of absorbed liquid previously measured to recalculate the new absorption coefficient, *a′*. By setting $\beta'=\beta\sigma$ the corrected force at time t takes then the following form:

$$F_{corr} = F_{original} - \alpha'\sqrt{t-t_A} + \beta'(t-t_A)\sqrt{t-t_A} \tag{3}$$

where the prime indicates that the absorption coefficient has been recalculated as explained and the evaporation coefficient incorporates the variation of evaporating surface.

5. WICKING ANALYSIS

Absorption measurements are often used for the determination of the contact angle of a porous material, notwithstanding the explicit limitations of the Washburn equation[3] and of the models derived from its application. This equation provides a kinetic model of the rise of a liquid in a cylindrical capillary and is written in the form:

$$h^2 = \frac{1}{2} r \frac{\gamma_{liq} \cos\theta}{\mu} t \tag{4}$$

with r the radius of the capillary, h the height of the meniscus at time t, γ_{liq} the surface tension, μ the dynamic viscosity of the liquid and θ the contact angle of the liquid on the solid. In the original paper of Washburn an equation for a porous system was obtained, simply considering the case in which the porous body "*can be taken as eqivalent to the penetration of n cylindrical capillary tubes of radii r_1,...r_n*"[3] otherwise, in the opinion of Washburn, the applicability of its equation "*could only be determ ined by experiments*"[3]. The finding that during the first part of the imbibition, not the whole porosity is used by the liquid[14] was an important improvement but as it is shown in this work other strong limitations still exist. Some experimental absorptions and contact angle measurements on two different calcareous and silicates porous media are here reported. Each absorption measurement with different liquids has been performed by collecting the mass of the sample and height of the liquid versus the imbibition time. The model applied is given by Washburn's equations for the mass absorbed and the height reached by the liquid as a function of time:

$$H_w(t) = h_1 \sqrt{t} \quad with \quad h_1 = \sqrt{\frac{\bar{r}\gamma_{liq} \cos\vartheta}{2\mu}} \tag{5}$$

$$M_w(t) = m_1 \sqrt{t} \quad with \quad m_1 = \sqrt{\frac{(\rho\varepsilon A)^2 \bar{r}\gamma_{liq} \cos\vartheta}{2\mu}} \tag{6}$$

Once the parameters m_l and h_l have been derived, the effective porosity and the equivalent capillary radius can be determined by means of the following relationships:

$$\varepsilon_{eff} = m_1 / \rho_{lq} \; A h_1 \tag{7}$$

$$r_{eq} = \frac{2{h_1}^2 \mu}{\gamma_{lq}} = \left(\frac{m_1}{\varepsilon A \rho}\right)^2 \frac{2\mu}{\gamma_{lq}} \qquad with \quad \cos\theta = 1 \tag{8}$$

The calculation of the contact angle θ_{nw}, for the non wetting liquid, can be done finally applying the simple relationship:

$$\frac{r_{eq-nw}}{r_{eq-w}} = \cos\theta_{nw} \tag{9}$$

where $r_{eq\text{-}w}$ and $r_{eq\text{-}nw}$ are the equilibrium radius calculated using a perfectly wetting and non wetting liquid and knowing that in the first case $\cos\theta_w=1$ ($\theta_w=0°$ for its definition). The values of effective porosity and equivalent radius for 4 different liquids found on Noto Stone are presented in Figure 5. The equivalent radius determined in this way turns out to be at least two orders of magnitude smaller than the average pore radius deduced from mercury porosimetry.

A qualitative explanation of this behaviour is that the meniscus spends the majority of time in the largest segments, where the capillary driving force is the smallest and the volume to fill is the greatest. On one hand this fact explains why only a part of the porosity is used, on the other hand it points out that this value does not have a physical meaning. As it is expected the final result reflects all the limitations of the too raw model and that the so determined contact angles of water on calcite are widely scattered and meaningless.

It is possible to note in Table 1 that the contact angle for water, found applying the Washburn procedure by using the n-Hexadecane radius is not so far from the same value found using the corrected Wilhelmy approach. Instead the results calculated starting from 1-Bromonaphtalene and Formamide radii are totally different: this can mean that the reference radius refers to a non-complete wetting fluid, but the measurements confirm that the contact angle of those liquids on Noto Stone is zero.

Table 1. Contact angle for Water/Noto Stone as calculated from the absorption results of other wetting liquids using the equivalent radius approach.

Employed liquid	Contact angle of water
n-Hexadecane	65.1±7
Formamide	0-23
1-Bromonaphtalene	0-27

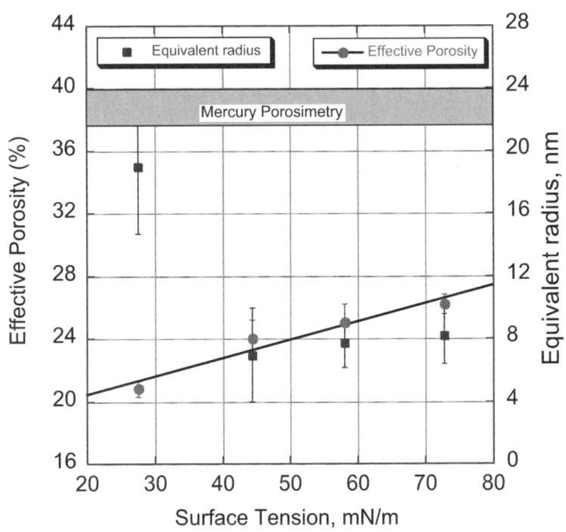

One of the reasons explaining why contact angle values are often meaningless, is the dependence of the parameters obtained by the application of the Washburn model on the sample shape. This limitation is due to the fact that in the Washburn's study the replaced fluid (air in this case) resistance is not considered; this approximation could be acceptable in a multi-capillary system, where the displacing liquid enters from a side and the air exits from the opposite side. On a porous system the situation is totally different because the porosity is a complex network and the removed fluid need to follow pathways different than just a simply capillary tube.

By applying Eqs.(5, 6) one must calculate the parameters h_l, m_l and ε_{eff} for samples of ceramic material having different transverse sections in water and in heptane. Notice that, the parameter m_l, increases linearly with the sample area (Figure 6a), as correctly indicated by the Washburn's model (see Eq.6). The parameter h_l shows instead a maximum (Figure 6b) or in any case a variation with the sample section whereas its mathematical model does not include such a dependence (see Eq.5).

The factors, which affect the absorbed mass in time and, respectively, the rate of propagation of the liquid front as measured by the front height, are therefore different. It is important to stress out that the weighted mass corresponds to a value independent on the size and shape of the sample, while the height is relative to the external faces of the parallelepiped sample. It is quite reasonable to assume that the height of the liquid front, as it is visible on the lateral surface, does not correspond to the analogous height reached by the liquid in the whole section of the sample, as supposed in the application of Washburn's model.

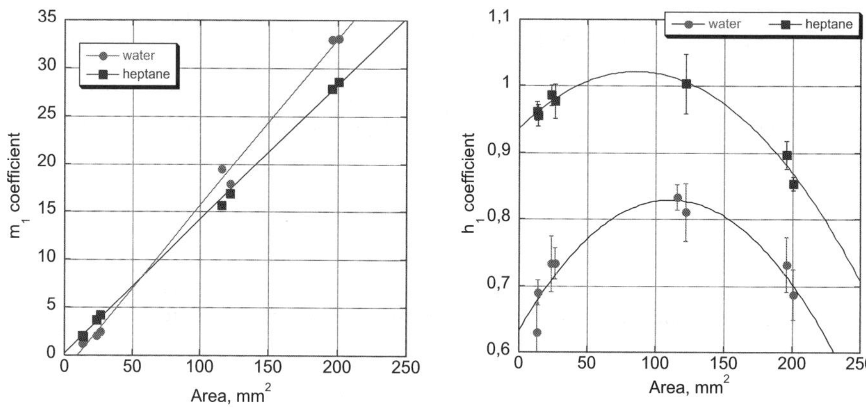

(a) (b)

Figure 6. (a) The dependence of the parameter m_l of the Washburn's equation on the size of the sample. (b) The dependence of the parameter h_l of the Washburn's equation on the size of the sample.

6. CONCLUSIONS

1. The use of the contact angle in the field of monument protection should be made with reference to its meaning on rough and heterogeneous surfaces; the advancing angle only is not able to show the ability of a protective agent but only to reveal its presence; receding angles or a more general view of the wettability properties of the material should be considered.
2. Obtaining contact angles of porous materials, as stones, through standard techniques is possible applying a suitable model to the obtained data; in the case of Wilhelmy experiment a model has been proposed to this direction.
3. The use of the Washburn equation to follow the liquid penetration in a porous material is certainly useful; the direct calculation of the contact angle from the results of a Wicking experiment, on the contrary, does not appear a good strategy.

REFERENCES

1. T. Young, *Philos. Trans.* **95**, 65 (1805).
2. P.S. De Laplace, in: *Mech. Celeste Suppl. Book* (1806).
3. E.W. Washburn, The Dynamics of Capillary Flow, *Phys.Review* **XVII**, 3 (1921).
4. J. Wilhelmy, Über die Abhängigkeit der Capilaritätsconstanten des Alkohols von Substanz und Gestalt des benetzten festen Körpers, *Annalen der Physik* **119**, 177 (1863).
5. M. Brugnara, E. De Gasperi, C. Della Volpe, D. Maniglio, A. Penati, S. Siboni, L. Toniolo, T. Poli, S. Invernizzi, and V. Castelvetro, The application of the contact angle in monument protection: new materials and methods, *Colloids Surf., A* **241**, 299-312 (2004)
6. B.H. Vos, Suction of Groundwater, *Studies in Conservation* **16**, 129-144 (1971).
7. G. Biscontin, La conservazione dei materiali lapidei: trattamenti conservativi, *La Pietra: Interventi conservaizone rest auro. Atti del Convegno Internaizonale Lecce 6-8 novembre 1981* (Congedo Editore, Galatina Italy, 1983).
8. L. Lazzarini and M. Laurenzi Tabasso, *Il restauro della pietra*, (CEDAM, Padova, 1986).
9. C. Della Volpe, D. Maniglio, S. Siboni, and M. Morra, An experimental procedure to obtain the equilibrium contact angle from the Wilhelmy method., *Oil & gas science and technology* **56**, 9-22 (2001).
10. C. Della Volpe, S. Siboni, M. Morra, and D. Maniglio, The determination of a "stable-equilibrium" contact angle on heterogeneous and rough surfaces., *Colloids Surf., A* **206**, 46-67 (2002).
11. R.D. Schulze, W. Possart, H. Kamusewitz, and C. Bischof, Young's equilibrium contact angle on rough solid surfaces. Part I. An empirical determination, *J. Adhes. Sci. Technol.* **3**(1), 1-78 (1989).
12. H. Kamusewitz, W. Possart, and D. Paul, The relation between Young's equilibrium contact angle and the hysteresis on rough paraffin wax surfaces, *Colloids Surf., A* **156**, 271-279 (1999).
13. R.E. Johnson Jr. and R.H. Dettre, Contact angle hysteresis III: Study of an idealized heterogeneous surface, *J. Phys. Chem.* **68**, 1744-1750 (1964).
14. L. Labajos Broncano, M.L. Gonzales, J.M. Bruque, C.M. Gonzales Garcia, and B. Janczuk, On the Use of Washburn's Equation in the Analysis of Weight-Time Measurements Obtained from Imbibition Experiments, *J. Colloid Interface Sci.* **219**, 275-281 (1999).

Chapter 9.3

THE CHALLENGE OF PROTECTING OUTDOOR EXPOSED MONUMENTS FROM ATMOSPHERIC ATTACK: EXPERIENCE AND STRATEGY

Tommaso Poli and Lucia Toniolo
CNR – ICVBC Sezione di Milano "Gino Bozza", P.za L. da Vinci 32, 20133 Milano, Italy

Abstract: Natural and synthetic waxes, acrylic and siloxane resins, perfluoropolyethers, fluorinated polyolefin and fluoroelastomers are some of the most used commercial products employed as protective coatings for stone materials. The application of these products is aimed to prevent the attack of the environmental pollutants and the penetration of water, with the connected risks of freezing and thawing cycles and the crystallization of salt solutions, in order to slow the stone degradation processes. The common and easiest way to pursue this goal is through the deposition, on the stone surface, and possibly deeper beyond the surface, of a coating layer able to modify the interaction of the stone with water, turning the partial hydrophilic character of the stone surface into a high hydrophobic interface. This modification should be achieved by affecting only the thermodynamic superficial potentials, but leaving unaltered, as long as possible, the superficial aspect, morphology, open porosity and roughness. Hence the challenge is avoiding the change of the substrate colour, the natural water vapour permeability and do not react with the stone material in order to respect its natural chemistry.The use of polymers, as protective coatings for stone materials, started in the sixties as a consequence of the expansion of macromolecular chemistry and the successful diffusion of synthetic resins in many different market fields but in most cases of conservation they have been applied without adequate knowledge of the properties of the polymer/stone system and, moreover, without a satisfactory optimization of the molecular structures for protection purposes. It is necessary to point out that the various commercial products rarely have been studied and developed expressly for applications in this field, but they have been merely transferred from different technological applications with higher economic advantages. The way these products are working, the reasons of some failures and the most common problems that can occur, have been reported here. For instance, the influence of the substrates' nature (open porosity, superficial roughness) on the protective performances and the behaviour of water dispersed systems for the

S.K. Kourkoulis (ed.), *Fracture and Failure of Natural Building Stones*, 553–563.

stone conservation have been investigated and reported. Stability and resistance to the photoxidative and thermal ageing of the different polymeric classes are other crucial problems: often a long term chemical integrity does not correspond to a similar protection efficacy. When a polymer is applied on a stone surface, physical rearrangements could occur to the protective layer, reducing the shielding efficacy but leaving the chemical structure unaltered. In a similar way the reversibility, even when assessed for a polymer in laboratory conditions, is practically lost when the protective is applied on a stone material with high open porosity and the product can penetrate deeply in the bulk of the system. This paper aims at being a short but complete overview of the issues that have to be considered and the problems that have to be faced when the protection of an outdoor exposed stone material is concerned.

Key words: stone protection; stone conservation.

1. INTRODUCTION

The last phase of a modern conservative intervention on an outdoor exposed building, monument or artifact is the important decision about the opportunity of a protective treatment application. This choice is crucial and decisive because to protect a stone surface, as well as consolidate and clean it, means to alterate irreversibly the surface[1]. Indeed, applying a protective treatment on a degradated stone material means, not only to shield the "stone system" with a "polymeric system" but to create a new and more complex stone/polymer system having durability and resistance, in the favourable cases, higher than the damaged and untreated stone. The main tasks of a treatment for the stone protection are, as well known, to avoid or at least reduce the water penetration and to protect the surface from dangerous interactions with any aggressive and reactive substances that can be easily found in a polluted environment. The risks connected to an outdoor exposition are well assessed and, as predictable, strictly correlated to the interactions of water with stone material and other pollutants dispersed in the environment[2-4]:

- Water imbibition, with the weight increase of the material inducing micro-structural damages.
- Salt crystallization; the precipitation and the different hydration state passages of salts can generate tensions in the porosity of stone which are able to induce micro-fractures, scalings and detachments.
- Chemical threats; the reaction of water with anhydrides or simply the dissolution of aggressive substances dispersed in the polluted environment, leads to the formation of acids or other compounds able to chemically attack stone materials, mostly in the case of limestones and marbles.
- Freezing and thawing phenomena in wet condition; the crystallization of water into the porous structure of the material induces mechanical stresses with consequent formation of micro-fractures.

- Biological threats; high humidity content promotes the growth of micro-organisms like bacteria, moulds, fungi, algae and musks, which could alterate and damage the stone substrates.
- Erosion; mechanical action of water flow and particulate matter transported by water could induce loss of material, after damages and detachments of superficial crystals.

2. TREATMENT EFFICACY AND STONE PROPERTIES

The above mentioned phenomena could be avoided or, at least, slowed down by the application of protective and water repellent treatment[5]. The common and easiest way to pursue this goal is through the deposition, on the stone surface, and, possibly deeper beyond the surface, of a coating layer able to modify the interaction of the stone with water, turning the partial hydrophilic character of the stone surface into a high hydrophobic interface. This modification should be achieved by affecting only the thermodynamic superficial potentials but leaving unaltered, as long as possible, the superficial aspect, morphology, open porosity and roughness. This means, it should not alter the substrate colour, the natural water vapour permeability and not react with the stone material[6]. Nevertheless, some silicon based commercial protectives can establish some binding interactions with silicatic substrates[5,7].

It is clear that the success of treatments strictly depends on the nature of the substrate and the type of interaction with stone materials. In fact, it is well assessed that the efficacy and the water-repellency of a treatment is directly correlated to the chemical and physical properties of the substrate like composition, open porosity and superficial roughness. The main challenge in conservation is preventing deterioration of limestones and marbles; many historical architecture and monuments are built with precious marbles for their high aesthetical value.

Marble is a quite critical substrate to protect, due to many intrinsic reasons. Its chemical nature leaves the stone exposed to the attack of acid pollutants, starting from the discontinuities among grains of the crystalline structure and does not exploit linking capabilities with the silicon based products[5-7]. The very low open porosity does not allow the penetration of treatments in the bulk of the material, with many obvious disadvantages like the formation of a very thin water repellent layer (stone/polymer) and the accumulation of the coating on the stone surface, directly exposed to chemical (aggressive pollutants), thermal, photochemical (the UV range of the sunlight spectrum and the radical species) and mechanical (washing and vibrations) stresses[8,9]. An excessive amount of accumulated polymer on the surface favour the soot deposition and particulate sticking (usually these polymers show T_g temperatures - glass transition or softening temperature - around

50°C, lower than the temperature of sun exposed surfaces). Consequently, the physical characteristics of coatings can be modified by the presence and accumulation of pollutants, layered in thick films[10]. Moreover, on a thick superficial film, the formation of polymer/polymer and stone/polymer discontinuities, where the biological growth is favoured by the water and soot prolonged retaining, is more likely. Finally, a thick surface coating modifies the gloss properties inducing an important aesthetic damage: the film changes the natural aspect of the stone giving it the typical "wet" or "greased" aspect. The usual low surface roughness, typical of the marble artefacts, reduces the specific surface and, therefore, the adhesion of the coating to the substrate.

Usually static contact angle measurements are used to assess the water repellence of a treated surface[11]: the higher the measured angles, the higher the hydrophobic effect of the coating (note that water repellent surfaces should have angles higher than 90°). The static angle value is not only influenced by the chemical nature of the product but by the surface roughness as well: the lower the roughness, the lower the angles, which tend to the value of the coating itself[12-14]. The polymers applied to low porosity and smooth substrates tend to remain over the surface, favouring the polymer mobility when energy, as for example heating, is supplied to the system.

The shielding efficacy of a protective treatment is usually tested through the measurement of water absorption by capillarity which evaluate the amount of water absorbed by a stone specimen per surface unit (absorbed water/surface area in contact with water, mg/cm^2) vs. time ($t^{1/2}$), before and after a treatment[15]. The trend of the obtained capillary absorption curves, the final amount of absorbed water and the relative capillary index[15] (ICr, the ratio of the capillary curves integrals before and after treatment), are the numeric instruments for the study of the behaviour of stone coatings. These parameters allow the comparison among the efficacy of different considered products.

In particular, the protective efficacy would be better understood if one takes into account the nature of the considered stone, for example the fact that only a low amount of water absorption is involved in all the experiments with marble and even relatively minor effects become meaningful. It is evident that, in the case of low porosity stones, it is more difficult for a hydrophobic coating to fully display its capability of modifying the surface properties of the substrate as it happens in the case of porous stones. The capillary absorption tests show that a fully satisfactory protective treatment on a high porosity stone, like a calcarenite, is able to grant a 95% reduction of water uptake, while on a low porosity stone, like marble, the reduction is only about 40%; nevertheless, it is important to consider that a protected high porosity stone still keeps absorbing twice as much water (mg/cm^2) than an untreated low porosity one. The elevated surface roughness (see above) and the large amount of permeable pores, in the case of high porosity materials, allow the formation of a very thick and efficient protection layer with quite good water-repellency and shielding properties.

The problems that have to be faced when the protection of a high open porosity stone, like calcarenite, is considered are specific for this kind of substrate and sometimes completely different from those of the low porosity stones. The high porosity and the pore size distribution allow a large amount of product to penetrate into the bulk of the stone with many advantages for the shielding efficacy (a thicker hydrophobic stone/polymer layer) and for the durability of products (most of the polymer is inside the stone and not exposed to sunlight and environmental aggressive pollutants)[6,16].

Some drawbacks arise from the large amount of product inside the stone that increases the weight of the substrate surface layer and favours the formation of areas with high protective concentration that can create discontinuity in the mechanical properties of the stone (risk of micro-fractures and scaling). The uptake of large amount of product, reduces dramatically the natural water vapour permeability, or breathability, of the stone, i.e. the ability of the material to get the hygro-thermal equilibrium of the environment[17, 18]. Finally, the distribution of the polymer inside the stone excludes any reversibility of the treatment, making dramatically complicate any removing operation. The elevated superficial roughness, typical of high porosity stones, enhances apparently the contact angle values, deceiving about the real shielding efficacy and water-repellency of the protective treatment[12-14].

Considering the differences among the stone materials, it is clear that the same protective agent would not work equally well for low and high porosity substrates. It should be necessary to tune the characteristics of the coating according to the specific requirements of each stone material that has to be protected. Starting from the remark that commercial products have rarely been developed specifically for conservation applications, but they have been simply transferred from other technological fields with higher economic advantages, it is absolutely necessary before treating an artefact to perform a thorough study concerning the efficacy of the selected products. The aim of this preliminary approach is the selection of the most suitable protective treatment and, above all, the optimization of the application conditions like solvents, concentration, temperature and application methodology.

3. THE PROBLEM OF SOLVENT AND OF APPLICATION METHODS

Indeed the choice of the solvent is decisive for an effective and successful protection treatment of stone materials. The solvent, as well known, is the medium that "transforms" the treatment from solid to liquid, carries the product into the bulk of the stone and, evaporating, allows the transformation from liquid to solid. The solvating power and the volatility are therefore the characteristics that should be carefully controlled in order to enhance the effectiveness of a protection treatment[19]:

- Maximum solvating efficacy. The solvent must have the highest possible affinity with the polymer.
- No interaction with the substrate. Solvents with acid or basic characteristics should be avoided.
- No toxicity. The solvent should not cause risks for the operators' health or damages to the environment, according to the EU standard requirements.
- Minimum solvent viscosity; the dissolution of a polymer usually increases the viscosity of the organic liquid phase, therefore it is necessary to employ low viscosity solvents in order to favor the penetration of the solution into the stone material.
- Tuned volatility. The solvent must completely evaporate after the treatment, in order to prevent unwanted reactions induced or catalyzed by compounds present in the substrate. The lower the solvent's affinity with the polymer is, the easier it can evaporate; but it should be emphasized, that a fast evaporation induces the formation of an inhomogeneous solid phase and a poor penetration of the polymer into the substrate[19]. A compromise should be carefully reached, pursuing the most suitable solvent or solvents mixture.
- No chromatographic effect. No preferential adsorptions by polymer or solvent in the stone matrix should happen in order to prevent the formation of a gradient of treatment's concentration. When a polymeric solution penetrates in a stone substrate, this latter tends to behave like a static phase of a chromatographic column, working both in size exclusion (discriminating by size) and in adsorption (discriminating by affinity). It is easy to comprehend that the small molecules of the solvent move faster than the polymer through the pores, so that rarely the solvent front corresponds with that of the polymer[20]. This gives birth to areas with different polymer/solvent ratio that will behave differently in the formation of the solid phase and then in the shielding efficacy.
- Polymerisation favouring; in the case of reactive pre-polymeric treatment, the solvent should grant the polymerization process and, successively, favour the evaporation of by-products;
- No reversed migration. The solvent, inside the porous material, migrates towards the external surfaces, for evaporation purposes; during these processes, it tends to carry the polymer with it[20]. In order to minimize this effect, it is necessary to reduce the interaction solvent/polymer (using a thermodynamically "poor" solvent) and to slow the evaporation rate.

A solvent or a blend of solvents, showing all these characteristics cannot exist for the evident contradictions of the required chemical physical properties[21]. A "good" solvent, i.e. with a great solvating effect and polymer affinity, favours the penetration, reduces the chromatographic effect and does not interact with substrate, but only a "bad" solvent completely and quickly evaporates without any "reversed migration" effect. Moreover, it is important to consider the undesirable interactions between polymer and solvent, such as the formation of intermolecular complexes, clusters or physical gels that can occur in particular conditions of temperature and pressure. Only a suitable experimental work before the treatment, allows one to choose the appropriate solvent or solvent mixture, considering the characteristics of the stone material and of the polymeric coating. It is important to evaluate the correlation between the protection efficacy and the different solvent systems.

In the last years, in order to reduce the toxicity and the environmental impact of conservation treatments, new formulations of protective products have been developed in water dispersion, through the use of surfactants and other additives. These dispersions, nowadays, still show many limits and drawbacks[22,23], mainly due to the particles size (the radius of particles ranges from 100 to 1000 nm and from 10 to 100 nm for the classic emulsion and the so called micro-emulsion, respectively) and to the permanence of different additives in the formed solid phase. Initiators, primary and secondary surfactants, if not completely removed after the film formation, can favour dangerous reactions (oxidations, hydrolysis etc.) of the protection treatment and of the substrate; they can also alter the optical properties (the visual aspect: opalescence, whitening, yellowing, etc.) of the coating. The presence of plasticizers and of thickeners, often employed to enhance the film quality, can also produce undesired similar effects.

Finally two major drawbacks are correlated with the use of water: first, it is a high boiling solvent with a quite good chemical affinity with stone substrates and, therefore, evaporates very slowly; second, it is able to dissolve the residues of soluble salts (even if present in very low amount) increasing locally the ionic force and producing, sometimes, the collapse of the dispersion[24,25].

4. THE PROBLEM OF DURABILITY

The other crucial problems in the protection of outdoor exposed monuments are the prediction and the evaluation of stability and resistance to the photoxidative and thermal ageing of the applied treatments: often a long term chemical integrity, both in laboratory and in situ, does not correspond to a similar protection efficacy.

As far as it is verified that the chemical degradation of a polymeric treatment corresponds to a loss of the shielding efficacy, the contrary is not necessarily true. With the progress of natural or artificial ageing, a polymer applied on a stone surface could lose the protection effectiveness because of some physical rearrangements of the macromolecules, maintaining the chemical structure unaltered[8,26]. For example, SEM observation of different acrylic copolymers[10] showed a common pattern of microfractures on the coating surface after 1500 hours of UV artificial ageing, while, during the same ageing time, no such pronounced chemical decay was assessed[27]. It is evident that the chemical nature definitely influences the aging stability and, hence, the durability of the polymer. Concerning the polymers employed in conservation, the acrylics and siloxanes are the more studied and applied classes of products; similarly their degradation processes and protective features are well known. The introduction of fluorine into classical acrylic polymers, like Paraloid B72 and B67, resulted in a class of partially fluorinated acrylic copolymers[8,16,17] showing higher chemical stability, enhanced water repellency and reduced sensitiveness to photooxidation. Nevertheless, the durability and long term effectiveness of these polymers applied on stone substrates are not yet completely satisfactory.

In the case of silicon based water repellent treatments, the polymer resistance is intrinsically granted by the higher stability of Si-C and Si-O bonds; similarly, these favorable characteristics are unable to maintain a long term durability of the shielding efficacy. Actually, some authors[28] have studied a large number of sites treated with silicon based water repellents, after different aging periods, demonstrating that the shielding effectiveness decreases in about 5 years and dramatically drops in 10 years. After this period of time, the stone substrate shows an inverted gradient of protective efficacy, i.e. the outermost part of surface completely loses the water repellency while the inner is maintaining this property. This condition, favouring water retaining induces hygric dilation processes, frost events and biological attacks, which take place preferably in this thin un-protected superficial layer.

It is important to highlight that the available test to evaluate in situ effectiveness of treatments (low pressure water absorption measurement or Karsten pipe) is not completely reliable, especially when low porosity stones are considered.

5. THE "UTOPIA" OF REVERSIBILITY

The idea of the reversibility of conservation treatments[29], since their introduction, seemed crucial for a modern and correct approach to the restoration issues. The term "reversibility" usually means the possibility of completely remove an applied product from the treated surface, recovering it in the condition preceding the treatment, in order to allow a new and, possibly, more effective intervention.

In the case of consolidation treatment the concept of reversibility is not applicable; actually, most of the treatments with inorganic products (silicates, fluosilicates or barium hydroxide), for example, are intrinsically irreversible, considering that they work through the precipitation of compounds necessarily more stable than the matrix that has to be consolidated.

As far as the protection treatment is concerned, it is necessary to clarify a common misunderstanding connected to the term reversibility: often the product solubility is confused with the product/substrate reversibility. The solubility, and the long term solubility, of a polymeric material is not the only requirement that can assure its complete real reversibility. The penetration of the polymer inside the porous micro-structure of the stone[30], the adhesion to the crystalline material and the interaction of the polymer with the substrate are the crucial factors which, usually, do not enable the removal of the product[30,31].

Some organic products, like waxes[32] and many synthetic polymers, easily soluble in the suitable solvents and then theoretically reversible, once applied on a stone substrate lose their reversibility, for many reasons. First of all, once the product penetrates deeply in the porous matrix of the stone it cannot be easily reached by the solvent and extracted; moreover, the extraction could be complicated by preferential adsorptions, interactions with the stone matrix and cross-linking of the polymeric chains occurring over time. Large amounts of solvent and hazardous operation would be necessary to partially remove the polymer, without taking care of the safety regulations.

Some difficulties in removing the treatment from low porosity substrate are also verified; actually the polymeric films, exposed to the environmental aggression, lose their solubility properties after aging (photooxidation, crosslinking, chain degradation, etc). Then, the assessed impossibility of completely removing a treatment from a real surface leads to consider a new approach to the conservation intervention: the re-treatableness. A new generation coatings should be developed in order to permit the application of another material with the minimum interference.

ACKNOWLEDGMENTS

The Authors are gratefully thanking the CNR Target Project "Safeguard of Cultural Heritage" for the research support on stone protection and Dr. Antonio Sansonetti for his careful reading of the paper.

REFERENCES

1. J. M. Teutonico, A. E. Charola, E. De Witte, G. Grasseger, R. J Koestler, M. Laurenzi Tabasso, H. R. Sasse, R. Snethlage, Group report: how can we ensure the responsible and effective use of treatments (Cleaning, Consolidation, Protection)? in: *Saving Our Cultural Heritage*, 293-313 (Wiley and Sons, Bognor Regis UK, 1997).
2. E. M. Winkler, *Stone in architecture*, (Springer-Verlag, Berlin, 1994).
3. A.A.V.V., Natural stone, weathering phenomena, conservation strategies and case studies, Ed. S. Siegesmund, T. Weiss, A. Vollbrecht, Geological Society Special publication **205** (2002).
4. G. G. Amoroso, V. Fassina, *Stone decay and Conservation* (Elsevier, Amsterdam, 1983).
5. A. E. Charola. Water repellents and other "protective treatments": a critical review, *International Journal for Restoration of Buildings and Monuments* **9**(1), 3-22 (2003).

6. J. Delgado, E. Charola, *General report on water repellents, Science and Technology for Cultural Heritage* **5**(1), 93-101 (1996).
7. E. S. Goins, G. Wheeler, M. T. Wypyski, Alkoxysilane film formation on quartz and calcite crystal surfaces, in: *Proceedings of 8th Int. Cong. on Deterioration and Conservation of Stone*, Berlin, 1255-1264 (1996).
8. T. Poli, L. Toniolo, O. Chiantore, M. Lazzari, V. Castelvetro, A. Manariti, Tailoring new fluorinated acrylic copolymers as protective coatings for marble, *Journal of Cultural Heritage* **3**, 309-316 (2002).
9. O. Chiantore, C. Colombo, R. Peruzzi, T. Poli, L. Toniolo, Effect of fluorinated groups on photooxidative stability of polymeric protectives applied on marble *Annali di Chimica,* **91**, 741-748 (2001).
10. L. Toniolo, F. Casadio, C. Colombo, T. Poli, SEM observation of polymers applied to natural stones, *Actes du congrès "Art et Chimie, les polymères"* CNRS Editions, Paris, 163-167 (2003).
11. H. R. Sasse, R. Snethlage, Methods for the evaluation of stone conservation treatments in: *Saving Our Cultural Heritage*, 223-243 (Wiley and Sons, Bognor Regis UK, 1997).
12. C. Della Volpe, A. Penati, R. Peruzzi, S. Siboni, L. Toniolo, C. Colombo, The combined effect of roughness and heterogeneity on contact angles: the case of polymer coating for stone protection, *J. Adhesion Sci. Technol.*, **14**, 273-299 (2000).
13. T. Poli, L. Toniolo, R. Peruzzi, M. Brugnara, C. Della Volpe, Partially fluorinated acrylic copolymers as coatings for stone protection: characterization and surface properties *MRS Proceedings Volume 712 - Symposium II Materials Issues in Art and Archaeology VI*, Ed. by P. B. Vandiver, M. Goodway, J. L. Mass, Boston, **II3.3** (2002).
14. M. Brugnara, E. Degasperi, C. Della Volpe, D. Maniglio, A. Penati, S. Siboni, V. Castelvetro, T. Poli, L. Toniolo, S. Invernizzi, The application of the contact angle in monument protection: new materials and methods, *Colloids and Surfaces A*, **241**, 299-312 (2004).
15. R. Peruzzi, T. Poli, L. Toniolo, The experimental test for the evaluation of protective treatments: a critical survey of the Capillary Absorption Index, *Journal of Cultural Heritage*, **4**(3), 93-96 (2003).
16. G. Alessandrini, M. Aglietto, V. Castelvetro, F. Ciardelli, R. Peruzzi, L. Toniolo, Comparative evaluation of fluorinated and unfluorinated acrylic copolymers as water repellent coating materials for stone, *Journal of Applied Polymer Science*, **76**, 962-977 (2000).
17. G. Alessandrini, L. Toniolo, C. Colombo, Partially fluorinated acrylic copolymers as coatings for calcareous stone materials in: Preprints of the 18th IIC International Congress *Tradition and Innovation. Advances in Conservation*, Edited by A. Roy and P. Smith 1-6 (Melbourne, 2000).
18. C. M. Hansen, Water transport and condensation in fluoro-polymer films *Progress in Organic Coatings* **42**, 167-178 (2001).
19. C. V. Horie, *Materials for Conservation. Organic Consolidants, adhesives and coatings*, (Butterworths, London, 1987).
20. W. Domaslowski, The mechanism of polymer migration in porous stone, *Wiener berichte über Naturwisenschaft in der Kunst*, Doppelband 4/5, Vendl A., Pichler B., Weber J., Banik G., 402-425 (Wien, 1988).
21. D. Lopez, H. Reinecke, C. Mijangos, Some consequences of the polymer-solvent interaction, *Macromol. Symp.*, **166**, 25-33 (2001)
22. J. Snuparek, Principles and limits of polymer latex tailoring, *Prog. Org. Coatings,* **29**, 225-233 (1996).
23. A.A.V.V. *Emulsion polymer technologies*, Paint Research Association, **12**, 9, 2001.
24. K. Wanli, Interaction between alkali/surfactant/polymer and their effects on emulsion stability, *Coll. Surf.*, **175,** 243-247 (2000).

25. I. Fortelny, Coalescence in polymer blends: solved and open problems, *Macromol. Symp.* **158,** 137-147 (2000).
26. T. Poli, L. Toniolo, Protection efficacy of fluorinated acrylic copolymers applied on historical italian marbles, *Material Research Society, Materials Issues in Art and Archaeology*, in press (2005).
27. O. Chiantore, M. Lazzari, Photo-oxidative stability of Paraloid acrylic protective polymers, *Polymer* **42,** 17-27 (2001)
28. E. Wendler, New Materials and approaches for the conservation of stone in: *Saving Our Cultural Heritage*, 181-196 (Wiley and Sons, Bognor Regis UK, 1997).
29. G. Urbani, *Intorno al restauro*, a cura di B. Zanardi, (Skira, Milano, 2000).
30. L. Toniolo, F. Casadio, F. Cariati, A key factor in modern protection of historic buildings: the assessment of penetration of water repellent polymers into porous stone materials, *Annali di Chimica,* **91**, 823-832 (2001).
31. F. Casadio, L. Toniolo, Polymeric treatments for stone materials conservation: the problem of penetration depth, *J. American Institute for Conservation,* **43**, 1-19 (2004)
32. K. Raft, A preliminary report on the possibility of using bleached beeswax to improve the resolubility of picture varnishes based on polycyclohexanones, *Studies in conservation*, **30**, 143-144 (1985)

Chapter 9.4

IN SITU POLYMERIZATION WITH ACRYLIC MONOMERS FOR STONE CONSOLIDATION AND PROTECTION

S. Vicini, E. Princi, E. Pedemonte, G. Moggi
Department of Chemistry and Industrial Chemistry, University of Genoa, Via Dodecaneso 31, 16146 Genova, Italy

Abstract: In this paper the in situ copolymerization of acrylic monomers, i.e. mixtures of ethyl methacrylate and methyl acrylate, is once again proposed as a remarkable technique in order to increase the amount of polymer present inside the stone and consequently to improve the consolidating properties of the substrate. To make better the protective properties of the copolymers two different possibilities are indicated: either to copolymerize a third fluorinated acrylic monomer, i.e. 2,2,2 trifluoroethyl methacrylate, or to dissolve in the copolymerizing mixture a fluorinated elastomer, i.e. Tecnoflon TN®. In both cases our results show that small amounts of fluorinated compounds are sufficient to improve remarkably the protective properties of the polymers.

Key words: In situ polymerization; acrylic copolymers; stone consolidation and protection.

1. INTRODUCTION

Water is commonly considered the main cause of stone degradation since it carries the pollutants which are present in the atmosphere, owing to the combustion processes i.e. carbon dioxide, sulphuric and nitric oxides. Water comes from rain or from condensation of moisture, due to thermal excursion between day and night and/or summer and winter.

The degradation level depends on the chemical nature of the rocks; actually, the limestones are damaged more than silicatic ones since calcium carbonate is easily transformed by acid rain in calcium hydrogen carbonate, sulphate and calcium nitrate. The penetration of water into the rocks is due

S.K. Kourkoulis (ed.), Fracture and Failure of Natural Building Stones, 565–575.

to the porosity of the stone[1,2]. From this point of view the pores size distribution is particularly important (micropores, mesopores, macropores), since inside the micropores ($\varnothing < 20$ Å) the capillary effect is more enhanced and the polluting solutions can reach higher depths.

The reaction of the chemical components of the rocks with the pollutants leads both to an increase of the average stone porosity and to a shift of the size distribution towards mesopores ($20 < \varnothing < 500$ Å) and macropores ($\varnothing >$ 500 Å). The increasing of porosity is associated to a decohesion of the material, that loses its intrinsic mechanical properties, and to a reduction of adhesion of the degraded layers to the substrate, i.e. the undegraded stone. The main purpose of the so called "consolidation" treatment consists in leading back the average porosity and the pore size distribution to the original values characteristic of the undegraded stone, improving both the adhesion of the altered layers to the substrate and the cohesion of the degraded material, in order to recover the original mechanical properties.

The consolidation can be performed with both inorganic and organic substances, introduced into the stone by capillary absorption of solutions or suspensions. Organic consolidants are polymeric compounds[3]; owing to the large dimensions of their molecules[4,5] in comparison with the pores sizes, they can not penetrate deeply into the degraded stone, but they often remain in the external layers. For this reason their consolidation properties are rather insufficient. On the other side, polymers can have water-repellent properties and therefore they can perform a protective action, forming a superficial barrier between stone and environment which hinders the penetration of liquid water. This means a peculiar characteristic, which puts polymeric compounds in a leading position if compared with inorganic consolidants, which do not exhibit hydrophobic properties.

Obviously one faces with two problems. First of all, the polymer must penetrate in the smallest pores and besides this the water repellence must be improved. The first topic can be settled with the in situ polymerization, carried out in the laboratory as described by Vicini et al.[5-8]. With this technique, the degraded stone is treated not directly with the polymer, but with the corresponding monomers: these molecules have small dimensions and therefore they can easily penetrate inside the micropores, where they are polymerized in a subsequent step following a usual radical mechanism.

As far as the water repellence is concerned, fluorinated polymers or monomers can be used[9-13]. It is well known that the introduction of fluorine atoms into polymer structures has the effect of improving their chemical, thermal, and photochemical stability, due to the stability of the C-F bond (bond energy: 116 kcal/mol). In addition, the formal substitution of hydrogen by fluorine atoms induces higher hydrophobicity as a consequence of the low surface energy brought by the fluorinated groups. The high cost of these products leads to minimize their amounts in the compounds used for the consolidation and protection. In this paper the results of in situ polymerization of acrylic monomers carried out in presence of small amounts of both fluorinated polymer and monomer, with the purpose of improving the performances of the restoration treatments, are reported.

2. EXPERIMENTAL

2.1 Materials

The monomers used in the experiments, i.e. ethylmethacrylate (EMA), methylacrylate (MA) and 2,2,2 trifluoroethylmethacrylate (TFEMA), were commercial products supplied by Aldrich. They contain an inhibitor (hydroquinone monomethylether) that has been removed before the polymerization, passing through an Aldrich *Inhibitor Removers* column. The pure monomers were subsequently stored at low temperature (4 °C) in the dark, in order to avoid polymerization.

The fluorinated monomer TFEMA has been chosen since it is comercially available at a reasonable low price. On the other hand, its conversion in homopolymer is rather high (86% in dioxane)[12] and it copolymerizes easily with other unfluorinated acrylic monomers. The homopolymer exhibits a quite high chemical, thermal and photochemical stability[14]. Acrylic monomeric mixtures containing 2.5% and 10% (mol/mol) of TFEMA were used, indicated in the following as Terpolymer 2.5 and Terpolymer 10. The polymerization initiator (2,2′ azobisisobutyronitrile, AIBN)[15] was supplied by Fluka.

The fluorinated polymer was Tecnoflon TN® supplied by Solvay Solexis; its composition is the following: vinylidene fluoride (64%), hexafluoropropylene (19%), tetrafluoroethylene (17%). The total amount of fluorine is 67% wt. The molecular weights are: Mw=495000, Mn=90000; the polydispersity index is 5.5; the glass transition temperature is -14 °C. This polymer is soluble in acrylic monomers and exhibits thermodynamic compatibility with acrylic and methacrylic resins[16-18]. Use was made of solutions in acrylic monomers of Tecnoflon TN® containing 2, 5 and 10 % w/w, indicated in the following as EMA/MA + Tecnoflon 2, Tecnoflon 5 and Tecnoflon 10, respectively.

Paraloid B72®, used for a comparative study since it is widely employed in stone restoration, was a commercial acrylic resin produced by Rohm and Haas, with a composition of 67 % EMA and 33 % MA[19]. The ratio between the repetition unit of EMA and MA in the copolymers used in the present series of tests has been purposely kept similar to the analogue Paraloid B72®. Solvents were laboratory grade products from Aldrich and they were used without any further purification.

Deionized water was employed throughout the work.

Obviously, specimens of deeply degraded stones coming from monuments of historical interest were not available in so large number to perform the whole investigation. For this reason one is obliged to choose a quarry rock, which has to satisfy two basic requirements: high porosity, to reproduce a stone weathered by physical and chemical agents and easy availability in standardized shapes, suitable for the experiments.

Two varieties of Finale stone, a calcareous sedimentary rock of biological origin, easily available in Liguria (Italy)[20] and with a fairly high porosity were used:

- the Mascia variety (porosity: 27%) for the experiments with the terpolymer EMA/MA/TFEMA (in the following experiments A).
- the Pale variety (porosity: 40%) for the experiments with Tecnoflon TN® (in the following experiments B).

2.2 In situ polymerization

The in situ polymerization is performed in three subsequent steps: absorption, polymerization and purification. In the first step, stones absorb by capillarity the different mixtures that must be polymerized, placing the sample on a thick layer of cotton soaked in the reaction mixture; the absorption time was standardized in 4 hours at 4 °C, in absence of light. Polymerization was carried out at 50 °C; this temperature is very close to that corresponding at the best performance of AIBN (the half period of AIBN is 25 hours at 60 °C and 50 hours at 50 °C); the reaction time was 24 hours. In these conditions the conversion is high. The purification involves the removing of the solvent and the unreacted monomers still present after the polymerization. The amount of AIBN was 2% wt of the monomer mixture.

The reactions have been performed in acetone (20% vol/vol of the monomer mixture), in order to avoid the large volume contraction which leads to fractures in the stone, when the polymerization is carried out in bulk.

The percentage of fluorinated compounds in the reaction mixture was intentionally limited to small values since the cost of these products is quite high; on the other side the solutions of Tecnoflon TN® in the acrylic monomer have a rather high viscosity, which hinders a good penetration into the stone.

To compare the results of the in situ polymerization with those obtainable from the traditional technique of application of consolidants, some stone specimens have been treated by capillarity for 4 hours with a 3% wt solution in acetone of Paraloid B72® or of a mixture Paraloid B72® / Tecnoflon TN® containing the 2% wt of the fluorinated elastomer (in the following indicated as commercial mixture).

The amount of polymer present in the stones (ΔM%), after the in situ copolymerization and the absorption of the preformed polymers, has been calculated using the simple equation:

$$\Delta M\,\% = [(P_f - P_i)/P_i] * 100$$

where P_f and P_i are the weights of the specimens after and before the treatments, respectively.

2.3 Consolidation and protection tests

To evaluate the consolidating and protective properties of the in situ polymerized copolymers, some tests have been carried out[21,22]. Each test has been performed on three specimens (5x5x2 cm) of the Finale stone samples, before and after each treatment.

The consolidating properties were evaluated with an original instrument[8], made up by a wood rail (100 cm of length x 5 cm of width) having a strip of sandpaper (granulometry: 60 mesh) on the top. The sample with the largest area surface (5x5 cm) is set on the rail and loaded with a weight of 2.0 kg; it runs for 30 times (equivalent to 30 meters of sandpaper) along the sandpaper with a constant driving force of 2.5 kg; every 5 meters it is weighed and the weight loss percent (WL%) is calculated. The test allows the calculation of the efficacy of superficial aggregation (EA) with the formula:

$$EA = [(WL_0 - WL_t) / WL_0] * 100$$

where WL_0 is the average value of the weight loss percent of three untreated stones after 30 meters and WL_t is the average value of the weight loss percent of three treated stones after 30 meters. The quantity of removed material is a function of the aggregation of the sample, and so it is different for treated and not treated samples.

The protective properties were evaluated by capillary water absorption and permeability to water vapor. The capillary water absorption test was carried out using the gravimetric sorption technique, according to the Normal protocol[23]. The stone specimen is laid on a filter paper pad, around 1 cm thick, partially immersed in deionized water, with the treated surface in contact with the pad. The amount of water absorbed by capillary force is determined by weighing the specimen after 10′, 20′, 30′ and 1, 2, 4, 6, 24, 48, 72 and 96 hours, to obtain the wet specimen mass M_i ($M \pm 0.0001$ g).

The amount of absorbed water Q_i, at the time t_i per surface unit, is defined as follows:

$$Q_i = (M_i - M_0) / S$$

where M_i is the specimen mass (g) at time t_i (second), M_0 is the dry specimen mass (g) and S is the contact surface (cm^2). The Q_i values (g/cm^2) are plotted against the square root of time ($t^{1/2}$) to give the capillarity absorption curve.

The angular coefficient of the first part of the curve enables one to evaluate the capillary absorption coefficient CA: this value should be reduced with treatments. The results can also be expressed as protective efficiency EP %:

$$EP\ \% = [(Q_0 - Q_t) / Q_0] * 100$$

where Q_0 is the average value of water absorbed by untreated stones series after 1 hour and Q_t is the average value of water absorbed by treated stones series after 1 hour.

The permeability to water vapor test was carried out according to the corresponding Normal protocol[24] on 5x5x1 cm specimens, using a measurement cell consisting in a cylindrical PVC chamber with open top fitted with an o-ring rubber seal, where the stone specimen is employed as the lid of the chamber; the chamber is sealed with the lid by means of an aluminium flange with an o-ring. The cell is partially filled with deionized water, therefore allowing measurement by gravimetry of the amount of water vapor that diffuses through the stone specimen with fixed thickness (1 cm) between two parallel surfaces. The test is carried out at constant temperature 20±0.5 °C, with the cells placed into a desiccator. The driving force for the diffusion of water vapor is, therefore, the constant difference between the water vapor pressure inside and outside the cell (in the presence of activated silica gel desiccant).

The permeability is monitored by determining the weight decrease per surface unit (S, m^2) in the unit time (24 h):

$$\Delta M_i = (M_i - M_{i-1}) / S$$

where M_i is the weight system (cell and stone) at i-day (g).

The cell is weighed (M ± 0.0001 g) and ΔM_i (daily weight variation) is calculated when a stationary condition (constant vapor flow through the stone) is reached; stationary flow was considered to be reached when:

$$(\Delta M_i - \Delta M_{i-1}) \times 100/\Delta M_i \leq 5\ \%$$

The permeability to water vapor, after the treatment, must be as high as possible and not too different from the value of the untreated material.

Besides the water vapor permeability, the reduction in permeability RP % due to the treatment was evaluated, also, according to the following equation:

$$RP\ \% = [(P_0 - P_t) / P_0] * 100$$

where P_0 is the permeability to water vapor of the untreated stones, used as reference and P_t the permeability to water vapor of the treated ones.

3. RESULTS AND DISCUSSION

The amount of polymer present inside the stones ($\Delta M\%$) is reported in Tables 1 and 2 for the experiments A and B respectively.

Table 1. ΔM % in the experiments A.

POLYMER	(ΔM %)
Paraloid B72®	0.10
EMA/MA copolymer	0.20
Terpolymer 2.5	0.30
Terpolymer 10	0.30

Table 2. ΔM % in the experiments B.

POLYMER	(ΔM %)
Paraloid B72®	1.6
EMA/MA copolymer	3.1
EMA/MA + Tecnoflon 2	2.4
EMA/MA + Tecnoflon 5	0.9
EMA/MA + Tecnoflon 10	0.3
Commercial mixture	1.1

The amount of polymer present in the stone in the experiments B is higher than in the experiments A, either if one applies the commercial copolymer Paraloid B72® or if one copolymerizes in situ the mixture EMA/MA. This is due to the different porosity of the stones used in the two series of experiments, 27% for the Mascia (A) and 40% for the Pale (B), respectively.

In both experiments, the in situ copolymerization of the mixture EMA/MA leads to ΔM% values double than those measured after the absorption of the Paraloid B72® solution. This result confirms what could be easily foreseen, namely that the absorption of the monomers by the stone instead of the preformed polymer sets up a real advantage since it allows introducing into the pores of the stone a larger amount of the consolidating and protective material.

In the experiments A, when the fluorinated monomer is present in the mixture copolymerized in situ, ΔM% reaches values of 0.3%. It is supposed that this result could be explained if one takes into account that the mixture with the fluorinated monomer reacts faster; since the polymerization time is standardized in 24 hours, the conversion yield is higher.

In the experiments B, the presence of Tecnoflon TN® mixed with the acrylic monomers reduces ΔM%, increasing in large amount the percentage of the fluorinated elastomer (from 2 to 10%). This result can be easily explained if one considers that Tecnoflon TN® is a preformed polymer; therefore, it increases the viscosity of the mixture applied to the rock that cannot penetrate deeply in the pores of the stone.

This explanation justifies the decreasing of ΔM% from 2.4% with the in situ copolymerized mixture EMA/MA + Tecnoflon 2 to 1.1% with the solution of the two polymers Paraloid B72®/ Tecnoflon TN® (commercial mixture with 2% of Tecnoflon TN®). Also in this case the in situ polymerization is favorable in comparison with the absorption of the preformed polymers.

The consolidating and protective properties related to the different treatments reflect the amount of polymer present in the pores of the stone.

Figures 1 and 2 show the results of abrasion and capillary absorption tests obtained from the experiments A.

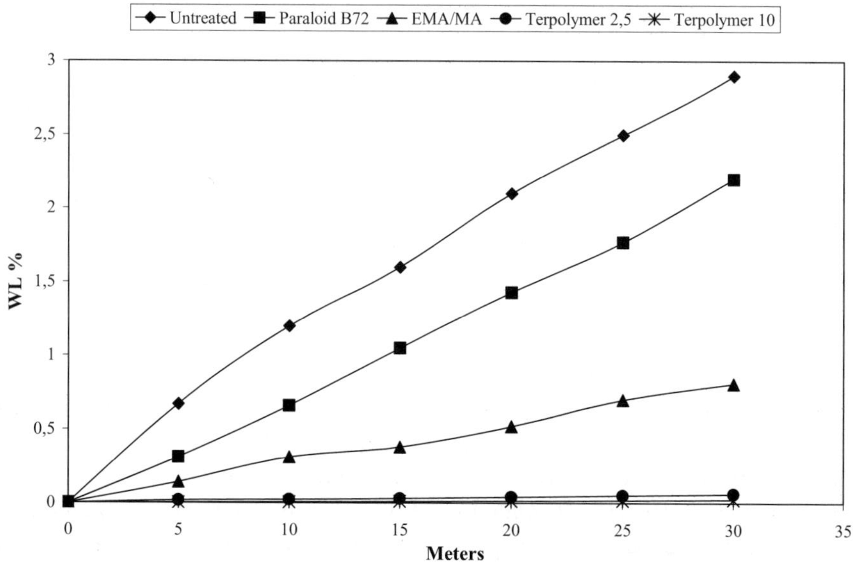

Figure 1. Results of abrasion test concerning the experiments A.

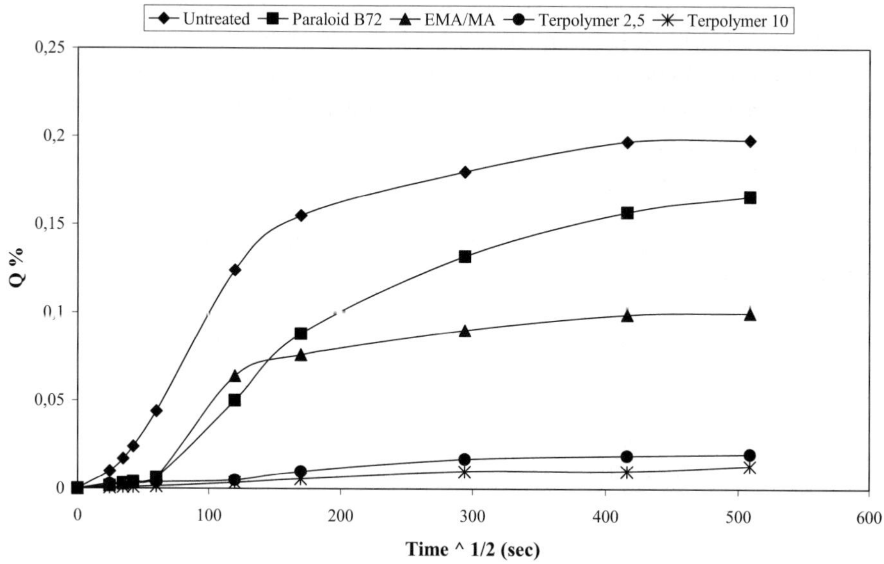

Figure 2. Results of capillary absorption test concerning the experiments A.

The aggregation efficiency is particularly high for the terpolymers (EA= 90%) which penetrate deeply in the stone and low for Paraloid B72® (EA= 30%), which forms just a thin polymer layer on the surface. For the copolymer EMA/MA polymerized in situ EA is good (50%), but it probably could reach higher values increasing the monomers conversion, since the reaction time is limited to 24 hours.

The capillary absorption test shows that the presence of TFEMA largely improves the water repellence of the in situ polymerized copolymer, owing not only to the higher amount of polymer in the stone pores, but also the presence of the fluorinated repetition unit.

Comparing the behavior of Paraloid B72® with that of the in situ polymerized copolymer, the latest shows better performances for long time of exposition to the liquid water. In the short time Paraloid B72® supplies good water repellence; actually it forms a continuous layer on the stone surface, which hinders the penetration of the liquid water, but the layer is too thin to hold out: after about 1 hour, water gets over the polymer and penetrates deeply into the stone.

Finally, it is underline that, both in the abrasion and in the capillary absorption tests, the behavior of the two terpolymers is similar. This remarkable result leads to the conclusion that, since the price of the fluorinated monomers is high, the amount of TFEMA in the terpolymer can be limited to 2.5%, without changing the performances of the polymeric material.

As far as the experiments B are concerned, attention is focused only on the capillary absorption test, since the main purpose of the addition of the Tecnoflon TN® to the copolymerizing monomers is to improve the resistance to water penetration. The data are reported in Table 3.

These results show that Paraloid B72® and EMA/MA copolymer polymerized in situ have the same behavior, independently of the amount of polymer present into the stone. Obviously one must realize that the capillary absorption coefficient and the corresponding protective efficiency refers to the initial period of exposure of the treated stone to liquid water. Later on the performance of the EMA/MA copolymer improves, as stated in the experiments A, and its capillary absorption becomes smaller than that of Paraloid B72®.

Table 3. Capillary absorption data for the experiments B.

TREATMENT	CA $g/cm^2\ sec^{1/2}$	EP %
None	4.2×10^{-3}	-
Paraloid B72®	2.4×10^{-3}	45
EMA/MA copolymer	2.4×10^{-3}	45
EMA/MA + Tecnoflon 2	1.1×10^{-4}	96
EMA/MA + Tecnoflon 5	4.3×10^{-4}	90
EMA/MA + Tecnoflon 10	2.8×10^{-4}	93
Commercial mixture	4.3×10^{-5}	99

The addition of Tecnoflon TN® doubles the protective efficiency. It is noteworthy that the behavior of the polymer does not improve increasing the amount of Tecnoflon TN® and therefore only few percent of fluorinated terpolymer are sufficient in order to have good protective characteristics.

Finally, the highest value of EP is obtained with the commercial mixture of Paraloid B72® and Tecnoflon TN®. This is probably due to the formation of a film on the stone surface which, owing to Tecnoflon TN®, is more adhesive and flexible than the one obtained with Paraloid B72® alone; it provides an excellent hindrance to the penetration of liquid water.

The last data, reported in Tables 4 and 5, concern the permeability test and the reduction of permeability after the application of the polymer. The permeability of the untreated stones follows their porosity: actually the Pale variety has porosity higher than that of Mascia variety (40% vs. 27%) and consequently the permeability increases from 28 to 56 $g/(m^2.24h)$. The treatment with the polymer reduces the permeability but the experimental values are still acceptable. The reduction is larger in experiments B due to the high porosity of the stone, whose pores are more easily filled up with the polymers.

It is worth to note that the largest reduction of permeability is found for terpolymer 10 in the experiments A and EMA/MA + Tecnoflon 10 in the experiments B. These results support the conclusion that only few percent of fluorinated compound are enough to give the best performances.

Table 4. Results of the permeability test for the experiments A.

TREATMENT	Permeability (g/m^2 24h)	RP (%)
None	28	-
Paraloid B72®	19	33
EMA/MA copolymer	20	29
Terpolymer 2.5	20	29
Terpolymer 10	12	57

Table 5. Results of the permeability test for the experiments B.

TREATMENT	Permeability (g/m^2 24h)	RP (%)
None	56	-
Paraloid B72®	28	50
EMA/MA copolymer	30	46
EMA/MA + Tecnoflon 2	23	59
EMA/MA + Tecnoflon 5	20	64
EMA/MA + Tecnoflon 10	18	68
Commercial mixture	22	61

4. CONCLUSIONS

The experimental results presented in this paper support once again that the in situ polymerization is a technique more favorable than the traditional one, since the polymerization of acrylic monomers inside the stone pores allows putting into the rock an amount of polymer larger than that obtainable by the application of a preformed polymeric material. This improves the consolidating and protective properties of the treatment.

On the other hand, it is well known that the protective efficacy of a polymer can be made better if fluorinated compounds are used. The present experiments show that either fluorinated monomers can be added to the co-polymerizing mixture or a fluorinated polymer can be dissolved in the acrylic monomers. In both cases the data lead to the conclusion that only few percent of fluorinated compound is enough to largely improve the protective properties. This is remarkable since it reduces the cost of the restoration.

REFERENCES

1. P. Rota Rossi Doria, *Bollettino dell'Arte* **41**, 37 (1997).
2. P. Rota Rossi Doria, in *Proceedings of the International Symposium on "Principles and applications of pore structural characterization"* (J. M. Haynes, P. Rossi Doria Eds., Arrowsmith, Bristol, 1985).
3. G. Amoroso, V. Fassina, *Stone decay and conservation* (Elsevier, Lausanne, 1983), ch.13.
4. J. P. Flory, *Principles of polymer chemistry* (Cornell Univ. Press, Ithaca, NY, 1962), ch.10.
5. S. Vicini, E. Princi, G. Moggi, E. Pedemonte, *La Chimica e l'Industria* **81**, 1013 (1999).
6. S. Vicini, S. Margutti, G. Moggi, E. Pedemonte, *J. Cultural Heritage* **2**, 143 (2001).
7. S. Vicini, S. Margutti, E. Princi, G. Moggi, E. Pedemonte, *Macromol. Chem. Phys.* **203**, 1413 (2002).
8. S. Vicini, E. Princi, E. Pedemonte, M. Lazzari, O. Chiantore, *J. Appl. Polym. Sci.* **91**, 3202 (2004).
9. D. M. Brewis, *Int. J. Adhes. Adhesives* **13**, 251 (1993).
10. J. Hopken, S. Shieko, J. Czech, M. Moller, *ACS Polym. Prepr.* **33**, 937 (1992).
11. V. Castelvetro, M. Aglietto, L. Montagnini di Mirabello, L. Toniolo, R. Peruzzi, and O. Chiantore, *Surface. Coat. Intern.* **11**, 551 (1998).
12. F. Ciardelli, M. Aglietto, L. Montagnini di Mirabello, E. Passaglia, S. Giancristoforo, V. Castelvetro, and G. Ruggeri, *Progr. in Org. Coat.* **32**, 43 (1997).
13. G. Alessandrini, M. Aglietto, V. Castelvetro, F. Ciardelli, R. Peruzzi, and L. Toniolo, *J. Appl. Polym. Sci.* **76**, 962 (2000).
14. 11. R.F. Brady, *Chem. Brit.*, **26**, 427 (1990).
15. G. Mohad, D. H. Solomon, Azo and peroxy initiators, in *Comprehensive Polymer Science* Vol. 3 (G. C. Eastmond, A. Ledwith Eds, Pergamon Press, UK, 1989, p. 97).
16. O. Chiantore, M. Guaita and M. Lazzari, *Int. J. Polym. Anal. Characterization* **2**, 395 (1996).
17. D.R. Paul, and J. Wbarlow, *J. Macromol. Sci. Rev. Macromol. Chem.* **C18**, 109 (1980).
18. J. Mijovic, H. L. Luo, and C.D. Han, *Polym. Eng. Sci.* **22**, 234 (1982).
19. www.rohmhaas.com/coatings/.
20. F. Carpene', Le meraviglie della Pietra di Finale, Ed. Bacchetta, 1997.
21. G. Alessandrini, A. Pasetti, Arkos, **14**, 26 (1991).
22. G. Alessandrini, A. Pasetti, Arkos, **21**, 50 (1993).
23. UNI 10859 Cultural Heritage – Natural and Artificial Stones – Determination of water absorption by capillarity, UNI Milan, Italy, 2000.
24. Normal Protocol 21/85, Water vapor permeability, ICR-CNR, Rome, Italy, 1986.

Chapter 9.5

THE DURABILITY OF HYDROPHOBIC TREATMENTS ON GOTLAND SANDSTONE

Malin Myrin[1] and Katarina Malaga[2]

[1]Stenkonservatorn Skanska, Department of Environmental Sciences, Göteborg University, Storängskroken 2, SE-115 42 Stockholm, malin.myrin@skanska.se; [2]SP Swedish National Testing and Research Institute, Box 857, SE-501 15 Borås, katarina.malaga@sp.se

Abstract: The main purpose of this study is to observe, record and evaluate changes in the physical and esthetical properties of Gotland sandstone after hydro-protective treatment and to evaluate the durability of treatments. The aim of water-repellent treatments is to prevent absorption of water into the stone by reducing the stone's capillary forces, and thus avoid the damaging effects caused by water. The treatment is usually not considered very durable and negative surface changes on treated stone objects have been observed. For this study four objects, treated respectively one, four, eleven and fifteen years ago with hydro-protective products, have been studied. Freshly quarried stones, having been treated and placed at field stations and, there, exposed to natural environment for up to three years, have been used as reference samples. The analysis has been made by inspection and by the use of Karsten's testing tube. Colour changes have been measured with spectrophotometer. All applied methods are suitable for usage in field. The results of the study show that untreated Gotland sandstone, freshly quarried as well as deteriorated, has a very rapid water uptake. The study implies that stone sculptures staying exposed to outdoor conditions benefit from hydro-protective treatment, e.g. reduced water uptake, if the treatment is made correctly. The durability of the treatment varies much depending on exposure. Observed negative effects, likely due to hydro-protective treatment, are colour changes, discolouring and loss of gloss.

Key words: hydro-protective treatment; Gotland sandstone; Karsten's testing tube; in-field treatments; field-testing.

1. INTRODUCTION

Water in its different states has an active and important role in all actions influencing the deterioration of building stone. This is especially true for porous and calcitic sandstones such as Gotland sandstone. Hydrophobic treat-

S.K. Kourkoulis (ed.), Fracture and Failure of Natural Building Stones, 577–590.

ments, when performed correctly, reduce the water uptake of the stone. The aim of water-repellent treatments is to prevent absorption of water into the stone by reducing the stone's capillary forces, and thus avoid the damaging effects caused by water. However, the treatment is usually not considered very durable and the durability varies much between different exposures, rendering maintenance planning difficult. Also stone conservators have observed negative surface changes on treated stone objects, e.g. patchy looking surface, colour changes and loss of gloss.

This paper focuses on results and evaluation of in-field hydro-protective treatments, with the aim to study changes in the physical and esthetical properties of Gotland sandstone after hydro-protective treatment and to study the durability of treatments:

- Four objects treated respectively one, four, eleven and fifteen years ago with hydro-protective products have been studied.
- Freshly quarried stone having been treated and placed at field stations, and, there, exposed to natural environment for three years, have been used as reference samples.
- Treated stone samples are presently being treated for achieving accelerated aging (UV-chamber).
- Two hydrophobic agents were used: Wacker BS 290, a silicone concentrate based on silane and siloxane diluted with organic solvents, and Wacker Elastosil E41, a silicone rubber, likewise diluted with organic solvents.

A further aim of the study is to establish methods for estimating the accurate point of time for re-treatment of stone and so to facilitate appropriate conservation work and active maintenance planning for sandstone objects. The necessity and effects of re-treatment projects have recently been presented by Meinhardt-Degen and Snethlage (2004).

The analysis of the treatment effects of the hydrophobic agents has been made by careful visual inspection and by the use of Karsten's testing tube. Colour changes have been measured with spectrophotometer. All applied methods are suitable for usage in field. The treatments' efficiency and durability were evaluated using a suitable treatment efficiency index (Nwaubani et al., 1999).

2. MATERIALS

2.1 Composition of Gotland Sandstone

Gotland sandstone is a soft, very porous stone with a fairly high degree of homogeneity. This makes it particularly easy to work with. Therefore it has become a popular material for sculptures and ornaments.

This sandstone traditionally is the predominantly used stone for outdoor sculptures and ornaments in Sweden (Löfvendahl et al., 1994). Gotland sandstone is relatively homogeneous as to its mineral composition. The stone's fine grains are weakly compacted (Wessman, 1995). The grain size varies between 2 and 0.06 mm with an average of 1 mm. The grains have a subangular shape and consist mainly of minerals, quartz and feldspars but also of small amounts of calcite (Lukaszewicz et al. 1995 and Löfvendahl, 1996). All these minerals have a light colour. A small content of the clay mineral glauconite gives the stone its slightly greenish hue. The stone usually contains small amounts of iron compounds, which give to the weathered Gotland sandstone a brown-yellow colour, whereas it is light grey when freshly quarried. Gotland sandstone has a weak binder made up of calcium carbonate, clay minerals, and silica (Lukaszewicz et al., 1995). The stone has a high degree of porosity, 5-30% by volume (Lukaszewicz et al., 1995), with a pore diameter of about 10 μm (Wessman 1995). The mechanical resistance, i.e. compressive strength, is low, 60 MPa (Lukaszewicz et al. 1995).

The suitability of the stone for carving decorative elements and sculptures and as a building stone varies much between different quarries and different blocks of stone (Löfvendahl 1996). Chemical, mechanical and biological actions, jointly or separately, all influence the weathering process of the stone. The main actions having an effect on the deterioration of Gotland sandstone are dissolution of calcium, swelling and dissolution of clay minerals, formation of gypsum, frost action, salt action and biological weathering.

Water in its different states, solid, liquid and gas, has in all actions mentioned above an active and important role in the weathering process of the stone. Gotland sandstone has, due to its high porosity, a strong and rapid capillary suction. Weathering of Gotland sandstone is visible primarily by sanding and exfoliation.

2.2 Hydro-protective Products

Two hydrophobic agents have been used for this study:

- The conventional hydrophobic product Wacker BS 290, which is a silicone concentrate based on silane and siloxane diluteable with organic solvents, produced by Wacker-Chemie GmbH. For treatment of Gotland sandstone the solvent is diluted with aliphatic hydrocarbons, e.g. White Spirit, in a weight ratio of 1:11 (9%).
- A silicone rubber product Wacker Elastosil E41, likewise dilutable with organic solvents and produced by Wacker-Chemie GmbH. For treatment of Gotland sandstone the solvent is diluted with White Spirit in a weight ratio of 1:67 (1.5%).

The silicon rubber product, E41, has been frequently used as a hydrophobic product in Poland since the middle of the 1980's. It was used by conservators in Sweden the first time in 1990 (the object treated at that time is included in this study). The suitability of the silicon rubber product as a hydro-protective product has been studied at the Nicholas Copernicus University, Torun, Poland by Domaslowski and Kesy-Lewandowska (1988).

2.3 Description of evaluated Stone Objects and their Treatments

In this study four objects treated with hydro-protective products have been observed and studied. The conservation of the objects took place between one and fifteen years ago. The conservation program for all objects consisted of the following interventions:

- Pre-consolidation
- Gluing
- Cleaning
- Desalination
- Consolidation
- Repair
- Hydro-protective treatment

The following objects have been studied:

- *Façade decorations, Storgatan 10, Stockholm (city centre.)*

The decorative sandstone elements on the façade of Storgatan 10 were conserved in 1990. The decorative elements are fairly protected from rain and snow. The conservation work was completed with a hydro-protective treatment using Wacker Elastosil E41.

- *Sculpture of Neptunus, Landbyska Verket, Birger Jarlsgatan 20, Stockholm (city centre).*

The Neptune sculpture is placed on the façade of the building situated at Birger Jarlsgatan 20. Conservation work was undertaken in 1994. The work included extensive reconstruction and hydro-protective treatment using Wacker Elastosil E41. The sculpture was dismantled and brought to a studio for conservation. The Neptune sculpture is, due to its protruding location on the façade, extremely exposed to water through direct rain and snow. This circumstance, together with the extent of the reconstruction work, was the reason for treating the object with a water-repellent product.

- *Column bases of the County Council administration building (Landstingshuset), Hantverkargatan 45, Stockholm (city centre).*

The columns were conserved in 2001. The column bases are very much exposed to water splashing from the ground during and after rain, therefore the conservation program included hydro-protective treatment of the bases but not of the whole columns. The bases were treated with Wacker BS 290 after consolidation and repair work.

- *Urns decorating the entrance staircase of the Supreme Court (Bondeska palatset), Riddarhustorget 8, Stockholm (old town).*

The façade of Bondeska palatset was restored in 2004. The two urns decorating the entrance staircase have a very exposed location and were in a bad condition. This was the reason for completing the conservation with a hydro-protective treatment. The urns were treated with Wacker BS 290.

It should be added that all elements studied are insulated from rising subsoil water and from water coming from walls and from the building foundations. All objects but one (Neptune) were conserved in situ. The method of wet-in-wet application was used for all objects, e.g. pouring the liquid onto the stone while brushing with full and flowing strokes keeping the surface wet until saturation is achieved.

2.4 Preparation of Stone Samples

Freshly quarried Gotland sandstone from Bushvik, Gotland, was used for the preparation of samples for the field stations. Two different sizes were prepared, cubes measuring 5x5x5 cm and panels measuring 50x40 cm of 3 cm thickness. The stone material was delivered from two different suppliers in order to widen the sampling. Two hydrophobic agents where used, Wacker BS 290 and Wacker Elastosil E41, both diluted with White Spirit as described in paragraph 2.2.

The panel samples were all treated using the method of wet-in-wet application with the aim to emulate in situ treatment. Half of the cube samples were treated using the method of wet-in-wet application and the remaining half of the cube samples were submerged in the hydro-protective product for slow absorption by free capillary forces until saturation was achieved. Thereafter all samples were left to air-dry (about 40% RH) until equilibrium was reached after 14 to 16 days.

The samples were placed at field stations. When the measurements for this study were made, the cubes had been exposed to natural environment for three years and the panels for three months. In addition several samples are presently being exposed to accelerated aging (UV-chamber) where they are exposed to 800 Watt. Measurements for this study were taken after 1400 hr in UV-chamber which is equivalent to one year natural out-door exposure.

3. METHODS FOR EVALUATION

The effects and the durability of the hydro-protective treatments were evaluated by means of visual inspection. Water repellence and absorption were detected by testing the stones wet-ability and measured by use of Karsten's testing tube, whereas colour changes were, when possible, detected by spectrophotometer. Karsten's tube is a conventional method for in situ absorption measurements and is considered to have a good reproducibility (Wendler and Snethlage, 1990). The mass of water absorbed per square meter in relation to time was calculated using 60-minute water absorption values when possible. The untreated stone samples had a very rapid absorption; many samples had absorbed the water in the tubes in less than 10 minutes. Therefore ten and one minute absorption values sometimes have been used.

All methods described are non-destructive and suitable for field-testing.

3.1 Treatment Efficiency Index (TEI)

The treatments' efficiency and durability were evaluated by using a treatment efficiency index (TEI) (Nwaubani et al., 1999).

$$TEI = [(Wo - Wt) / Wo] \times 100 \quad (1)$$

where Wo is the water absorption of the untreated stone and Wt is the water absorption of the treated stone.

The index is useful for evaluating the extent of improvement in surface performance achieved by hydro-protective treatments and is valuable in practical conservation work, for evaluating the effectiveness of treatments as well as for estimating the accurate point of time for re-treatment. The magnitude of the treatment efficiency index shows the extent of improvement in surface performance achieved by a given treatment (Nwaubani et al. 1999). A TEI value of 100 means that the tested stone did not absorb any water at all after treatment.

4. RESULTS AND DISCUSSION

4.1 Evaluation of Stone Objects

The measurements are summarized in Table 1. The objects were all measured on the same day (June 2005). The weather was cloudy and dry with sunny intervals, with a temperature of about 15°C.

- *Façade decorations, Storgatan 10.*

The hydro-protective treatment of the decorations was made in 1990 using the Wacker product E41. The decorations on the façade are protected from direct rain. The decorations show no further signs of weathering, e.g. sanding, since treatment. Large parts of the stone surface is still water-repellent, especially the upper parts of the decorations, which are most protected against rain. These parts had no water absorption at all measured after one hour. The lower parts of the decorations (less protected) absorbed

0.29 kg/m^2 in one hour. No colour changes or other negative effects of the treatment could be observed. Re-treatment of the object is not yet considered necessary (Figure 1).

- *Sculpture of Neptunus*

The hydro-protective treatment of the sculpture was made in 1994 using the Wacker product E41. The sculpture is extremely exposed to direct rain and snow. Evident signs of sanding were observed on a few especially exposed areas. When testing the wet-ability of the sanding areas it was observed that the areas had an almost water-repellent surface. The less exposed areas showed no signs of sanding; these areas had a virtually water-repellent surface. Two sampling areas were measured, one area showing no signs of deterioration, which had no water absorption at all after one hour measuring, and one area showing signs of superficial sanding, which absorbed about 0.14 kg/m^{2}h (Figure 2) .

Evident dark traces from water running from the trident were observed. These dark traces are most probably due to the fact that the water-repellent agent changed the surface tension and forced the dirty water to run off from the surface in small concentrated veins. No other colour changes or negative effects of the treatment could be observed. The conservation work included extensive reconstruction work. The reconstructions apparently were in a good condition. However, smaller repairs, e.g. technical mending, need to be exchanged. The conclusion of the evaluation is that the sculpture is in need of re-conservation (Figure 2).

Figure 1. Water uptake by means of Karsten's tube, Storgatan 10, Stockholm.

Figure 2. Neptunus, Landbyska Verket, Stockholm.

- *Column bases (Landstingshuset)*

  The columns were conserved in 2001. The bases are very exposed to water, whereas the higher parts of the columns are well protected from water. Only the bases, not the whole columns, were treated with hydro-protection Wacker BS 290. The columns show no signs of further weathering, e.g. sanding, since treatment. The column bases have a water-repellent surface and had no water absorption at all after one hour measuring. The higher parts of the columns, which have been conserved and consolidated but not treated with hydro-protection, absorbed 1.14 kg/m^2 in one hour. No colour changes or other negative effects of the treatment could be observed. Re-treatment of the object is not yet considered necessary (Figure 3).

- *Urns (Bondeska palatset)*

  The hydro-protective treatment of the decorations was made one year ago (2004) using Wacker E41. The two urns are very exposed to rain and snow and signs of sanding were observed only one year after conservation. The urns had a relatively high absorption of water, 0.71kg/m^{2}h and 0.57 kg/m^{2}h. Colour changes were observed, in the form of thin, white veins running in the direction of the rainwater and covering the stone surface. The conclusion of the evaluation is that the urns are not yet in need of re-treatment but should be checked regularly for signs of further weathering. (Figure 4)

Observations of and measurements on the four treated objects show that the durability of the treatment varies much depending on exposure. The façade decorations on Storgatan, in a rain-sheltered location are in excellent condi-

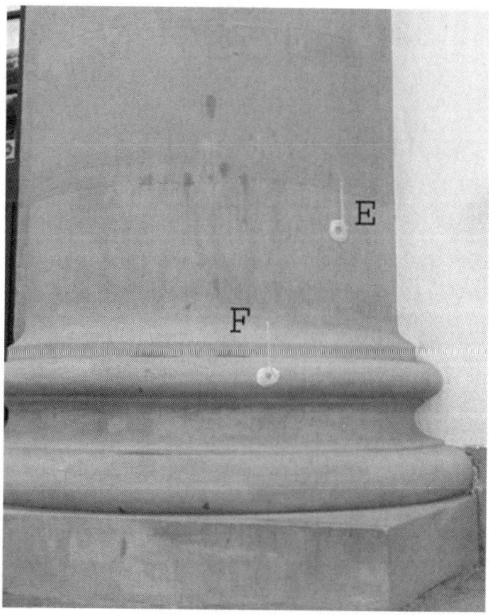

Figure 3. Column bases, Landstingshuset, Stockholm.

Figure 4. Urns at Bondeska palatset, Stockholm.

tion although the treatment dates 15 years back. The urns with a very exposed location had the highest absorption although the treatment was made only one year ago.

Discolouring, likely due to the hydro-protective treatment, could be observed on two of the objects, the urns and Neptune. Both objects have locations very exposed to rain. On Neptune the discolouring has gradually been caused by dirty water running from the trident. The discolouring of the urns on the other hand appeared shortly after the treatment and seems to be directly caused by it. The pattern of the discolouration is small veins of dirt running in the same direction as the rain does. These veins probably dried more slowly than the rest of the sculpture and therefore were more susceptible to dust/dirt that could attach to them directly after the treatment.

As there are no records of measurements taken before treatment it is not possible to calculate TEI for the treated stone objects (the freshly quarried sandstone was not considered relevant for comparison). However, as in the case of Landstingshuset, measurement could be made on stone consolidated and hydro-protected as well as on stone only consolidated and it was possible to calculate TEI for the treatment of the columns based on the differences between conservation with and without hydro-protection. As the parts only consolidated had a water absorption of 1.14 kg/m^2h and the parts consolidated and hydro-protected of 0.00 kg/m^2h, TEI for the treatment is 100.

Table 1. Measurements of stone objects

Name of object:	**Absorption:**	**Product:**
Storgatan (1990)	A. 0 kg/m^2h	E 41
	B. 0,29 kg/m^2h	E 41
Neptune (1994)	C. 0 kg/m^2h	E 41
	D. 0,14 kg/m^2h	E 41
Landstingshuset (2001)	E. 1,14 kg/m^2h	Consolidation
	F. 0 kg/m^2h	BS 290
Bondeska palatset (2004)	G. 0,71 kg/m^2h	E 41
	H. 0,57 kg/m^2h	E 41
Description of sampling area:		
Storgatan (1990)	A. Sound stone, no signs of sanding	
	B. Sound stone, no signs of sanding	
Neptune (1994)	C. Sound stone, no signs of sanding	
	D. Sound stone, signs of superficial sanding	
Landstingshuset (2001)	E. Sound stone, no signs of sanding	
	F. Sound stone, no signs of sanding	
Bondeska palatset (2004)	G. Sound stone, signs of superficial sanding	
	H. Sound stone, signs of superficial sanding	

Consolidation treatment without hydro-protection significantly decreases, also, the water-absorption of both freshly quarried and deteriorated stone, as shown by Malaga et al. (2004). Therefore it is appropriate to study the results of the measurements taken on the columns showing how effectively the hydro-protective treatment reduces the stone's water uptake. The fact, that the bases were in a much worse condition than the other parts of the columns before conservation work started, should also be taken into account when comparing the measurements.

No differences in results, effects or durability between the two different hydro-protection products can be established based on the results of the evaluation. The correlation between the results of the measurements with Karsten's tube and the interpretation of the visual inspections is not always obvious. Stone with signs of sanding can have a very low water-absorption and vice versa as shown in the evaluation. With regard to maintenance planning the results from both methods need to be considered for evaluating a treatment's effects and durability.

4.2 Evaluation of Stone Samples

The measurements are summarized in Table 2.

The freshly quarried untreated sandstone samples were measured in laboratory environment. The samples had an instant and rapid water absorption during the first ten minutes. A few samples absorbed the water in the tubes in less than ten minutes. The quickest water absorption was 1.39 kg/m^2min and the slowest was 0.07 kg/m^2min. An average was calculated by excluding the two samples with the quickest absorption and the two samples with the slowest absorption. Thereby it was possible to calculate the mass of water absorbed per square meter using 60-minute water absorption values. The average water-absorption during one hour for untreated unexposed samples was calculated to 7.25 kg/m^2.

After treatment the water uptake was considerably reduced. The stone samples treated with Wacker BS 290 did not absorb any water during one hour of testing. Samples treated with Wacker E41 had an average absorption of 0.28 kg/m^2h. TEI for treatment with BS 290 is 100, at comparison to before treatment. For treatment with E41 TEI is 96. The measurements were taken in laboratory environment with an RH of 50%. So soon after treatment no colour changes at comparison to the untreated samples could be detected.

The panel samples were measured after exposure to natural environment for three months. The measurements were taken in field on a sunny and dry day with a temperature of about 25 °C. The water uptake for the untreated stone varied much between the different samples. One sample absorbed 14.28 kg/m^2 during the first ten minutes. The slowest water absorption was 2.06 kg/m^2h. An average of 3.24 kg/m^2h was calculated (excluding the sample

Table 2. Measurements of stone samples

Water absorption of untreated stone (laboratory), panels 500x400x30 mm		**TEI**
Quickest absorption:	1,39 kg/m²min	
Slowest absorption:	0,07 kg/m²min	
Average absorption:	*7,25 kg/m²h*	
Water absorption of treated stone (laboratory), panels 500x400x30 mm		
Average BS 290:	*0 kg/m²h*	100
Average E41:	*0,28 kg/m²h*	96
Water absorption of untreated stone after 3 months outdoor exposure (in-field), panels 500x400x30 mm		
Quickest absorption:	14,28 10 min	
Slowest absorption:	2,06 kg/m²h	
Average:	***3,24 kg/m²h***	
Water absorption of treated stone after 3 months outdoor exposure (in-field), panels 500x400x30 mm		
Average BS 290:	*0,14 kg/m²h*	96
Average E41:	*0,71 kg/m²h*	78
Water absorption of treated stone after 1400 h in UV 800W (laboratory), panels 500x400x30 mm		
Average BS 290:	*0 kg/m²h*	100
Average E41:	*0,82 kg/m²h*	62
Water absorption of un-treated stone after 1400 h in UV 800W (laboratory), panels 500x400x30 mm		
Average:	*2,18 kg/m²h*	
Water absorption of treated stone after 3 years outdoor exposure (laboratory), cubes 50x50x50 mm		
W - 290-A	0,14 kg/m²	
W - 290-B	0,71 kg/m²h	
W - 290-C	0,57 kg/m²h	
W - 290-D	0,14 kg/m²h	
Average BS 290 :	*0,39 kg/m²h*	95
W E 41 – A	3,57 kg/m²h	
W E 41 – B	1,57 kg/m²h	
W E 41 – C	5,43 kg/m²h	
W E 41 – D	0,86 kg/m²h	
Average E41:	*2,86 kg/m²h*	61

with the quickest absorption and the sample with the slowest absorption. All untreated samples had a slower water uptake after three months exposure than before exposure.

The treated samples had all increased their water uptake after three months of natural exposure compared to before exposure. Samples treated with BS 290 had an average water uptake of 0.14 kg/m²h (TEI 96) at comparison to exposed untreated samples. Samples treated with E41 had an average of 0.71kg/m²h (TEI 78).

After three months no colour changes could be observed when comparing the treated samples to the untreated samples. However, it was observed that the treated samples had a loss of gloss, more lustreless surface, compared to untreated samples. These differences were noticeable on moist days or when the stone surface was wet whereas on dry and sunny days no difference between treated and untreated samples could be observed.

It should be pointed out that the quality of Gotland sandstone varies not only between different quarries but also within the same quarry. Differences within the same block of stone can be observed. It has been shown that deteriorated stone samples, collected from different objects, have a considerable lower absorbing capacity than freshly quarried stone samples (Malaga et al., 2004). Samples from this study, having been exposed only for three months to outdoor condition, show a decrease in water absorption from an average value of 7.25 kg/m²h to an average value of 3.24 kg/m²h. The differences are most likely due to the differences in the relative humidity. The unexposed samples had been stored at 50% RH while the outdoor RH is usually higher.

Also deposition of dirt particles on the stone surface has probably resulted in a reduction of the absorption capability of the untreated samples.

Panel samples were placed in UV-chamber for accelerated aging where they are exposed to 800 Watt. Measurements were taken after 1400 h, equivalent to one year natural out-door exposure. No colour changes could be observed on the panels samples placed in UV-chamber when comparing the treated samples to the untreated samples. Samples treated with BS 290 had no water uptake at all, after one hour, and the TEI value was 100 in comparison to UV-exposed untreated samples. Samples treated with E41 had an average of 0.82 kg/m²h, and the TEI value was 62 in comparison to UV-exposed untreated samples. The average water uptake for UV-exposed untreated samples was 2.18 kg/m²h.

The *cube samples* were measured in laboratory environment after three years of natural exposure. The average absorption for naturally exposed and treated samples was for BS 290 0.39 kg/m²h (TEI 95 compared to before treatment) and for E41 2.86 kg/m²h (TEI 61 compared to before treatment).

All samples, treated as well as untreated, had a darker, more brownish colour after three years of exposure, as confirmed by spectrophotometer measurements. Discolouring could be observed on the treated samples after two years. The samples had a patchy looking and glossless surface. This was first observed on samples treated with BS 290, but later also on samples treated with E41. It was observed that there is an evident correlation between very low water uptake and discolouring. Samples treated with BS 290, with a much lower water uptake (0.39 kg/m^2h) than samples treated with E41 (2.86 kg/m^2h), had a stronger, more evident discolouring.

No differences between the two different methods of treatment, wet-in-wet application and submerging, could be detected with regard to water uptake, durability and discolouring.

5. CONCLUSIONS

The results of the study show that untreated Gotland sandstone, freshly quarried as well as deteriorated, has a very rapid water uptake. Treatment with BS 290 effectively reduces the water uptake of the stone. Also E41 reduces the water uptake of the stone but to a considerably lower extent.

Observations of and measurements on the four treated objects show no differences with regard to efficiency, durability and colour changes between the two different products. The durability of the treatment varies much, depending on the exposure. Negative colour changes, likely due to the hydrophobic treatment, were observed on the two most exposed objects. No other negative effects, e.g. exfoliation or other damages, could be found due to hydrophobic treatment.

Studies in laboratory show that Wacker BS 290 reduces the water uptake more effectively than Wacker Elastosil E41. Studies of samples at field stations, based on observations after up to three years of natural outdoor exposure, also show that treatment with Wacker BS 290 is more effective and durable than Wacker Elastosil E41. Studies at field stations indicate that samples treated with BS 290 have a stronger more evident discolouring than samples treated with E41.

The study implies that stone sculptures staying exposed to outdoor conditions benefit from hydro-protective treatment, e.g. have an evident reduction in water uptake, if the treatment is made correctly at the completion of an adequate conservation program, including consolidation treatment. The negative effects observed, likely due to hydro-protective treatment, are slight colour changes and loss of gloss.

Objects treated with hydrophobic agents should be monitored by means of visual inspection and Karsten's tube, in order to estimate the correct point of time for re-treatment. The correlation between results from measurement with Karsten's tube and interpretation of visual inspections is not always obvious. The results from both methods need to be taken into account.

REFERENCES

Domaslowski W., Kesy-Lewandowska M. 1988. Investigation on Applicability of Silicone Caoutchoucs to Hydrophobing Sandstones, in: *Proceedings of the VIth Int. Congress on the Deterioration and Conservation of Stone.* Torun (Poland), 12-14 September 1988, pp. 577-602.

Lukaszewicz J.W., Kwiatkowski D., Klingspor M. 1995. Consolidation of Gotland Stone in Monuments, in: *Methods of Evaluating Products for the Conservation of Porous Building Materials in Monuments.* ICCROM, Rome, Italy 19-21 June, pp. 179-186.

Löfvendahl R. 1996. Sador och Orsaker, in: *Stenen i tiden, från 1000-talet till 1940,* Friberg G., Sundnér B. (Eds) Riksantikvarieämbetet och Statens historiska museer, pp. 116-131.

Malaga K., Myrin M., Lindqvist J.E. 2004: Consolidation of Gotland Sandstone, in: *Proceedings of the 10th International Congress on the Deterioration and Conservation of Stone.* Stockholm 2004, pp. 439-446.

Meinhardt-Degen J., Snethlage R. 2004. Durability of Hydrophobic Treatment of Sandstone Façades - Investigations of the Necessity and Effects of Re-treatment, in: *Proceedings of the 10th International Congress on the Deterioration and Conservation of Stone.* Stockholm (Sweden) 2004, pp. 347-354.

Nwaubani S.O., Mulheron M., Tilly G.P., Schwamborn B. Pore-Structure and Water Transport Properties of Surface-Treated Building Stones, in: *Materials and Structures*, Vol. 33, April 2000, pp. 198-206.

Wendler E., Snethlage R. 1990. Durability of Hydrophobing Treatments of Natural Stone Buildings, in: *Advanced Workshop "Analytical Methodologies for the Investigation of Damaged Stones"* Pavia (Italy) 14-21 September 1990.

Wessman L. 1995. *Expansioner och avskalningar vid frysning av Gotländsk sandsten och Öländsk kalksten i NaCl och Na2 SO4- lösningar.* Rapport TVBM-7089, Lunds Tekniska Högskola.

Index